The Handbook of Environmental Chemistry

Founded by Otto Hutzinger

Editors-in-Chief: Damià Barceló • Andrey G. Kostianoy

Volume 65

Advisory Board:
Jacob de Boer, Philippe Garrigues, Ji-Dong Gu,
Kevin C. Jones, Thomas P. Knepper, Alice Newton,
Donald L. Sparks

More information about this series at http://www.springer.com/series/698

Terrestrial and Inland Water Environment of the Kaliningrad Region

Environmental Studies in the Kaliningrad Region

Volume Editors: Vladimir A. Gritsenko · Vadim V. Sivkov · Artem V. Yurov · Andrey G. Kostianoy

With contributions by

A. V. Aldushin · N. S. Belov · T. A. Bernikova · E. S. Bubnova · R. Capell · B. Chubarenko · A. R. Danchenkov · V. V. Danishevskij · D. Domnin · E. E. Ezhova · I. V. Frizina · M. A. Gerb · V. A. Gritsenko · O. V. Kocheshkova · T. B. Kolesnik · A. G. Kostianoy · E. G. Kropinova · E. K. Lange · E. A. Masyutkina · G. Mikhnevich · N. S. Molchanova · N. N. Nagornova · M. G. Napreenko · T. V. Napreenko-Dorokhova · E. Nesterova · J. J. Polunina · E. A. Romanova · D. V. Sergeeva · T. V. Shaplygina · S. V. Shibaev · M. N. Shibaeva · V. Sivkov · V. V. Sivkov · A. V. Sokolov · N. A. Tsoupikova · K. V. Tylik · O. L. Vinogradova · O. V. Vinogradova · I. I. Volkova · A. A. Volodina · A. V. Yurov · S. K. Zaostrovtseva · V. A. Zhamoida

Editors
Vladimir A. Gritsenko
Immanuel Kant Baltic Federal University
Kaliningrad, Russia

Artem V. Yurov
Immanuel Kant Baltic Federal University
Kaliningrad, Russia

Vadim V. Sivkov
P.P. Shirshov Institute of Oceanology
Kaliningrad, Russia

Andrey G. Kostianoy
P.P. Shirshov Institute of Oceanology
Moscow, Russia

S.Yu. Witte Moscow University
Moscow, Russia

ISSN 1867-979X   ISSN 1616-864X   (electronic)
The Handbook of Environmental Chemistry
ISBN 978-3-319-72164-4   ISBN 978-3-319-72165-1   (eBook)
https://doi.org/10.1007/978-3-319-72165-1

Library of Congress Control Number: 2017964380

© Springer International Publishing AG 2018
This work is subject to copyright. All rights are reserved by the Publisher, whether the whole or part of the material is concerned, specifically the rights of translation, reprinting, reuse of illustrations, recitation, broadcasting, reproduction on microfilms or in any other physical way, and transmission or information storage and retrieval, electronic adaptation, computer software, or by similar or dissimilar methodology now known or hereafter developed.
The use of general descriptive names, registered names, trademarks, service marks, etc. in this publication does not imply, even in the absence of a specific statement, that such names are exempt from the relevant protective laws and regulations and therefore free for general use.
The publisher, the authors and the editors are safe to assume that the advice and information in this book are believed to be true and accurate at the date of publication. Neither the publisher nor the authors or the editors give a warranty, express or implied, with respect to the material contained herein or for any errors or omissions that may have been made. The publisher remains neutral with regard to jurisdictional claims in published maps and institutional affiliations.

Printed on acid-free paper

This Springer imprint is published by Springer Nature
The registered company is Springer International Publishing AG
The registered company address is: Gewerbestrasse 11, 6330 Cham, Switzerland

Editors-in-Chief

Prof. Dr. Damià Barceló
Department of Environmental Chemistry
IDAEA-CSIC
C/Jordi Girona 18–26
08034 Barcelona, Spain
and
Catalan Institute for Water Research (ICRA)
H20 Building
Scientific and Technological Park of the
 University of Girona
Emili Grahit, 101
17003 Girona, Spain
dbcqam@cid.csic.es

Prof. Dr. Andrey G. Kostianoy
P.P. Shirshov Institute of Oceanology
Russian Academy of Sciences
36, Nakhimovsky Pr.
117997 Moscow, Russia
kostianoy@gmail.com

Advisory Board

Prof. Dr. Jacob de Boer
IVM, Vrije Universiteit Amsterdam, The Netherlands

Prof. Dr. Philippe Garrigues
University of Bordeaux, France

Prof. Dr. Ji-Dong Gu
The University of Hong Kong, China

Prof. Dr. Kevin C. Jones
University of Lancaster, United Kingdom

Prof. Dr. Thomas P. Knepper
University of Applied Science, Fresenius, Idstein, Germany

Prof. Dr. Alice Newton
University of Algarve, Faro, Portugal

Prof. Dr. Donald L. Sparks
Plant and Soil Sciences, University of Delaware, USA

The Handbook of Environmental Chemistry Also Available Electronically

The Handbook of Environmental Chemistry is included in Springer's eBook package *Earth and Environmental Science*. If a library does not opt for the whole package, the book series may be bought on a subscription basis.

For all customers who have a standing order to the print version of *The Handbook of Environmental Chemistry*, we offer free access to the electronic volumes of the Series published in the current year via SpringerLink. If you do not have access, you can still view the table of contents of each volume and the abstract of each article on SpringerLink (www.springerlink.com/content/110354/).

You will find information about the

– Editorial Board
– Aims and Scope
– Instructions for Authors
– Sample Contribution

at springer.com (www.springer.com/series/698).

All figures submitted in color are published in full color in the electronic version on SpringerLink.

Aims and Scope

Since 1980, *The Handbook of Environmental Chemistry* has provided sound and solid knowledge about environmental topics from a chemical perspective. Presenting a wide spectrum of viewpoints and approaches, the series now covers topics such as local and global changes of natural environment and climate; anthropogenic impact on the environment; water, air and soil pollution; remediation and waste characterization; environmental contaminants; biogeochemistry; geoecology; chemical reactions and processes; chemical and biological transformations as well as physical transport of chemicals in the environment; or environmental modeling. A particular focus of the series lies on methodological advances in environmental analytical chemistry.

Series Preface

With remarkable vision, Prof. Otto Hutzinger initiated *The Handbook of Environmental Chemistry* in 1980 and became the founding Editor-in-Chief. At that time, environmental chemistry was an emerging field, aiming at a complete description of the Earth's environment, encompassing the physical, chemical, biological, and geological transformations of chemical substances occurring on a local as well as a global scale. Environmental chemistry was intended to provide an account of the impact of man's activities on the natural environment by describing observed changes.

While a considerable amount of knowledge has been accumulated over the last three decades, as reflected in the more than 70 volumes of *The Handbook of Environmental Chemistry*, there are still many scientific and policy challenges ahead due to the complexity and interdisciplinary nature of the field. The series will therefore continue to provide compilations of current knowledge. Contributions are written by leading experts with practical experience in their fields. *The Handbook of Environmental Chemistry* grows with the increases in our scientific understanding, and provides a valuable source not only for scientists but also for environmental managers and decision-makers. Today, the series covers a broad range of environmental topics from a chemical perspective, including methodological advances in environmental analytical chemistry.

In recent years, there has been a growing tendency to include subject matter of societal relevance in the broad view of environmental chemistry. Topics include life cycle analysis, environmental management, sustainable development, and socio-economic, legal and even political problems, among others. While these topics are of great importance for the development and acceptance of *The Handbook of Environmental Chemistry*, the publisher and Editors-in-Chief have decided to keep the handbook essentially a source of information on "hard sciences" with a particular emphasis on chemistry, but also covering biology, geology, hydrology and engineering as applied to environmental sciences.

The volumes of the series are written at an advanced level, addressing the needs of both researchers and graduate students, as well as of people outside the field of "pure" chemistry, including those in industry, business, government, research establishments, and public interest groups. It would be very satisfying to see these volumes used as a basis for graduate courses in environmental chemistry. With its high standards of scientific quality and clarity, *The Handbook of Environmental Chemistry* provides a solid basis from which scientists can share their knowledge on the different aspects of environmental problems, presenting a wide spectrum of viewpoints and approaches.

The Handbook of Environmental Chemistry is available both in print and online via www.springerlink.com/content/110354/. Articles are published online as soon as they have been approved for publication. Authors, Volume Editors and Editorsin-Chief are rewarded by the broad acceptance of *The Handbook of Environmental Chemistry* by the scientific community, from whom suggestions for new topics to the Editors-in-Chief are always very welcome.

Damià Barceló
Andrey G. Kostianoy
Editors-in-Chief

Contents

Introduction 1
Vladimir A. Gritsenko, Vadim V. Sivkov, Artem V. Yurov,
and Andrey G. Kostianoy

Mineral Resources of the Kaliningrad Region 13
Vladimir Zhamoida, Vadim Sivkov, and Elena Nesterova

Amber Deposits in the Kaliningrad Region 33
Vadim V. Sivkov and Vladimir A. Zhamoida

**Composition of Pre-Quaternary Surface and Quaternary Sediments
Allocation on the Territory of the Kaliningrad Region** 45
Galina Mikhnevich

Pleistocene Deposits in the Kaliningrad Region 81
T.B. Kolesnik

Modern Landscapes in the Kaliningrad Region 97
Elena A. Romanova, Olga L. Vinogradova, and Irina V. Frizina

**The History and Pattern of Forest and Peatland Formation
in the Kaliningrad Region During the Holocene** 121
T.V. Napreenko-Dorokhova and M.G. Napreenko

Eolian Coastal-Marine Natural Systems in the Kaliningrad Region ... 147
I.I. Volkova, T.V. Shaplygina, N.S. Belov, and A.R. Danchenkov

**Factors and Patterns of Current Development of Territorial Units
in the Kaliningrad Region** 179
Elena A. Romanova, Olga V. Vinogradova, and Darja V. Sergeeva

Specific Features of Urban Geosystems in the Kaliningrad Region ... 201
Elena A. Romanova, Olga L. Vinogradova, Vladislav V. Danishevskij,
and Irina V. Frizina

Environmental Features of Watercourses in the Kaliningrad Region ... 223
T.A. Bernikova, N.N. Nagornova, N.A. Tsoupikova, and S.V. Shibaev

Formation and Re-Distribution of the River Runoff in the Catchment of the Pregolya River ... 269
Dmitry Domnin, Boris Chubarenko, and Rene Capell

Hydrobiological Characteristics of Water Bodies in the Kaliningrad Region ... 285
M.N. Shibaeva, E.A. Masyutkina, and S.V. Shibaev

The Structure and Composition of Biological Communities in the Pregolya River (Vistula Lagoon, the Baltic Sea) ... 317
E.E. Ezhova, E.K. Lange, M.A. Gerb, O.V. Kocheshkova, J.J. Polunina, and N.S. Molchanova

Phytoplankton Community of Small Rivers of the Pregolya River Basin ... 373
E.K. Lange

Flora and Vegetation of the Small Rivers of the Pregolya River System in the Kaliningrad Region ... 385
A.A. Volodina and M.A. Gerb

The Protection Conditions of the Groundwater Against Pollution in the Kaliningrad Region ... 411
Galina Mikhnevich

Current Status of the Lake Vistytis in Kaliningrad Region ... 441
S.V. Shibaev, A.V. Sokolov, K.V. Tylik, T.A. Bernikova, M.N. Shibaeva, E.A. Masyutkina, N.N. Nagornova, A.V. Aldushin, and S.K. Zaostrovtseva

Specially Protected Natural Areas of the Kaliningrad Region ... 481
I.I. Volkova, T.V. Shaplygina, and E.S. Bubnova

The Reduction in the Beach Area as the Main Limiting Factor for Sustainable Tourism Development (Case for the Kaliningrad Oblast) ... 497
Elena G. Kropinova

Rare and Protected Macrophytes and Semiaquatic Plants of Flora of the Kaliningrad Region ... 513
M.A. Gerb and A.A. Volodina

Conclusions ... 527
Vladimir A. Gritsenko, Vadim V. Sivkov, Artem V. Yurov, and Andrey G. Kostianoy

Erratum to: Mineral Resources of the Kaliningrad Region 539
Vladimir Zhamoida, Vadim Sivkov, and Elena Nesterova

Erratum to: Specially Protected Natural Areas of the Kaliningrad Region 541
I.I. Volkova, T.V. Shaplygina, and E.S. Bubnova

Index 543

Introduction

Vladimir A. Gritsenko, Vadim V. Sivkov, Artem V. Yurov, and Andrey G. Kostianoy

Abstract This introductory chapter briefly describes the content of the book, which has 22 chapters including the present Introduction and Conclusions written by Volume Editors of the book. This book entitled *Terrestrial and Inland Water Environment of the Kaliningrad Region* is the first one in the series of four volumes which will be published in the coming years under the general title *Environmental Studies in the Kaliningrad Region*. This first volume is devoted to the physico geographical and bio-geo-ecological conditions, as well as to environmental problems of the Kaliningrad Region focusing on terrestrial and inland water environment.

Keywords Bioecology, Geoecology, Inland water, Kaliningrad Region, Terrestrial environment, The Southeastern Baltic Sea

Contents

1 Introduction 2
2 Geological Structure and Mineral Resources 5
3 Landscapes 7
4 Water Bodies 8
5 Hydrobiology of Water Bodies 8
6 Nature Management and Protection 9
7 Conclusions 10
References 11

V.A. Gritsenko (✉) and A.V. Yurov
Immanuel Kant Baltic Federal University, 14, A. Nevskogo Str., Kaliningrad 236041, Russia
e-mail: gritsenko_vl@mail.ru

V.V. Sivkov and A.G. Kostianoy
P.P. Shirshov Institute of Oceanology, Russian Academy of Sciences, 36, Nakhimovsky Pr., Moscow 117997, Russia
e-mail: sivkov@kaliningrad.ru; kostianoy@gmail.com

1 Introduction

Kaliningrad Region (Oblast) is a federal subject of the Russian Federation located on the coast of the Southeastern Baltic Sea (Fig. 1). This is the westernmost region of the Russian Federation with the city of Kaliningrad as the administrative center (467,000 people). The territory was formerly a part of East Prussia (the Kingdom of Prussia was the leading state of the German Empire after its creation in 1871), which was divided between Russian SFSR, Lithuanian SSR (both former the USSR republics), and the People's Republic of Poland (Fig. 2) after a defeat of Nazi Germany in 1945 with the end of WWII. The capital city Königsberg was renamed to Kaliningrad in 1946. Today this is exclave of the Russian Federation with total area of about 15,100 km^2 and population of about 986,261 (2017). In the south it borders with Poland, in the north and east with Lithuania, and in the west it is washed by the Baltic Sea and its bays – the Curonian and Vistula Lagoons (Fig. 3). The exclave (from Latin "ex" – "from" and "clavis" – "key") is an untrusted region, separated from the main territory of the country and surrounded by other states (one or several). The exclave, which has access to the sea, is called a semi-exclave, because it is formally linked to the main territory by the sea. Thus, correctly, the

Fig. 1 Geographic location of the Kaliningrad Region

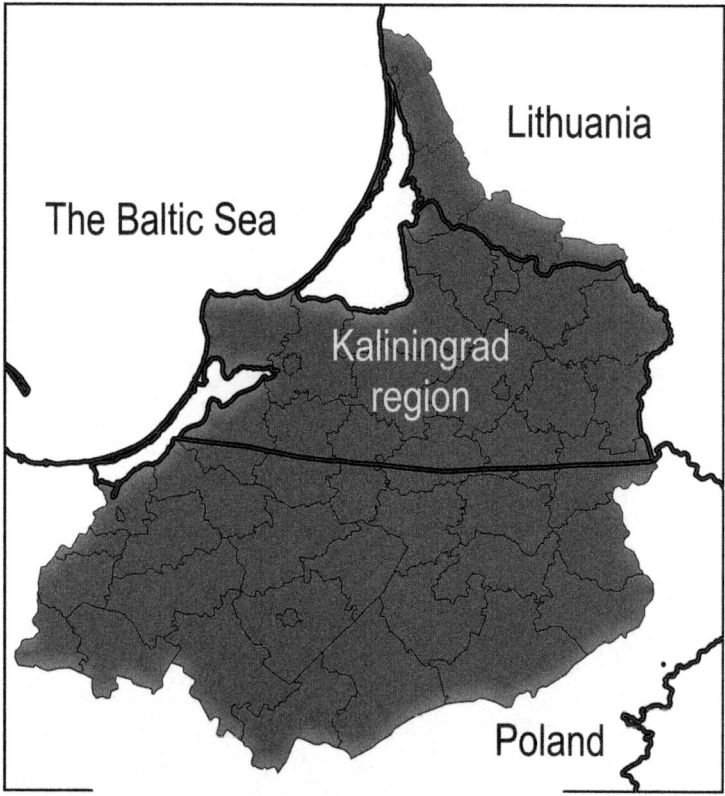

Fig. 2 Territory of East Prussia (green color) on the modern map of Lithuania, Poland, and Russian Federation (Kaliningrad Region)

Kaliningrad Region is semi-exclave of the Russian Federation. In the world there are a few examples of semi-exclave with the most known Alaska which is a semi-exclave of the United States and semi-enclave of Canada.

Increased interest in the nature of the Kaliningrad Region is due to several reasons. Firstly, this territory has a significant landscape, geoecological and biodiversity. The region includes a variety of landscapes, inland waters, rivers, marine waters of the Southeastern Baltic Sea, and two sandy Curonian and Vistula Spits which separate the Curonian and Vistula lagoons from the sea (Fig. 3). Since 2000, the Curonian Spit is a UNESCO World Heritage Site shared by Russian Federation and Lithuania. Secondly, it is located within the Central Europe and differs markedly from neighboring countries by the legal and economic conditions of nature management. Thirdly, it is in the Kaliningrad Region that the world's largest amber deposits are concentrated, which ensure de facto world monopoly of Russia for the extraction of amber raw material. This natural feature of the region gives grounds to call it "amber area." And, finally, in the world scientific literature, there is a lack of information about the nature of the Kaliningrad Region and environmental studies

Fig. 3 Map of the Kaliningrad Region (https://en.wikipedia.org/wiki/Kaliningrad_Oblast)

performed by research and educational organizations. Despite a significant international collaboration between the Kaliningrad Region institutions and research and educational organizations in Lithuania, Poland, Germany, and other European countries, a little is published in western editions in English, and most of the environmental research results are published in local Russian editions hardly accessible to western readers [1–27]. These references show a little part of the books only published in Russian editions about Kaliningrad Region since 2000.

The idea of this book project was born in December 2013 at the meeting between all Volume Editors in Immanuel Kant Baltic Federal University (Kaliningrad, Russia). Several research organizations and universities located in Kaliningrad, Moscow, and St. Petersburg are involved in the environmental studies of the Kaliningrad Region (including the Southeastern Baltic Sea). Among them are Immanuel Kant Baltic Federal University (BFU, Kaliningrad), Atlantic Branch of P.P. Shirshov Institute of Oceanology of Russian Academy of Sciences (AB IORAS, Kaliningrad), P.P. Shirshov Institute of Oceanology of Russian Academy of Sciences (IORAS, Moscow), Kaliningrad State Technical University (KSTU, Kaliningrad), Atlantic Research Institute for Fishery and Oceanography (AtlantNIRO, Kaliningrad), and A.P. Karpinsky Russian Geological Research Institute (VSEGEI, St. Petersburg). The leading specialists in different fields of science (ecology, geoecology, bioecology, geology, biology, botany, forestry, watercourses, groundwater, limnology, mineral resources, urban geosystems, and tourism) were invited to contribute to this book project entitled "Environmental Studies in the Kaliningrad Region" which will have four volumes. Here we introduce the first volume *Terrestrial and Inland Water Environment of the Kaliningrad Region* which will be followed in the coming years by "Oceanography of the Southeastern Baltic Sea," "Geoecology of the Southeastern Baltic Sea," and "Bioecology of the Southeastern Baltic Sea."

This book does not pretend to a comprehensive description of environmental conditions and problems of the Kaliningrad Region. It consists of a number of independent articles (chapters) in which, in the form of generalizations or special studies, certain issues relevant to the nature of the region are revealed.

2 Geological Structure and Mineral Resources

Over millions of years of its geological history, numerous changes in the marine, lagoon, lacustrine, and continental climatic conditions, including periods of glaciation, land uplifts, and descents, occurred on the territory of the region; the sea was advancing and receding. All these processes have formed the geological basis of the modern environment in the region. The modern general description of the geological structure of Kaliningrad Region area can be found in several chapters.

The modern general description of the geological structure and mineral resources of Kaliningrad Region area can be found in Zhamoida et al. [28]. Two structural stages are distinguished in the geological section of the Kaliningrad Region: the lower one is composed of gneisses, crystalline schists, and amphibolites of the *Archaean-Proterozoic age* (platform basement) and the upper one is the *Phanerozoic platform* cover, represented by poorly dislocated and slightly metamorphosed sediments. The presence of major breaks and stratigraphic unconformities in the Lower Cambrian, at the boundary of the Lower and Middle Devonian, in the Lower Permian, between the lower Devonian and the upper Permian, allows to distinguish the structural complexes formed in the Caledonian, Hercynian, and Alpine cycles of tectogenesis.

Within the Kaliningrad Region in accordance with lithological and facial features of the *Quaternary sediments*, two zones are distinguished: South Kaliningrad and North Kaliningrad. The South Kaliningrad zone occurs within the boundaries of neotectonic basin and is characterized by the completeness of the section. The total thickness of the Pleistocene formations here is 270 m. The North Kaliningrad zone is characterized by the absence of the deposits of the lower and middle stages of Pleistocene. The average total thickness of the Pleistocene sediments here, as a rule, does not exceed 60–70 m. During the Pleistocene the territory was glaciated three times; accordingly, it is possible to find deposits of three glacial and three interglacial periods, as well as preglacial formations. The polygenetic sediments of the Holocene are sufficiently investigated. It is possible to distinguish different types of marine and alluvial sediments, as well as aeolian, boggy, etc.

Actually all economic minerals discovered within Kaliningrad Region are spatially and genetically associated with the deposits of the platform cover. They are represented by oil, drinking and mineral underground water, amber, and building raw materials which are actively exploited. Potassium-magnesium salts, numerous occurrences of which are known here, are of the greatest economic interest among undeveloped types of other useful minerals. The occurrence of rock salt, zeolite, lignite, peat, ferrous, and nonferrous metals should also be mentioned. The deposits of the main mineral of the Kaliningrad Region – amber – are examined in more detail in Sivkov and Zhamoida [29]. For a generalization of modern ideas about the conditions for the formation of various amber accumulations, the geological structure of the world's largest Primorsky amber deposit is given in this chapter.

The description of quaternary sediments as the geological basis of modern natural conditions on the territory of the region is given in Mikhnevich [30]. It carried out the regionalization of the territory of the Kaliningrad Region according to the nature of the surface of pre-quaternary sediments. The most interesting forms of the relief of the pre-quaternary surface and the main mechanisms of its transformation during the Cenozoic time (exacerbation, water-ice erosion, tectonic movements) are considered in detail.

Despite the fact that the Pleistocene has been studied for more than a dozen years, there are different opinions on the history of development of the Pleistocene deposits. The sections that would have formed during the Pleistocene are not there, which complicates the study of sediments and is one of the main causes of disagreements. A wide spectrum of deposits participates in the structure of the quaternary section, the interpretation of the genesis of which is also not always unambiguous. In Kolesnik [31], a description of a number of outcrops of the northern coast of the Sambia Peninsula is given, the structure and genesis of deposits showing unexplained contradictions within the framework of the glacial theory.

3 Landscapes

It is well known that landscapes are formed under the great influence of climatic and geological conditions. The winds of the western quarter dominate in the Kaliningrad Region, which determine the transfer of warm and humid masses from the Atlantic Ocean and, therefore, the characteristic features of the temperate marine climate: relatively small fluctuations in the average monthly air temperature, high humidity and cloudiness throughout the year, and significant amount of precipitation.

The relief of the region (Fig. 3) was influenced by the three stages of migration of the Valdai Glacier – the southern, middle, and north Kaliningrad stages. It is believed that the formation of hilly-moraine ridges along the border with Poland is associated with the southern Kaliningrad stage and with the middle Kaliningrad one the formation of lowlands within the periglacial water bodies and the finely moraine ridges of the Kaliningrad (Sambia) Peninsula, as well as a chain of hills along the right slope of the Pregolya and Instruch (River) Valley. The formation of the hilly-moraine ridges of the coastal part of the Sambia Peninsula and the chain of hills are associated with the north Kaliningrad stage, on which the Vistula and Curonian Spits were subsequently formed.

In the southwest of the region, there is the Warmian Upland, and in the southeast is the Vishtynetskaya Upland, which is shared by the Lava River Valley. On their northern slopes, many rivers of the region originate. Pregolya Lowland, formed on the site of the lake-glacial reservoir and the ancient hollow of the runoff of melted glacial waters, is adjacent to the hilly-morainic hills from the north. The northeastern part of the region is occupied by the Sheshupe glaciolacustrine plain, on which individual hills rise. From the west to this plain adjoins the Instruch-Sambia hilly-moraine ridge, which is stretched by an arc almost across the entire area along the valleys of the rivers of Instruch and Pregolya, and on the Sambia Peninsula, it divides into several branches. Features of modern landscapes of the region are described in Romanova et al. [32].

A period of recent environment formation started on the territory of the Kaliningrad Region in the postglacial time during which present-day features of climate, soils, and vegetation developed together with ecosystem alterations caused by human impact. Since peatlands are considered as a natural record of vegetation dynamics, they play a great role for investigations of ecosystem formation which can be transcribed by means of paleobotanical researches (including pollen analysis and study of the botanical composition of peat) and radiocarbon dating. There are a number of landscape types in Southeastern Baltic Region that, having formed here after the ice sheet ablation, the largest among them are glaciolacustrine plains in the central part of the region and coastal lowlands which comprise a vast area along the Curonian Lagoon and in the Neman Delta. This can be found in Napreenko-Dorokhova and Napreenko [33].

An important role in shaping the external appearance of the region was played by the movements of the earth's crust and fluctuations in the Baltic Sea level after the retreat of the Valdai Glacier. The slow descent of the coastal part of the land in the last 6,000 or 7,000 years led to the appearance of the Curonian and Kaliningrad (Vistula) lagoons, together with two sandy spits, the Curonian and the Baltic (Vistula). The Curonian Spit extends almost 100 km from the southwest to the northeast, of which just about a half is located on the territory of the Kaliningrad Region. The length of the Baltic (Vistula) Spit is 65 km, of which 35 is in the Kaliningrad Region. A description of the state and evolution of aeolian coastal-marine natural complexes of the region is given in Volkova et al. [34].

The most important feature of landscape complexes in the Kaliningrad Region is determined by the high level of influence on them of economic activity. All of them can be attributed to anthropogenic landscapes what is discussed in Romanova et al. [35]. So-called agricultural landscapes have appeared: fields, orchards, meadows, pastures, and water anthropogenic landscapes (ponds, water reservoirs, canals). On the coast of the region is widely represented recreational (resort) landscape. Residential landscapes are widespread: the territories of towns and villages, where natural landscapes are heavily altered or replaced by buildings of residential and industrial buildings, with their entire complexes. The small size of the Kaliningrad Region provoked the process of "introduction" of urban systems in the nature of the region which is well demonstrated in Romanova et al. [36].

4 Water Bodies

Low and flat terrain, the prevalence of clay and loamy rocks on the surface contributed to the formation of many rivers and lakes. A general description of the water resources of the region is given in Bernikova et al. [37]. Rivers belong to the plain type. Almost all of them are deep-watered with a slow current and carry their waters to the Curonian and Kaliningrad (Vistula) lagoons of the Baltic Sea. Only a few small rivers of the Sambia Peninsula flow directly into the Baltic Sea. The total number of rivers in the region exceeds 4,000, and most of them belong to the category of small rivers with a length of 10 to 25–50 km. In the region there are only six rivers that can be classified as medium. The length of Pregolya River is 123 km, and together with its tributary of Angrapa (169 km), their total length reaches 292 km. The area of the Pregolya River Basin exceeds 13,000 km^2, and most of it is in Poland. The processes of formation and redistribution of river flow in the basin of the Pregolya River are discussed in Domnin et al. [38].

5 Hydrobiology of Water Bodies

Several chapters deal with various aspects of hydrobiological research. The results of a long-term study of zooplankton and zoobenthos complexes are given in Shibaeva et al. [39]. It may serve as a basis for ecological mapping of water bodies in the region, because their bio-indication properties reflect the state of the environment not only at the current time but also in the future.

Ezhova et al. [40] summarizes the data on the lower and middle reaches of the Pregolya River, the catchment area of which covers almost the entire Kaliningrad Region. The composition, structure, seasonal and interannual dynamics, and productivity of phytoplankton, zooplankton, and zoobenthos are described, water flora is characterized, and information on ichthyofauna is provided. It is shown that, despite the remaining high level of anthropogenic impact, in recent decades the ecological state of the river ecosystem has gradually improved, which led to the growth of biodiversity and the restoration of biological communities in the anthropogenically disturbed lower reaches of the Pregolya River. Seasonal changes in the species composition, structure of the phytoplankton, its dominant complex, and abundance indicators were also studied in the lower reaches of the Pregolya River in Lange [41]. An assumption is made about the introduction into the river and further to the Curonian Lagoon of the cyanobacterium *Planktothrix agardhii* from the Mazury Lakes of Poland. For the first time, information has been obtained on the structure and quantitative development of the phytoplankton of the tributaries of the Pregolya River. In Volodina et al. [42], the results of floristic surveys of small rivers of the Pregolya River system in the Kaliningrad Region are presented for the first time.

6 Nature Management and Protection

Groundwater protection against pollution is one of the important problems in the Kaliningrad Region, which is studied since 1980. There are different sources of groundwater pollution. Among the main criteria of protection from pollution, the study considers power, lithological composition, and permeability of rocks composing the upper waterproof and the correlation of the level of pressure and groundwater. An attempt to quantify assessment of groundwater protection is taken in Mikhnevich [43]. This research defined the regularities of the spatial differentiation of the protection conditions of the Moscow-Valdai aquifer in the Kaliningrad Region and isolated areas characterized by varying degrees of protection of the upper intermoraine aquifer. The chapter defines the areas of potential water-use conflicts in the oil-extraction areas and in the areas of construction material excavations.

Lakes in the Kaliningrad Region are numerous and, with the exception of Lake Vistytis, are small in size. The lake basins have a glacial and/or erosional origin. In the valleys of the Neman and Pregolya are floodplain lakes. The only large lake in the region, Lake Vistytis, has an area of almost 18 km^2 and a depth of 47 m. It is located in the extreme southeast of the region, among the hills of the Vistytis Upland. On the territory of the region, there are several hundred marshes with a total area of more than a thousand square kilometers. The natural bogs are located in the interfluves and in the Pregolya Valley, as well as in other parts of the region. Lake Vistytis environment is described in Shibaev et al. [44].

As is known, the main form of conservation of biological and landscape diversity is the creation and development of specially protected natural areas. For these territories, a special regime for the protection and use of natural resources is established, including full or partial and permanent or temporary restriction of the use of nature and in some cases the use of special methods of nature management or restoration. Protected areas in the Kaliningrad Region are discussed in Volkova et al. [45]. In the Kaliningrad Region, there is a National Park "Curonian Spit," established in 1987 with the aim of preserving a unique natural complex, which has a special ecological, historical, and aesthetic value. In 2000, the Curonian Spit was included in the list of UNESCO World Cultural and Natural Heritage sites in the category of cultural landscapes.

The problems of rational nature management and nature protection due to small size of the territory, the vulnerability of natural resources, high economic development, and significant population density are of the greatest importance in the Kaliningrad Region. One of the factors of sustainable development of the region is the use of tourist and recreational potential. The problem of shrinking beaches as the main limiting factor in the development of the tourism industry is discussed in Kropinova [46].

The description of rare and protected macrophytes and coastal-aquatic plants in the flora of the Kaliningrad Region is given in Gerb et al. [47]. The list of rare and protected water and water plants is presented. The identification of the modern floristic composition of water bodies in the Kaliningrad Region and the documentation of the distribution of various aquatic and near-water species are relevant in connection with their poor knowledge.

We summarized the main results described in the book in Gritsenko et al. [48].

7 Conclusions

We hope that all the abovementioned research topics presented in the respective chapters, despite their fragmentation, will still give a clear impression of the environmental conditions in the territory of the Kaliningrad Region. We also hope that readers will be interested in the previously unknown information published in this book, on the one hand, indicating the unquestionable originality of the Russian coastal exclave and, on the other hand, naturally supplementing the overall picture of the environment of the entire Baltic Region.

Acknowledgments V. A. Gritsenko was partially supported by Immanuel Kant Baltic Federal University Program "Integrated geographical surveys of the Kaliningrad Region and the Baltic Region." A. G. Kostianoy was partially supported by the Russian Science Foundation Grant N 14-50-00095. As Volume Editors of this book, we would like to thank Immanuel Kant Baltic Federal University for investing in our book project by funding translation of most of the chapters.

Also, we are very grateful to Springer-Verlag and *The Handbook of Environmental Chemistry* book series, in particular, for their support on our four-volume book project *Environmental Studies in the Kaliningrad Region*.

References

1. Zagorodnykh VA, Dovbnya AV, Zhamoida VA (2001) In: Kharin GS (ed) Stratigraphy of the Kaliningrad Region. KGU, Kaliningrad, 226 pp
2. Orlenok VV, Barinova GM, Kucheryavyy PP, Ulyashev GL (2001) Lake Vištytis: nature, history, ecology. Kaliningrad State University Press, Kaliningrad, 212 pp
3. Fedorov GM (2001) Population of the Kaliningrad Region. Demographic conditions for substantiation of the territorial scheme for town planning development of the Kaliningrad Region and its individual areas. KSU, Kaliningrad, 111 pp
4. Orlenok VV (ed) (2002) Geographic atlas of the Kaliningrad Region. KSU, Kaliningrad, 276 pp
5. Barinova GM (2002) The Kaliningrad Region. Climate. Yantarnyi Kray, Kaliningrad, 413 pp
6. Volkova II, Korneevets VS, Fedorov GM (2002) In: Fedorov GM (ed) The Vistula (Baltic) Spit. The potential of possibilities. KSU, Kaliningrad, 72 pp
7. Kaliningrad Printed Yard (2002) Report on the state and protection of the environment of the Kaliningrad Region in 2001. Kaliningrad Printed Yard, Kaliningrad, 160 pp (in Russian)
8. Committee for Natural Resources of the Kaliningrad Region (2003) The state of environment in the Kaliningrad Region in 2003. Committee for Natural Resources of the Kaliningrad Region, Kaliningrad, 160 pp
9. Orlenok VV (ed) (2004) Geography of the Amber region. Yantarnyy Skaz, Kaliningrad, 416 pp
10. Tsybin YA (ed) (2004) Nature protection scheme of the Kaliningrad Region. TENAX MEDIA, Kaliningrad, 136 pp
11. Amber Tale (2004) Report on the state and protection of the environment of the Kaliningrad Region in 2003. Amber Tale, Kaliningrad, 216 pp (in Russian)
12. Zagorodnykh VA, Kunaeva TA (2005) Geology and minerals of the Kaliningrad Region. Yantarny Skaz, Kaliningrad, 208 pp
13. Orlyonok VV, Fedorov GM (2005) Regional geography of Russia. KSU, Kaliningrad, 259 pp
14. Regional Geography of Russia (2005) Kaliningrad Region: a manual for students enrolled in the geographical specialties. Immanuel Kant Baltic Federal University, Kaliningrad, 259 pp
15. Gusev AL, Maslov VN (eds) (2006) History of agriculture in the Kaliningrad Region: 1945–2006. IP Mishutkina I.V, Kaliningrad, 464 pp
16. Orlenok VV (ed) (2007) Geography of the Kaliningrad Region. Geographical fieldwork: a textbook. IKSUR, Kaliningrad
17. Naumenko EN (2007) The zooplankton of the Vistula Lagoon. KGTU, Kaliningrad, 210 pp
18. O. Yu. Shmidt IFZ RAN (2008) Complex seismological and tectonic studies for the seismic danger evaluation of the territory of Kaliningrad in 2008. Scientific and technical report, 306 pp
19. Kulakov VI et al (2008) The Curonian Spit. A cultural terrain. Yantarny Skaz, Kaliningrad, 432 pp
20. Kadzhoyan YS, Kasimov NS (eds) (2008) Oil and environment of the Kaliningrad Region. V. I. Terrestrial ecosystems. Yantarny Skaz, Kaliningrad, 360 pp
21. Tylik KV, Shibaev SV (eds) (2008) Lake Vistytis. SP Mishutkina, Kaliningrad, 144 pp
22. Shibaev SV, Khlopnikov MM, Sokolov AS et al (2008) Fishery inventory of transboundary waters of Russia (Kaliningrad Region) and Lithuania. SP Mishutkina, Kaliningrad, 200 pp

23. Dedkov VP, Grishanov GV (eds) (2010) The Red Book of the Kaliningrad Region. Immanuel Kant University, Kaliningrad, p 334
24. The Government of the Kaliningrad Region (2013) Report: on environmental state of the Kaliningrad Region in 2012. The Government of the Kaliningrad Region, Kaliningrad, 164 pp
25. Medvedev VA, Alexeev FE (eds) (2013) Kaliningrad Region nature. Landscapes. Specially protected natural areas. Istok, Kaliningrad, p 192
26. Kropinova EG (ed) (2015) Natural, historical and cultural crossroads in Baltic region. Pictorica, Kaliningrad, 144 pp
27. Fedorov GM (ed) (2016) Kaliningrad Region. Natural conditions and resources: rational use and protection. Immanuel Kant Baltic Federal University, Kaliningrad, 224 pp
28. Zhamoida VA et al (2017) Mineral resources of the Kaliningrad Region. Handb Environ Chem. doi:10.1007/698_2017_115
29. Sivkov VV, Zhamoida VA (2017) Amber deposits in the Kaliningrad Region. Handb Environ Chem. doi:10.1007/698_2017_116
30. Mikhnevich GS (2017) Composition of pre-quaternary surface and quaternary sediments allocation on the territory of the Kaliningrad Region. Handb Environ Chem. doi:10.1007/698_2017_87
31. Kolesnik TB (2017) Pleistocene deposits in the Kaliningrad Region. Handb Environ Chem. doi:10.1007/698_2017_88
32. Romanova EA et al (2017) Modern landscapes in the Kaliningrad Region. Handb Environ Chem. doi:10.1007/698_2017_86
33. Napreenko-Dorokhova TV, Napreenko MG (2017) The history and pattern of forest and peatland formation in the Kaliningrad Region during the Holocene. Handb Environ Chem. doi:10.1007/698_2017_89
34. Volkova II et al (2017) Eolian coastal-marine natural systems in the Kaliningrad Region. Handb Environ Chem. doi:10.1007/698_2017_96
35. Romanova EA et al (2017) Factors and patterns of current development of territorial units in the Kaliningrad Region. Handb Environ Chem. doi:10.1007/698_2017_90
36. Romanova EA et al (2017) Specific features of urban geosystems in the Kaliningrad Region. Handb Environ Chem. doi:10.1007/698_2017_91
37. Bernikova TA et al (2017) Environmental features of watercourses in the Kaliningrad Region. Handb Environ Chem. doi:10.1007/698_2017_108
38. Domnin DA et al (2017) Formation and re-distribution of the river runoff in the catchment of the Pregolya River. Handb Environ Chem. doi:10.1007/698_2017_97
39. Shibaeva MN et al (2017) Hydrobiological characteristics of water bodies in the Kaliningrad Region. Handb Environ Chem. doi:10.1007/698_2017_99
40. Ezhova EE et al (2017) The structure and composition of biological communities in the Pregolya River (Vistula Lagoon, the Baltic Sea). Handb Environ Chem. doi:10.1007/698_2017_107
41. Lange EK (2017) Phytoplankton community of small rivers of the Pregolya River Basin. Handb Environ Chem. doi:10.1007/698_2017_100
42. Volodina AA et al (2017) Flora and vegetation of the small rivers of the Pregolya River system in the Kaliningrad Region. Handb Environ Chem. doi:10.1007/698_2017_95
43. Mikhnevich GS (2017) The protection conditions of the groundwater against pollution in the Kaliningrad Region. Handb Environ Chem. doi:10.1007/698_2017_93
44. Shibaev SV et al (2017) Current status of the Lake Vistytis in Kaliningrad Region. Handb Environ Chem. doi:10.1007/698_2017_112
45. Volkova II et al (2017) Specially protected natural areas of the Kaliningrad Region. Handb Environ Chem. doi:10.1007/698_2017_98
46. Kropinova EG (2017) The reduction in the beach area as the main limiting factor for sustainable tourism development (case for the Kaliningrad Oblast). Handb Environ Chem. doi:10.1007/698_2017_113
47. Gerb MA et al (2017) Rare and protected macrophytes and semiaquatic plants of flora of the Kaliningrad Region. Handb Environ Chem. doi:10.1007/698_2017_106
48. Gritsenko VA et al (2017) Introduction. Handb Environ Chem. doi:10.1007/698_2017_118

Mineral Resources of the Kaliningrad Region

Vladimir Zhamoida, Vadim Sivkov, and Elena Nesterova

Abstract Mineral resources of the Kaliningrad Region are associated with the deposits of the platform cover: Paleozoic, oil, salt, and nonferrous and rare metals; Mesozoic, marls, limestones, and phosphorites; and Cenozoic, amber, brown coal, aggregates, peat, etc.

Keywords Geology, Kaliningrad Region, Mineral resources

Contents

1	Introduction	14
2	Stratigraphy	16
	2.1 Archean and Proterozoic	16
	2.2 Paleozoic	17
	2.3 Mesozoic	18
	2.4 Cenozoic	19

The original version of this chapter was revised. The erratum to this chapter is available at DOI 10.1007/698_2017_197.

V. Zhamoida (✉)
A.P. Karpinsky Russian Geological Research Institute (VSEGEI), 74, Sredny Prospect, St. Petersburg, Russian Federation

St. Petersburg State University, Universitetskaya Embankment, 13B, St. Petersburg 199034, Russian Federation
e-mail: vladimir_zhamoida@vsegei.ru

V. Sivkov
P.P. Shirshov Institute of Oceanology RAS, Nakhimovsky Pr., 36, Moscow 117997, Russia

Immanuel Kant Baltic Federal University, 14, A. Nevskogo Street, Kaliningrad, Russia
e-mail: sivkov@kaliningrad.ru

E. Nesterova
A.P. Karpinsky Russian Geological Research Institute (VSEGEI), 74, Sredny Prospect, St. Petersburg, Russian Federation

3 Mineral Resources 20
   3.1 Fossil Fuels 20
   3.2 Metallic Minerals 23
   3.3 Nonmetallic Minerals 24
4 Building Materials 25
5 Sapropel and Therapeutic Mud 26
6 Groundwater 27
7 Conclusions 28
References 29

1 Introduction

Since the middle of the nineteenth century to 1945, the territory of the former Eastern Prussia was studied by German geologists. Among the most significant works of the nineteenth century, it is possible to mention papers devoted to the geological structure of the Western Sambia [1] and study of Tertiary lignite and glauconitic formations by [2]. Essential part of the geological investigations carried out in the beginning of twentieth century in particular papers of [3, 4] was devoted to amber-bearing Paleogene deposits. A review of geological studies and papers published during the period of 1918–1945 was gathered in several reviews [5–7].

The next period of geological research began in 1945. Even during the military operations, the Soviet Military-Geological detachment no.1 collected a large number of data on geomorphology and geology. Immediately after the Second World War, gathering of available geological materials was broadened. It is possible to mention the map of Quaternary deposits of the Kaliningrad Region and adjacent Polish areas (1:1,000,000 scale) which was compiled by [8]. In the same year, a geological map of the Kaliningrad Region of scale 1:1,000,000 (G.V. Bogomolov, I.V. Danilevsky) was published. In 1947 R.P. Teush and T.V. Hakhonin made a hydrogeological map of the Kaliningrad Region and adjacent areas of Poland (1: 1,000,000 scale). In 1951–1952, geological mapping (1:200,000 scale) was carried out for the western part of the Kaliningrad Region [9].

Since 1957, the Kaliningrad expedition of the North-West Production Geological Association based in Gusev carried out geological, geophysical, and hydrogeological survey [6].

The study of the territory of the Kaliningrad Region by geophysical methods (aeromagnetic and gravimetric surveys at a scale of 1: 200,000 and electrical prospecting at a scale of 1:500,000) began in 1957. In 1961 the Baltic expedition of the Trust "Spetzgeofizika" started seismic exploration. The Sambia Peninsula and some other areas were covered by gravimetric and magnetic survey at a scale of 1:50,000 [10].

Since 1956, drilling of exploration and production wells for water began. To date, it was drilled with more than 2100 wells. In the 1960s, different geological and hydrogeological surveys of 1:50,000 and 1:200,000 scales in different districts of the Kaliningrad Region were carried out [11–15]. At that time, prospecting works for various types of minerals were also very active. Primorskoye amber deposit was discovered and then explored in detail. Preliminary exploration of the Plyazhevy area of the Palmnicken amber deposit was completed in 1971 [16]. The exploration work on brown coal was carried out in the north-west part of the Sambia Peninsula in 1968–1976; as a result Grachevskoe economic deposit was identified. In 1976–1977 the same expedition revealed a Mamonovskoe deposit of brown coal located in the southwest of the region [17]. According to the results of geological surveys in 1961–1967, several geological maps (1:200,000 scale) of the northern part of the territory were published in 1983.

A great contribution to the study of the geological structure of the Kaliningrad Region was made by Lithuanian geologists [18–21].

In 1969–1970 the Atlantic Branch of P.P. Shirshov Institute of Oceanology together with the Kaliningrad State University (today Immanuel Kant Baltic Federal University) conducted research of the outcrops of amber deposits in the coastal zone of the Sambia Peninsula, which allowed mapping the area of the amber-bearing layer of the so-called blue earth under a thin layer of sandy-silty deposits [22].

In 1970, the volume XLV of *Hydrogeology of the USSR* (Kaliningrad Region) was published; it described the engineering and geological conditions of the region on the basis of the actual material received by different scientific organizations [23].

The exploration work was also carried out for quartz sands [24], keramzite clays [25], sand, and gravel [26]; in the years during hydrogeological investigations, a large volume of drilling was performed in Baltiysk and Mamonovo [27, 28].

As a result of geological and geophysical studies followed by drilling of parametric, prospecting, and exploration wells in 1960–1970, 11 oil deposits were discovered in the Cambrian deposits.

Information about deep geological structure of the area was also obtained as a result of drilling of structural wells, aimed to assess the oil prospects of the Devonian sequence [29] and parametric wells up to the rocks of crystalline basement [30, 31].

Systematic oil exploration work in offshore area started in 1976 by the international company Petrobaltic. In 1982, exploratory drilling was started on the shelf, and in 1983 the largest Kravtsovskoye oil field in the region was discovered. In 1991, compilation of forecasting maps for oil and solid mineral resources of pre-Quaternary rocks was completed [32].

In 1995–2002 VSEGEI carried out a geological study of Quaternary and pre-Quaternary formations with consideration of the issues of stratigraphy, lithological-facial and structural features of the region, as well as patterns of mineral resource allocation and their predictive assessment [33]. In 2011 VSEGEI published the set of geological maps (1:1,000,000 scale) with explanatory note [17]). The set includes geological maps of pre-Quaternary and Quaternary formations, maps of mineral resources, hydrogeological and environmental geological maps, a lithological map of the bottom of the Baltic Sea, a forecast map for oil resources, and a volumetric model of the geological environment.

At the last period of time, most ambitious oil-prospecting works are conducted in the Kaliningrad Region by Lukoil-Kaliningradmorneft LLC and OAO

Kaliningradneft jointly with All Russia Petroleum Research Exploration Institute (VNIGRI). As a result of these works, new oil fields were discovered and promising local structures were identified.

2 Stratigraphy

The mostly full modern general description of the geological structure of the Kaliningrad Region area can be found in several publications [34, 6, 17]. These publications were used as a base for short general information about geological sequence in the Kaliningrad Region that is outlined below.

Two structural stages are distinguished in the geological section of the Kaliningrad Region: the lower one is composed of gneisses, crystalline schists, and amphibolites of the Archaean-Proterozoic age (platform basement) and the upper one is the Phanerozoic platform cover, represented by poorly dislocated and slightly metamorphosed sediments.

Sediment rocks of all geological systems, with the exception of Carboniferous, are represented in the sequence of the platform cover. They are characterized by relatively gentle bedding and completeness of stratigraphic sequence. The presence of major breaks and stratigraphic unconformities in the Lower Cambrian, at the boundary of the Lower and Middle Devonian, in the Lower Permian, between the lower Devonian and the upper Permian, allows to distinguish the structural complexes formed in the Caledonian, Hercynian, and Alpine cycles of tectogenesis.

2.1 Archean and Proterozoic

The crystalline basement, composed of Archean and Proterozoic rocks, occurs at a depth of 1,100–3,350 m with a general subsidence to the west. The most ancient of them are represented by Archaean and Early Proterozoic (Karelian) metamorphic stratified and intrusive complexes that form several structures.

The structure of the crystalline basement is a combination of linear, annular, and block elements. According to the degree of metamorphism of the basement rocks, two large zones are distinguished: the Western zone, within which the rocks of the amphibolite facies are developed, and the Eastern, where the rocks of the granulite facies of metamorphism predominate [17]. Within the zones, magmatized gneisses with interlayers and lenses of crystalline schists and amphibolites are common. Relict structures and petrochemical diagrams of gneisses allow assuming primary volcanogenic and volcanogenic-sedimentary nature of these rocks.

The intrusive formations of the Lower Proterozoic are combined into gabbroid, granodiorite, plagiogranite, and granite complexes, which form linear and annular structures. Intrusions are represented by large interstratal bodies, stock-like massifs, and dikes.

2.2 Paleozoic

Cambrian System Marine terrigenous Cambrian deposits form the bottom part of the Baltic syneclise section. Deposits of the Lower and Middle Cambrian are most widespread and lesser the deposits of the Upper Cambrian. The depth of occurrence of the roof of the Cambrian rocks on land varies from 1,250 m.b.s.l. in the eastern to 2,600 m.b.s.l. in the southwestern part of the region, and the thickness varies from 120 m to 270 m, respectively. Cambrian deposits unconformably overlap Archaean-Proterozoic basement rocks and underlie Ordovician formations with the intraformational interruption.

Ordovician System The Ordovician system is represented by terrigenous-carbonate marine deposits distributed within the whole Kaliningrad Region. The depth of the Ordovician roof occurrence varies from 1,150 m.b.s.l. in the east of the region to 2,616 m.b.s.l. in offshore area. The general monoclinic dipping of the structure in the western direction (Baltic monocline) is complicated by a transverse uplift, where the thickness of Ordovician formations is minimal (less than 60 m).

Silurian System Silurian deposits are distributed throughout the area of the Kaliningrad Region and represented by marine terrigenous and carbonate rocks. It is possible to mark predominance of relatively deep-water carbonate mudstones and clays differing by the color of rocks, carbonate content, and content of organic matter. Structural contour lines of Silurian formation roof vary from −700 m in the east to −1,400 m in the western part of the region in offshore area. In the same direction, the thickness of the sediments increases from 400 to 1,446 m. Silurian sediments are deposited transgressively on the eroded surface of Ordovician formations and unconformably overlapped by Devonian deposits.

Devonian System Devonian deposits occur only in the northern part of the region. They are located in the southwest margin of the Main Devonian Field of the Russian Platform and represented by all three series. The depth of the roof is traced at absolute elevations from 500 m.b.s.l. in the northeastern part to 1,200 m.b.s.l. in the west. They are represented by marine, lagoon, and continental facies and overlie transgressively on the Silurian formations with a change in the structural plan. The direction of the layers dipping and increasing of their thickness is replaced by the northern one in comparison with the western one for Silurian formation. The lower Devonian contact is fixed by the change of gray and multicolored deposits and the upper one by the unconformity with the Lower Permian deposits, which complete the Hercynian tectonic cycle.

Permian System Permian deposits, represented by two series, form the southeastern part of the Baltic Syneclise. They are confined to two structural levels – Lower Permian, formed at the final stage of Hercynian tectogenesis, and Upper Permian, belonging to the Alpine cycle. Lagoon-marine deposits predominate within Permian section. At the basement, Permian deposits are represented by gravels and sandstones with carbonate cement, replacing upward the section by carbonate rocks. The upper part of section is composed of evaporites of the Pregolya suite (anhydrite, rock salt with interlayers of potassium-magnesium salts), alternating with limestones and dolomites. The depth of the Permian roofing in the northeast of the region is 500 m.b.s.l. In offshore area in the northern part of Russian economic zone, the Permian deposits are exposed on the pre-Quaternary surface and pinch out near the slope of the Gdansk depression. The thickness of Permian deposits onshore achieves 360 m. Permian deposits with a stratigraphic break and disagreement overlie the Devonian and Silurian deposits and, in turn, after interruption in deposition during the Upper Tatar time overlap Triassic formation.

2.3 Mesozoic

Triassic System Triassic sediments cover erosional surface of the Upper Permian formation and overlap with erosion Jurassic sediments. Triassic deposits are represented by Lower and Upper series. The Upper Triassic is distributed only in the southern part of the region. Formation of a relatively thick Triassic sequence took place under arid climate conditions in the lagoon-sea basin, as well as in the subcontinental brackish and freshwater basins (regressive and emersion stages of the region's development). In the sole is a thin layer of sandstone. Above the section there is a stratum of variegated and red-colored clays of montmorillonite and montmorillonite-hydromicaceous composition with an admixture of fine crystalline dolomite and with interlayers of marls and oolitic limestones with a thickness up to 34 m. The position of roof of the Triassic formations varies from 280 m.b.s.l. in the northeastern part of the region to 610 m.b.s.l. in its southwestern part, and thickness of sequence varies from 202 to 520 m, respectively. The Triassic deposits are exposed at the pre-Quaternary surface in the northern part of offshore area of the Russian sector.

Jurassic System Jurassic deposits are included with Triassic and Cretaceous deposits in the structure of large Alpine structure – the Polish-Lithuanian basin. The Jurassic deposits overlie unconformably on different erosional horizons of Triassic and overlap unconformably with erosional contact by Cretaceous sediments. Lithologically and genetically, Jurassic deposits are separated into two large strata. From the middle of the Early Jurassic up to early Callovian, terrigenous continental and coastal-continental sediments accumulated with some interruptions. They are represented by weakly cemented sandstones with interlayers of kaolinite-hydromica clays and calcareous silts, with rare inclusions of charred plant remains and carbonaceous clays. The thickness of the strata reaches 135–200 m. From the Middle Callovian to the end of the Jurassic, due to the downwarping of the region, a stratum of shallow marine sediments was formed. The thickness of the second stratum is 130–340 m. The depth of occurrence of the roof surface of Jurassic deposits varies from 95 m.b.s.l. in the northeast part of the region to 550 m.b.s.l. in the southwest. Jurassic deposits are exposed at the pre-Quaternary surface in the central part of offshore area of the Russian sector.

Cretaceous System The Cretaceous sequence is characterized by numerous internal breaks. These deposits transgressively overlie the Upper Jurassic sediment rocks. Commonly Cretaceous deposits are overlapped by Quaternary formations, and only in the southern and western parts of the region, they are covered by Paleogene sediments. Lower Cretaceous (Aptian-Albian) terrigenous deposits are composed of greenish-gray glauconite-quartz sands, silts, and weakly cemented siltstones. The content of glauconite in these rocks can reach 30–50%. In the upper part of the Lower Cretaceous section, the nodules of phosphorites occur. The Upper Cretaceous (Cenomanian-Maastrichtian) terrigenous-siliceous-carbonate sequence is composed of glauconite-quartz sands, chalk-like marls, silts, and weakly cemented siltstones containing a small admixture of gravel material. The lenses of strongly silicified rocks and pyrite concretions are found. At their base, a thin (0.5–2.0 m) layer of medium- and coarse-grained glauconite-quartz sandstones with siliceous-phosphate calcite cement (the "phosphorite slab") is deposited. The maximum thickness of Cretaceous deposits is found in the southwest of the region, where it reaches 320 m. The depth of their occurrence varies from 10 m.b.s.l. in the valley of the Neman River to 277 m.b.s.l. in the southeastern part of the region.

2.4 Cenozoic

Paleogene System Paleogene deposits are well studied due to their unique amber content. They are developed only in the southern part of the region and most widely in the southwestern part. They lie on erosional roof of the Upper Cretaceous sediment rocks and overlap sediments of the Neogene and Quaternary formations. The thickness of Paleogene sequence varies in the range of 80–190 m. Paleogene deposits are represented by shallow water thin-grained clastic rocks, sometimes slightly calcareous.

Two structural-facies zones are distinguished in the area of Paleogene deposit development: the Western and Southeastern zones, differing mainly in the completeness of the section with insignificant differences in the composition of the deposits. In the Southeastern zone, there are no deposits of Oligocene, Upper, and Middle Eocene. The Western zone is distinguished by a much more complete sequence, consisting of a successive series of seven suites representing all stages Paleogene development except Lutetian. Lutetian time was everywhere the time of a break in the Paleogene sedimentation.

The Paleogene sediments are sometimes exposed on the surface of the Sambia Peninsula, and they are uncovered by quarries for the extraction of amber. Within the rest of the territory, they are overlaid by Neogene and Quaternary sediments and occur at a depth of up to 280 m.b.s.l.

Neogene System Neogene deposits occur only in the southwestern part of the region – on the Sambia Peninsula and to the south from the Vistula Lagoon. Their thickness reaches 70 m. They lie on erosional roof of the Eocene sediments and overlap with Quaternary formations. They are represented by lacustrine-boggy and river sediments, including sand, clay, lignite, and brown coal.

Quaternary System Within the Kaliningrad Region in accordance with lithological and facial features of the Quaternary sediments, two zones are distinguished: South Kaliningrad and North Kaliningrad. The South Kaliningrad zone occurs within the boundaries of neotectonic basin and is characterized by the completeness of the section. The total thickness of the Pleistocene formations here is 270 m.

The North Kaliningrad zone is characterized by absence of the deposits of the lower and middle stages of Pleistocene. The average total thickness of the Pleistocene sediments here, as a rule, does not exceed 60–70 m. The most complete Quaternary sequence can be found only within the deep palaeovalleys located in the mouth area of the Neman River valley and on the Sambia Peninsula.

During the Pleistocene, the territory was glaciated three times; accordingly, it is possible to find deposits of three glacial and three interglacial periods, as well as preglacial formations.

The polygenetic sediments of the Holocene are sufficiently investigated. It is possible to distinguish different types of marine and alluvial sediments, as well as aeolian, boggy, etc.

3 Mineral Resources

Actually all economic minerals discovered within the Kaliningrad Region are spatially and genetically associated with the deposits of the platform cover (Figs. 1 and 2). They are represented by oil, drinking and mineral underground water, amber, and building raw materials which are actively exploited. Potassium-magnesium salts, numerous occurrences of which are known here, are of greatest economic interest among undeveloped types of other useful minerals. The occurrence of rock salt, zeolite, lignite, peat, and ferrous and nonferrous metals should also be mentioned. Mostly, the full description of mineral resources can be found in several publications, which were used for short general compilation that is outlined below [35, 6, 17].

3.1 Fossil Fuels

Oil and Gas The oil fields which are distributed in the Kaliningrad Region and adjacent offshore areas of the southeastern Baltic Sea occur within the Baltic oil-bearing area and are mainly associated with the Deimenian suite of the Middle Cambrian (Fig. 1). The productive section of the suite is represented by a stratum of interbedded sandstones and siltstones, with a subordinate number of intercalations of mudstone and clay. The oil deposits are confined to local anticlinal uplifts and

Mineral Resources of the Kaliningrad Region

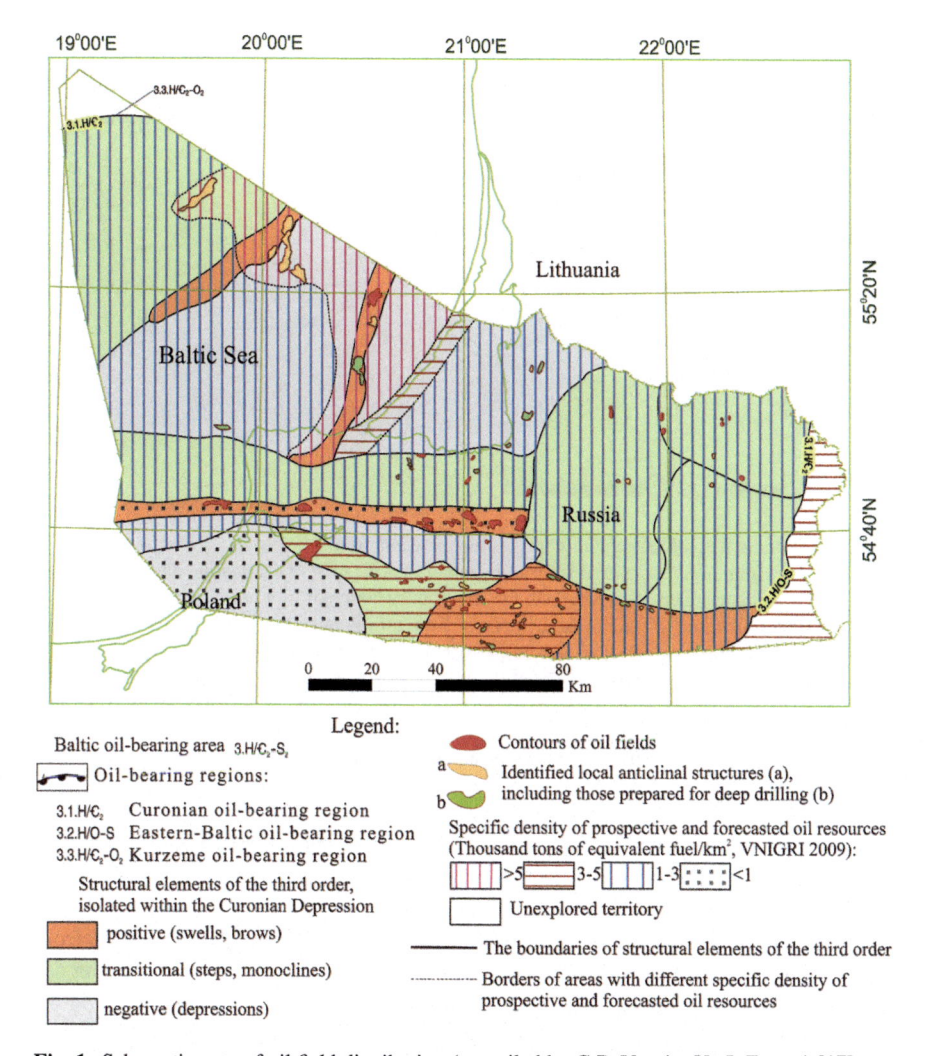

Fig. 1 Schematic map of oil field distribution (compiled by G.P. Vargin, Yu.I. Zytner) [17]

located at depths from 1,500 m.b.s.l. to 2,500 m.b.s.l. These deposits are characterized by structural and, to a lesser extent, combined types of traps. The main types of oil traps are structural – reservoir, arched, disjunctive-shielded, and less often structural–lithological combination. Rarely oil occurrence is found in Ordovician and Silurian carbonate rocks. The presence of oil traces (liquid and solidified) were fixed in Permian sandstone, anhydrite, and limestone. Thirty-seven oil fields have been discovered within the Kaliningrad Region to 2011. All these fields are concentrated in the Curonian oil-bearing region. Their reserves make up more than 70% of total oil reserves of the entire Baltic oil-bearing area. Two of them are located in offshore area, the rest in the mainland [36, 37, 17].

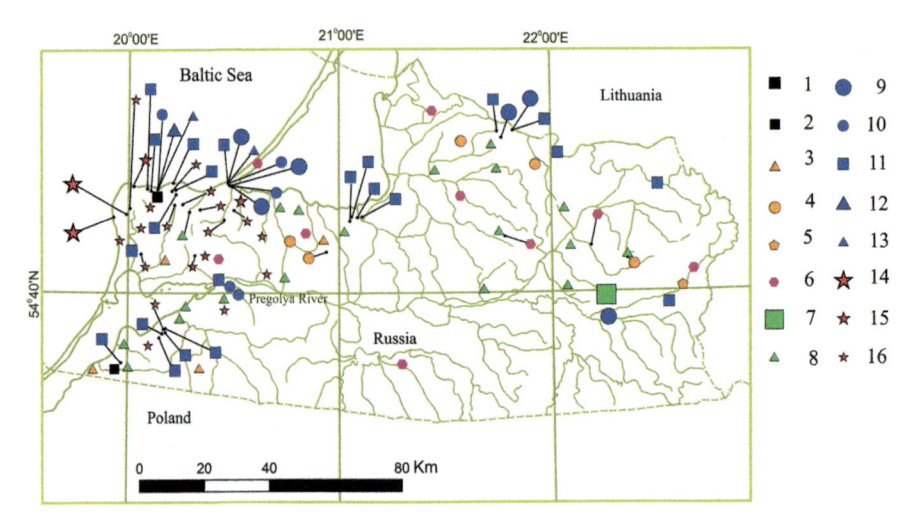

Fig. 2 Schematic map of selected mineral resources (economic deposits and site of occurrence) in pre-Quaternary deposits (simplified from Lukyanova et al. 2011): (*1*) economic deposit of brown coal, (*2*) site of occurrence of brown coal, (*3*) site of occurrence of iron ore, (*4*) site of occurrence of Pb-Zn mineralization, (*5*) site of occurrence of Zn-Cu mineralization, (*6*) site of occurrence of zeolites, (*7*) economic deposit of rock salt, (*8*) site of occurrence of potassium-magnesium salts, (*9*) medium deposit of underground mineral water, (*10*) minor deposit of underground mineral water, (*11*) deposit of fresh underground water, (*12*) medium deposit of underground mineral bromine medicinal water, (*13*) minor deposit of underground mineral bromine medicinal water, (*14*) large amber-bearing deposit, (*15*) minor amber-bearing deposit, (*16*) site of occurrence of amber-bearing deposits

The oils of the Kaliningrad Region are mainly of methanoneften type, light (density up to 0.855 g/m^3), low-sulfur, low-resinousness (0.78–9.71%), low-viscous, and paraffinic (2.02–12.3%).

The geological reserves of individual oil fields are measured in the range from 11 to 5,300 thousand tons onshore and 15,800 thousand tons offshore. The recoverable reserves of open deposits are measured in the range from 7–10 thousand tons to 10 million tons [17].

Pure gas fields in the region have not been found. Associated gas in the total amount of hydrocarbons in the Kaliningrad Region is 3% [38]. The gas is extracted from 23 oil fields located on land. The volume of associated gas of the exploited oil fields is estimated approximately at 140–150 million m^3 [35]. The gas factor of oil is 10–50 m^3/t.

Brown Coal Neogene deposits of brown coal belong to the Paleogene-Neogene epoch of coal formation of the Russian platform [39]. Exposures of brown coal-bearing Neogene deposits are observed on the Baltic Sea coast. They are developed over an area of about 600 km^2 and reach a thickness of 50 m. Coal-bearing deposits of lacustrine-alluvial genesis are distributed as isolated bodies separated by ancient valleys. The predicted resources of Neogene brown coal do not exceed 100 million tons.

Furthermore layers of brown coal up to 10 m thick were discovered in Jurassic deposits [40].

Shale Oil The layers of shale oil are found in the terrigenous-carbonate formation of the Upper Permian. Usually the content of organic matter in shaly carbonaceous-silt-carbonate rocks is 10–20%. There are interlayers in which the content of organic matter can reach 85% [41]. The small thickness of the shale strata (1–2 m) and considerable depth of there occurrence (up to 700–800 m) does not allow considering them to be useful for industrial extraction. The high content of some nonferrous and rare metals was found in shale [41, 42].

Peat About 300 peat deposits are known in the Kaliningrad Region. They occupy more than 4% of its territory. The lowland type of peat accounts to about 65% of the area of peatlands, top peat about 30%, and transitional type about 5%. Peat ash content varies from 3 to 50%, average of 15–20%. The top peat is characterized by the lowest ash content (up to 10%). The calorific value of peat varies from 10.5 to 21.6 megajoule. Peat resources are currently estimated at 238.6 million tons (dry weight) [6, 17].

3.2 Metallic Minerals

Iron Ore Several types of iron ore are found in the Mesozoic (Lower Triassic, Middle and Upper Jurassic) and Holocene deposits [6, 17].
The Triassic ore is represented by brown sandstones with thin siderite interlayers. Cement of sandstones is composed of goethite-leptochlorite. Jurassic ores are represented by sandstones and limestones containing oolites. Oolites are composed of shamosite and hydrogelite with siderite-clayey cement. The average content of total iron for different types of ore varies from 17.6 to 20.0%. The predicted resources of iron ores of the Mesozoic are estimated to be 5.0 billion tons.

Bog ores associated with the Quaternary sediments lie in the form of lenticular deposits. Most commonly they are found in the southern part of the Sambia Peninsula and in the valleys of Niemen and Pregolya rivers. The Fe_2O_3 content in bog ore varies from 45.1 to 52.1%.

Nonferrous and Rare Metals Four stratiform occurrences of nonferrous and rare metals have been found in the Upper Permian carbonaceous shale. The summary content of nonferrous metals (Pb, Zn, Cu) varies from 10.4 to 26.5 kg/t, whereas V, Mo, Co, Cd, etc. from 4.5 to 5.5 kg/t. Some types of ore of nonferrous metals (Cu, Zn, Pb) are associated with the Devonian and upper Silurian deposits.

The group of nonferrous metal ore also includes alumina-containing rocks represented by allites with Al_2O_3 content of up to 16.3%. The rocks containing alumina are confined to the Upper Permian, Silurian, and Ordovician deposits. The most promising area for identifying bauxite deposits is Nesterovsky District.

Special exploration studies for nonferrous and rare metals were not carried out, so the region's prospects for these metals require further justification [41, 42]. Nevertheless, it is possible to predict identification of stratiform deposits of nonferrous and rare metals lying at depths of more than 1,300 m.

3.3 Nonmetallic Minerals

The group of nonmetallic minerals includes rock salt, potassium-magnesium salts, amber, zeolites, strontium ores, quartz sands, phosphorites, and glauconite [43, 6, 44, 17].

Rock Salt Rock salt was formed during two cycles (Verra and Stassfurt), and its deposits are located within the Kaliningrad salt basin. This basin covers an area of about 9,000 km^2 within the Kaliningrad Region. The contour of the basin has a rather complicated configuration. The depth of the salt bed occurrence increases from northeast to southwest (from 530 to 1,230 m). Stone salt is confined mainly to the deposits of the Pregolya suite of the Upper Permian (Verra cycle). The thickness of the salt layer reaches 214 m. Halite, associated with the sediments of the Aystmar suite (Stassfurt cycle), is found in the southwestern part of the Kaliningrad Region.

In the northern part of Kaliningrad Region, rock salt occurs at the depths of 530–600 m.b.s.l., which are quite accessible for mining. At present, underground storage facilities for liquefied gas in the rock salt layer are worked out in the area of Romanovo village located on the Sambia Peninsula. Resources of rock salt achieve more than 1,500 billion tons.

Potassium-magnesium Salts A large number of potassium-magnesium salts are found among the halogen stratum of the Pregolya suite of the Upper Permian (Verra cycle). They are represented by non-sulfate (sylvite, carnallite), sulfate (polygalite, kieserite), and mixed double salts (kainite).

The second type of potassium-magnesium salts is associated with the subsalt anhydrite of the Pregolya suite and represented by polygalite. These salts lie either directly beneath the rock salt stratum or are separated from it by a thin layer of anhydrite (3–10 m). Currently, more than two dozen of such salt occurrences are known.

Resources of potassium-magnesium salts in the region are estimated at 250 million tons [45].

Strontium Ores Celestine, containing strontium, occurs almost throughout the entire section of the sulfate-halogen-carbonate sequence of Zechstein (upper Permian). According to the spectral analysis, the strontium content exceeds 6.0%. Celestine is represented by small (0.3–0.7 mm) spherulites and lamellar aggregates filling the leaching cavities of 10–25 mm in size. According to the visual definition, the content of celestine can reach 20%. The thickness of the layers enriched by it varies from 1.5 to 3.0 m.

Apatite Apatite is found in granite-biotite schists of the crystalline basement (depth 1,690 m). The content of apatite reaches 20%. The content of phosphorus pentoxide is 2.5–3.0%.

Phosphorites Phosphorites are found in the deposits of the Paleogene, Upper and Lower Cretaceous, and Upper Jurassic. The deposits of the Paleogene and Upper Cretaceous are most studied for phosphorite content. The content of phosphorus pentoxide in the nodular phosphorites varies from 5.0–26.7%.

Glauconite There are high prospects for glauconite deposits, which can be used as permutite, mineral pigment, and potassium fertilizer. The highest concentrations of glauconite (30–40%) are found in the Paleogene (Prussian suite) and Cretaceous deposits.

Amber Amber is associated with the deposits of the Prussian suite of the Upper Eocene, represented by glauconite-quartz sands and clay siltstones of the sea (delta) genesis, which have been called "blue earth." Industrial concentrations of amber are controlled by the area of distribution of the deposits of the Prussian suite. The world's largest Primorsky amber deposit can ensure the smooth operation of the mining enterprise for more than 100 years. Two new amber fields (Filinsko-Shatrovskoe and Pokrovskoye) contribute to a further increase of reserves. The Palmenniken field, developed in the past on land, found its offshore continuation.

The Quaternary glacial moraine deposits locally include erratic blocks of Paleogene rocks containing an amber-bearing "blue earth." Nadezhdinsky amber deposit is a conglomeration of such erratic blocks.

Zeolites The Jurassic, Upper Cretaceous, and Paleogene deposits are often characterized by increased content of zeolites. This area belongs to the Baltic zeolite-bearing region, which is part of a large East European zeolite-bearing province [40]. The highest (more than 30%) content of zeolites (clinoptilolite) is confined to the deposits of the Upper Jurassic (Lermontov suite). Recently, prospects for the zeolite resources of the Paleogene deposits (Lubawa suite), in which the content of clinoptilolite can reach 20%, have been positively evaluated.

Quartz Sand Sands with the highest content of silica (for ceramic and refractory raw materials) are confined to the Neogene deposits overlying amber-bearing Paleogene rocks and to aeolian sediments of Holocene located on the right bank of the Neman River [46].

4 Building Materials

Sandy-gravel Aggregates Sandy-gravel mixture is associated mainly with the fluvioglacial deposits of the last glaciation. They are also found in alluvial and rarely in glaciolacustrine sediments. Totally 28 economic deposits of sandy-gravel aggregates were explored. Their total resources exceed 143 million m^3. The largest deposits are situated in the central part of the Kaliningrad Region and traced along the ancient glacial drain trough, which is located in a modern valley of river Pregolya. Deposits of sand and gravel occur at the depths from a few meters to 27 m under a layer of peat, sand, sandy loams, or loams. Useful thickness of deposits varies from the first meters to 26.9 m; its length for individual deposits achieves 4 km. The average content of gravel (grains larger than 5 mm) is 30–35%. The sandy component is represented by coarse- and medium-grained particles of quartz, feldspar, and carbonate rocks. The petrographic composition of gravel is formed by quartz, limestone, granite, sandstones, and metamorphic rocks [6, 17].

Building Sand Sand separated from sand-and-gravel deposits is most often used as building sand. However, this sand is mostly suitable only as fine filler for concrete. Sand, suitable for plastering and masonry mortars, is confined to the glaciolacustrine and fluvioglacial sediments of the last glaciation. Two deposits of sand are suitable for production of silica brick and porous concrete. Totally 23 economic deposits of sandy-gravel aggregates were explored.

Brick Clay Totally 14 deposits of clays are suitable for production of building bricks, tiles, and drainage pipes. A useful clay stratum is confined to the glaciolacustrine deposits of the last glaciation. This clay is characterized by medium to high plasticity with the inclusion (up to 0.3%) of coarse-grained carbonate particles. Manufacture of ceramic products using this clay requires addition of fine-grained sand, which is used as emaciating material. A general lack of clay is changing composition of coarse-grained carbonate inclusions and nonuniform plasticity.

Keramzit Clays Two deposits of keramzite clay were explored (Yablonovskoe and Lermontovo). Useful rock strata are confined to the glaciolacustrine deposits of the last glaciation. In the natural state, the clays are not swelled well enough. Only with the addition of 0.5–1.0% of diesel oil, it is possible to get expanded clay gravel. Yablonovskoe deposit was exploited. Currently, production of keramzite clay sharply decreased.

Raw Materials for the Production of Binders Chalk marl with a high content of CaO can be considered as raw materials for the production of binders. Limestone and anhydrite can be used as raw material for the production of Portland cement and gypsum binder material. However, due to the deep occurrence of these rocks, they do not represent practical interest at the moment.

5 Sapropel and Therapeutic Mud

The Sapropel In the Kaliningrad Region, there are several explored deposits and numerous locations of the sapropel. According to the conditions of occurrence, it is possible to distinguish three groups of this sediment: occurring at the bottom of the modern lakes, occurring under the peat, and buried (older) deposits of lacustrine sapropel. Deposit of modern lacustrine sapropel was explored in detail in the southern part of the region. In the north of the region, in the Slavsky District, a deposit of buried sapropels – Verhny Biser – was discovered. The deposit is confined to the delta sediments (meander lake facies), and sapropels occur here at a depth of about 3 m. Overlying sediments are represented by loam and sandy loam. Sapropels, lying under peat bogs, were practically not studied.

Therapeutic Muds Only one deposit (Lake Goreloe), located to the south of Svetlogorsk, has been studied as therapeutic muds.

6 Groundwater

The Kaliningrad Region is located in the central part of the Baltic Artesian Basin, where fresh, mineral, and thermal waters, as well as commercial brines, have been identified in various hydrogeological levels.

Fresh Underground Water The main source of underground fresh water is the Quaternary aquifer complex. The source of fresh water supply in the western and locally in the southern part of the region is the Paleogene water-bearing complex and in the north the Upper Cretaceous aquifer. Total reserves of fresh water in the Kaliningrad Region by various authors are estimated at 300,000–380,000 m^3/day [23, 35, 6, 17].

Mineral Water Practical use of mineral waters was started in the region in the late 1960s, with the Albian-Cenomanian aquifer as a source of healing table water. High-pressure waters of the Pliensbachian-Callovian aquifer with a mineralization of 10–50 g/dm^3 can be used for the preparation of medicinal baths, as well as diluted water of the Devonian aquifer.

Thermal Waters Due to the active oil exploration within the region, numerous geothermal anomalies have been identified (Fig. 3). Within the oil fields confined to the Cambrian deposits, the water below the water-oil contact has a temperature from 60°C (Gusevskaya area) to 96°C (Ladushkinskaya area) [47, 17]. The zone of constant temperatures, where the annual amplitude does not exceed 0.5°C, is fixed at a depth of about 20 m. The average value of the geothermal stage is 30–31 m; the geothermal gradient is 1–5°C/100 m. Within the region, the temperature rises in the south direction. If in the northern part of the region, the temperature at depths of 1,450–1,550 m (Slavsk town) is 41°C; then in the vicinity of Gusev town, it already reaches 54°C [17].

Industrial Brines Paleozoic hydrogeological levels are characterized by presence of waters with total mineralization up to 203.5 g/dm^3. In these brines, the maximum bromine content can reach 1,750 mg/dm^3, boron 135 mg/dm^3, and iodine 16 mg/dm^3 [17].

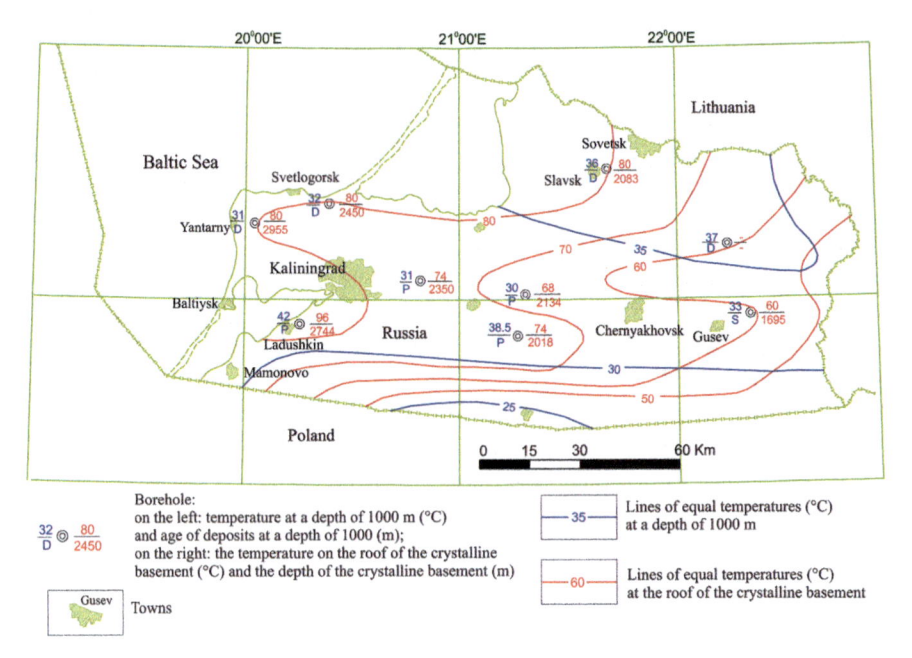

Fig. 3 Schematic map of groundwater temperature distribution (compiled by N.G.Nikutina) [17]

7 Conclusions

The Kaliningrad Region is characterized by wide variety of mineral resources. All of them are associated with the deposits of the platform cover. To date, the mineral resources of the Kaliningrad Region are represented by 37 oil fields, 3 deposits of amber, a salt deposit, deposits of fresh (25 deposits) and mineral (6 deposits) water, numerous deposits of building materials, promising manifestations of potassium-magnesium salts, amber, and other economic minerals.

Oil, the extraction of which is actively conducted in the region, is mainly confined to the Middle Cambrian deposits. Rock and potassium-magnesium salts are found in Permian deposits. Relatively high concentrations of different nonferrous and rare metals are fixed in Ordovician, Silurian, Devonian, and Permian deposits. There are no significant minerals in the Mesozoic stratum, except marls, limestones, and phosphorite, which can be used for production of building materials, cement, and phosphate fertilizers. Cenozoic in the region is represented by sediments of the Paleogene, Neogene, and Quaternary systems. Paleogene coastal marine facies are associated with such minerals as amber, phosphorite, and glauconite. Neogene sediments are characterized by the presence of glass sands and brown coal. The glacial complex provides the region with building materials, such as sandy-gravel aggregates and ceramic clay. Peat, sapropel, and healing mud are connected with the modern sedimentation.

References

1. Berendt GC (1866) Erläuterungen zur geologischen Karte des West-Samlandes. Schriftenreihe der physikalisch-ökonomischen Gesellschaft zu Königsberg 7:131–144
2. Zaddach E (1867) Das Tertiargebirge des Samlands/Schriften der Physik.-ökonom. Gesellschaft VIII, Königsberg, pp 85–194
3. Jentzsch A (1903) Verbreitung der bernsteinführenden blauen Erde. Z Dtsch Geol Ges, Monatsberichte 55:122–130
4. Tornquist A (1914) Die Wirkung der Sturmflut von 9 bis 10 Januar 1914 auf Samland und Nehrung. Schriften der Physik.-ökonom. Gesellschaft, pp 241–256
5. Geological study in the USSR (1970) Volume 6. Kaliningrad Region of RSFSR. The period 1918–1945, issue 1. Printing Works, Vilnius, p 129 [in Russian]
6. Zagorodnykh VA, Kunaeva TA (2005) Geologiya I mineral'nye resursy Kaliningradskoy oblasti (Geology and mineral resources of the Kaliningrad Region). Yantarny Skaz, Kaliningrad, p 171. [in Russian]
7. Mikhnevich GS (2014) Rol' nemetskih uchenykh v isuchenii chetvertichnoy geologii Kaliningradskoy oblasti (The role of German scientists in the study of the quaternary geology of the Kaliningrad Region). Actual Probl Hum Nat Sci 10:459–463 [in Russian]
8. Vereisky NG (1946) Poyasnitel'naya zapiska k karte chetvertichnykh otlozheniy Kaliningradskoy oblasti (Explanatory note to the geological map of quaternary deposits of the Kaliningrad Region. RSFSR and adjacent areas of Poland), 1: 100,000 scale. Unpublished report. Trust "Specgeo", Moscow, p 302 [in Russian]
9. Valuev GA, Osipova AI, Kirshin Yu N (1952) Otchet o kompleksnyh geologicheskih i gidrogeologicheskih pochvennyh issledovanijah v masshtabe 1:200,000, proizvedennyh v zapadnoj chasti Kaliningradskoj oblasti RSFSR (Report on complex geological and hydrogeological soil studies, 1: 200,000 scale, conducted in the western part of the Kaliningrad Region). Unpublished report. Fifth Geological Administration, Leningrad, p 248 [in Russian]
10. Izraileva EN (1983) Otchet o rezul'tatah gravimetricheskoj i magnitnoj s'emok masshtaba 1:50,000 na territorii Kaliningradskoj oblasti (Sambijskaja ploshhad') za 1981–1982 gg (Report on the results of gravimetric and magnetic surveys at a scale of 1:50,000 on the territory of the Kaliningrad Region (Sumbian Area) in 1981–1982. North-West Geological Association). Unpublished report [in Russian]
11. Zharkov MP, Polferova VM, Ivanova ZK (1964) Geologicheskoe stroenie i gidrogeologicheskie uslovija juzhnoj chasti Kaliningradskoj oblasti. Otchet o geologicheskoj s'emke m-ba 1:200,000 (Geological structure and hydrogeological conditions of the southern part of the Kaliningrad Region. Report on geological survey in scale 1:200,000 on sheets N-34-XIV, XV, XVI, XVII). Gusev, Kaliningrad geological and hydrogeological expedition. Unpublished report, p 1109 [in Russian]
12. Lasberg IK, Zhao S, Yeltsina GN (1965) Otchet o kompleksnoj geologo-gidrogeologicheskoj s'emke masshtaba 1:50,000, provedennoj v predelah Kaliningradskogo poluostrova v 1962–64 gg (Report on an integrated geological and hydrogeological survey of a scale of 1:50,000, conducted within the Kaliningrad Peninsula in 1962–64). State Geological Committee of the Lithuanian SSR, Kaliningrad Expedition. Unpublished report [in Russian]
13. Kazanov Yu V, Makarova EP, Ivanova ZK (1967) Otchet po geologo-gidrogeologicheskoj s'emke listov N-34-VIII-IX m-ba 1:200 000 za period 1961-66 gg (Report on the geological and hydrogeological survey of sheets N-34-VIII-IX on a scale of 1: 200,000 for the period 1961-66). Gusev, Kaliningrad geological and hydrogeological expedition. Unpublished report [in Russian]
14. Kazanov Yu V, Makarova EP, Ivanova ZK (1983) Gosudarstvennaja geologicheskaja karta SSSR (State geological map of the USSR. Scale 1: 200,000. Series Pribaltiyskaya. Sheet N-34-VIII-IX. Explanatory note). Moscow, p 116
15. Eryukhin VA, Vakorin VF (1971) Otchet po geologo-gidrogeologicheskoj s'emke masshtaba 1:50,000, provedennoj v jugo-zapadnoj chasti Kaliningradskoj obl. 1969–71 gg (The report on

geological and hydrogeological survey at a scale of 1:50,000, held in the southwestern part of the Kaliningrad Region in 1969–71). Gusev, Kaliningrad geological and hydrogeological expedition. Unpublished report [in Russian]

16. Nakhabtsev Yu S, Bibikov BI, Zharkov MP (1972) Otchet o geologorazvedochnyh rabotah, provedennyh na Pljazhevom uchastke Pal'mnikenskogo mestorozhdenija jantarja v Zelenogradskom rajone Kaliningradskoj oblasti v 1970–71 gg (A report on geological exploration carried out on the beach area of the Palmnicken amber deposit in the Zelenogradsky district of the Kaliningrad Region in 1970–71). Kaliningrad geological and hydrogeological expedition. Unpublished report [in Russian]

17. Lukyanova NV, Bogdanov Yu B, Vasileva OV, Vargin GP, Verbitskiy VR, Gorbatsevich NR, Zhamoida VA, Zytner Yu I, Kirikov VP, Maksimov AV, Nikutina NG, Semenova LR, Sivkov VV, Fenin GI (2011) Gosudarstvennaja geologicheskaja karta Rossijskoj Federacii. Masshtab 1:1000000 (tret'e pokolenie). Serija Central'no-Evropejskaja. List N-(34) (State geological map of the Russian Federation. Scale 1:1000000 (third generation). Series Central European. Sheet N-(34)) – Kaliningrad. Explanatory note. VSEGEI Cartographic Factory, St. Petersburg, p 226 [in Russian]

18. Suvezdis PI (1965) K voprosu o vozraste "nemoj" tolshhi terrigennyh podcehshtejnovyh obrazovanij Juzhnoj Pribaltiki (Concerning the age of the "barren" stratum of terrigenous under-Zechstein formation in the Southern Baltic). Geology and oil content of the Paleozoic of the Southern Baltic. In: Proceedings of the institute of geology, issue 1. Vilnius, pp 84–91 [in Russian]

19. Kondratene OP, Eryuhin VA (1974) Novye razrezy butenajskogo (lihvinskogo, gol'shtejnskogo) mezhlednikov'ja Kaliningradskoj oblasti (New sections of the Butenay (Likhvinian, Holsteinian) interglacial of the Kaliningrad Region). In: Problems of studying of quaternary deposits of Lithuania. Proceedings of LitNIGRI, vol 27. Mintis, Vilnius, pp 123–136 [in Russian]

20. Kisneryus Yu L (1976) Stratigrafija triasovyh otlozhenij Pribaltik (Stratigraphy of the Triassic deposits of the Baltic States Materials on the stratigraphy of the Baltic States). Vilnius, pp 84–85 [in Russian]

21. Grigelis AA, Kaplan AA (1976) Stratigrafija paleogenovyh otlozhenij Pribaltiki (Stratigraphy of the Paleogene deposits of the Baltic States materials on the stratigraphy of the Baltic States). Vilnius, pp 101–108 [in Russian]

22. Blazhchishin AI, Boldyrev VL, Moroshkin KV (1978) Jantarenosnye otlozhenija paleogena i uslovija ih zaleganija na podvodnom sklone Sambijskogo poluostrova (The amber-bearing deposits of the Paleogene and conditions of their occurrence on the underwater slope of the Sambian Peninsula). Tectonics and Minerals of Belarus and the Baltic States, Kaliningrad, pp 119–127 [in Russian]

23. Gidrogeologija SSSR. Tom XLV. Kaliningradskaja oblast' (Hydrogeology of the USSR. Volume XLV. Kaliningrad Region) (1970) In: Grigelis AA, Iodkazis VI, Kondratas AR (eds). Nedra, Moscow, p 158 [in Russian]

24. Zhitkova EA (1962) Otchet o rezul'tatah poiskovo-razvedochnyh rabotah na formovochnye peski v Kaliningradskoj oblasti v 1960–1961 (Report on the results of exploration work on molding sands in the Kaliningrad Region in 1960–1961). Unpublished report. North-West Geological Administration, 703 pp [in Russian]

25. Zagorodnykh VA (1977) Otchet o rezul'tatah poiskovyh rabot na keramzitovye gliny v rajone g. Kaliningrada i detal'noj razvedki Jablonevskogo mestorozhdenija glin, provedennyh v 1970–76 gg (A report on the results of prospecting for keramzite clays in the Kaliningrad Region and the detailed exploration of the Yablonevskoye clay deposit carried out in 1970–1976). Unpublished report. North-West Geological Administration, p 1108 [in Russian]

26. Gurskaya TV, Galvinets EN, Rusmanova LA (1988) Otchet o rezul'tatah poiskov i poiskovo-ocenochnyh rabot na vyjavlenie mestorozhdenij gravijno-peschanogo materiala v drevnealljuvial'nyh otlozhenijah na severo-zapade Kaliningradskoj oblasti v 1985–88 gg (Report on the results prospecting and evaluation works to identify deposits of gravel-sand

material in ancient alluvial deposits in the southeast part of Kaliningrad Region in 1985–1988). Unpublished report. "Sevzapgeologiya", p 366 [in Russian]
27. Stoyanova SV, Nisnevich AK, Andreeva LP (1979) Otchet o razvedke podzemnyh vod dlja vodosnabzhenija g. Mamonova Kaliningradskoj obl., provedennoj v 1978–79 gg (Report on the exploration of groundwater for water supply in the town of Mamonovo, Kaliningrad Region, conducted in 1978–1979). Kaliningrad geological and hydrogeological expedition. Unpublished report [in Russian]
28. Trufanova LF, Tiunov Yu V, Andreeva LP (1984) Otchet o rezul'tatah predvaritel'noj razvedki podzemnyh vod dlja vodosnabzhenija goroda Baltijska, provedennoj v 1982–84 gg (Report on the results of preliminary groundwater exploration for the city of Baltiysk water supply, conducted in 1982–1984. Kaliningrad geological and hydrogeological expedition). Unpublished report, p 295 [in Russian]
29. Shustov VA, Fadeev NS, Fadeenkova LI (1974) Otchet o rezul'tatah rabot po ocenke perspektiv neftegazonosnosti devonskih otlozhenij Kaliningradskoj Oblasti (Report on the results of assessment of petroleum potential of the Devonian deposits of Kaliningrad Region). Kaliningrad geological and hydrogeological expedition. Unpublished report, p 295 [in Russian]
30. Kagramanyan NA (1973) Otchet o rezul'tatah burenija parametricheskoj skvazhiny 1-Mamonovo v Kaliningradskoj oblasti (Report on the results of drilling 1-Mamonovo parametric well in the Kaliningrad Region). Unpublished report. Trust "Yaroslavnefterazvedka" [in Russian]
31. Kagramanyan NA, Sednin Ya V (1984) Otchet o rezul'tatah burenija Pervomajskoj parametricheskoj skvazhiny 1 v Kaliningradskoj oblasti RSFSR (Report on the results of drilling of the 1-Pervomayskaya parametric well in the Kaliningrad Region of the RSFSR). Unpublished report. Volgokaskgeologiya, p 110 [in Russian]
32. Zagorodnykh VA, Vorobiev Yu N, Boykova OV (1991) Sbor i obobshhenie materialov i sostavlenie prognoznyh kart territorii Kaliningradskoj oblasti na neft' i tverdye poleznye iskopaemye (Data collecting and summarizing of materials and compiling forecast maps of the territory of the Kaliningrad Region for oil and solid minerals (scale 1:200,000)). Gusev, Kaliningrad geological and hydrogeological expedition. Unpublished report, p 375 [in Russian]
33. Zagorodnykh VA, Dovbnya AV, Zhamoida VA (2002) Proizvodstvo geologicheskogo, gidrogeologicheskogo doizuchenija, geologo-jekologicheskih issledovanij i kartografirovanija territorii Kaliningradskoj oblasti masshtaba 1:200,000, kartosostavitel'skih rabot masshtaba 1:200,000 na akvatorii Juzhnoj Baltiki (v predelah jekonomicheskoj zony Rossii, prilegajushhej k Kaliningradskoj oblasti) i podgotovka k izdaniju komplektov Gosgeolkarty-200 (Production of geological, hydrogeological additional exploration, environmental geological studies and mapping of the territory of the Kaliningrad Region at a scale of 1:200,000, cartographic works of scale 1:200,000 of the Southern Baltic (within the economic zone of Russia adjacent to the Kaliningrad Region) and preparations for the publication of sets state geological maps-200 (new series) of sheets N-34-II-III, VIII, IX, X, XI, XIV, XV, XVI, XVII (all partially) in 1993–2002). Unpublished report. VSEGEI, St. Petersburg [in Russian]
34. Zagorodnykh VA, Dovbnya AV, Zhamoyda VA (2001) Stratigrafija Kaliningradskogo regiona (Stratigraphy of the Kaliningrad Region). Kaliningrad, p 225 [in Russian]
35. Yeltsina GN (1995) Минерально-сырьевая база СЭЗ "Янтарь" (Mineral and raw materials base of FEZ "Yantar"). In: Problems of physical and economic geography of Kaliningrad Region. Kaliningrad University, Kaliningrad, pp 26–33 [in Russian]
36. Otmas AA, Meshchersky AA, Desyatkov VM (2010) Neftegazonosnost' Kaliningradskogo regiona (Oil and gas potential of the Kaliningrad Region). Mining J (Gorny Zhurnal) 3:25–28 ("Ore and Metals" Publishing House [in Russian])
37. Desyatkov VM, Otmas AA, Syryk SI (2006) Neftegazonosnost' Kaliningradskogo regiona (Oil and gas potential of the Kaliningrad Region). Geol Geophys Dev Oil Gas Fields 8:24–30 [in Russian]
38. Adamov VG (2007) Neftegazonosniy potenzial Kaliningradskogo regiona I novye yfghfvleniya poiskov (Oil and gas potential of the Kaliningrad Region and new directions of searches). In:

Materials of the international scientific and practical conference, VNIGRI, St. Petersburg, 3–5.10.2007, pp 226–229 [in Russian]
39. Zagorodnykh VA, Cheremnykh GV (2010) Neogenovye i jurskie burye ugli Kaliningradskogo regiona (Neogene and Jurassic brown coals of the Kaliningrad Region). Mining J (Gorny Zhurnal) 3:31–33 ("Ore and Metals" Publishing House [in Russian])
40. Zagorodnykh VA (1993) Diageneticheskie ceolity Juzhnoj Pribaltiki (Diagenetic zeolites of the Southern Baltic region). In: Scientific and technical achievements and best practices in the field of geology and mineral prospecting: scientific and technical information collection, vol 2. Geoinformmark, pp 25–31
41. Zagorodnykh VA (1996) Poleznye iskopaemye verhnepermskih otlozhenij Kaliningradskoj oblasti (Minerals of the upper Permian deposits of the Kaliningrad Region). Lithol Miner Resour 1:97–105
42. Zagorodnykh VA (2000) Metallonosnost' osadochnogo chehla Severo-Zapadnoj okrainy Russkoj plity (Metal-bearing capacity of the sedimentary cover of the North-Western margin of the Russian plate). Lithol Miner Resour 5:500–511 [in Russian]
43. Gurevich GI, Shustov VA, Vostryakova NV (1979) Kaliningradskij solerodnyj bassejn i perspektivy ego kalienosnosti (Kaliningrad salt basin and prospects of its potassium content). In: New in the geology of the Northwest deposits of the RSFSR. Leningrad, pp 116–120 [in Russian]
44. Vishnyakov AK, Zagorodnykh VA, Rudenko DG (2010) Kamennye i kalijno-magnievye soli Kaliningradskoj oblasti (Stone and potassium-magnesium salts of the Kaliningrad Region). Mining J (Gorny Zhurnal) 3:28–30 ("Ore and Metals" Publishing House [in Russian])
45. Shaporev AA, Vostryakova NV (1982) Otchet o rezul'tatah poiskovyh rabot na kalijnuju sol' v predelah Mamonovskogo progiba Kaliningradskogo solenosnogo bassejna v 1980–82 gg (Report on the results of prospecting works for potassium salt within the Mamonovsky trough of the Kaliningrad salt basin in 1980–1982). In: Kaliningrad geological and hydrogeological expedition. Unpublished report [in Russian]
46. Nikutina NG (1998) Otchet o rezul'tatah poiskovyh rabot na stekol'nye peski, provedennyh v 1992–93 gg. v Kaliningradskoj oblasti (Stekol'nyj ob'ekt) (Report on the results of prospecting works on glass sand, carried out in 1992–93 in the Kaliningrad Region (Glass object)). In: Kaliningrad geological and hydrogeological expedition. Unpublished report [in Russian]
47. Zagorodnykh VA, Terskih AV, Laskevich AI (2010) O vozmozhnostjah osvoenija resursov Kaliningradskoj geotermal'noj anomalii (On the possibilities of developing the resources of the Kaliningrad geothermal anomaly). Mining J (Gorny Zhurnal) 3:33–35 ("Ore and Metals" Publishing House [in Russian])

Amber Deposits in the Kaliningrad Region

Vadim V. Sivkov and Vladimir A. Zhamoida

Abstract Amber is one of the most valuable mineral resources of the Kaliningrad Region. It is mined by several methods, both open and closed, on sea and on land; the vast majority of this type of amber world reserves is located on the Sambian Peninsula and its continental slope. Nevertheless, geological research of amber deposits still did not carry out up to sufficient rate.

Keywords Amber, Blue earth, Glacier, Neogene, Paleogene

Contents

1 Introduction	34
2 Paleogene Sea-Beach Amber Placers	35
3 Neogene and Quaternary Amber Placers	41
4 Conclusions	42
References	43

V. Sivkov Vadim (✉)
P.P. Shirshov Institute of Oceanology RAS, 36, Nakhimovsky Pr., Moscow 117997, Russia

Immanuel Kant Baltic Federal University, 14, A. Nevskogo Street, Kaliningrad, Russia
e-mail: sivkov@kaliningrad.ru

A. Zhamoida Vladimir
A.P. Karpinsky Russian Geological Research Institute (VSEGEI), 74, Sredny Prospect, St.-Petersburg, Russia
e-mail: vladimir_zhamoida@vsegei.ru

1 Introduction

The world's largest industrial deposits of amber (succinite) are located in the Kaliningrad Region (Russian Federation). About 90% of all the world's proven reserves of this "sunstone" are located in the Sambian Peninsula. Amber was mined and processed within the territory of the modern Kaliningrad Region even in the Neolithic period – 3,500–1,700 BC [1]. The manuscript *Chronicle of the Prussians* [2] reports that during the Pompeian Wars in 47 BC, amber was recovered from the "white sand mountains." Since ancient times, amber was gathered at the coast and within shallow water areas. According to the available estimates [3, 4], over the last three millennia, about 100 thousand tons of amber was collected on the Baltic coast.

Free amber gathering continued until the thirteenth century, when the Teutonic Order declared amber deposits as its property. Since the fourteenth century, open-cast mining of amber began to develop. There are numerous evidences of amber mining on the sixteenth to seventeenth century's maps of the Baltic Sea southern coast (Fig. 1). The first mention of land amber extraction is referred to the middle of the sixteenth century [5]. In 1781, the first mine was created on the west coast of the Sambian Peninsula, but just 7 years later, work has stopped due to unprofitability. In the nineteenth century, there were attempts to extract amber on the coastal slopes of the Sambian Peninsula. Since the middle of the nineteenth century, the largest

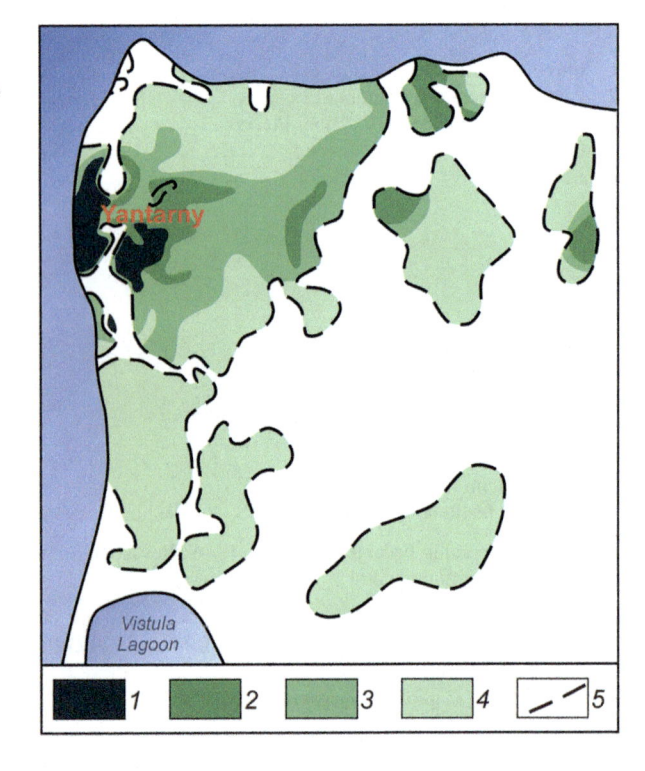

Fig. 1 Contents of amber in the "blue earth" [24] (kg/m^3): *1*, >1; *2*, 0.5–1; *3*, 0.2–0.5; *4*, <0.2; *5*, "blue earth" location

amber firm of East Prussia *Stantien und Becker* had successfully used divers and steam dredger. The work of this company had begun after the amber deposits discovery during the deepening of the channel in the Curonian Lagoon's northern part. In 1873 and in 1883, two mines were founded in the Palmnicken area (current Yantarny settlement), but by 1922 they were closed. In 1912 a large quarry was developed in the Palmnicken area, where amber was mined by open-pit method until 1944. Before World War II, this site produced about 400 tons of raw amber per year.

K. Andree [6] calculated the total amount of amber extracted on the Sambian Peninsula from 1,876 to 1935 as 9–524 tons per year. During this period totally 16,168 tons of amber was mined from sea and from land using opencast and underground mining methods.

Currently on the amber deposit Primorskoye with balance reserves (at the beginning of twenty-first century), about 166 thousand tons is developed. This amber deposit located near the Yantarny settlement was found in 1952–1955.

The amber formation can be described as a process of resin maturing, comparable to carbonification – from plants through brown to stone coal [7]. The variety of resins is explained by different botanical nature, as well as the conditions of diagenesis (fossilization). Amber is found on all continents except Antarctica. Its varieties are found both in the coals of the Upper Carboniferous and in the deposits of the Early Mesozoic. However, the actual "amber period" is Cenozoic, to which most of the known fossil resins belong [7]. Distribution of flora and position of climatic zones during amber formation, the types, age, and distribution of its clusters are connected to two large amber provinces – the Eurasian and American [8]. Both of them have subprovinces, regions, and areas characterized by significant differences in the ratio of primary and secondary aggregations and their age, the composition and properties of fossil resins, and the presence and scale of industrial deposits. The Baltic-Dnieper subprovince is located in the central part of the Eurasian amber province and associated with the distribution of so-called Baltic amber. The main feature of the Baltic-Dnieper amber subprovince is the location within its boundaries of ancient (Paleogene) buried marine placers. Small Neogene alluvial and lacustrine placers, Pleistocene fluvioglacial and lacustrine glacial, and modern (Holocene) coastal-marine (beach) placers are also widely developed in this subprovince.

In this chapter, modern ideas about the formation of concentrations (deposits and outcrops) of amber on the Sambian Peninsula are summarized.

2 Paleogene Sea-Beach Amber Placers

In the Lower Eocene (56–49 million years ago) during the "climatic optimum", the highest air temperature for the entire Cenozoic era was reconstructed [9]. Subtropical and tropical conditions, apparently, spread beyond the boundaries of actual climatic zones 15° south and north [7]. Since there is a correlation between air temperature and gum exudation, V. Katinas [10] associated the occurrence of increased resin yield of "amber conifers" precisely with the beginning of the Eocene time. The primary deposits of the Baltic amber and, consequently, the vast area of the "amber-bearing" forests growth were located to the northwest and north of the present Kaliningrad Region [10, 12]. The transformation of resin into amber (succinite), accompanied by its chemical transformations with succinic acid discharge, took place in a slightly alkaline conditions, with the participation of oxygen-containing, potassium-rich muddy waters [3, 4, 10, 13].

For a long time, it was believed that only pine *Pinus succinifera* is the initial plant for amber production [12, 14, 15]. However, still an extremely limited amount of needle-foliage has been found in amber [6]. At the same time, the infrared spectra of the Baltic amber have much more similar characteristics with the resin of the Araucarius (*Agathis australis*) recently growing in New Zealand, rather than with the resin of modern pines [16]. Resins of the modern cedar (*Cedrus atlantica*) of the Atlas Mountains (North Africa) also show a great similarity with succinite. By the beginning of the Late Eocene (about 37 million years ago), the relative climate cooling led to a complete disappearance of "amber coniferous".

Spatial distribution and frequency of occurrence of inclusions of various fossil organisms found in amber give important information on paleoecological relationships in the biotopes of the "amber forest" and, as a consequence, on the paleoclimate [16]. A modern idea of the occurrence rate of certain taxonomic groups of inclusions in Baltic amber was formed on the basis of collection studies in the Earth Museum in Warsaw [17, 18]. This collection is very representative, but there is no reliable information about the primary sources of its acquisition. Therefore, it remains an open question how distorted the natural occurrence of various taxa became during the collection formation (seizure and loss of individual samples, exclusion of small fractions of amber, etc.).

Thus the paper [19] is advantageous, based on the collection study, which belongs to LLC "Marine Venture Bureau" (Kaliningrad). This collection was gathered in 2001–2006, with the assistance of the Kaliningrad Amber Combine and the Atlantic Branch of the P.P. Shirshov Institute of Oceanology of RAS. During the furrow testing of the Primorsky and Plyazhevy open-pit mines (136 furrows in total), several amber samples containing inclusions were selected for taxonomic studies. The path of each amber sample was tracked by the authors during all stages – from sampling to examining inclusions under a microscope, allowing minimizing the distortion of the initial information. For amber samples sized from 0.4 to 10 cm, the number of inclusions was counted, and taxonomic composition (orders, even suborders) of at least 0.1 mm size was determined (Table 1). Plant residue designation did not take into account stellate hairs of oak, which are found in amber everywhere. The total number of studied amber for inclusions was 5,857 pieces.

The main difference in occurrence rate obtained in [19] from the similar data from the collection of the Earth Museum in Warsaw is very high maintenance of ticks (Acari). Apparently, due to the small sizes (from 0.1 mm to several mm), ticks in the Warsaw collection were simply not taken into account. For the same reason,

Table 1 Occurrence rate (%) of fossil fauna taxa in Baltic amber (Based on materials of furrow sampling in the Kaliningrad Amber Combine)

Order	Suborder		Occurrence rate (%)		Occurrence rate (%)
Diptera	Mosquitos –Nematocera		37.5		**45.09**
	Flies – Brachycera		6.68		
	Non. defined Diptera		0.93		
Collembola					7.47
Homoptera	Plant lice – Aphidinea		4.29		**5.11**
	Others	Coccinea	0.43	0.82	
		Cicadinea	0.28		
	Hymenoptera	Aleyrodinea	0.11		
Arachnida	Ticks – Acari		21.71		**26.05**
	Spiders – Araneae		4.25		
	Pseudoscorpiones		0.09		
Hymenoptera	Ants – Formicidae		2.03		**4.82**
	Others		2.79		
Coleoptera					**3.00**
Flora					1.05
Others	Caddis flies – Trichoptera		0.59		**3.05**
	Maggots larvae		0.52		
	Thrips – Thysanoptera		0.48		
	Dust lices – Psocoptera		0.36		
	Bugs – Heteroptera		0.18		
	Wood louses – Isopoda		0.14		
	Centipedes – Myriapoda		0.12		
	Butterflies – Lepidoptera		0.11		
	Thysanura		0.09		
	Caddis flies – Plecoptera		0.07		
	Cockroaches – Blattodea		0.07		
	Termites – Isoptera		0.07		
	Worms – Nematoda		0.07		
	Dayflies – Ephemeroptera		0.05		
	Daddy longlegs – Opiliones		0.05		
	Lacewings – Neuroptera		0.04		
	Pincher bugs – Dermaptera		0.02		
	Dragonflies – Odonata		0.02		

the occurrence of such taxa as Collembola, Thysanoptera, Aphidina, and some other Homoptera (Coccinea, Aleyrodinea) also exceeds the corresponding values of the collection of the Earth Museum.

Despite the fact that the imprints of organisms in inclusions are perfectly preserved, they still do not show the true balance of the small-sized fauna of the past. The point is that different organisms had a different probability to get into the resin and to remain in it, i.e., had a different "fossilization potential" [20]. Higher probability to remain in fossil form was for very active animals, such as dipterans (mosquitoes, flies). There are also vast of organisms, which lived on bark of trees or on forest soil, such as ants, beetles, or termites. The organisms of open landscapes are rare in inclusions, since they became victims of sticky resin only by chance.

Flora and fauna of the amber forest, which has survived till the present day in the form of "inclusions," as modern biogeography sees, represent an accidental mix of temperate, subtropical, and tropical forms of life. This phenomenon is explained by the limitation of the actuality principle, which allows modern biogeographical criteria to be transferred to the past for paleobiological reconstructions. It is assumed that the flora and fauna of the Paleogene amber forest cannot be judged by modern biogeographical criteria, which extend only to the Quaternary period (1.8 million years ago) [6]. Another explanation may be the vastness of the amber forest covering various natural zones, and the effect of high altitude natural zonality, when the change of natural zones occurs in a limited area. Thus, aquatic insects found in Baltic amber testify that the area of resinous trees was a mountainous terrain with fast mountain streams and rivers [21].

Despite the huge area of amber forest, the number of proved deposits of Baltic amber is relatively small, since the processes of resin masses washing from forest soil and transferring them to conditions where they would be protected from weathering (oxidation and dehydration) occurred only on very limited area of amber forest.

When studying amber occurrence on the Sambian Peninsula, it was initially assumed that the erosion of the primary amber deposits was carried out by the transgressing Late Eocene (or Early Oligocene) sea, and then washout amber was deposited in its coastal sediments [5, 22].

V. Katinas [10] supposed that the river transfer was crucial for amber outwashing from its primary deposits and following accumulation in rivers estuaries of the Paleogene sea basin. In his opinion, if river carried amber, its burial occurs at the estuary, in organic- and glauconite-rich sediments. Organic matter decomposition changes weak oxidizing conditions of initial amber deposition to reducing conditions with hydrogen sulfide presence. Resin reduced under such conditions, resulting in specific properties that distinguish it from the same compounds of modern resins. Polish geologists, relying on the "estuarine" concept of V. Katinas, showed the uniformity of amber accumulation on the Sambian Peninsula with Eocene amber accumulations on the Baltic coast of Poland [5].

Despite the "estuarine" theory, V. Trofimov [13, 23] suggested that the amber-rich sediment layer enriched with glauconite has a marine genesis, and the demolition of material in the Oligocene took place from the south where the continent (or a series of islands) existed. The delivery of amber to the sea was carried out not only by rivers but also with the general transgression of the sea to low marshy banks covered with plants, producing resin. Accumulation of amber occurred near the river delta, but only in relatively calm parts of the basin, deep enough that waves had no effect on the bottom.

S.G. Krasnov and A.A. Kaplan [24] proposed another theory, in our opinion, the most probable. According to this theory, the formation of Paleogene amber accumulations occurred at a distance from estuaries and was associated with the facies of the open sea shelf.

Nowadays the prevailing opinion is that in the Oligocene (34–23 million years ago), the erosion of the amber-bearing Lower-Middle Eocene sediments was carried out by the transgressing sea; as a result amber was spread all over the paleobasin of the Tethys Ocean (North Atlantic) [10].

Geological research of amber on the Sambian Peninsula began in the middle of the nineteenth century by the work of K. Thomas [25]. Lithostratigraphic terms, for example, such names as "wild earth," "blue earth," etc., introduced by E. Tsaddakh [26, 27], are firmly established in the scientific literature in the designation of local Tertiary deposits. The first geological maps of the Sambian Peninsula, containing notes about amber, were compiled by G. Berendt [28, 29]. Studies of E. Tsaddah and G. Berendt created theoretical basis for amber mining. Other important results of amber research were presented by A. Jencz [30, 31]. The map of A. Jencz with information about amber distribution was developed by F. Kaunhoven [32].

The amber-bearing Prussian Formation of the Priabonian Eocene Stage on the Sambian Peninsula (Fig. 1) has a thickness of 15–20 m and unites three horizons (from the bottom to the top): (1) "wild earth," (2) "blue earth," and (3) "upper drift sand" [10].

The basal horizon of the "wild earth" on the seabed near the Sambian Peninsula is distributed everywhere. It lies on the eroded Alkian sediments, the roof of which is located at the depths from 7–9 m.b.s.l. at Cape Taran to 20–25 m.b.s.l. in Pokrovskaya Bay (the western coast of the peninsula), where the thickness of the "wild earth" reaches 5–6 m. Wild earth represents a sandy-silty heterogeneous grain of brecciated form with clumps of sandy clay and numerous nodular phosphorites. Mineral composition is characterized by a high content of glauconite (up to 80%) and – in severe subfraction – authigenic phosphates, sulfides, and siderite.

The "blue earth" becomes "wild earth" gradually. The "blue earth" is glauconitic-quartz clayey sand with an admixture of silt [3] and has this name due to the greenish-blue hue, which is caused by the presence of a significant amount of glauconite – iron-aluminum silicate. The glauconite absolute dating by the potassium-argon method by Z. Ritzkovsky [33] allowed to conclude the "blue earth" was formed in the Middle Eocene (44–47 million years ago), much earlier than it was commonly believed [10].

On the Sambian Peninsula coast, the basis of the "blue earth" is located quite high (at of sea depth of 4–8 m), while its roof is sometimes open to the beach zone and to the base of the coastal cliffs (Fig. 2).

"Upper drift sand" is uneven-grained glauconite-quartz sands up to 35 m thick in the northern part of Sambia [10]. The main difference between underlying "blue earth" and these sands is the smaller content of clayey material and glauconite [3]. The total content of amber is reduced to 143 g/m^3, but the share of large amber fractions (up to 50%) increases [10].

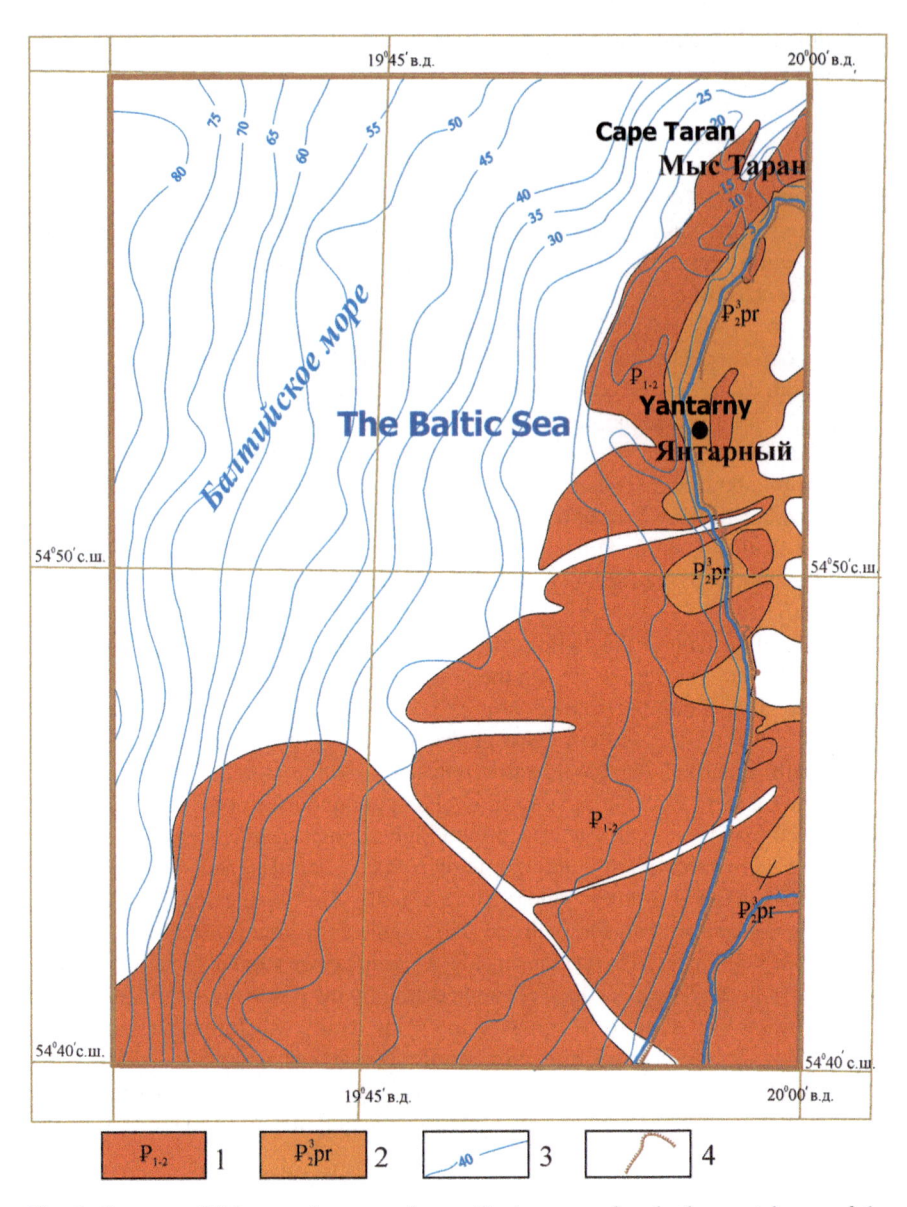

Fig. 2 Outcrops of Paleogene layers on the pre-Quaternary surface in the coastal zone of the western coast of the Sambian Peninsula: *1*, undivided Paleogene; *2*, amber deposits of the Prussian suite; *3*, isobaths; *4*, abrasion cliffs

3 Neogene and Quaternary Amber Placers

The southern Baltic area was almost completely characterized by terrestrial conditions in the Neogene. Starting in the Miocene, large river systems transferred clastic sediments to the south and southwest from the places of amber Paleogene accumulation. Most amber outcrops of the Neogene time are associated with continental alluvial or lacustrine deposits. Findings of amber are associated with lenses of carbonaceous sands. It is believed that brown coal sands have received amber from the reworked Paleogene layers and come along with this amber in the third occurrence [34]. As the main argument against the version of the existence of Miocene "amber" forests, the fact of the identity of the flora and fauna of Bitterfeld (Central Germany) and Baltic amber is given, which can be explained only by the identity of the paleoclimatic conditions [35].

Significant redeposition of the Baltic amber, undoubtedly, took place in the Pleistocene. It is no coincidence that the boundaries of the region of amber distribution roughly coincide with the boundaries of glacial deposits. Large inland glaciers coming from Scandinavia carried masses of detrital rocks, including amber-bearing deposits, to the south. If the glaciers stopped or retreated, as it was in interglacial periods, the role of meltwater increased, reprocessing the displaced loose rocks and redepositing them in the new areas in accordance with their grain size and specific gravity. In glacial (morainic) sediments, amber is found either in the form of sporadic inclusions or erratic blocks of original amber-bearing deposits. Not only these erratic mass but also the greenish-gray color of morainic loams, which is due to the presence of processed glauconite, testifies the intensity of the glacial erosion of the original amber-bearing layers [10]. There are significant differences between the amber resources in the Early and Late Pleistocene deposits, which are explained by the fact that most ancient glaciers have captured more tertiary material with amber than the subsequent ones.

The erosion of the bedrock amber-bearing deposits by glacier meltwater is confirmed by the extensive development of the Quaternary river network on the Sambian Peninsula. Most of the traced valleys intersect the amber-bearing sediment layers. Amber lies in nests in alluvial sands containing wood and, apparently, representing floodplain areas of rivers.

Despite the existence of quarries with production volumes of hundreds of tons of amber, locals continue to collect amber on the seashore, i.e., the modern beach placers of amber have not lost their value (Fig. 3). According to existing estimates, on the average about 38 tons of amber are annually washed ashore by the sea in the Kaliningrad Region [36]. In the Holocene, up to the present time, amber was eroded by rivers or the sea from older deposits and redeposited.

The underwater slope of the Sambian Peninsula, where the outcrops of the "blue earth" are found, can serve as an example of littoral (bottom) amber placers development [37]. According to S.S. Savkevich [38], erosion of the "blue earth" outcrops results in passing of 50 tons of amber to modern sediments annually. An additional contribution to the formation of bottom placers can be produced by bared

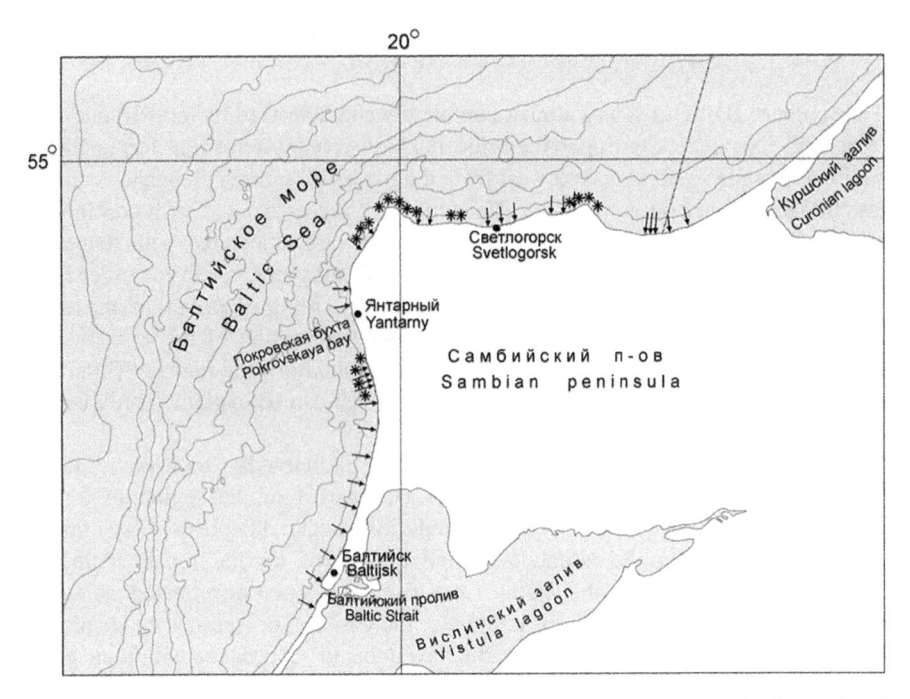

Fig. 3 Basseting of glacial sediments, where high occurrence of redeposited amber is (stars) and places of catching and collecting amber on the shore (arrows) at the present time

glacial ("blue earth" detached masses) and postglacial deposits on the seabed, in particular sediments associated with the ancient coastal lines of the Baltic Sea.

Due to low weight and the presence of voids, amber is easily transported in the seawater. It is obvious that amber species (yellow, white, foamy) with a density lower than the density of the Baltic Sea water are thrown on the shore. It can be assumed that denser transparent amber can stay in the sea, gradually moving to the extreme depths of the wave action zone on the seabed (20–30 m) and accumulate in local depressions. Amber, brought by the sea, is collected on the Baltic coast in Germany, Denmark, Lithuania, Latvia, Poland, and Estonia.

4 Conclusions

For nearly 150 years, Baltic amber has been an object of intensive research in various scientific directions. One might assume that after so many years of research, there are only a few unresolved issues of fundamental importance. A large number of publications in recent years still prove the opposite. Fundamental topics, such as the question of resin sources or the precise temporal classification of Baltic amber and the position and period of the legendary "amber forest," are still a material for discussion. The question of underwater deposits of amber is also open.

Unfortunately, purposeful and comprehensive studies of amber in Russia have not been carried out since the 1970s of the last century. The existence of the world's largest amber deposit on the territory of the Kaliningrad Region, nevertheless, allows us to hope for the resumption of such research.

References

1. Klebs R (1882) Der Bernsteinschmuck der Steinzeit. In: Beiträge zur Naturkunde Preussens 5, Königsberg, 1 – III
2. Chronik der Preussen (1626) Wie das Land erfunden und gebaut ist vom Glauben, Sitten und Gebrauch der Einwohner von Anfang des Ordens und seiner Regierung und wie sie das Land verlohren, Ms. Universität Torun, Archiv, Nr. 60. Kopie der Berliner Chronik des Heinrich von Reden aus dem 16. Jahrhundert
3. Savkevich SS (1970) Amber. Nedra, Leningrad, 192 pp [in Russian]
4. Srebrodolsky BI (1984) Amber. Hauka, Moscow, 112 pp [in Russian]
5. Kosmowska-Ceranowicz B (1996) Bernstein – die Lagerstätte und ihre Entstehung. In: Bernstein – Tränen der Götter. Dt. Bergbau-Museum, Bochum, pp 161–168
6. Andree K (1937) Der Bernstein und seine Bedeutung in Natur- und Geisteswissenschaften, Kunst und Kunstgewerbe, Technik, Industrie und Handel. Gräfe & Unzer Verlag, Königsberg
7. Weitschat W (2001) Baltischer Bernstein. Herkunft – Lagerstätten – Einschlüsse. Ribnitz-Damgarten, 120 pp
8. Ivanova AM, Smirnov AN (2008) Mineral genesis of shelf Trudy VNIIOkeangeologii. T. 214. VNIIOkeanologiya, St. Petersburg, 108 pp [in Russian]
9. Buchardt B (1978) Oxygen isotope paleotemperatures from the tertiary period in the North Sea area. Nature 275:121–123
10. Katinas V (1971) Amber and amber deposits of Southern Baltic. Mintis, Vilnius, 156 pp [in Russian]
11. Heer O (1869) Miozäne baltische Flora. Beitr Natur Preußens Königsberg 2:104
12. Conwentz H (1890) Monographie der baltischen Bernsteinbäume. Engelmann, Leipzig, S. 151
13. Trofimov VS (1974) Amber. Nedra, Leningrad, 184 pp [in Russian]
14. Göppert HR, Berendt GC (1845) Der Bernstein und in ihm befindlichen Pflanzenreste der Vorwelt. In: Berendt GC (ed) (1845–1956): Der Bernstein und die in ihm befindlichen Pflanzenreste der Vowelt. 1, vol I. Abth., Berlin
15. Arnold V (1998) Vergessene Einschlüsse – Blütenstaub in Baltischem Bernstein. Milt Geol Paläont Inst Univ Hamburg 8:269–282
16. Poinar GO (1992) Life in amber. Stanford University Press, Stanford
17. Krzeminska E, Krzeminski W, Haenni J-P, Dufour C (1992) Les Fantômes del'Ambre. Insectes fossils dans l'ambre de la Baltique, Neuchâtel
18. Krumbiegel G. (1996) Bernstein (Succinit) – Die Bitterfelder Lagerstätte. In: Ganzelevski M, Slotta R (eds) Bernstein – Tranen der Goötter. Bochum
19. Bukanova YV, Borisenko VV, Sivkov VV, Lukashina NP, Gaydyukov AA (2007) New data in Baltic amber fossil fauna occurrence rate. Baltiyskiy yantar', pp 7–13 [in Russian]
20. Seilacher A (1995) Fossile Kunst. Albumblätter der Erdgeschichte. Weinstadt
21. Weitschat W, Wichard W (1998) Atlas der Pflanzen und Tiere im Baltischen Bernstein. Verlag Dr. Friedrich Pfeil, Munich, 254 pp
22. Conwentz H (1897) Über englischen Bernstein und Bernstein im Allgemeinen. Naturw Rundschau 12

23. Trofimov VS (1965) Abmer placers and their genesis – placers geology. Moscow, pp 77–97 [in Russian]
24. Krasnov SG, Kaplan AA (1976) Paleogene amber deposits genesis according to lithostratigraphic data. Litologiya i poleznyye iskopayemyye 4:95–106 [in Russian]
25. Thomas K (1847) Die Bernsteinformation des Samlands. In: Preussische Provinzblätter 3. Königsberg, pp 241–245
26. Zaddach EG (1860) Über die Bernstein- und Braunkohlenlager des Samlands. In: Schriften der physikalisch-ökonomischen Gesellschaft zu Königsberg 1. Königsberg, Taf. I–IV, pp 1–44
27. Zaddach EG (1867) Das Tertiärgebirge Samlands. In: Schriften der physikalisch-ökonomischen Gesellschaft Königsberg 8. Königsberg, Taf. VI–XVII, pp 85–194
28. Berendt GC (1866) Erläuterungen zur geologischen Karte des West-Samlandes. Verbreitung und Lagerung der Tertiär-Formationen. In: Schriften der physikalisch-ökonomischen Gesellschaft zu Königsberg 7. Königsberg, pp 131–144
29. Berendt GC (1867) Lagerung und Verbreitung des Tertiar-Gebirges im Bereich der Provinz Preussen. In: Schriften der physikalisch-ökonomischen Gesellschaft zu Königsberg 8. Königsberg, pp 65–84
30. Jentzsch A (1877) Beiträge zur Kenntnis der Bernsteinformation. In: Schriften der physikalisch-ökonomischen Gesellschaft zu Königsberg 17. Königsberg, pp 101–108
31. Jentzsch A (1903) Verbreitung der Bernsteinführenden blauen Erde. Z Dtsch Geol Ges, Monatsberichte 55:122–130
32. Kaunhowen F (1914) Bernstein. In: Die nutzbaren Mineralien 2, 446 pp
33. Ritzkowski S (1997) K-Ar – Altersbestimmungen der bernsteinführenden Sedimente des Samlandes (Palaogen, Bezirk Kaliningrad). In: Metalla. Sonderheft zum Symposium "Neue Erkenntnisse zum Bernstein", Veröffentlichungen aus dem Deutschen Bergbau-Museum Nr. 66, pp 19–23
34. Andree K (1951) Der Bernstein. Das Bernsteinland und sein Leben. Kosmos, Stuttgart, p 96
35. Weitschat W (1997) Bitterfelder Bernstein – ein eozäner Bernstein auf miozäner Lagerstätte. In: Sonderheft zum Symposium "Neue Erkenntnisse zum Bernstein", Veröffentlichungen aus dem Deutschen Bergbau-Museum Nr. 66, pp 71–84
36. Metel'skiy GV (1969) Amber shore. Mysl', Moscow, p 86 [in Russian]
37. Blazhchishin AI, Boldyrev VL, Moroshkin KV (1978) Paleogene amber deposits and its storage conditions at Sambian peninsula continental slope. In: V kn.: Tektonika i poleznyye iskopayemyye Belorussii i Pribaltiki. Kaliningrad, pp 119–127 [in Russian]
38. Savkevich SS (1973) Perspectives and exploration degree of amber in USSR. Sov Geologiya 5:48–55 [in Russian]

Composition of Pre-Quaternary Surface and Quaternary Sediments Allocation on the Territory of the Kaliningrad Region

Galina Mikhnevich

Abstract The chapter studies the relief peculiarities of pre-Quaternary sediment surfaces (PQS) and the consistent pattern of the distribution of the Quaternary sediments thickness. The most notable forms of pre-Quaternary surface relief – denudation elevations and paleoincisions – are studied in detail. It was described the stages of PQS development, as well as the main mechanisms of its transformations during the neogen, described the polygenic nature of PQS and demonstrated the results of the effect on the erosive and denudation basis of exaration, fluvioglacial erosion and the tectonic factor. The focus of this chapter is on the analysis of thickness of Quaternary sediments and the peculiarities of their allocation on the territory of the region.

Keywords Paleoincisions, Pre-Quaternary surfaces, Relief, Sediment

Contents

1 The History of Studies of Pre-Quaternary Relief and Quaternary Sediments 46
2 Pre-Quaternary Sediments, Their General Profile and Surface Zonation 54
3 Characteristic Features of Paleoincision Structure and Origin 61
4 Allocation of Quaternary Sediments Thickness on the Territory of the Kaliningrad Region 71
5 Pre-Quaternary Relief Influence on Quaternary Sediments Allocation and the Contemporary Relief of the Kaliningrad Region 73
6 Conclusions 76
References 77

G. Mikhnevich (✉)
Institute of Environmental Management, Urban Development and Spatial Planning, Immanuel Kant Baltic Federal University, 14, A. Nevsky Street, Kaliningrad, Russia
e-mail: mi78galina@mail.ru; GMikhnevich@kantiana.ru

1 The History of Studies of Pre-Quaternary Relief and Quaternary Sediments

The studies of sediments and the general palaeogeography of the Quaternary Period on the territory of Eastern Prussia (Fig. 1) began as early as in the nineteenth century. However, the dealluvial hypothesis of the sediments surface mantle formation, which prevailed in the first half of the nineteenth century, didn't allow for the correct understanding of the formation processes. Despite the fact that the origin

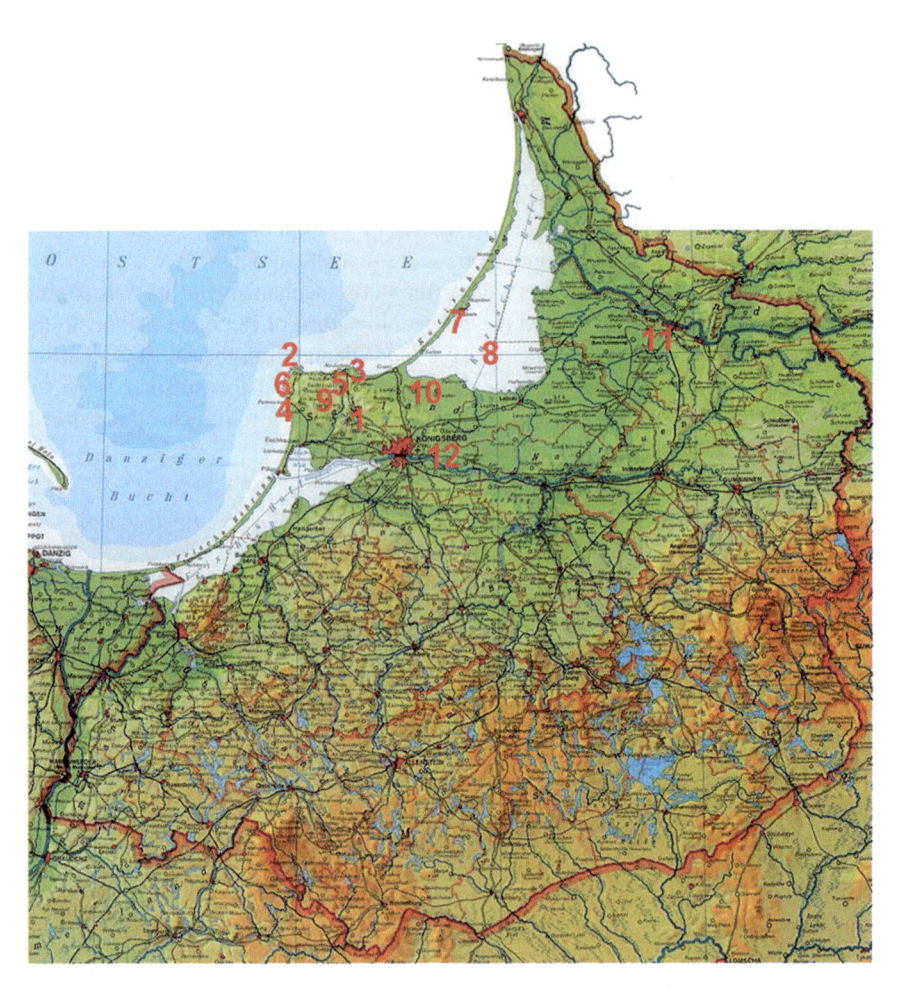

Fig. 1 Overview map of East Prussia (www.familienforschung-rimek.de/ostpreussen). The numbers indicate *1*, Sambian (Kaliningrad) Peninsula; *2*, Cape Brusterort (Taran); *3*, settlement Rantau (Zaostrov'e); *4*, settlement Nodems (Okunevo); *5*, Neukuhren (Pionerskij); *6*, settlement Dirschkeim (Donskoe); *7*, Curonian Spit; *8*, Curonian Lagoon; *9*, settlement Nortiken; *10*, settlement Powunden (Khrabrovo); *11*, Tilsit (Sovetsk) town; *12*, Königsberg (Kaliningrad) town

of the majority of Quaternary sediments was defined incorrectly, a detailed description of the nature of sediments and the character of their allocation proves to be of great interest. In the territory of the modern Kaliningrad Region, it was its western part, especially the Sambian (Kaliningrad) Peninsula (Zamland, Zemland), the territory adjacent to Königsberg (now Kaliningrad), as well as the north-east of the region near Tilsit (now Sovetsk), that was studied the most. Explorations of the region's geology were fuelled by the practical interest in Palaeogene amber-bearing deposits as well as by the search of reliable sources of water supply for the numerous villages and small towns.

In the coastal zone, the explorers concentrated mostly on the territory of the northern Zamland shore – starting from Brusterort (Mys Taran) to settlement Rantau (settlement Zaostrov'e), where amber quarrying and mining were centred in the nineteenth century. It was only in the last quarter of the nineteenth century that the scientific studies that involved boring and experimental excavations were carried out on the western coast of Zamland, starting from Brusterort (Mys Taran) to settlement Nodems (settlement Okunevo). The fundamental role in the research of the geology of Zamland coast played the works of Gustav Zaddach "On the Amber and Coal Deposits of Zamland" [1] and "The Tertiary Rocks of Zamland" [2]. Both works were dedicated to Palaeogene and Neogene sediments, and they discussed in detail the relief of their surfaces as well as the character of dealluvial (Pleistocene) sediments and their occurrence. It was for the first time that geological sections were made in this region, which made it possible to see the structure of the shore of Zemland. These sections also demonstrated rather peculiar deviations in the horizontal deposits of Palaeogene, Neogene and Quaternary sediments. As the amber production had been done in the open-cast mines, the most attractive areas for mining were those, where the marks of the roof of amber-bearing sediments were at their minimum (e.g. near Neukuhren, now Pionerskij; Mys Kupal'nyj; settlement Dirschkeim – now Donskoe), and the ones that showed minimal overburden thickness of Neogene and Pleistocene rocks. It was suggested that geotectonical factors and fracture tectonics influenced the formation of clay diapirs, which went through the Tertiary rocks, and fold dislocations in Mesozoic and Caenozoic rocks. Numerous erratic blocks of Tertiary rocks in Pleistocene sediments are characterized, and the younger age of amber deposits, which are connected with the erratic blocks, is emphasized when compared with autochthonous deposits. At the same time, their genetic similarity is emphasized on the basis of the study of vegetation and insect remains, found in amber. The graphs in this article often demonstrate cross beddings, including those connected with the appearance of paleoincisions, e.g. near settlement Kleinkuhren (settlement Filino), the Rosenort bay (to the south of Mys Taran), etc.

One of the first significant works was the geological map of Quaternary sediments of the Prussian province, made in scale 1:100,000, funded by the Royal Physics and Economics Society and created in Königsberg in 1866–1888. The map consisted of 41 sections, 12 of which (3–9, 13–17) cover the territory of the modern Kaliningrad Region. The data has mostly become out of date, e.g. the abruptions of Cretaceous and Tertiary rocks, which were depicted as autochthonous, were identified as erratic blocks later. The map does not provide a clear distinction between the contours of drift clay, end moraine or non-boulder clay. Significant areas are identified as soil and not geological formations. The map cites monographs of German scientists, e.g. the work of G. Berendt – *Geology of the Curonian Lagoon and the Surrounding Areas, and Explanatory Notes to Sections 2, 3 and 4 of the Geological Map of Prussia*, published in 1869 [3]. This work describes in detail the sediments of the lagoon, the spit and the surrounding areas (the northern part of the modern Kaliningrad Region); it also provides the age and genetical differentiation of sediments, suggests the hypotheses of the formation of the Curonian Spit and determines the evolution mechanisms of the spit, including those connected with the fluctuations in the Baltic Sea level; forecasts of the state of the spit and the lagoon are given.

The development of well-boring, which expanded greatly in the second half of the nineteenth century, was vastly significant for the evolution of views on the geology of the region. Around 4/5 of wells of the prewar period were bored by the firm Pöpcke-Anclam, better known at the end of the nineteenth century under the name of E. Bieske. There were private entrepreneurs engaged in drilling – R. Quäck and L. Dost – as well as state institutions, the Royal Regional Construction Inspectorate and the Royal Road Workshop. The rock samples from wells, as well as the drilling journals, were delivered to the Museum of the East Prussian Province. Massive contribution to the processing of the material was made by the head of the Museum – A. Jentzsch. In his work "Geological Study of the Province of Prussia in 1877 with Detailed Consideration for all the North German Plain", we find indications of specific formations, which existed in the surface of pre-Quaternary sediments and are characterized by the valley-like shape (near Nortiken, at Katzengründe slope, in Sassau, etc.). The author believes that they were formed due to erosion by water during or before Pleistocene [4]. A. Jentzsch includes several images of paleo-incisions. Many things were borrowed by A. Jentzsch from the earlier works of Berendt [3] and Zaddach [2].

In 1899 "Report on Managing the Museum of the East Prussian Province in 1896 - 1898" were interesting sections of wells and hydrogeological data presented as well as stratigraphic division of sediments into systems [5]. On the basis of the further generalizations of data on numerous wells bored in Königsberg, in the article "Deep Foundation of Königsberg in the Context of Water Supply of the City", A. Jentzsch describes nine main water-bearing horizons, including four horizons of Quaternary age and their dynamic parameters [6]. He provides detailed description of the nature of Quaternary sediments, gives the statistics of the thickness of pre-Quaternary sediments and absolute marks of pre-Quaternary surface and characterizes roughness of PQS relief. In the work, there is the first description of a giant paleoincision that stretches over the territory of Königsberg, and the incline of the borders of the paleoincision is established. There are numerous graphs and maps in this work, including a map of boreholes in Königsberg and the surrounding areas, as well as profiles, a geological map of pre-diluvial foundation of the city, a scheme of sub-Quaternary relief surface, a scheme of the high-level position of the roof of Cretaceous formation, etc. [6]. At the same period of time (1899–1900), A. Jentzsch created a geological map of the pre-Quaternary surface of East Prussia (scale 1:1,000,000) and published a work in which he described the pre-Quaternary foundation of the province – "Pre-Diluvial Foundation of the North-Eastern German Plain" [7].

The beginning of the tenth century was marked by interest in Quaternary history of the region. If earlier the data concerning the youngest geological sediments had existed as a necessary part of a sedimentary thickness description or as the data concerning water-bearing material, from that time on, glacial sediments and relief became the subject of close scrutiny. In 1882, one of the first papers on the problem of glaciodislocation formation comes out. Its author – F. Wahnschaffe – explained the appearance of such formations by the rocks being squeezed out from under the lip of a glacier under the weight of ice [8]. In 1905 the paper by P.-G. Krause "On End Moraines of the Western Part of Zamland" is published. In this paper, P.-G. Krause singled out "end moraines of Zamland", no analogues to which were found in Prussia [9].

Speaking of papers that summarize the structure of the coast of Zamland, we must mention "Geological Pictures of the Coast of Zamland" ("Geologische Bilder von der samlandischen Küste") written by E. Schellwien, the Königsberg University professor. It is a separate reprint from *The Writings of Physical and Economic Society* [10]. This small book with 54 illustrations (including photographs) sums up the results of numerous works of G. Zaddach, A. Jentzsch, G. Behrendt and other researches. In simple terms, E. Schellwien described the main geological and geomorphological features of the coast of Sambia. He studies geological age of the layers forming steep banks and the disruptions of the layers occurring on the abrasive shores. The connection is made between the formation of a complex pre-Quaternary surface relief and two components of a glacial geological activity: a destructive one, which deepens and widens depressions, formed by fluid Pliocene waters and, a creating one, which fills the old valleys with ground moraine or with stratified fluvioglacial sediments, brought by thaw water to the ice-covered areas as well as to the grounds freed from ice. Schellwien points out that the breaks in Tertiary rocks, which are filled with glacier formations, originate not from the result of tectonic processes but rather from superficial destruction of Tertiary layers. He also points to the extremely deep occurrence of the Tertiary rocks roof near Warnicken (settlement Lesnoe). Deep drilling, carried out in the Warnicken manor, showed that glacial material occurs there as deep as 68 m below the sea level. Tectonic processes are used as explanation in this case [10].

In 1910, a consolidated regional paper of A. Tornquist "Geology of East Prussia" was published, in which he summarizes the materials on both autochthonous and Quaternary sediments [11]. It was the first time that the assumption of East Prussian pre-Quaternary rocks declivity being a complex glacial groove. A. Tornquist introduced the term "Zamland plinth" to describe the most elevated pre-Quaternary substrate surface on the Sambian Peninsula [12].

For further research of glaciation processes and glacial geomorphology, the works of E. Mayer – "Deviations in the North-Western Part of the Sambian Peninsula (the Sheet of Gross-Dirschkeim)" and "On Water Bodies and Coating

Clay Formation in the Remote Parts of Königsberg in East Prussia" [13, 14] – are of great importance. Works of Hess von Wichdorff dedicated his works – "Zamland End Moraines Motion and Continuation in East Prussia" and "Masurian Interglacial" [15, 16] – to the problems in studies of glacial morphology. A paper of great importance is the work of H. von Wichdorff on the structure and evolution of the Curonian Spit [17].

In the end of the nineteenth–beginning of the twentieth century (1891–1921) – there were geo-agronomic maps published in scale 1:25,000, which covered only the western part of the region. The unprecedented range of production well drilling served as a solid basis for the map (2,000–5,000 of sounding wells per mapboard) [12]. Geo-agronomic maps and resulting maps were of great importance for the further exploration of characteristic patterns of geological sediment distribution and for determining the hydrogeological conditions of Quaternary water-bearing horizons.

E. Kraus is the author of the following papers on palaeogeography and tectonics of the Quaternary Period in Prussia and the East Baltic: "Quaternary Tectonics of East Prussia", "Geological Overview Map of East Prussia" and "Tertiary and Quaternary Periods of the East Baltic" [18–20]. This series of maps is continued by the earlier published maps, i.e. a geological map of Prussia and adjoining federative lands (the sheet of Memel) and the explanatory note written by H. von Wichdorff in 1917, and a geological map of pre-Quaternary rocks in scale 1:500,000, made by the Polish researchers J. Lewinski and J. Samsonowicz in 1918 [21]. Several aspects of Quaternary geology were studied in the articles of Gagel [22, 23], Andree [24] and Beurlen [25]. B. Körnke devotes his papers to palaeogeography and the study of aspects of the late ice-cap degradation in general and to the history of the river system formation in the northern part of East Prussia and the adjacent regions, in particular [26].

In 1929 the Prussian Department of Geology publishes a geological overview map of Königsberg neighbourhood (scale 1:100,000) edited by Kaunhowen in 1926 [27]. The map shows distribution of sediments of different origin of the last and the penultimate glaciation, including ground and end moraines, lake and glacial sediments and fluvioglacial sediments; it also describes the composition of those sediments. Holocene sediments are mostly described in terms of structure; their genetic characteristics are hardly mentioned. The map also shows the abruptions of pre-Quaternary rocks. In some cases erratic blocks were depicted as autochthonous: smaller ones near Warglitten (now settlement Lyublino) and of significant size between Schulstein (settlement Volnoye) and Powunden (settlement Khrabrovo) [27].

The paper by P. Woldstedt on the genesis and stratigraphy of the Quaternary sediments in the North German Plain – "On Marginal Sediments of Last Glacification in Poland and Northern Germany" [28] concludes this stage of the development of notitions about Quaternary geology. On the basis of the received data, P. Woldstedt put together an overview geomorphological map of the North German Plain in scale of 1:1,500,000 and an explanatory note to it [29, 30]. The last map, created when the war had already started, was a geological map of East Prussia in scale 1:50,000, put together by I. Della in 1942 [12].

Among the aforementioned works, another outstanding contribution to the geological studies was an archive of a drilling firm of E. Bieske. In the period of its existence from 1880 to 1944, the extensive actual material included descriptions of drilling logs, dewatering logs, chemical analyses of water and notes on water supply. The archives comprise data on 10,000 drilling wells, presented in chronological order in 50 volumes, and the additional 10 volumes of chemical analyses of water. Many postwar papers were written on the basis of the data, obtained from the firm.

The research, which had been carried out by many scientists for more than 50 years and summarized the data on geological and geomorphological materials, resulted in the scientific basis for the development of ideas concerning the nature and patterns of continental glacification. The most indicative forms of glaciogenic relief, such as end moraines, were studied and mapped; their strike, composition and structure were specified. Other forms of glacial genesis, typical for the marginal glacier zones (oses, kames, sandres, glacial plains, meltwater runoff plains and drift-dammed lakes), were studied and specified. The composition, genesis and age of the rocks, underlying Quaternary sediments, were well explored, and maps and schemes of pre-Quaternary surface relief were created. Thus even a brief review of studies which had been carried out by German scientists for a century before the 1940s in the territory of East Prussia testifies to their significance and complexity.

Despite the hardships of the postwar time, the studies resumed straight after the Soviet troops arrived to the Kaliningrad Region. During battle actions, Military Geological Detachment No. 1 managed to gather a lot of data concerning geological morphology of the region. The first 5 years were marked by examination, restoration and reprocessing of the materials. In 1946, I. I. Rodionov, who worked in trust "Spetsgeo", comprised geological and geomorphological maps in scale 1:500,000 based on Sheet N-34 [31].

During the first postwar years, there was a complex expedition of USSR Academy of Science working on the territory of the region with the aim of collecting and summarizing the data on natural resources. The head of the geological and geomorphological party was prof. N. N. Sokolov. In the course of these works, in 1946 a catalogue of drillholes was made up (by E. F. Chanovskij and R. P. Teush), and it listed the data on 2,130 drillholes from Biske's archive, 1,190 of those in the territory of the Kaliningrad Region (Gidrogeologiya SSSR [32]). In 1946, on the basis of the catalogue, N. G. Verejskij compiled a map of Quaternary sediments of the Kaliningrad Region and the adjoining areas (scale 1:100,000), which was the first summary geological map of the region [31].

The first monograph, describing the hydrogeology of the region, comes out of print in 1947 (by R. P. Teush, T. M. Khokhonkina). The monograph comprised medium-scale maps (1:100,000) of water-bearing Quaternary and pre-Quaternary rocks, as well as hydrogeological sections accompanied by obligatory notes. There were a few maps in the monograph, which deserve special attention, the geological map of pre-Quaternary and Quaternary age, the map of Quaternary sediments thickness, as well as maps of geological exploration. About the same time, a geological map in scale 1:1,000,000 (Sheet N-34) is published, and it also includes the territory of the region (G. V. Bogomolov, I. V. Danilevskij). In 1951–1952, there was a complex geological and hydrogeological survey of the territory if the Kaliningrad Region carried out by the 5th Geological Department. In the course of works, a geological map of Quaternary sediments, as well as a geomorphological map, and for the very first time a map of water-bearing capacity of Quaternary rocks of the Kaliningrad Region (by N. A. Valuev, Z. B. Kostolomova, G. G. Shumskaya) were drawn [31, 32].

In 1958, an Integrated Geological Survey Party (IGSP) was founded in the North-West Geological Department in Leningrad. During 1958–1968 IGSP was conducting geological and geomorphological survey by large and medium scale. The survey was accompanied by test and structure drilling. While route surveys did not contribute much to glacial morphological studies, drilling provided plenty of new material for understanding the structure of Pleistocene and the geology of pre-Quaternary thickness. The survey was carried out in compliance with some sheets from State Geological Map of the USSR. Authors and time of completion differed from sheet to sheet. Geomorphological maps and maps of Quaternary sediments were completed by E. P. Makarova, M. P. Zharkov and I. K. Lasberg [12].

In 1964 IGSP finished its work on the report on the geological survey in scale 1:200,000 (M. P. Zharkov, the responsible researcher). In 1968 E. P. Makarova completed a review map of the region on the basis of the state geological survey materials. It was only in 1983 that sets of maps of the northern part of the region were published (sheets N-34-VIII, IX, X, XI). Only sheets N-34-VIII, IX and X were accompanied by explanatory notes. Geological and hydrogeological maps of Quaternary and pre-Quaternary sediments of the southern part of the region (sheets N-34-XIV, XV, XVII) were considered substandard and were not published [31]. The extended materials and preliminary research, together with geological survey, made it possible to specify the geological and hydrogeological structure of the region. For the first time ever, Jurassic, Cretaceous and Quaternary sediments were separated from each other, while the latter were further subdivided into three glacial and two interglacial complexes. Glacial, fluvioglacial and postglacial relief groups were identified together with the related mineral deposits.

At present, the survey in scale 1:50,000 covers 70% of the territory of the region [31]. The survey started in 1960 in the vicinity of Kaliningrad and was aimed at finding raw materials for building material production, as well as studying the engineering and geological basis (I. K. Lasberg, Z. K. Ivanova). Later, geological and hydrogeological survey of the same scale was carried out in the western part of the Sambian Peninsula (I. K. Lasberg, G. N. Yeltsina). For reclamation purposes, in 1974–1991, there were complex hydrogeological and engineering-geological surveys carried out in scale 1:50,000, and special attention was paid to Quaternary sediments. The southeastern part of the region remained the less researched area, just like in prewar years. A complex hydrogeological and engineering-geological survey in scale of 1:50,000 was conducted in 1987–1991 in the southeast of the Kaliningrad Region for the purposes of land reclamation construction (I. F. Dell, the responsible researcher). The conducted works resulted in a set of maps in scale 1:50,000 (11 maps) and 3 sketch maps in scale 1:200,000. The thickness of Quaternary sediments was studied in great detail.

In the Kaliningrad Region, Gusev Building and Construction Department had been drilling water wells since 1950, and from the mid-1960s, the biggest contribution to the research of the basis of the old networks and reconstruction of the new water conservation networks had been made by Leningrad Trust of Engineering and Construction Research (LenTISIZ) and by Zapgidrovodkhoz Institute. Several research and economical organizations of the Kaliningrad Region, such as Kaliningrad Integrated Geological Survey Party, Kaliningrad Oil Exploratory Expedition, the branch of Rosgiprovodkhoz Institute, Promburvod, LenTISIZ, SMU-4, etc., played a significant role in accumulating actual material on Quaternary geology and palaeogeography as well as on geomorphology of the region. Yet, the obtained materials cannot be accessed freely, are poorly summarized and are stored in reserves without being published. This is one of the reasons why there are still no review papers on Quaternary geology, geomorphology and palaeogeography of the Kaliningrad Region.

Some general data concerning the geology and geomorphology of the Kaliningrad Region can be found in the papers of Lithuanian scientists. For example, in 1966 P. P. Vajtekunas studied the marginal glacier formations as well as the stratigraphy of the Quaternary thickness in the Sambian Peninsula. In his article "Marginal Glacier Formations and Deglaciation Patterns on the Territory of the Kaliningrad Region and Adjoining Regions", P. P. Vajtekunas published the scheme of anthropogenic substrate surface of the region [33]. In the monography of V. K. Gudelis "Quaternary Sediments of the Baltic" (1973), there is a map of sub-Quaternary relief of the Soviet Baltic territory [34]. Multiple schemes of pre-Quaternary rock surfaces, schemes of pre-Quaternary surface isolines of the Southern Baltic as well as the schemes of buried paleoincisions can all be found in the only book on paleoincisions of Lithuania and the Kaliningrad Region, edited by A. Gaigalas, named "Buried Paleoincisions of pre-Quaternary Rocks in the Southern Baltic" [35]. Hence, during the Soviet period (1945–1990), there were maps and schemes of pre-Quaternary relief drawn for several parts of the Kaliningrad Region. Many of them came out of date and provide very general understanding of the nature of PQS, and it does not help that these maps are stored in reserves with no free access.

During the post-Soviet period, there were further works on the territories of the Kaliningrad Region which were previously surveyed in scale 1:200,000. These works resulted in sets of geological maps of Quaternary and pre-Quaternary formations combined with mineral maps in scale 1:200,000. The maps also have indications of pre-Quaternary surface isohypses. Only four sheets out of ten that covered the territory of the Kaliningrad Region were approved by the scientific and editorial board of the Russian Ministry of Natural Resources [31]. Works were carried out with the aim of creating a State Geological Map (scale 1:1,000,000) of the third generation. The maps were approved and published in 2011 [36], and to some extent, the gap in the open access cartographic information concerning the Kaliningrad Region was filled in with the help of the pre-Quaternary relief map (in scale 1:500,000) by G. N. Yeltsina, the map of Quaternary sediments by V. V. Orlenok and A. N. Efimov and the geomorphological map by G. N. Yeltsina, which were all created for the Geographical Atlas of the Kaliningrad Region [37].

2 Pre-Quaternary Sediments, Their General Profile and Surface Zonation

On the basis of the available data on more than 1,500 drillholes made by different agents (IGSP, "Zapgiprovodkhoz", LenTISIZ, Promburvod), the author of the article succeeded in completing the maps of pre-Quaternary relief and Quaternary sediments thickness in scale 1:200,000; simplified versions of those maps are presented in Figs. 2 and 10. The maps demonstrate the nature of PQS and agree with the earlier works on the matter. The methodology of pre-Quaternary surface maps compilation can differ, and the first arguments in favour of particular methods were attempted at in the early 1970s in Lithuania, which territory is very similar to the territory of the Kaliningrad Region [35, 38]. Even today pre-Quaternary surface maps made by different researchers demonstrate different approaches [39]. Some researchers interpolate all the data concerning the points on the map, extreme

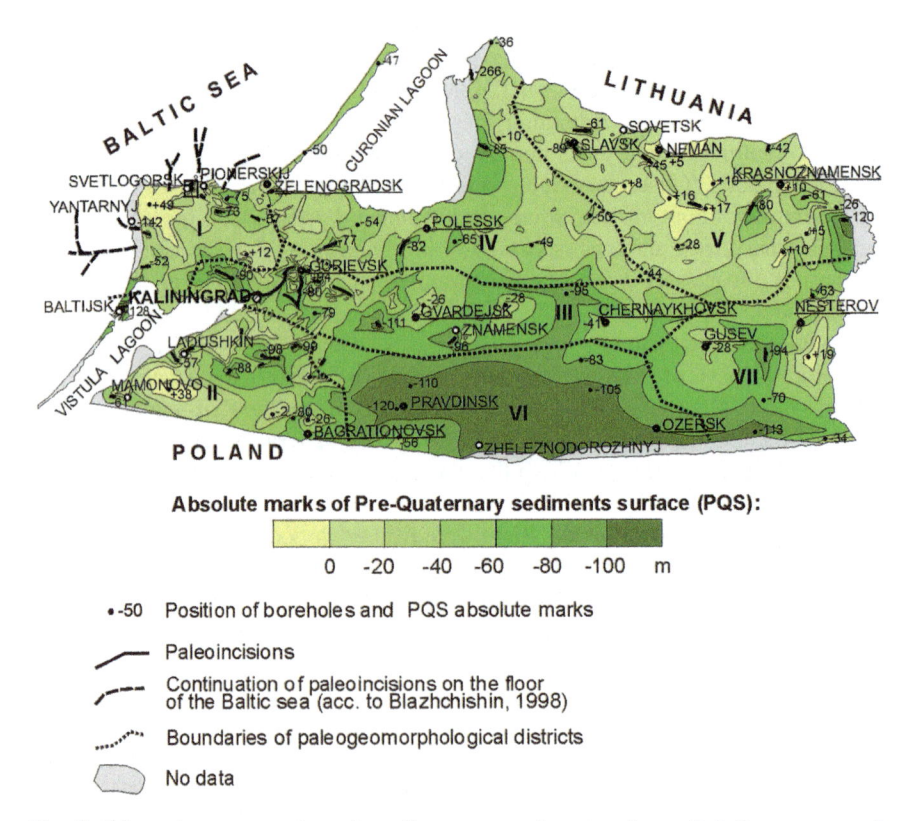

Fig. 2 Schematic representation of pre-Quaternary sediment surface relief. Roman numerals stand for *I*, Sambian (Zamland) plinth; *II*, south-west elevation; *III*, the Pregolya zone of local depressions; *IV*, Curonian lowland; *V*, north-east elevation; *VI*, south depression; *VII*, southeast undulating plain. The *rectangle* shows the region which is depicted in Fig. 5

depressions and "elevations" alike. The maps presented here demonstrate the same principle. Other researchers produce pre-Quaternary surface relief maps, ignoring the extra-deep incisions, and sketch paleoincisions over the ready maps afterwards. In both cases the maps of pre-Quaternary surface have their advantages and disadvantages. When using the discrete drilling data, the results of the former method might include a series of narrow oblong or rounded depressions instead of unified elongated negative landforms. Determining the true long axis orientation of such landforms might sometimes become problematic. Justification of such interpretation of the data or the possibility of combining several landforms into one valley-like form is only possible if additional drilling is carried out in order to obtain more information or by other geophysical methods, which is very rare. On condition that the data on paleoincisions and other extreme depressions in PQS relief is excluded, the latter method allows us to "ignore" the data concerning the upper part of the depressions, which differs insignificantly from the surface of "elevations". Alternatively, we can "assign" a new position to some local depressions with the amplitude of 10–20 m. This method doesn't solve the problem of identifying the direction of paleoincisions, either.

The formation of sub-Quaternary surface relief is usually regarded as the result of an exaration process overlaying the Pliocene and Eo-Pleistocene relief [35]. However, the surface of pre-Quaternary sediments (PQS) in the region was not smooth or flattened out, instead it was characterized by significant roughness and larger amplitudes than at present. At least two peculiarities of PQS morphology can be singled out: presence of over-deepened incisions (paleoincisions, palaeovalleys), which are usually regarded as river valleys after the exaration, and of residual elevations of pre-Quaternary surface, located above sea level.

Pre-Quaternary sediments that form PQS belong to three systems: Cretaceous, Palaeogene and Neogene. The most common sediments are Campanian and Maastricht sediments in the upper part of the Cretaceous system. The sediments were opened in the drillholes in the north, east and southeast of the region, as well as in the lowerings of paleoincisions in other districts, including those on Sambian Peninsula (Fig. 3). There were Cenomanian and Cenomanian-Albian sediments of the Lower Cretaceous found in the north-east of the region, between the Neman and the Sheshupe, and also in paleoincisions in the east and the north of the region (settlement Mysovka). There were also sediments of the Lower Cretaceous and Middle and Upper Jurassic periods found in between Neman and Sheshupe rivers (Fig. 3). Palaeogene sediments are spread over the larger part of the Sambian Peninsula, south-west of the region and only sporadically in the south of the region. Neogene sediments (Miocene ones, mostly) can be found only in the Sambian Peninsula and the south-west of the region. The patches of Palaeogene and Neogene sediments are separated by deep depressions, the bottoms of which incise into Cretaceous sediments.

The sediments that overlay PQS date back to Pleistocene. Submoraine lacustrine and alluvial sediments of the Lower Neo-Pleistocene were discovered in the south of the region, and we expect that they can be found in paleoincisions in other parts of the region [31]. As a rule, the oldest sediments that fill deep depressions and

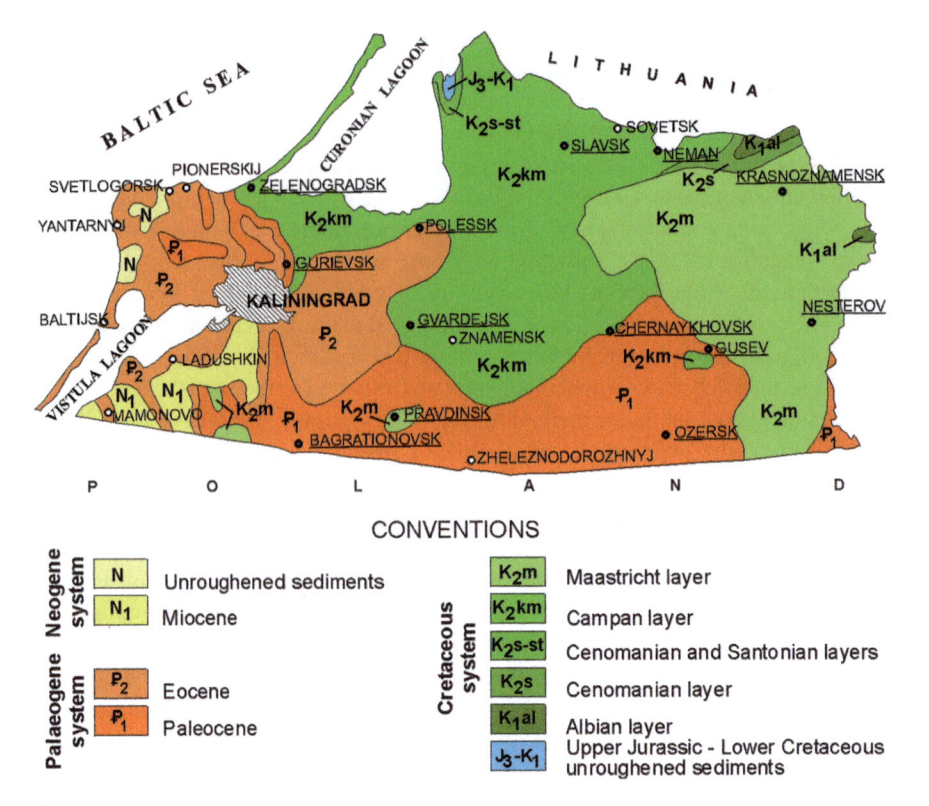

Fig. 3 Schematic representation of pre-Quaternary sediments in the Kaliningrad Region (http://atlaspacket.vsegei.ru/_Files/Северо-Западный%20ФО/KALININGRAD/04_geol1000_KALININGRAD.jpg)

paleoincisions of the Kaliningrad Region belong to the Lower and Middle Neo-Pleistocene. Genetically speaking, those are the complexes of glacier, lacustrine or sea origin. It is only in the north and north-east of the region that the oldest sediments that fill deep depressions belong to the Middle Neo-Pleistocene or sometimes to the Upper Neo-Pleistocene and Holocene (in the delta of the Neman river). However, even here the extra-deep depressions are filled with the whole range of Neo-Pleistocene sediments.

On the whole, in the central part of the region, the roof of pre-Quaternary sediments deepens from the north and north-west towards the south and southeast. In reality the relief of PQS is much more complex and alternates between depressions and elevations; the absolute marks of the pre-Quaternary floor vary significantly over short distances, and the matters are made worse by the depth of valley incisions (Fig. 2). The roof of pre-Quaternary surface lies in the Sambian Peninsula (settlements Shatrovo, Krasnotorovka, Otradnoe) on the absolute mark +30–40 m, in the south-west of the region; on the absolute mark +20–30 m (to the south of settlement Pyatidorozhnoe), to the north of Kaliningrad (near settlement Kholmogorovka and settlement Lermontovo); and on the absolute mark +10–15 m. The maximum mark for PQS, at our estimation, is +49.5 m (the Sambian Peninsula, near settlement Krasnotorovka) (Figs. 2 and 4a). The deepest depression (−110–120 m) is situated in the south of the region (Pravdinsk, settlement Sosnovka) (Fig. 2). According to V. A. Zagorodnykh, the deepest paleoincision was opened in the upper part of the Neman river (settlement Mysovka): absolute marks of the roof of Jurassic sediments reach −266 m [31, 40]. Thus, the amplitude of heights for PQS is 315 m, when the modern relief amplitude is less than 250 m.

Besides the aforementioned elevation of PQS, also called denudation remnants, there are other local elevations of pre-Quaternary relief with absolute marks of higher than 0 m and with relative height of 20–30 m. The majority of those are concentrated in the north-east and the east of the region (Figs. 2 and 4). Elevations in the district of Neman occupy the largest area, settlement Bolshoe Selo, settlement Kanash, settlement Pushkino (up to +8 m) (Fig. 4b) and near settlement Lunino-Zabrodino-Ulyanovo (up to +17 m), to the north-east of settlement Zarechnoe (+7 m) and near settlement Kalachevo (+10 m), to the west of settlement Dobrovolsk (+10 m) and near settlement Pokryshkino (+19 m).

The remnants of the western part of the region are formed by Palaeogene and Neogene sediments (Fig. 4a); the eastern "elevations" are formed in the Upper Cretaceous formations (Fig. 4b). The existence of denudation remnants was proved on the example of the Sambian Peninsula back in the early twentieth century by the German scientist A. Tornquist. He was the one who named the elevated pre-Quaternary surface of the Sambian Peninsula "the Sambian plinth" [11].

The mechanism of formation of these elevations is unclear; it cannot be explained by a selective exaration impact on the rocks, which have various resistances to destruction: the main part of the highest "remnants" consists of Neogene sands and clays and Palaeogene sands and aleurites, while deep depressions were formed in aleurolites, sandstone, marls, etc. The peculiarities of the dynamics of the continental glaciers and/or the tectonic influence are the only available explanations for the formation of the remnants made of easily destructible rocks. There is an opinion that the north-west part of the Sambian Peninsula near Mys Taran is actually a horst uplift; and back in the nineteenth century, there was an opinion that the orthogonal shape of Zamland is the result of splits [1, 2].

The elevated areas in the west are separated by a deep depression (−80–140 m), which corresponds to the modern mouth part of the Pregolya river and the adjoining parts of the Kaliningrad bay (Figs. 2 and 9). Further to the east, along the valley of the present-day Pregolya, there is an area of an extremely complex and rough pre-Quaternary relief, particularly near settlement Isakovo, settlement Vasilkovo, Gurievsk and up to settlements Ushakovo and Malinovka (Fig. 2). Within 500 m the absolute marks may change by 70 m (from −94 to −25 m), which corresponds to 7° gradient. Maximum gradients of PQS are noted in the territory of Kaliningrad: the difference in the marks of the PQS at a distance of 300 m can be up to 100 m (slope 18°). To the west of the Deima river, near Polessk, the relief is characterized by absolute marks of −30–50 m with deeper depressions (settlement Bayevka, settlement Nekrasovo – less than −80 m). To the south of the Pregolya, near settlements

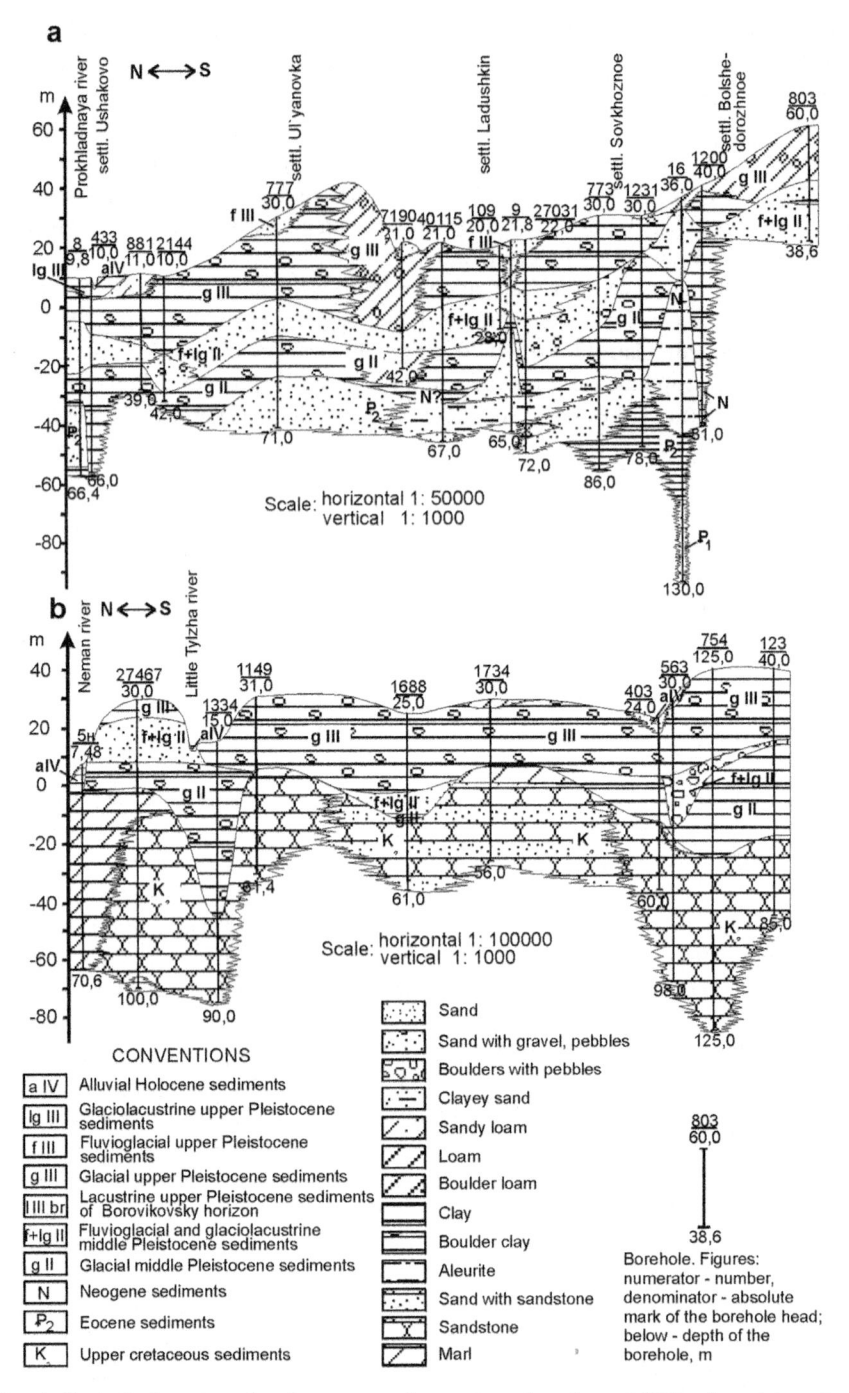

Fig. 4 Geological sections that demonstrate the remnant elevations of PQS: *above* (**a**) section along the line of settlement Ushakovo, settlement Bolshedorozhnoe (the south-west of the region); *below* (**b**) section along the line of Neman sity, settlement Zhilino (the north-east of the region)

Ozerki and Znamensk, there are two large depressions (−111 m and −90 m, respectively). Further to the south, there is an extensive depression with absolute marks that vary from −70 to −120 m. Minimum marks are found in Pravdinsk region (Fig. 2). A. Tornquist called the central part of the Kaliningrad Region (from the Couronian Lagoon to settlement Zheleznodorozhnyj) the East Prussian lowland due to the minimum marks of the pre-Quaternary surface [11]. This low is traditionally regarded as a glacial stream route [33, 41].

The area to the south of the Deima river and the east coast of the Curonian Lagoon forms a single structure with a low to the west of the Deima river and the south coast of the Curonian Lagoon. This is the area of pre-Quaternary lowland relief with absolute marks of −40–60 m with elevations up to −26 m (settlement Zaborie), −27 m (settlement Krasnyj Bor) and depressions down to −85 m (settlement Zapovednoe), −65–70 m (settlement Saranskoe) and −266 m (settlement Mysovka) (Fig. 2). The north-east of the region represents a slightly elevated surface. Average marks are −10–30 m. The maximum absolute marks are higher than 0 m; the minimum can reach −80–110 m (Fig. 2). The elevations serve as a kind of watersheds, and numerous depressions represent ancient valley-like lows which might have been connected with the river flows (the proto-Neman and in Holocene times – the Neman and its migrating influxes). The erosive activity of the watercourses became the reason for the partial or total destruction of Pleistocene sediments in the Low Neman lowland and in the north-east of the region.

There are minimum absolute marks of pre-Quaternary relief found in the southeast part of the region, which is in fact a single entity with the area to the south of the Pregolya valley. Generally speaking, the roof of pre-Quaternary surface deepens from the north towards the south from −40–50 m to −113 m. Near Gusev, at settlements Pokryshkino and Babushkino, there are local elevations (−28 m, +19 m, −6 m, respectively) (Fig. 2). So, according to the nature of pre-Quaternary surface relief, we can single out, conceptually, the following areas:

1. The Sambian (Zamland) plinth
2. South-west elevation
3. The Pregolya zone of local depressions
4. Curonian lowland
5. Northwest elevation
6. South depression
7. Southeast undulating plain (Fig. 2)

When reconstructing the history of PQS development in the pre-Quaternary times, we can briefly present it as follows. The nature of the sediments in the lower part of Cretaceous system shows that the climate conditions in the territory of the region were mostly coastal, but the continental conditions still prevailed. The territory must have been elevated and was denudation prone, as can be seen by the small thickness of the early Cretaceous sediments – 20–50 m – and the local nature of their distribution. Late Cretaceous transgression came to the relatively flat surface from the west and represented the result of a sharp redesign of a structural plan. The depth of the sea basin in the territory of the Southern Baltic gradually grew deeper. At the same time, there were constant tectonic movements of different kinds. This is evidenced by the uneven character of late cretaceous sediments and the incomparability of thickness of different parts of sections [42]. In the process of late Cretaceous movements (Campan and Maastricht), the east part of the region, which had become dry land, got some linear tectonic structures: sublateral, marked out by rivers Neman and Pregolya, and sub-meridional, marked out by rivers Lava, Deima and Pissa. There is an opinion that the elongated in sub-meridional dimension depression of Lake Vištytis is of tectonic origin [42].

In the late Cretaceous, the marine climate, which was typical for the time period, changed into the continental one in the east and north-east parts of the region (north-east elevation, the north of Curonian lowland). It could be back in pre-Quaternary age that bodies of remnants became isolated and the deepest valley incisions appeared and significantly weakened the thickness of Upper Cretaceous sediments or eliminated them completely (territory between Neman and Sheshupe rivers, east part of the region; near settlements Pobedino and Dobrovolsk, north part of the region near settlement Mysovka (Fig. 3)). These formations were partially completed by exaration, and in the postglacial period, it was done by the river system of Neman.

Throughout the Palaeogene in the west part of the region, there were marine-like conditions for sediment accumulation, while the central and south parts (the Pregolya zone, south depression and southeast plain), where the characteristic feature was the sea basin regression back in Paleocene, start their development in continental conditions. The most intact Palaeogene sections are in the west part of the region, which is characterized by the complex nature of pre-Quaternary surface and its elevated location. The roof of Palaeogene sediments must have been located even higher if we take into account denudation in Neogene and exaration in Pleistocene. In the end of Palaeogene, the sea basin draws back from the west part of the region, where continental climate conditions had set in [31]. Neogene sediments show that there used to be a vast undulating plain; lakes, swamps and river system were formed (Sambian plinth, south-west elevation). It is possible that accumulation of lake and swamp sediments, as well as of alluvial sediments, took place in the depressions in the east part of the region, which were destroyed later by the process of exaration the same way the Pliocene sediments had been destroyed at Sambia.

With the beginning of the Pleistocene, glacial and fluvioglacial processes became dominant in the transformation of PQS. It had undergone the exaration processes several times, which were at their highest in the early and middle Pleistocene; pre-Quaternary depressions and paleoincisions sunk in the interglacial and Holocene periods (the Neman river delta). Washing out of the surface in the course of interglacial marine transgressions is not precluded. The central and northern parts of the region were subjected to glacial erosion of the greatest extent. It shows in the absence of lower sediments and mid-Pleistocene sediments as well as Maastricht sediments in the structure of this territory (Fig. 3). These are the reasons why Curonian lowland is also regarded as a depression formed by glacial exaration (Fig. 2). There is a similar decrease of glacier exaration in the southeast of the region. The line that shows the limits of the area of Paleocene sediment occurrence is drawn roughly along the longitude of Gusev. To the west of this line, there were documented Maastricht formations buried under Quaternary sediments. However, further to the east in the close vicinity from Lake Vištytis, pre-Quaternary surface is formed by Paleocene rocks (Fig. 3). The "gap" in Paleocene sediments is characterized by minimum absolute marks (−70–94 m) in comparison with "Paleocene" elevations (−38.0–28.0 m and −33.0 to +19.0 m) (Fig. 2).

These examples demonstrate the significant transformation of post-Pliocene surface as the result of glacial processes and emphasize the impossibility of discovering the interconnected network of Neogene rivers and determining the alluvial origin of paleoincisions (taking into consideration the absence of alluvium in the bottom part of paleoincisions). The surface of pre-Quaternary sediments gives no information on the relief that existed at the beginning of glaciation: the map shows not the late Pliocene relief but the result formed on the basis of pre-Quaternary surface by various processes, and tectonic activity is not the least important one among them. It is known that during Mesozoic and Cainozoic periods, transformations of the structural level of the region had been taking place. The evidence to that can be seen in the folded deformations of the corresponding layers found in Sambian Peninsula, in the absence of sediments in some Upper Cretaceous layers under deep paleoincisions (we allow the possibility of vertical movements under them), etc. [31, 40, 43]. At the same time, the load created by the glacier cover on the underlying surface had to activate the movements along the snap lines, and we also cannot ignore the glacial isostatic movements.

Thus the surface, where glaciation came to, had a long and complex history of development. We can divide it into late Cretaceous, late Palaeogene, Pliocene and ancient Quaternary marks that were characterized by a radical change of physical and geographical conditions and processes, which transformed the surface of pre-Quaternary sediments. The data, provided by boreholes, made it possible to reconstruct the PQS, and it demonstrates significant roughness and peculiar forms of relief such as denudation remnants and paleoincisions. PQS can also be divided into several palaeogeomorphological areas.

3 Characteristic Features of Paleoincision Structure and Origin

Identification of paleoincisions is carried out with a help of a number of geological methods, lithological, palaeontological, palaeogeomorphological, structural, geological, etc., and by applying the data of gravimetric, seismic and electrical surveys [35, 44]. Paleoincisions are of polygenic nature: their origin is usually linked to preglacial, fluvioglacial or exaration processes or a complex of processes (modified valleys) [35].

The territory of the Baltic region is characterized by the presence of a dense network of ancient erosional incisions. When interpreting the material on spreading of paleoincisions, it is necessary to take into consideration the history of the region development, existence of sediment accumulation basins, which level determines the basis of paleohydrographic network. Paleoincision is supposed to be able to form a unified system, generally caused by Pliocene and ancient Pleistocene paleohydrographic network in a combination with other geological factors [45]. The watershed of the paleohydrographic network could have existed in the area of Mesozoic and Cainozoic sediments to the south of the depression of the modern Baltic Sea, and it was much lower than the modern sea level. Thus, the paleoincisions found on the continent continue within the limits of the Baltic Sea water area. Paleoincisions of maximum depth were found on the coast of the Baltic [46]. The maximum depth of paleoincisions in Poland is −341 m (Żarnowieckie Lake) and −324 m (Karvia, to the west of Hel Peninsula). In the mouth of the Neman river, the depth of an incision is −266 m (settlement Mysovka), in the Jūrmala District − −282 min. Other paleoincisions can reach the depth of 158 m (Ventspils), 146 m (Palanga) and 139 m (Šventoji). Many paleoincisions on dry land and at the coast can be found at absolute depth of −80 and −100 m. There are several levels of paleoincisions (340–320 m, 280–260 m, 160–140 m and 100–80 m) which correspond to the existent levels of erosion basis of ancient water bodies. The varying depth of paleoincisions is usually attributed to changes in the level of erosion basis during their formation and to their position in the system of paleohydrographic network (valleys of different types). Processes of glacial exaration played a significant part in the changes of paleoincisions profile. Some areas in the valleys expanded, and abnormally deep zones appeared. It explains the absence of gradual level decrease in the transition to the basis of erosion. In separate areas paleoincisions could have been destroyed. The role of glacial water washing out in the transformation of paleoincisions at the sea floor is not yet clear. The changes in paleohydrographic network profile can be linked to tectonic activity, though this factor cannot be assessed accurately. In certain areas the resemblance between the modern relief and pre-Quaternary surface relief can be traced back through all the sediment layers, and paleoincisions are clearly conditioned by deep tectonic processes.

The territory of the region holds numerous deep paleoincisions (Fig. 2 and Table 1), but they are yet to be studied in detail [35, 40, 47]. Most of palaeovalleys are situated in the west and north parts of the region (Fig. 2), probably due to the minimal thickness of Quaternary sediments, which made it a more research-friendly object. The Table 1 shows the characteristics of some paleoincisions.

The main problems are usually associated with definition of age, genetic type and morphological and morphometric characteristics of paleoincisions. The paleoincision near Svetlogorsk, which was discovered with the help of boreholes back in the 1870s, can serve as an example [4]. This paleoincision was studied in detail in the 1960s–1970s in the course of hydrogeological works in this district. It can be seen on the plan that it stretches from settlements Salskoe and Zori to settlement Rybnoe (Fig. 5) and continues further into the Baltic Sea water area

Table 1 Examples of the parameters of paleoincisions in the Kaliningrad Region

Position of the incision	Absolute mark of the incision floor (m)	Age and nature of sediments in the bottom part of the incision	Relative elevation of the lips of the incision over the thalweg (m)	Nature of the filling material	Age of the filling material	Length/width of the incision (km)	Alignment
Slavsk	−89	Sandstone, K_2	75	Boulder clays, sand and gravel sediments	Q_1–Q_3	2.5/1.0	NW-SE
Kaliningrad	−140	Marl, K_2	100	Boulder clays, loams, sand	Q_1–Q_3	15(?)/3.0	SW-NE
Baltiysk	−128	Clays, aleurites, marls K_2	80	Sand	Q_1–Q_2	3.0/1.0	W-E
Settlement Yantarnyj	−142	K_2	145	Boulder clays	Q	3.0/2.0	W-E
Settlement Sosnovka (Zelenogradsk)	−86	Alevrolit, K_2	70	Sand, sandy loams, clays, boulder loams	Q_1–Q_3	4.0/1.0	W-E
Settlement Mysovka	−266	Aleurites, alevrolites, J_2	240	–	$Q_{1?}$	10.0/5.0	NNW-SSE
Settlement Medovoe	−88	Clays, Pg_1–Pg_2	40	Boulder loams and clays, sand	$Q_{1(?)}$–Q_3	2.5/1.0	SW-NE
Settlement Nivenskoe	−99	Sand, K_2	60	Boulder clays and loams, sand	Q_1–Q_2	4.0/2.0	SW-NE
To the southeast of Krasnoznamensk	Less than −80	K_2	50	Boulder loams, sand	Q_1–Q_2	2.5/1.5	SW-NE
Settlement Nekrasovo	Less than −82	K_2	50	Boulder clay	Q_1	2.0/1.0	SW-NE

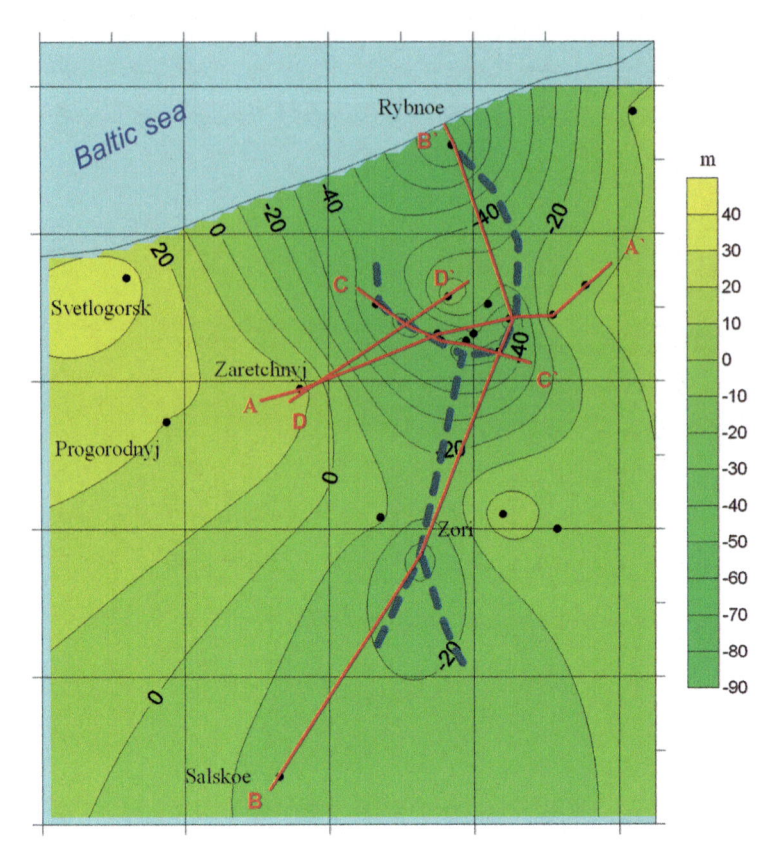

Fig. 5 Pre-Quaternary sediment surface relief. The *dotted line* shows the thalweg lines of the paleoincision. Grid size in the figure – 1 × 1 km

(Fig. 2) [45]. The paleoincision is aligned from south-west to north-east (Fig. 5). The floor of the incision is on absolute marks from −20 to −80 m (Fig. 5). The edge-to-edge width of the paleoincision is about 0.5–1.5 km (Figs. 5 and 6). However, when studied in greater detail, this paleoincision shows a number of peculiarities in its structure.

The age of incisions is usually only roughly estimated by using the method of age marks. The lower age mark is determined by the age of the newest sediments that form the lips of a paleoincision, and the highest mark is determined by the age of the most ancient sediments that fill this paleoincision (filler rocks). This is how an "age fork" can appear. For example, for the paleoincision near Svetlogorsk, it was calculated that the time of its formation – judging by the age of the sediments, found in the boreholes, that had been drilled in the lips and in the middle of that paleoincision – fits the interval between the early Neogene and the early Pleistocene (according to the archive materials of Integrated Geological Survey Party). The incision is about 4.5–5.0 km long, and the cross sections that cut across its length

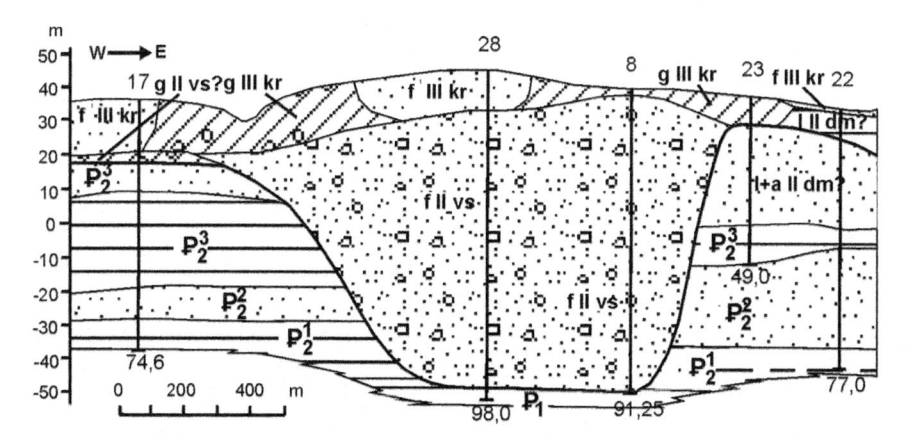

Fig. 6 Cross profile of the paleoincision near Svetlogorsk along A–A line. Profile position is shown in Fig. 4. Vertical scale 1:1,000, horizontal scale 1:10,000. Conventions – see Fig. 7

show filling formations of different age – from lower to middle upper Pleistocene; and the age of the newest rocks from the lips of the valley can vary from upper Eocene to Miocene and sometimes to mid-Pleistocene (Fig. 5). It can be explained by destruction of mid-Pleistocene sediments on some territories during the late Pleistocene, rather than by different stages of paleoincision formation.

Estimation of width and length of paleoincisions can also be problematic. The width of paleoincisions in the Kaliningrad Region is usually about 1.0–2 km (Figs. 5, 6 and Table 1), and it is rarely exceeded. Estimation of the parameters of paleoincisions is based on the data obtained from drilling, and the boreholes are often sparsely positioned, which makes it difficult to determine the nature of the floor and lips of the valley. For example, there are three boreholes, one of which detected a paleoincision in what supposedly is its central part, and two other boreholes fix the position of the "valley wall". Figure 6 demonstrates the impossibility of determining the position of the west incision lip by using solely the borehole data as the distance between boreholes 17 and 28 is about 1 km. Such situation makes it difficult to estimate the width of the incision, its slopes, etc. Sometimes we are speaking not of a single paleoincision but of a system of narrow, more complex formations. The paleoincision we chose as our example turned out to be a system of intersecting V-shaped depressions: not far from Svetlogorsk-Prigorodny, there is a branch of the incision that goes in the north-west direction. The side area of the paleoincision can reach −48–52 m in depth (Figs. 5 and 7). Among the sediments that fill the paleoincision, the greatest thickness is of fluvio-glacial sand and sand with gravel. These are the sediments that constitute the water-bearing horizon of Lake Vištytis, which is one of the sources for Svetlogorsk water supply (Figs. 6 and 7). At the same time, the cut clearly shows that the side branch of the paleoincision in settlement Prigorodny is formed by glacial sediments – boulder loams and clays (Fig. 7).

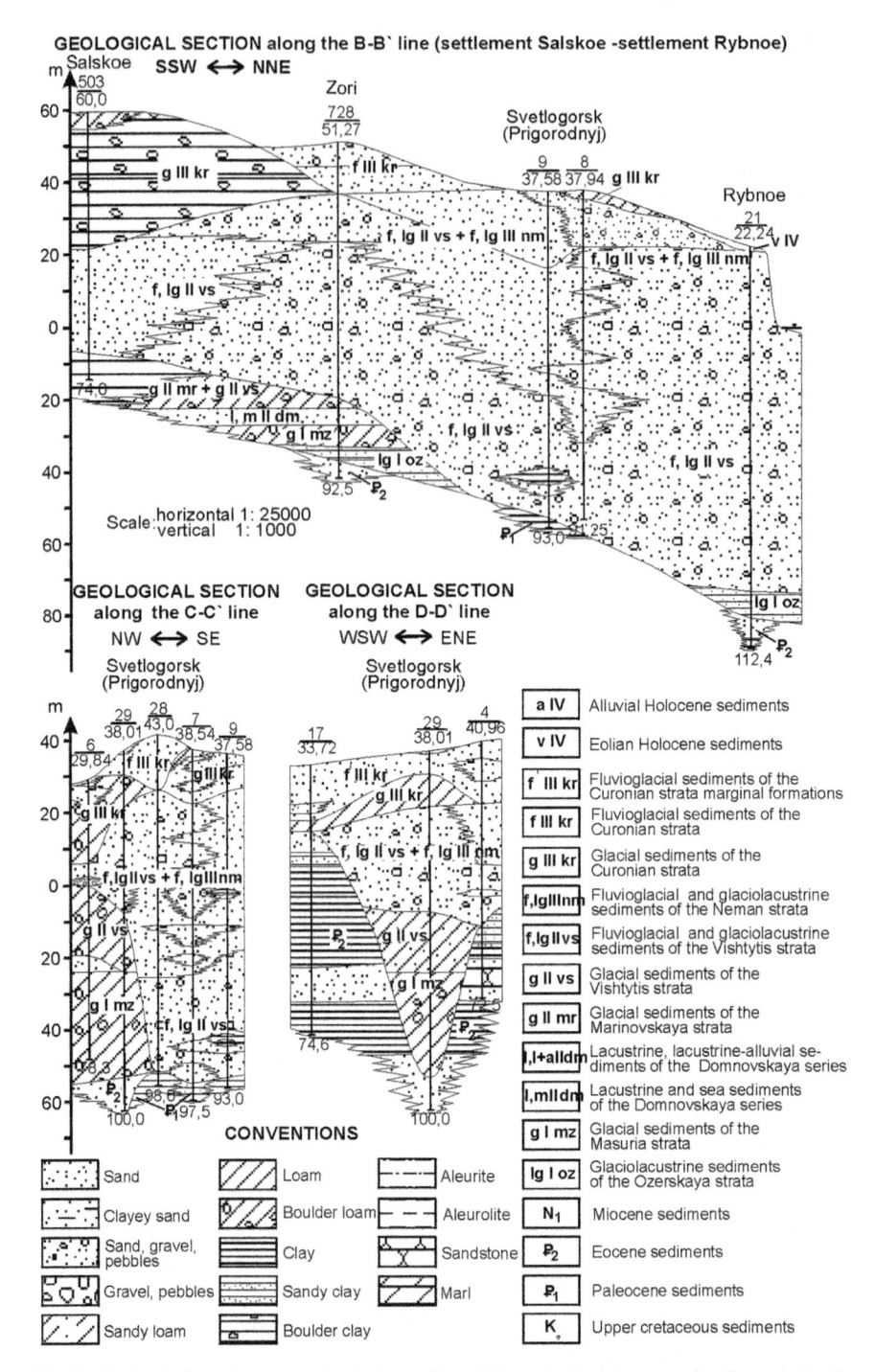

Fig. 7 Geological section along the thalweg line of the paleoincision near Svetlogorsk (to the *right*) and vertical and cross sections of its branch (to the *left*). Section positions – see Fig. 5

In the process of adding the results of gravimetric works, which had been carried out in 2012–2013, to the existing data from boreholes, we came up with an assumption that in the south part (near settlement Zori), there is a branch in the south-west direction. In favour of complex nature of pre-Quaternary surface speaks the fact that in close proximity from the incision in question, there is another paleoincision, aligned in submeridian direction from settlement Grachevka, across settlement Majskij and up to Svetlogorsk. The upper part of this incision is revealed in the sea cliff to the west of Svetlogorsk. The floor mark of the incision in settlement Majskij is on −40 m; its age is of late Neogene-early Pleistocene. It serves as another proof that PQS relief studies must take into consideration geophysical studies and not only the data obtained from boreholes. Further to the west near settlement Primorie, there is a 50 m deep paleoincision aligned in sublatitude direction. It was discovered in 1987–1991 as a result of terracing the slope of the Filinskaya Bay.

There is evidence that allows us to say that some paleoincisions were conditioned by deep tectonic processes (e.g. near settlement Yantarnyj, settlement Mysovka) [40, 47, 48]. Such evidence is the presence of fault deviations in the lips of paleoincisions and their filling and the fact that paleoincisions belong to either Pregolya, Neman or Yantarnyj fault zone [49]. The cyclic nature of glaciation and deglaciation could contribute to constant activation of tectonic ruptures and repeated recurrence of erosive incisions [35, 50]. The tectonic influence can be direct or mediated, through the increased fracturing of pre-Quaternary rocks in fault zones and through the decreased destruction resistance of rocks.

One of the most famous is the paleoincision in settlement Yantarnyj. The floor of the palaeovalley is on the absolute mark of −142 m (Fig. 8). Thickness of Quaternary rocks that fill the incision exceeds 160 m. The floor of the palaeovalley reveals the rocks of Upper Cretaceous. Formation of the paleoincision could be preceded by submersion of separate areas long before the beginning of the Quaternary Period. The increased thickness of upper Eocene amber-bearing horizon up to 20 m, which can be seen on the sections of Primorskoye amber deposit, is a consequence of syndepositional negative movement (boreholes 412, 650, 684 in Fig. 8a). Another manifestation of tectonic activity is the faults that preceded the formation of paleoincisions. Fault amplitude can reach up to 20 m. For example, absolute marks of the roof of the amber-bearing horizon in the south part of the section (Fig. 8b) are 20 m lower in comparison with the northern part. The same types of faults were detected in palaeovalleys near settlements Yantarnyj, Svetlogorsk and Baltiysk [40]. Here, active manifestation of denudation processes was predetermined by the tectonic factor.

In his works, V. A. Zagorodnykh notes that formation of super-deep palaeovalleys (incisions are 150–270 m deep) as the result of exaration only is highly unlikely, even if we take into consideration the areas weakened by tectonic activity [31, 40]. In separate cases, the origin of paleoincisions is associated with long-lived tectonic disturbances of Archean-Proterozoic foundation that penetrate the sedimentary cover-up to Quaternary sediments. He was also the one who pointed out the possibility of palaeovalley formation within riftogene depressions, which had

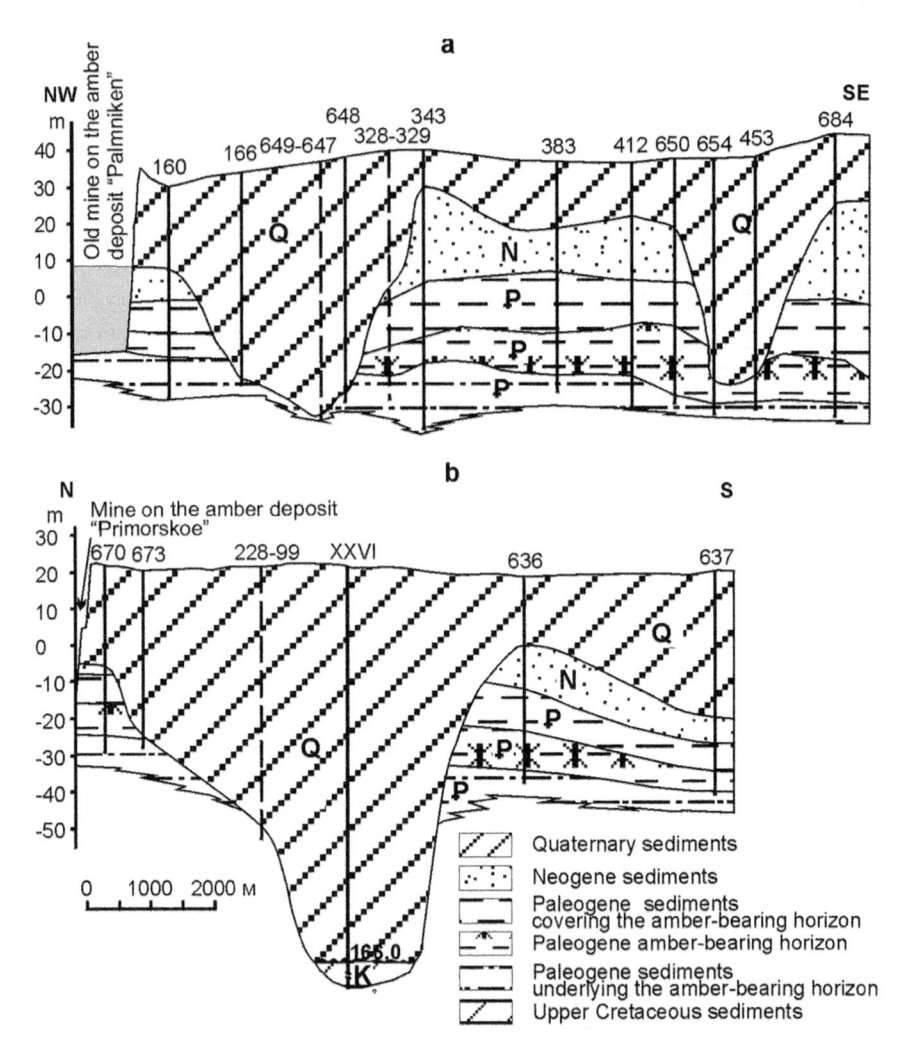

Fig. 8 Schematic section of the western part of the Sambian Peninsula near settlement Yantarnyj: (**a**) NW-SE alignment of section; (**b**) N-S alignment of section [47]

undergone some inversions at separate periods of time, and it serves as the reason for the breaks in sediment accumulation (e.g. absence of Turonian and Santonian sediments, as well as upper Campanian sediments). This fact confirms the possibility of vertical movements in different directions in these tectonic zones [31, 40].

There was another big paleoincision found near Baltiysk at the west coast of the Kaliningrad Peninsula (Fig. 2); it stretches across the Baltic Spit from north-west to southeast. Its width varies from 1 to 1.5 km, its length in the studied area is 5 km and the surface of the incision is on marks of -120–130 m. The floor of the incision consists of Upper Cretaceous rocks (aleurites and alevrolites), while the lips consist of aleurites and Palaeogene sand. The lips are elevated above the floor at 70–80 m.

The paleoincision is filled with a thick layer of Quaternary sediments which form the alteration of glacial and interglacial horizons. Interglacial sediments, which consist of sand and sand with gravel, form two water-bearing horizons that are actively used for centralized water supply of Baltiysk.

The analysis of palaeovalleys location showed that some of them belong to tectonic faults, detected by scientists from Kaliningrad [37] and from Institute for Earth Physics of RAS [49]. Long axes of depressions and elevations of pre-Quaternary relief lay nicely on flexural and rupture zones – Balakinskaya and Melnikovskaya. Pregolya snap zone is responsible for the formation of paleoincisions that stretch from Kaliningrad to Znamensk and for the modern valley of the Pregolya river. It is remarkable that the snap lines in the south part of the region "frame" the minimum marks of pre-Quaternary surface, while in near Gusev, the snap lines "frame" elevations [47].

The fact that palaeovalleys are connected with tectonic deviations is supported by some hydrogeological data. Some paleoincisions are filled with water-saturated sediments and serve as an origin for underground drinking water deposits. Variations in the quality of underground waters (increased mineralization, high concentration of chlorine ion, bromine ion, silica, etc.) can be indicators of water movements from one horizon to another or of the waters rising up through the snaps if the hydrodynamic conditions are favourable. For example, when mineralization and chlorine concentration at Baltiysk water intake increased in the mid-1990s from 0.6 to 2.0–2.5 g/L and from 90 to 350–500 mg/L correspondingly, it could be the result of the rise of mineralized waters [47]. In the process of decoding the images made from space, there was detected a light-rimmed ellipsoid dark mark at the extension of the paleoincision, 1.0 km away from the shore. The object was about 700 m long. The appearance of such a mark can be a result of thermal saline water intrusion along the snap [36]. A similar object was detected in the water area of the Curonian Lagoon near its south-west coast, and again it was not far from a paleoincision on dry land [36].

Increased mineralization of underground waters upper horizons is often the case with paleoincisions directly underneath. Underground waters near Chernyakhovsk show the increase of SiO_2 up to 2.36 MPC, Br^- up to 5.65 MPC and Cl^- up to 1.52 MPC and near settlement Znamensk – SiO_2 up to 3.52 MPC and Br^- up to 6.25 MPC [47, 48]. Salinization of waters in the Upper Cretaceous horizon is observed in the north of the region [32]. It is remarkable that paleoincisions near settlements Znamensk and Mysovka are situated within the limits of East-Curonian tectonic dislocation, which was first identified by Zagorodnykh [40] and which crosses the region from south to north. V. A. Zagorodnykh indicated that in near settlement Gremyach'e (Chernyakhovsk district), there is a Quaternary water-bearing horizon with mineral water located at a relatively shallow depth [36]. Some hydrochemical anomalies are observed here: waters with mineralization of 5.0–15.0 g/dm^3 are detected at the depth of 20–40 m. There were also ascension springs found near settlement Krasnooktyabr'skoe [32]. The appearance of such springs can be explained as follows. Due to the topographic low from the Vištytis (Masuria) elevation towards the Pregolya-Instruch valley, there is drainage of water-bearing horizons, which leads to the increase of pressure difference and can potentially be the reason for the increased water filtration. The most favourable conditions for deep water-bearing horizons drainage can be observed in the valleys of the contemporary relief with deep pre-Quaternary depressions, filled with water-permeable Quaternary sediments. Such depressions serve as discharge zones for deep waters. Mineralized waters (up to 6–8 g/L) are to be expected in the river valleys at the bottom of elevations (Chernyakhovsk and Gusev districts).

There is data concerning hydrothermal low-temperature solutions rising up from deeper horizons to upper layers of rocks. For example, in the north-east of the region the core, which had been brought up from the depth of 93.5 m from the borehole drilled within a deep paleoincision and consisted mostly of loosely cemented fractured sandstone, showed an excessive amount of pyrite (weight content exceeds 12%). Pyrite covers the fractures with a thin coating (0.3–0.5 mm) of microcrystals. It is often that aggregations of such crystals fill out the cavities and start a concretion. Single microcrystals of chalcopyrite can be found as well. The sulphides described earlier are new formations, and it could have been deep-earth gas of low-temperature solutions that move along the snaps and became the building material for the sulphides [36].

Usually, paleoincisions in the Kaliningrad Region can be traced up to several kilometres of their length, though sometimes we can assume that they are much longer. For example, a paleoincision (or a system of paleoincisions), that was discovered in Kaliningrad and that stretches from the mouth of the Pregolya river across the north part of the city and in the north-west direction towards Gurievsk, has a minimum length of 15 km (Figs. 2 and 9) and a branchy structure. At the same time, speaking of such lengthy structures in the context of PQS can be premature, as it is not uncommon when such incisions turn out to be a series of shallow hollows separated by insignificant elevations (10–30 m while the incision is 110–120 m deep). Even small incisions (as the one near Svetlogorsk) have uneven floors with numerous shelves (Fig. 7). As the Kaliningrad Region is situated on the coast of the Baltic Sea, some paleoincisions continue into its water area. It is true not only for the paleoincisions near Svetlogorsk but also for the paleoincisions near settlement Yantarnyj and Baltiysk [45]; that is why their total length in the continental and water areas can reach up to double figures.

There can be different filling materials: fluvioglacial (sand, sand and gravel sediments, tape clays) and moraine sediments (boulder clays and loams) (Table 1). Only several cases show alluvial or lacustrine sediments in the bottom part of paleoincisions. On the territory of the Kaliningrad Region, there is no regular dependence between the extension of a paleoincision and its filling material, though such dependence had been established for the territory of Lithuania [50].

Coming to a decision on the appearance of these hollows is quite problematic. Even if the initial nature of the paleoincision was that of erosion, the material could have been destroyed in the process of exaration and fluvioglacial erosion. Alternatively, they can be regarded as a result of the erosive impact of outwash under the glacier level. This is essentially the theory of how super-deep channel-like (valley-like) depressions had been formed, which was formulated in the end of the nineteenth century by A. Jentzsch and is very popular in Lithuania, the Netherlands

Palaeogene system. Eocene

$P_2^3 pr$	Prussian series. Sand, aleurite with fragments of amber
$P_2^2 al$	Alkskaya series. Sand, aleurite with phosphorites
$P_2^1 sm$	Sambian series. Clay, aleurolite with phosphorites

Paleocene

$P_1^2 lb$	Lyubava series. Sandy and micaceous aleurite
$P_1^1 cz$	Chistoozerskaya series. Clayey marl

Cretaceous system. Upper stage

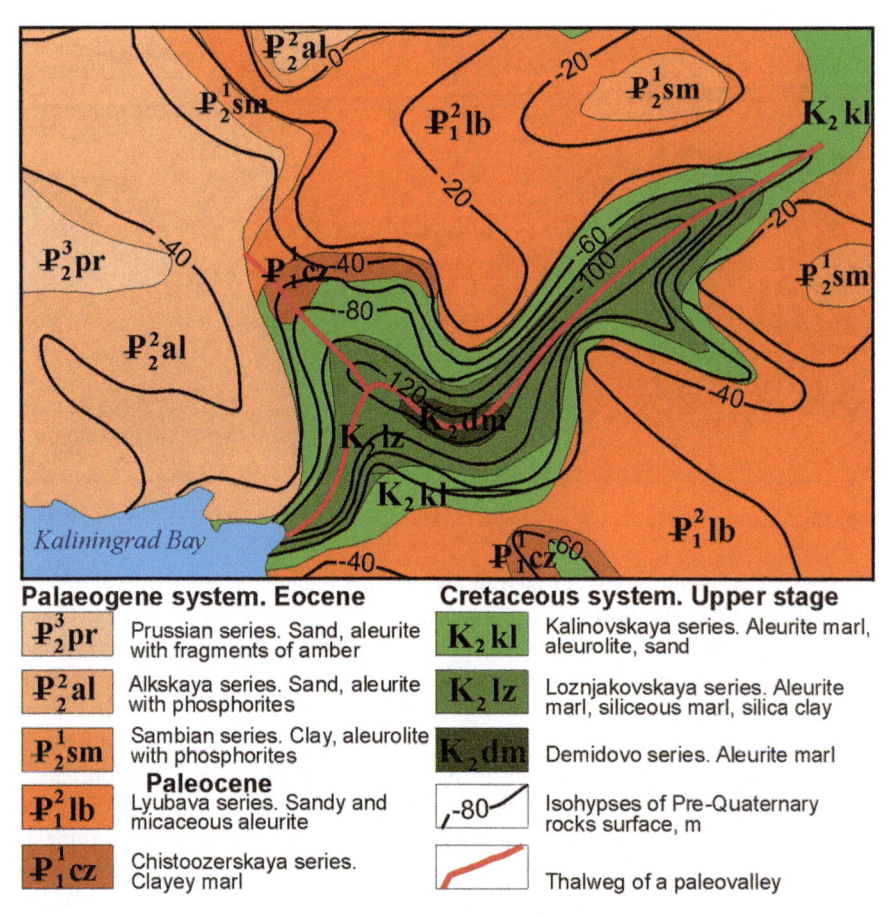

$K_2 kl$	Kalinovskaya series. Aleurite marl, aleurolite, sand
$K_2 lz$	Loznjakovskaya series. Aleurite marl, siliceous marl, silica clay
$K_2 dm$	Demidovo series. Aleurite marl
−80	Isohypses of Pre-Quaternary rocks surface, m
	Thalweg of a paleovalley

Fig. 9 Paleoincision within the boundaries of Kaliningrad

and Germany [39, 44, 51, 52]. According to this theory, paleoincisions are, in fact, underglacier tunnels which served as a means of outwash discharge under high hydrostatic pressure. The theory explains the existence of deep elongated hollows as well as significant differences between the depression floor marks along the thalweg. A considerable number of shallow paleoincisions might have similar origin.

4 Allocation of Quaternary Sediments Thickness on the Territory of the Kaliningrad Region

Quaternary sediments can be found everywhere in the Kaliningrad Region. Their thickness may vary from singular numbers to 277.5 m (settlement Sosnovka) (Fig. 10). There are several exceptions to that: areas on the Sambian Peninsula

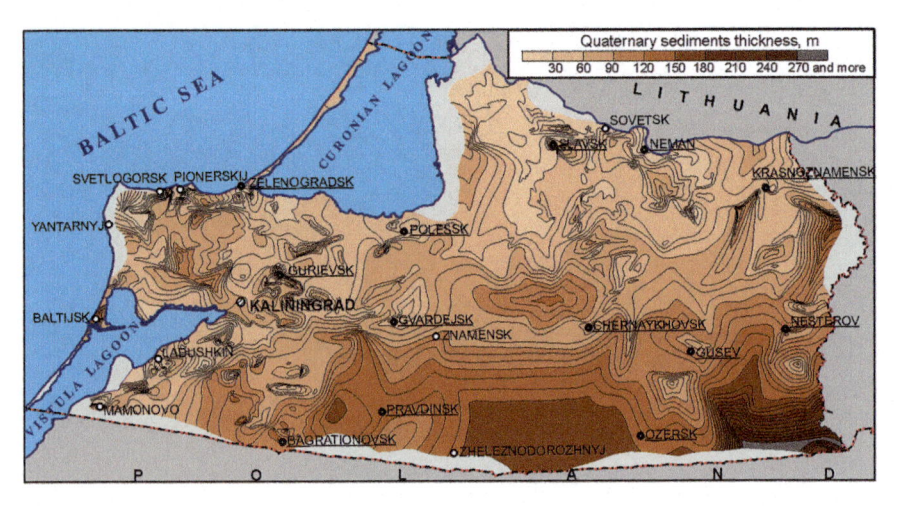

Fig. 10 Allocation of Quaternary sediments thickness on the territory of the Kaliningrad Region

(Neogene sediments on the surface or the surface is covered with thin Holocene sediments) and in the north-east part of the region near the Neman river (Upper Cretaceous sediments underlying Holocene alluvium). When describing Quaternary sediments, we usually say that their thickness increases from north towards south (from 25 to 150 m on average) [33, 34]. It is mostly correct but requires some stipulations. Firstly, the thickness of Quaternary sediments should reach its maximum not "in the south part" but in the areas of predominant accumulation, i.e. in areas of peripheral glacier formations, especially of the final Baltic stage of Valdai glaciation. These are the territories of Warmian elevation (Quaternary sediments thickness about 130 m), Vištytis elevation (up to 277.5 m), Sambian moraine plateau (up to 120 m), Instruch end moraine range (up to 90 m) and Dobrovolskaya range (up to 80 m) (Fig. 10).

Secondly, the map clearly shows the areas that do not belong to peripheral formations but still demonstrate considerable thickness of Quaternary sediments: near Baltiysk (up to 142 m), Svetlogorsk (up to 100 m), mouth of the Pregolya river (100–130 m), Zelenogradsk (up to 82 m), settlement Konstantinovka (91 m), settlement Nekrasovo (96 m), settlement Zapovednoe (86 m), Slavsk (more than 100 m), settlement Zelenodol'e (74 m), Krasnoznamensk (more than 100 m), settlement Pobedino (163 m), settlement Gremyach'e (120 m), settlement Majskoe (more than 100 m), settlement Nivenskoe (90 m), etc. (Fig. 10). In most cases, significant thickness of Quaternary sediments appeared as a result of deep ancient incisions being filled (Fig. 2). These "azonal" and local multimetre layers bring certain chaos to the general pattern of thickness allocation.

Thirdly, the areas of minimal Quaternary sediments thickness can be found not only in the north and north-east of the region but also in other parts. Sometimes, as in Svetly or settlement Pribrezhnyj, it is the result of sediments washing out during the postglacial transgression. But it happens more often that abnormal areas inherit the rise of pre-Quaternary relief. It is clearly visible in the Sambian Peninsula (settlement Prislovo −9 m, settlement Krasnotorovka −23 m, settlement Otradnoe −6.5 m), in the south-west of the region (settlement Yablonevka −10.8 m, near settlement Golubevo −17 m, near settlement Pyatidorozhnoe −24 m, near settlement Medovoe −9.7 m), in the north-east and in the east of the region (Neman and settlement Bolshoe Selo, from 3–6 to 20 m, near settlements Kanash and Pushkino −22–27 m, settlement Zabrodino −7 m, settlement Zarechnoe −16 m, settlement Pokryshkino −60 m), i.e. in the areas of denudation remnant formation.

The aforementioned points, which are related to the glacier activity and to PQS undulation, do not fall into the category of "flat dip" and create a complex figure of Quaternary sediments isopachytes. Correlation of pre-Quaternary relief, Quaternary sediments thickness and contemporary relief is not direct, especially if we take into consideration the existence of areas with inversive relief. These areas usually have matching contemporary maximum absolute marks and minimum absolute marks of PQS. As a result, these areas are bound to have maximum thickness of Quaternary sediments. The largest area of inversive relief is situated near Vištytis elevation. This is the area of the highest contemporary absolute marks of the relief (242 m), maximum thickness of Quaternary sediments (277.5 m) and minimum absolute marks of pre-Quaternary surface (−113 m) [53]. Less extreme, but still rather significant, parameters are typical for the area situated to the east of Pravdinsk and to the south of the Pregolya valley (60 m, 120 m and −120 m, respectively). Warmian elevation is a combination of a direct relief, which matches the pre-Quaternary remnants (settlement Il'ichevka − absolute pre-Quaternary surface mark +38 m, contemporary relief 109 m), and an inversive relief: near settlement Mamonovo (−43 m and 37 m, respectively), settlement Novoselovo (−40 m, 69 m) and settlement Domnovo (−100 m, 100 m). The complex relief of Sambian Peninsula also has areas of inversive character: near settlement Lyublino-Pereslavskoe (−60–100 m, 62–110 m), near the base of the Curonian Spit (−50 m, 20 m). Inversive relief can also be found on the eastern riverbank of the Deima river, near the Instruch range, in Sheshupskaya plain (−61 m, 33 m), etc.

5 Pre-Quaternary Relief Influence on Quaternary Sediments Allocation and the Contemporary Relief of the Kaliningrad Region

It is common knowledge that the speed of a glacier cover movement is a function with many variables. Some of these variables are common for all large areas of the glacier, e.g. glacier cover feed intensity, ice temperature and how it changes with depth. Climatic conditions influence the dynamics of the glacier lips on a regional scale but become less significant on smaller areas due to certain inertness of the glacier cover: short-term climate changes could not affect the dynamic potential as much as another factor did − the nature of subglacial bed. Among the underestimated parameters, we can often find the following: orographical position of separate areas, magnitude of relative elevations, degree of subglacial bed roughness, slope direction and gradient and the alignment of elongated PQS forms against the direction of glacier movement [54].

At the margin of the glacier cover, the nature of PQS relief affects the dynamics more actively and contributes to the formation of certain complexes of glacier relief forms on particular territories. This is why the marginal glacier zone is associated with the formation of glacial dislocations and paleoincisions (depressions). For a long time, it was extremely difficult to estimate the influence of subglacial surface due to the lack of actual material, necessary for creating large-scale and medium-scale maps, but the sufficient data accumulated with time.

The influence of PQS on the allocation of Quaternary sediments thickness and on the formation of the contemporary relief is both significant and complex. Unlike the climatic factor, the nature of subglacial bed is local and is subject to frequent changes even in small distances. Differences in activity of separate areas of the glacier cover on a relatively small territory like the Kaliningrad Region, and allocation of stagnant ice areas or outflows in the marginal part of the glacier in particular can only be explained by specific factors like peculiarities of the subglacial relief. A question might arise – how can layers of ice, which are kilometres thick, be affected by relatively small (up to 100 m) height differences? Especially the definition of inland ice states that a glacier is not influenced by the nature of the underlying surface. Therefore, we must mention the fact that the territory of the Kaliningrad Region, which is situated in the south of the Baltic, remained a marginal zone in the course of multiple glaciations. It is due to multiple glaciations that Quaternary sediments thickness reaches up to 300 m, and it is also the reason for the increased influence of pre-Quaternary surface on the dynamics of a degrading glacier which is burdened with clastic material.

Glacier sheet is to expand along the path of least resistance; therefore we expect the highest speed of glacier movement for the areas where the bed slopes are directed towards the line of such movement. In this case accumulation of clastic material is small. It is the reason why such partially transformed depressions became a path for glacial flows of several glaciations [41, 54]. At present, these are the territories with relatively insignificant thickness of upper Pleistocene sediments. In relief they usually take the form of valleys, e.g. Polesskaya valley.

The areas with a higher orographical position of the subglacial bed slow down the glacier, which gets even slower as the ice sheet thickness diminishes. At the same time, a large quantity of fractures appears. Slow motion and ice fragmentation stimulate the accumulation of the material carried by the glacier. Glacier scales and fractures facilitate the covering of the subglacial bed and the formation of dislocations and are quite common for the Sambian Peninsula. Clastic material accumulation area is irregular: it mostly accumulates in ice fractures and cavities that get filled with sediments brought by meltwater and with material embossed from beneath as a result of great pressure difference at the base of the glacier [55]. Thus, ice fracturing leads to undulating relief which is the most typical elevation element. This could be the way of formation for such elevations as Sambian and Warmian, as they were based on a combination of significant elevations and deep depressions.

The fact that remains unclear is how Neogene sediments survived the exaration. Neither lithological composition nor orographical position promised long life for Neogene formations. The same is true for some areas in Lithuania. For example, Turgelyay ledge of the East Lithuanian elevation is also associated with a sandy-argillaceous ledge of Neogene sediments [34].

Ice fracturing speeds up its stagnation on the elevated parts of a subglacial bed. These areas become ice divides and condition the appearance of dynamically isolated blade-shaped ice sheets which belong to a lower level of pre-Quaternary surface. Later, blade outlines might become sharper as they are sensitive to changes in the marginal parts of a glacier cover [55]. Repeated adjustment of the dynamic plan of the glacier cover marginal part causes formation of new areas of increased fracturing and, consequently, increased indulation of the contemporary relief. They appear on the contact points of stagnant and active ice and in interaction zones of glacial flows of varying degrees of activity. The influence of subglacial bed in such areas is insignificant. It explains the discrepancies between the contemporary elevated relief and the pre-Quaternary lowered surface, i.e. it explains the existence of inversive relief on the territory of the Sambian Peninsula, Warmian elevation, the Instruch range and the north-eastern part of the region [54]. In these territories, there are depressions with inversive relief which can be found in the shadow areas of elevated pre-Quaternary relief aligned against the direction of the glacier movement. Elevations served as ice divides with stagnant ice, while active ice flows went through depressions of the relief. By contacting with each other, they created favourable conditions for accumulation of glacial material and formation of inversive relief.

As for the south and southeast of the region, which is associated with the Vištytis elevation, it is a well-known fact that its position matches the position of the north margin of a large pre-Quaternary depression. This depression spans from sublatitude areas of the Pregolya river, the Angrapa river and the Pissa river to the south of the Masuria elevation. Suwalki elevation, Masuria elevation and Suduv elevation are situated above this depression. Here is an opinion that these elevations could have been a result of a powerful glacial accumulation when the glacier had been held up by Sokulko-Grodno and Olsztyn-Ciechanów main rock plinths [34]. Ledges on the northern margin of the Vištytis elevation – Gusev (-28 m), Pokryshkino (+18 m), Krasnooktyab'rskij (-27.5 m) and Gordovskij (-33 m) – could also have contributed to the formation of thick glacial sediments accumulations in their shadow area and the formation of an indulated relief.

Unfortunately, it is rather difficult to analyse the influence of valley-like forms of the subglacial bed. Some of these forms, which were aligned alone the direction of the glacier movement, could serve as convenient paths for the ice, where it could move faster and further if compared to the body of the glacier. These depressions were filled with ice which gradually melted and prevented the accumulation of thick fluvioglacial sediments and must be therefore clearly visible in the relief. They could serve as hollows for meltwaters flow or lack the traces of erosive activity altogether. Other valleys, which were aligned perpendicular to the direction of the glacier movement, are often buried with a complex of glacial sediments with occasional remnants in them.

Until recently, the role of subglacial water flows was not taken into consideration, and the existence of and the difference between "wet" and "dry" slides was not discussed. Extreme discharge of subglacial water stimulated faster movement of surges – separate blades of the ice sheet (even in the areas of degradation). An outburst or a discharge in talics or in the areas of no congealed water stand, conditioned by the presence of a water-resistant glacier cover and permafrost rocks, resulted in channels, basins or even "wells". Such a scenario could take place only if the rocks were tractable enough, and for that they had to consist of sand, aleurites, loosely cemented sandstones, marls or chalk stones, i.e. soft and highly porous rocks and never of dolomites and limestones. In foreign literature, N-channels are opposed to R-channels, where accumulation of coarse deposits leads to the formation of oses. Paleoincisions and oses are genetically similar formations which appeared due to the differences in nature of subglacial bed rocks.

6 Conclusions

PQS on the territory of the Kaliningrad Region is characterized by complex relief with extreme cases presented by denudation elevations, i.e. "remnants", and paleoincisions. The character of PQS (predominant absolute marks, their amplitudes, prevalence of extreme marks of heights) allows us to speak of several separate areas, such as Sambian Plinth, south-west elevation, Pregolya zone of local depressions, Curonian lowland, north-east elevation, south depression and southeast undulating plain. There are late Cretaceous, late Palaeogene and Pliocene and ancient Quaternary marks that were characterized by a radical change of physical and geographical conditions and processes, which transformed the surface of pre-Quaternary sediments. PQS is a complex polygenetic formation: erosive and denudation pre-Quaternary relief was transformed by glacial and fluvioglacial processes. It is difficult to find any regularities in the allocation of paleoincisions, but the incisions of several kilometres long, 1–2 km wide and up to 260 m deep are the most common. Detailed research reveals a branchy system of incisions in the north and western part of the Sambian Peninsula that continues further into the Baltic Sea floor. Allocation of Quaternary sediments thickness, mainly Pleistocene, corresponds to the principle "the further to the south, the greater the thickness" only on a regional scale. Locally, the value of thickness depends on peculiarities of the pre-Quaternary relief, dynamics of the glacier cover and Holocene denudation processes. Along the elevations and depressions inherited from the pre-Quaternary surface, we identified some inversive forms of the contemporary relief which were formed as a result of repeated adjustment of the dynamic plan of the glacier cover marginal part under the influence of abrupt fluctuation of absolute marks of the subglacial bed.

References

1. Zaddach G (1860) Ueber die Berstein- und Braunkohllager des Samlandes. Erste Abhandlung. In: Schriften der Königlichen physikalisch-ökonomischen Gesellschaft zu Königsberg. Erster Jahrgang. Erste Abteilung. Königsberg, pp 1–44. http://archive.org/details/schriftenderkn13kn. Accessed 20 Feb 2012
2. Zaddach G (1867) Das Tertiärgebirge Samlands. In: Schriften der Königlichen physikalisch-ökonomischen Gesellschaft zu Königsberg. Acher Jahrgang. Erste und zweite Abteilung. Königsberg, pp 84–194. http://archive.org/details/schriftenderkn79kn. Accessed 20 Feb 2012
3. Berendt G (1869) Geologie des Kurischen Haffes und seiner Umgebung zugleich als Erläuterung zu Sektion 2, 3 und 4 der geologischen Karte von Preußen. In: Schriften der Königlichen physikalisch-ökonomischen Gesellschaft zu Königsberg, vol 9. Königsberg, pp 131–238. www.archive.org/details/schriftenderkn79k
4. Jentzsch A (1878) Die geognostische Durchforschung der Provinz Preußen in Jahre 1877 mit eingehender Berücksichtigung des gesamten norddeutschen Flachlandes. In: Schriften der physikalisch-ökonomischen Gesellschaft zu Königsberg. Königsberg, Jahrgang 18, pp 185–257
5. Jentzsch A (1899) Bericht über die Verwaltung des Ostpreußischen Provinzialmuseum in den Jahren 1896, 1897, 1898. In: Schriften der Königlichen physikalisch-ökonomischen Gesellschaft zu Königsberg, vol 40. Königsberg. www.archive.org/details/schriftenderkn40kn
6. Jentzsch A (1900) Der tiefere Untergrund Königsbergs mit Beziehung auf die Wasserversorgung der Stadt. In: Jahrbuch der Königlich Preußischen geologischen Landesanstalt und Bergakademie zu Berlin für das Jahr 1899, vol XX, T. II. Berlin, pp 1–172. www.archive.org/details/geologischesjah00landgoog
7. Jentzsch A (1900) Der vordiluviale Untergrund des Nordostdeutschen Flachlandes. In: Jahrbuch der Königlich Preußischen geologischen Landesanstalt und Bergakademie zu Berlin für das Jahr 1899, vol XX, T. II. Berlin, pp 266–285. www.archive.org/details/geologischesjah0landgoog
8. Wahnschaffe F (1882) Über einige glaciale Druckerscheinungen im norddeutschen Diluvium. Z Dtsch Geol Ges Berlin 34(3):562–601. www.archive.org/details/zeitschriftderd341882deut
9. Krause P-G (1905) Über Endmoränen im westlichen Samland. In: Jahrbuch der Königlich Preußischen Geologischen Landesanstalt und Bergakademie für das Jahr 1904, vol 25, H. 2. Berlin, pp 369–383. www.archive.org/details/geologischesjah17landgoog
10. Geologische SE (1905) Bilder von der samländischen Küste. Königsberg in Pr.: verl. von Wilh. Koch, p 43. http://archive.org/details/geologischebild00schegoog
11. Tornquist A (1910) Geologie von Ostpreussen. Berlin
12. Iz istorii kartografirovaniya territorii Kaliningradskoj oblasti (History of map compilation in the Kaliningrad Region) (1969) Kaliningradskaya oblast'. Ocherki prirody. Kaliningrad, pp 193–205
13. Mayer E (1914/1916) Die Storungen in nordwestlichen Samlande, auf Blatt Gross-Dirschkeim. Jahrb Preuss Geol Landesaustalt Berlin XXXV(Teil II)
14. Mayer E (1916/1919) Über Staubecken und Decktonbildung in der weiteren Umgebung von Königsberg in Preussen. Jahrb Preuss Geol Landesanst Berlin XXXVII(T. II)
15. Wichdorff H (1914/1916) Das masurische Interglatial. Jahrb Preuss Geol Landesanst Berlin XXXV(Teil II)
16. Wichdorff H (1914) Fortsetzung und Verlauf der sämlandischen Endmoränen in Ostpreussen. Z Dtsch Geol Ges LXVI
17. Wichdorff H (1919) Geologie der Kurischen Nehrung. Abhandlungen der Preußischen Geologischen Landesanstalt, Berlin, H. 77, pp 1–181
18. Kraus E (1924) Die Quartertecktonik Ostpreussens. Jahrb Preuss Geol Landesanst Berlin XXXXV
19. Kraus E (1926) Eine geologische Übersichtskarte von Ostpreussens. Geol archiv 4(H. 1)
20. Kraus E (1928) Tertiär und Quartär des Ostbaltikum. In: Die Kriegschauplatze 1914–1918 geologisch dargestellt. Berlin, H. 10

21. Mikhnevich GS (2014) Rol' nemetskikh uchenykh v izuchenii chetvertichnoj geologii Kaliningradskoj oblasti (Contribution of the German scientists to the studies of Quaternary geology of the Kaliningrad Region). Aktual'nye problemy gumanitarnykh i estestvennykh nauk. Sci J 10:459–463. Strategic Research Institute, Moscow
22. Gagel C (1923) Bericht über einige Ergebnisse der Aufnahmearbeiten im nördlichen Ostpreussen. Jahrb Preuss Geol Landesanst Berlin XLIV
23. Gagel C (1925) Die Beschaftenheit und die Herkunft des roten ostpreussisen Decktones. Jahrb Preuss Geol Landesanst Berlin XXXXVI
24. Andree K (1926) Der geologische Bau Ostpreußens. Ostpreußen. In: Land und Leute in Wort und Bild. Aufl. 3. Gräfe und Unzer o. J. Königsberg, pp 14–23. www.polona.pl/item/1791367/12/
25. Beurlen K (1927) Diluvialtektonik und diluvialstratigraphie. Beitrage und Ergebnisse aus Nordostdeutschland. Fortschr Geol und Palaeontol VI(H. 18)
26. Körnke B (1930) Letzglazialer Eisabbau und Flußgeschichte im nördlichen Ostpreussen und seinen Nachbargebieten. Z Dtsch Geol Ges 82
27. Kaunhowen F (ed) (1929) Geologische Übersichtskarte der Umgebung von Königsberg (Hochschul-Exkursionskarte Nr.2). Pr. Geol. Landesanstalt, Berlin
28. Woldstedt P (1931) Über Randlagen der letzten Vereisung in Ostdeutschland und Polen. Jahrb Preuss Geol Landesanst Berlin 52:59–67
29. Woldstedt P (1935) Erläuterungen zur Geologisch-morphologischen Übersichtskarte des norddeutschen Vereisungsgebietes, 1:1 500000. Hrsg. Preußische Geologische Landesanstalt, Berlin, 33 pp
30. Woldstedt P (1935) Geologisch-morphologische Übersichtskarte des norddeutschen Vereisungsgebietes
31. Zagorodnykh VA, Dovbnya AV, Zhamoida VA (2001) Stratigrafiya Kaliningradskogo regiona (Stratigraphy of the Kaliningrad Region) (Kharin GS, ed). Kaliningrad, p 226
32. Gidrogeologiya SSSR. Kaliningradskaya oblast' RSFSR (Hydrogeology of the USSR. The Kaliningrad Region of the RSFSR), T. XLV (1970) Nedra Publishers, Moscow, p 158
33. Vajtekunas PP (1972) Kraevye lednikovye obrazovaniya i zakonomernosti deglyatsiatsii territorii Kaliningradskoj oblasti i prilegayushhikh rajonov (Marginal glacier formations and deglaciation patterns on the territory of the Kaliningrad Region and Adjoining Regions). In: Kraevye obrazovaniya materikovykh otlozhenij. Nauka, Moscow, pp 7–19
34. Gudelis VK (1973) Rel'ef i chetvertichnye otlozheniya Pribaltiki (Relief and Quaternary sediments of the Baltic). Mintis Publ, Vilnius
35. Gaigalas AA (ed) (1976) Pogrebennye paleovrezy poverkhnosti dochetvertichnykh porod Yuzhnoj Pribaltiki (Buried paleoincisions of the Pre-Quaternary rocks surface). Vilnius, p 140
36. Gosudarstvennaya geologicheskaya karta RF (State geological map of the RF) (2011) Scale 1: 1000000 (the third generation). A Central-European series, Sheet N-(34), Kaliningrad. Explanatory note. VSEGEI Kartfabrika, Saint-Petersburg, 226 pp
37. Orlenok VV (ed) (2002) Geograficheskij Atlas Kaliningradskoj oblasti (Geographic atlas of the Kaliningrad Region). KSU Press, Kaliningrad, p 276
38. Chepulite V (1974) Nekotorye voprosy sostavleniya kart rel'efa paleogeomorfologicheskikh poverkhnostej plejstotsena Litvy (Problems of map compilation in the context of palaeomorphological Pleistocene surfaces of Lithuania). In: Voprosy izucheniya chetvertichnykh otlozhenij Litvy, vol 27. LitNIGRI, Vilnius, pp 33–53
39. Bitinas A (1999) Some remarks on distribution and genesis of palaeoincisions in the East Baltic area. Geol Q 43(2):183–188
40. Zagorodnykh VA (1999) Paleovrezy v dochetvertichnuyu poverkhnost' na yugo-zapade Pribaltiki (Paleoincisions in the Pre-Quaternary surface in the south-west of the Baltic). Geoinformatika 4:33–37
41. Tarvidas RI (1967) Dinamika neoplejstotsenovogo lednika na territorii Litvy i kaliningradskoj oblasti (Dynamics of a non-Pleistocene glacier on the territory of Lithuania and the Kaliningrad Region). In: Voprosy geologii i paleogeografii chetvertichnogo perioda Litvy. Vilnius, pp 161–181

42. Yeltsina GN (1999) Geologiya i mineral'nye resursy (Geology and mineral resources). Ocherki prirody. Yantarny Skaz, Kaliningrad, pp 8–36
43. Kharin GS, Eroshenko DV, Kharin SG (2011) Geologicheskie kriterii ustojchivosti i slabye uchastki Kaliningradskogo poberezh'ya (Geological criteria of stability and weak areas of the Kaliningrad coast). Problemy izucheniya i okhrany prirodnogo i kul'turnogo naslediya Natsional'nogo Parka "Kurshskaya kosa". Collection of works, vol 7. IKBFU Press, Kaliningrad, pp 183–200
44. Kluivert SJ, Bosch JHA, Ebbing JHJ, Mesdag CS, Westerhoff RS (2003) Onshore and offshore seismic and lithostratigraphic analysis of a deeply incised Quaternary buried valley-system in the Northern Netherlands. J Appl Geophys 53:249–271
45. Blazhchishin AI (1998) Paleografiya i ehvolyutsiya pozdnechetvertichnogo osadkonakopleniya v Baltijskom more (Paleography and evolution of late Quaternary sediment accumulation in the Baltic Sea). Yantarny Skaz, Kaliningrad, p 200
46. Lithuanian Geological Institute, USSR (1991) Geologiya i geomorfologiya Baltijskogo morya (Geology and geographical morphology of the Baltic Sea). In: Grigyalis AA (ed) Explanatory note to geological maps on scale 1:500000. Nedra Publishers, Leningrad, pp 229–234
47. Mikhnevich GS (2013) Perspektivy vyyavleniya aktivnykh geologicheskikh struktur na territorii Kaliningradskoj oblasti na osnove issledovaniya drevnikh ehrozionnykh vrezov (Prospects of activ geological stuctures identification on the territory of the Kaliningrad Region based on the studies of ancient erosive incisions). In: Nauka i obrazovanie v XXI veke: collection of scientific papers. Tambov, Part 4, pp 100–104
48. Mikhnevich GS (2011) Rol' izucheniya paleodolin v vyyavlenii aktivnykh geologicheskikh struktur na territorii Kaliningradskoj oblasti (The importance of studies of palaeovalleys in identification of active geological structures on the territory of the Kaliningrad Region). In: Aktual'nye voprosy monitoringa geologicheskoj sredy i bezopasnosti urbanizirovannykh territorij: Outline reports of the 1st International conference (Kaliningrad, December 12-15, 2011). IKBFU Press, Kaliningrad, pp 30–32
49. Kompleksnye sejsmologicheskie i sejsmotektonicheskie issledovaniya dlya otsenki sejsmicheskoj opasnosti territorii g. Kaliningrada v 2008 godu (Complex seismological and tectonic studies for the seismic danger evaluation of the territory of Kaliningrad in 2008) (2008) Scientific and technical report. IFZ RAN im. O. YU. SHmidta, p 306
50. Vaitonis VP, Iodkazis VI (1978) Rol' tektoniki v formirovanii mestorozhdenij podzemnykh vod chetvertichnykh otlozhenij yuzhnoj Pribaltiki (Role of tectonics in the formation of underground water deposits of Quaternary sediments of the South Baltic). In: Tektonika i poleznye iskopaemye Belorussii i Pribaltiki: collected works. KSU Press, Kaliningrad, pp 59–63
51. Piotrowski JA (1997) Subglacial hydrology in north-western Germany during the last glaciation: groundwater flow, tunnel valleys and hydrological cycles. Quat Sci Rev 16:169–185
52. Piotrowski JA (1994) Tunnel-valley formation in northwest Germany – geology, mechanisms of formation and subglacial-bed conditions for the Bornhöved tunnel valley. Sediment Geol 89:107–141
53. Yeltsina GN, Gimbitskaya LA, Mikhnevich GS (1999) Chetvertichnaya geologiya Vishtynetskoj vozvyshennosti (Quaternary geology of Vištytis elevation). Ekologicheskie problemy Kaliningradskoj oblasti i Yugo-Vostochnoj Baltiki. KSU Publ, Kaliningrad, pp 30–36
54. Mikhnevich GS (2003) Vliyanie dochetvertichnogo rel'efa Kaliningradskoj oblasti na raspredelenie moshhnostej chetvertichnykh otlozhenij i sovremennyj rel'ef (Influence of Pre-Quaternary relief of the Kaliningrad Region on the allocation of Quaternary sediments thickness and the contemporary relief). In: Uchenye zapiski RGO (Kaliningradskoe otdelenie), vol 2. KSU, RGO, Kaliningrad, pp 3B-1–3B-12
55. Danilans NY (1972) Vliyanie podlednikovoj poverkhnosti na protsessy deglyatsiatsii i lednikovogo morfogeneza. Kraevye obrazovaniya materikovykh otlozhenij (Influence of subglacial surface on the processes of deglaciation and glacial morphogenesis. Marginal formations of mainland sediments). Nauka, Moscow, pp 19–22

Pleistocene Deposits in the Kaliningrad Region

T.B. Kolesnik

Abstract Although Pleistocene has been studied for many decades, there is still no unanimous opinion regarding the history of Pleistocene deposits. There are no geological features dating back to the Pleistocene, which complicates studying and analysing deposits and constitutes a major cause of disagreement. Another contentious issue is the stratigraphy of the Mid- and Late Pleistocene, namely, the identification of different-aged moraines of the Dnieper and Moscow Glaciation, as well as the Valdai glacial formations and their edges. Alongside boulder sandy clays, clays and loamy sands associated with the Pleistocene, the Quaternary deposits also include a wide range of rocks, whose origin is not always clear. Based on field observations, this article describes a number of denudations on the northern coast of the Sambia Peninsula, whose structure and genesis cannot be accounted for by the glacial theory.

Keywords Ice-marine sediments, Pleistocene, Quaternary glaciation

Contents

1 Introduction 82
2 An Overview of Pleistocene Studies 82
3 Stratigraphy and Geology of Pleistocene Deposits in the Kaliningrad Region 84
4 Description of the Geological Structure of Selected Denudations on the Northern Coast of the Sambia Peninsula, Based on Field Data 85
5 Some Problems of the Origin of Pleistocene Deposits 88
6 Conclusions 93
References 95

T.B. Kolesnik (✉)
Institute of Environmental Management, Urban Development, and Spatial Planning, I. Kant Baltic Federal University, Kaliningrad, Russia
e-mail: Kolesnik.Tatiana@mail.ru

1 Introduction

For the first time, the hypothesis about large Quaternary glacier masses was put forward 150 years ago by European naturalists – Louis Agassiz, Ignaz Venetz, William Buckland, Jens Esmark, James Croll, Otto Torell and James Geikie. This hypothesis was based on an attempt to establish the causes of crystalline rock boulder accumulations on the European plain.

The hypothesis that glaciers covered vast territories of the northern hemisphere in the Quaternary is widely accepted in the academic community. Most researchers agree that the whole layer of Pleistocene deposits is a continental formation composed of glacial, aqueous-glacial and interleaving alluvial facies.

Today, the glacial theory is widely used in basic and applied science. For the north-west of the Eastern European Plain, it serves as a basis for palaeogeographical, stratigraphic and geomorphological models and stratigraphic charts.

According to this theory, the territory of the Kaliningrad Region was more than once covered by up to 3–4 km thick layers of continental ice [1]. The last Valdai Glaciation (75 Ka to 10–12 Ka), which destroyed the formations of earlier glacial and interglacial periods observed beyond this layer, left traces in the form of well-preserved terrain dominated by flat and rolling plains and Quaternary deposits. However, there are certain contentious issues relating to the origin and age of Quaternary deposits.

2 An Overview of Pleistocene Studies

Although the geology of Pleistocene emerged as an academic discipline in the first half of the twentieth century, the first data on Pleistocene deposits dates back to the seventeenth/eighteenth centuries. However, these data were rather fragmented and reported by nonspecialists. In the eighteenth/nineteenth centuries, first qualified reports were drawn up by travellers. Naturalists visited different remote areas, for instance, the north of the Russian Plain and the lower reaches of the Rivers Ob and Yenisei.

Pleistocene studies became an actively developing discipline as the Moscow Society of Naturalists was established in 1805, followed by the Geographical Society (1845) and the Geological Committee (1882). However, for a long time, information on Pleistocene deposits was limited. First systematic data were provided by geologists and naturalists in the mid-nineteenth century. K.F. Rouillier and G.E. Shurovsky (1852–1855) believed that the formation of boulder deposits in European Russia was caused by marine rather than continental ice. In 1856, G.E. Shurovsky published a map, where the southern border of continental ice was drawn close to Voronezh, i.e. he supposed that most of the Russian Plain had been covered by ice. The ideas about the glacier's genesis proposed by these scholars became widely accepted. Of special importance is the research carried out by P.A. Kropotkin (1876). Based on the data on European Russia, Siberia and North America, he conducted a comprehensive analysis of boulder deposits and their distribution. His data served as a basis for the theory of boulder deposit origins, which was gaining increased popularity.

However, some researchers supported the idea of the marine origins of boulder sandy clays. Some of them assumed that large-scale transgression had taken place when the Arctic Ocean and the Aral-Caspian Basin conjoined. This idea was supported by such prominent naturalists as A.F. Middendorff and G.G. von Helmersen. The hypothesis about the wide spread of marine boulder deposits was advocated by I.S. Polyakov in the case of the north of Western Siberia and F.N. Chernyshev and A.A. Stuckenberg in the case of European North.

The first polyglacialist ideas were developed in the 1880s–1890s and in the early twentieth century by such scholars as A.P. Pavlov, K.M. Feofilaktov, V.P. Amalitsky, W. Ramsay and K.A. Vollosovich. The hypothesis about several glaciation centres was also put forward at the time.

A qualitative shift in Pleistocene studies took place in the first years of Soviet rule due to the efforts of research and production associations. The 1930s witnessed the development of stratigraphic charts and palaeographic maps, dissemination of polyglacialist ideas and the emergence of new areas in Pleistocene studies.

The history of geological studies of the Kaliningrad Region dates back several decades. Until 1945, the geology of the Kaliningrad Region was studied by German specialists. They paid special attention to upper pre-Quaternary deposits, namely, amber-rich Palaeogene and Quaternary deposits rich in construction materials. However, the most comprehensive studies were carried out in the region after the war in the Soviet period. Exploration and geological surveys were conducted alongside numerous geophysical studies.

In the pre-WWII period, a significant contribution to Quaternary studies was made by L.S. Berg, I.P. Gerasimov, V.I. Gromov, V.A. Obruchev, N.N. Urvantsev, S.A. Yakovlev, and others.

During World War II, the number of studies reduced to reach even higher levels in the post-war years. Laboratory facilities improved. Pleistocene researchers devised a number of new methods, for instance, physiochemical ones, including the method of absolute chronology and palaeontological methods, the most widespread being palynological techniques. Geological studies started to use aeroplane photography.

All of the above had a profound effect on the development of theoretical palaeogeographical problems. Newly proposed theories differed significantly from earlier ones. They concerned the quantity, size, nature, and age of glaciation, features of vegetation of glacial and interglacial periods, etc.

Over the years, different organisations have carried out a significant number of Pleistocene studies in the Kaliningrad Region. A significant contribution has been made by the Lithuanian geologists J. Dalinkevičius, A. Grigelis, P. Suveizdis, O. Kondratienė, J. Kisnėrius, A. Šliaupa, V. Narbutas, V. Joudkazis, A. Klimašauskas, R. Tarvydas, P. Vaitiekunas, and others. Stratigraphic issues were tackled by V. A Zagorodnykh and the staff of the Karpinsky Russian Geological Research Institute – A.V. Dovbnya and V.A. Zhamoida.

3 Stratigraphy and Geology of Pleistocene Deposits in the Kaliningrad Region

In the Kaliningrad Region, Pleistocene deposits account for most of Quaternary deposits. Pleistocene sediments of different genesis are deposited on Cretaceous, Palaeogene and Neogene rocks. The base of Quaternary deposits is found at depths ranging from 47.0 m in the north-west and 20–40 m in the north and 60–100 m in the south to −260 m in lowland areas. Quaternary deposits are best developed in the regions' south. The northern districts are dominated by a thin layer of boulder sandy clay dating back to the late Valdai Glaciation.

A geological overview of Quaternary deposits and underlying rock denuded by erosion on the northern coast of the Sambia Peninsula is given below. The overview is based on data presented in [2]–[5].

Palaeogene deposits are well studied, since they are unprecedentedly rich in amber. They are developed in the south-west of the region, exposed on the Sambia Peninsula, found at a depth of up to −280 m and covered by Neogene and Quaternary deposits on the rest of the territory, including the Baltic Sea area.

Deposits rest erosively on the upper Cretaceous layer covered by Neogene and Quaternary deposits. Their thickness reaches 190 m. Most of Palaeogene deposits are marine shallow-water medium and fine clastic quartz and glauconite sand and aleurites, clay and marlstones with silicate bands and phosphate, siderite and amber concretions.

Neogene rocks are found in the western part of the Sambia Peninsula and in the region's south-west along the coast of the Vistula Lagoon. They consist of continental lacustrine-palustrine deposits composed of quartz and quartz-feldspathic sand, aleurites and light-grey and dark-grey carbonaceous clay. These rocks include lignite bands with fragments of weakly carbonised wood. Lignite is overlain by thin quartz sand followed by an up to 3 m thick grey loamy sand layer. The thickness of Neogene deposits ranges from 20 to 30 m, sometimes reaching up to 91 m.

Quaternary deposits top the platform cover, resting with a stratigraphic hiatus on pre-Quaternary rocks. They are observed throughout the region's territory.

The differentiation of formations is carried using the regional stratigraphic chart of Quaternary deposits approved by the Northwestern Russian Regional Interdepartmental Stratigraphic Commission in 1998 and the legend of the Central European series of the National Geological Map at a scale of 1:1,000,000 [5]. Local regional units were identified by V.A. Zagorodnykh in a comprehensive geological study of the region at a scale of 1:200,000 [3]. The differentiation is based on the genetic and climatostratigraphic principle. According to the approved compound legend of the local stratigraphic chart of Kaliningrad regional Quaternary deposits, neo-Pleistocene deposits are divided into the lower, mid- and upper stages and the Holocene.

A preglacial or interglacial period is identified within each stage. Lacustrine, marine, alluvial and palustrine deposits developed during these periods. Moraine rocks – dark-grey and brown boulder clay, sandy clay and loamy sand with fragments of well-abraded gravel – were deposited by glaciers. The thickness of coastal deposits ranges from 3 to 5 m.

The studied territories demonstrate outcrops of upper neo-Pleistocene deposits observed in natural littoral denudations. This stage is associated with the last Late Quaternary (Valdai) Glaciation, which affected the whole territory of the Kaliningrad Region. V.A. Zagorodnykh [3] classified these deposits as the Baltic superhorizon. On the Baltic chart, formations of the same age were named the Neman superhorizon. On the East European Platform and in the north-west, it is identified as the Valdai superhorizon.

In terms of the petrographic composition of cobbles and boulders, moraine deposits associated with the Brandenburg (Early Baltic) and Pomeranian (Late Baltic) Glaciations are identified within Baltic superhorizon deposits. The superhorizon base is composed of moraine formations consisting of bluish dark-grey sandy clay and loamy sand. The upper part of moraine rocks is brown in colour. Most recent Quaternary deposits (Holocene) consist of alluvial, lacustrine and palustrine deposits usually composed of loamy sand, silty clay, peat, silts and sapropels. The thickness of the Quaternary cover is very variable. The maximum thickness is observed in valley cuts in pre-Quaternary rocks and island and marginal uplands (up to 250–300 m in the region's south-east within the Baltic Ridge). The minimum thickness is associated with lowlands (10–15 m in the valley of the River Neman).

4 Description of the Geological Structure of Selected Denudations on the Northern Coast of the Sambia Peninsula, Based on Field Data

Selected natural denudations that, in our opinion, raise questions as to the origin of constituting deposits are described below. Observation sites are situated on the northern coast of the Sambia Peninsula stretching from Cape Kupalny through the coast of Svetlogorsk and village of Otradnoe to the villages of Lesnoe and Filino (Fig. 1).

The first denudation is found on the left bank of the River Svetlogorka. The river cliff contains outcrops of deposits consisting of unstratified bluish dark-grey sandy clay including cobbles and gravel dating back to the Early Baltic Glaciation, according to the regional stratigraphic classification. The roof of the sandy clay is found at a height of 2.5 m above sea level, followed by a 20 cm layer of boulder

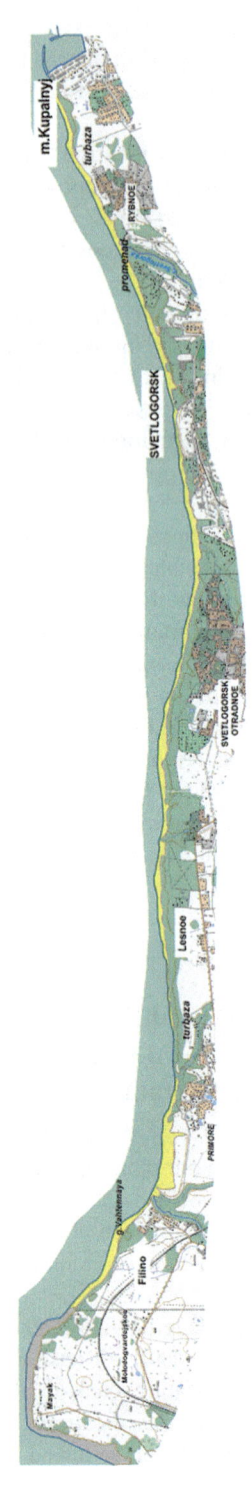

Fig. 1 North coast of Sambia Peninsula (Fragment of the topographic map of the Kaliningrad Region, 2001)

sand, which is covered by boulder sandy clay resting at the base of the cliff and associated with the Late Baltic Glaciation. The denudation contains a boulder of a diameter of approximately 40 cm and iron-rich sand bands and pockets of a thickness of up to 20 cm. The height of the denudation is approximately 4 m above sea level.

The second observation site (Fig. 2) is situated approximately 500 m north of the first one. Approximately 9 m high cliff contains bluish dark-grey boulder sandy clay with 2–3 cm cobbles, which is associated with the Early Baltic Glaciation. The roof of the sandy clay is found at a height of 2.7 m above sea level. It is followed by a horizontal level of red loamy sand containing numerous clay fragments of a diameter of 2–3 cm. The contacts between the loamy sand and upper and lower layers are pronounced and filamentous. The lower contact between the loamy sand and the sandy clay contains a lens of inequiregular iron-rich sand with cobble and gravel inclusions. It is approximately 1 m long and 50 cm thick. The roof of the lens is in contact with brownish quartz silty sand. The sand is well sorted and abraded. The thickness of the layer reaches 70–80 cm. The layer has an indistinct contact with the overlaying greenish-grey fine-grain sand. At a height of 5 m, there is a

Description of the second observation site

Light grey medium-grained quartz sand, medium abrasion - (j)
Dark brown small-grained quartz sand, medium abrasion - (i)
Greenish light brown small-grained quartz sand,
medium abrasion - (h)
Dark brown fine-grained sand with brown fragments - (g)
Pinkish-brown small-grained well-abraded quartz sand with cobbles and gravel - (f)
Greenish-grey small-grained well-abraded quartz sand - (e)
Brown silty quartz sand, well-sorted and abraded - (d)
Brown loamy sand with clay (c)
Inequigranular iron-rich sand with cobbles and gravel - (b)
Dark grey boulder sand clay with cobbles and gravel, Early Baltic Glaciation (a)

Fig. 2 Observation site 2. The *bottom* of the picture shows *bluish dark-grey* boulder sandy clay of the Early Baltic stage with cobble and gravel inclusions (indicated with *red arrows*). It is followed by gravel and cobble deposits and interleaving fine sands of different colours containing glauconite

layer of pinkish-brown fine-grained quartz sand. The layer is 60 cm thick. It is covered by fine-grained dark brown sand with occasional brown inclusions. At a height of 7.5 m, there is greenish light brown fine-grained quartz sand replaced 0.8 m higher by dark brown fine-grained sand. At a height of 9 m, there is a layer of light-grey medium-grained medium-abraded sand. Iron hydroxide films are observed on surfaces of quartz granules.

Deposits are considered as marine facies since quartz granules are well sorted and abraded. Moreover, under the microscope, all sands without exception proved to contain glauconite, which is indicative of sediment accumulation in the marine coast conditions and at shelf depths.

The third denudation of a height of 27 m is situated 2.5 km away from the River Svetlogorka, 150 m west of the end of the Svetlogorsk Promenade. The denuded Neogene rocks consist of continental lacustrine-palustrine facies (Fig. 3).

The denudation is not completely open for observation, since its bottom is covered behind accumulations of sliding sand. The slope has an uneven profile – its top is steep, almost vertical, and, at a height of 13 m, the slope becomes gentler (with a slope angle of 32–35°). As a result of sand constantly sliding down the slope, thick accumulations are found at the bend of the slope, which makes studying the bottom part of the denudation impossible. The accumulation is followed by 2.5 m thick horizontal dark-grey clay, whose roof contains a 1.5 m lens of iron-rich sand. The sand is covered by interleaving laminated fine-grained light-grey quartz sands, low-plastic dark-grey clays with lenses of coarse-grained clay sands with a high content of carbonaceous substances and fine-grained sands and a 60 cm lignite layer with a high content of well-preserved wood. The lignite layers are followed by dark-grey dense sandy clays, whose roof contains laminated light-grey quartz sands of a thickness of up to 3 m. Similarly to the first denudation, the upper layers consist of horizontal brown boulder sandy clays associated with the Late Valdai Glaciation. It is impossible to study the boulder sandy clay materials because of the danger of subsidence. However, in the sections where subsidence did take place, the sandy clay contains cobbles and small boulders of a size of up to 20 cm, which were intercepted in the gentler parts of the slope, which are composed of accumulated loose materials.

A similar structure of denudations is observed across the coast from the village of Otradnoe to the Filino Cove (the village of Primorye). Here, horizontal Neogene rocks are overlain by not only brown or light brown boulder sandy clays and loamy sands of the Early Baltic Glaciation (Fig. 4) but also bluish dark-grey boulder sandy clays of the Early Baltic stage (Fig. 5).

5 Some Problems of the Origin of Pleistocene Deposits

The fact that dark-grey and brown sandy clays with cobble, gravel and boulder inclusions rest on light-grey laminated Neogene sands of lignite rocks (Figs. 3, 4, and 5) raises certain questions. The horizontal bedding of sandy clays is indicative

Fig. 3 A denudation west of the Svetlogorsk promenade. *Brown* sandy clay rests on horizontal Neogene rocks

of undisturbed sediment accumulation. Literature describes these sandy clays as moraines [3, 6] and associates them with the Late Baltic Glaciation. According to the glacial theory, huge ice masses of a thickness of 3–4.5 km pushed the platform cover at a depth of hundreds of metres reaching the crystalline basement and ploughed deep hollows in the bedding rock. They ripped off and moved crystalline erratics and boulders hundreds and thousands of kilometres. Under their weight, huge stones were rounded and polished. Thus, it is highly unlikely that they would neatly rest on fine-grained loose Neogene sands. The inconsistency is evident.

The National Geological Map of the Russian Federation [7] shows limnoglacial deposits of the Baltics glacial lake in this area. However, it is not clear how a lake could be located in a watershed area (absolute denudation marks at 25–27 m and higher), when adjacent denudations (see observation sites 1 and 2) are at a much

Fig. 4 *Brown* boulder sandy clays of the Late Baltic Glaciation (**a**), clays resting on Neogene lignite (**b**)

lower hypsometric level (3–10 m). Another peculiar fact is the agglomeration of large boulders of a diameter of 2–3 m on the capes of the Sambia Peninsula, in particular, on the nameless cape 700 m east of the village of Rybnoe (Fig. 6).

Adherents of the glacial theory would say that these agglomerations are a product of the abrasive destruction of boulder sandy clays forming coastal cliffs and containing boulders. However, the number of boulder agglomerations on a small section of the beach is so great that the moraine sandy clays have to be stuffed

Fig. 5 Horizontal Early Baltic moraines (**a**), Late Baltic formations (**b**), formations resting on lignite (**c**). The picture shows that the "moraines" contain boulder cobble material (indicated by *red arrows*)

with boulders. However, this is not the case. There are no boulders in the coastal cliff of the cape. Moreover, boulders are rarely found in coastal cliffs anywhere along the coast of the Sambia Peninsula. Obviously, the cause of boulder agglomerations in certain areas of the Sambia Peninsula has to be associated with a different geological event.

Another fact that cannot be explained by the glacial theory is the so-called ice mass "injection" penetrating the enclosing rocks of the coastal cliffs. In the area

Fig. 6 Boulder agglomeration on the nameless cape 700 m north of the village of Rybnoe (http://baltic-dolphin.livejournal.com/116732.html?thread=2067196)

stretching from the villages of Otradnoe and Filino through Cape Taran to the western coast of the peninsula, one can observe bluish dark-grey sandy clays with cobbles and boulders between the layers of horizontally bedded loose Neogene rocks (Fig. 7).

Such phenomena are rather common south of the village of Donskoe on the western coast of the Sambia Peninsula. This gives rise to the question as to how a huge ice mass, thick enough to produce metres of high moraine deposits, could invade horizontally bedded loose Neogene rocks without disturbing them. Another seemingly obvious fact is glaciodislocation. One can easily notice its instances near the cleft (Figs. 8 and 9).

Clefts, as well as rivers, are associated with tectonically weak rocks and tectonic faults. Probably, the dislocated rock sections can be a result of tectonic movements rather than glaciodislocations. In this case, the folding of rocks often observed along the coast of the Sambia Peninsula can be easily explained by common fault and pull-apart tectonics, since it is impossible to explain the selective and limited nature of the push moraine from the perspective of its effect on plastic rocks.

Fig. 7 One of the instances of glacier "injection": boulder sandy clays of the Early Baltic stage (**a**), lignite Neogene rocks (**b**), boulder sandy clays identical to boulder sandy clays of the Early Baltic Glaciation (**c**), *brown* boulder sandy clays of the Late Baltic stage (**d**)

6 Conclusions

A widespread opinion is that the whole layer of Pleistocene sediments is a continental formation consisting of glacial, aqueous-glacial and interleaving alluvial facies. However, some researchers believe that most Pleistocene deposits accumulated in

Fig. 8 Instances of "glaciodislocation." We believe them to be manifestations of Neotectonics

Fig. 9 Neotectonic dislocations

the sea under the impact of icebergs and fast ice [8–11]. The geological facts discussed in this article cannot be accounted for by the glacial theory:

1. It is highly probable that the cobble/loamy sand/sand sediments and underlying bluish dark-grey "moraine" sandy clays associated with the Early Baltic Glaciation [3, 4] (observation point 2) are a single marine layer composed of facies varieties of Upper Quaternary coastal marine deposits. This can be proven by well-sorted and abraded quartz granules and the presence of glauconite in sands.
2. The presence of boulder sandy clays horizontally bedded on fine Neogene sands is indicative of undisturbed sediment accumulation, which can be explained by an influx of coarse material into the sediment at small depths under the influence of floating iceberg of fast ice particles stranded in shallow waters.
3. We believe that the boulder accumulation in the area of the nameless cape and other capes is a facies of glaciomarine deposits, which could form as a result of an influx of coarse gravel and cobble material and boulders from icebergs and fast ice.
4. As to the ice "injection" into the enclosing rocks, the observed connection between the sands and embedded boulder sandy clays is probably indicative of aqueous sediment facies rather than the moraine origin of boulder layers.
5. We believe that the formation of so-called glaciodislocations is accounted for by different tectonic movements. Probably, these structures are a result of Late Cainozoic seismic activity, since they have a pronounced relief and, thus, their formation was caused by the most recent tectonic cycle. The deformation of loose sediments interpreted as glaciodislocations are most probably a result of recent tectonic dislocations.

These facts raise doubts over the existence of a Pleistocene ice sheet on the territory of the Kaliningrad Region. Our field observations correspond to the conclusions of the "Nature of the Shelf and Archipelagos of the European Arctic" 8th National Conference, which was held in Murmansk in 2008. A new glacial theory was developed based on the findings of an international research expedition organised by the Murmansk Marine Biology Institute in collaboration with researchers from Rostov and the USA. The theory suggests that the Pleistocene Glacier was so thin and inert that it did not account for significant destruction. Therefore, the tectonic factor had a more profound effect on the relief than the glacier did.

References

1. Koronovskiy NV (2006) Obshchaya geologiya. Moscow State University Press, Moscow
2. Zagorodnykh VA, Kunaeva TA (2005) Geologiya i poleznye iskopaemye Kaliningradskogo regiona [Geology and minerals of the Kaliningrad region]. Yantarny skaz, Kaliningrad
3. Zagorodnykh VA, Dovbnya AV, Zhamoyda VA (2002) Informatsionny otchet [An informational report]. Karpinsky Russian Geological Research Institute Press, Saint Petersburg

4. Zagorodnykh VA, Dovbnya AV, Zhamoyda VA, Kharin G (ed) (2001) Stratigrafiya Kaliningradskogo regiona [Stratigraphy of the Kaliningrad region]. Kaliningrad
5. Lukyanova NV, Bogdanov Yu B, Vasilyeva OV, Vargin GP et al (2011) National geological map of the Russian Federation at a scale of 1:1000000 (3rd generation). *Central Europe. Sheet N-34 (Kaliningrad)*. Karpinsky Russian Geological Research Institute Press, Saint Petersburg
6. Orlenok VV (ed) (2007) Geografiya Kaliningradskogo regiona. Polevaya obshchegeograficheskaya uchebnaya praktika: uchebnoe posobie [Geography of the Kaliningrad region. Geographical fieldwork: a textbook]. Kaliningrad, IKSUR Press
7. Maksimov AV, Semenova LR, Zhamoyda VA (2011) Karta chetvertichnykh obrazovaniy [A map of quaternary sediments]. *National geological map of the Russian Federation at a scale of 1:1000000 (3rd generation). Central Europe. Sheet N-34 (Kaliningrad)*. Karpinsky Russian Geological Research Institute Press, Saint Petersburg
8. Danilov ID (1962) Pleystotsenovye otlozheniya vostoka Bolshezemelskoy tundry i usloviya ikh obrazovaniya [Pleistocene sediments in the east of Bolshaya Zemlya tundra and the conditions of its formation]. Izvestiya Acad Sci Geogr 6:74–80
9. Kulik NA (1926) O severnom postpliotsene [On northern post-Pliocene]. Geologicheskiy Vestnik 5(1–3):1–12
10. Popov AI (1963) Pleystotsenovye otlozheniya v nizhnem techenii r. Pechory. [Pleistocene sediments in the lower reaches of the River Pechora]. *Kaynozoyskiy pokrov Bolshezemelskoy tundry* [*Cenozoic cover of the Bolshaya Zemlya tundra*]. Moscow University Press, pp 24–49
11. Chuvardinskiy VG (1985) Geologo-geomorfologicheskaya deyatelnost pripaynykh ldov (po issledovaniyam v Belom more) [Geological and geomorphological activities in fast ice: the case of the White Sea]. Geomorfologiya 3:70–77

Modern Landscapes in the Kaliningrad Region

Elena A. Romanova, Olga L. Vinogradova, and Irina V. Frizina

Abstract This article summarises the findings of studies into the Kaliningrad Region's landscapes and their components tracing the development of natural features of modern landscapes – the paleogeography of postglacial and Quaternary deposits. Based on field and laboratory studies, the authors conduct a component analysis of territorial complexes. Genetic types and varieties of regional landscapes are identified, their spatial and temporal structure described. The landscape structure of the Kaliningrad Region is characterised by the following genetic types of landscapes – landscapes of glacial and fluvial origin and sea and marine landscapes. In their turn, they are divided into the following types of natural landscapes: ground moraine plains, terminal moraine uplands, glaciolacustrine plains, coastal landscapes, ancient delta landscapes, valley landscapes, and aeolian ancient alluvial plains. The authors consider key trends in natural landscape transformations resulting from their economic use. The article proves that modern landscapes are polygenetic systems consisting of the areal and network components of the environment. A new concept of modern landscape is proposed. The authors describe and substantiate the idea of landscape metachronicity and secondary succession in the region.

Keywords Land use, Landscape, Landscape genesis, Secondary succession, Settlement system

Contents

1 Introduction 98
2 Theoretical Background 98
3 Methodology 101

E.A. Romanova (✉), O.L. Vinogradova, and I.V. Frizina
Institute of Environmental Management, Territorial Development and Urban Planning,
Immanuel Kant Baltic Federal University, Kaliningrad, Russia
e-mail: alberta63@mail.ru

4	Research Results and Discussion	102
	4.1 Characteristics of Landscapes' Major Natural Components: Relief, Climate, Soils, and Vegetation	102
	4.2 Metachronicity of Regional Landscapes	105
	4.3 Genetic Typology of the Landscape's Natural Basis: The Territory's Landscape Structure	106
	4.4 Characteristics of Modern Landscape's Anthropogenic Components: The Impact of Agricultural Cultivation of Regional Territories	109
	4.5 Landscape-Dependent Settlement in the Kaliningrad Region	110
	4.6 Modern Landscapes as Polygenetic Systems of Areal and Network Environmental Components	113
5	Conclusions	115
	References	116

1 Introduction

Any modern landscapes of a single climate zone develop under the influence of two factors – natural evolution and the type of land use, which is closely connected with the settlement system. The natural component of a modern landscape is largely determined by its relief and Quaternary deposits, as well as current geological processes – endogenous (tectonically active areas) and exogenous ones – which can be of both natural and anthropogenic origin. The region's landscape structure has characteristics common for areas affected by the last Valdai glaciation and shows features pertaining to its coastal position. The transformation of the territory's landscape environment started as early as the thirteenth century. Relatively intact natural landscapes can be found on the coasts of the Baltic Sea and the lagoons and in natural reserves and large forests. The appearance of the other modern landscapes has been strongly affected by human use.

2 Theoretical Background

Modern Russian geography distinguishes between natural, anthropogenic-natural, and anthropogenic landscapes. Different schools of geographical thought attach different meanings to these notions.

Without polemicising with adherents of different schools of thought, let us consider the possibilities of employing different scientific approaches to studying modern landscapes on a certain territory through uniting them into common paradigms.

Natural Paradigm The natural approach is a 'classic' of landscape studies. In the USSR, the physiographical zoning was based on identifying natural landscapes, which holds true for the current Russian practices. V. L. Kagansky believes that classical 'natural' landscape studies focus on the 'anatomy' of landscape paying little attention to the functional aspects. Since actual natural complexes do not exist in most regions of the world, 'natural' landscape studies deals with the 'reconstruction' of non-existent landscapes [1]. The term 'anthropogenic' denotes landscapes 'transformed by humans'. Human impact on landscapes is interpreted as a sum of changes in natural conditions. Many studies give priority to identifying functions pertaining to the economy and land use to describe the types of anthropogenic transformation of landscape. The term 'cultural landscape' denotes 'good' anthropogenic landscapes, where natural components are in harmony with artificial ones meeting the needs of human society [2, 3]. An impotent landmark in the development of this approach was the influential work of E. P. Romanova *Modern landscapes of Europe* [4]. Its advantages are attention to detail and a logical approach to classifying and mapping landscapes. By the transformation degree and age, landscapes are divided into relatively indigenous (primary), natural, natural-anthropogenic (secondary and human-modified), and anthropogenic.

Cultural Paradigm Another major research area is 'cultural' landscape studies. It focuses on isolated cultural components superimposed on the continuous natural basis. Advocates of this approach use the term 'cultural landscape'. However, they understand it as individual historically or culturally significant elements [5–8]. Cultural elements are often attributed a special meaning; sometimes, they are consecrated.

In this framework, cultural landscape develops as natural and cultural heritage. In 2004, UNESCO adopted the Code of the Cultural and Landscape Heritage including the following classification: landscapes designed and created intentionally by human beings (e.g. parks and memorials), organically evolved landscapes (created through a society's spontaneous activities), and associative landscapes (associated with certain events or personalities) [9–11]. This aspect is of major practical significance, since it pertains to preserving valuable natural and cultural objects. However, it is not suitable for the comprehensive zoning of modern landscapes. A special case is the works of R. F. Turovsky dedicated to not only the zoning of cultural landscapes (including by linguistic, confessional, and natural characteristics) but also regional identity.

An interesting instance of this approach is ethnocultural landscape studies yielding valuable results for research on landscape characteristic of settlement areas of traditional or relict ethnic groups. An important element is studying not only the material but also mental and semantic layers – the reflection of an area's features in songs, literary works, and oral lore [12, 13].

Theoretical Paradigm The third area of research on modern landscapes is the theoretical geographical approach represented by Kagansky and Rodoman [14, 15]. This approach treats natural and cultural components as equal and interconnected. They are considered similarly – as areas, networks, and districts and different 'layers' of the same territory. Theoretical geography (primarily, the works of B. B. Rodoman) provides plenty of material. However, it would be premature to call it a completely suitable approach to studying landscapes of concrete territories. In particular, B. B. Rodoman considers cultural landscapes a non-existent ideal object. In his

2007 work *Cultural landscapes in post-Soviet Russia*, V. L. Kagansky stresses that since studying cultural landscapes empirically is complicated and not always necessary, they have to be considered within theoretical geography.

The key difference of international studies from Russian researches is the absence of division into 'natural' and 'cultural landscapes'. The landscape is an object of cultural geography. The founder of American landscape studies C. O. Sauer wrote, 'the cultural landscape is an amalgam of physical and cultural forms: Culture is the agent, the natural area is the medium, the cultural landscape the result' [16]. However, another influential US geographer R. Hartshorne did not consider the natural landscape as an independent object (and an element of the cultural landscape) viewing it as a background of human activities [17]. Today's international geography offers different approaches to studying modern landscapes that are very similar to those developed in Russia.

One of the advocates of the natural paradigm is the British researcher J. Lowton [18]. He stresses, cultural landscapes 'do not include some of the most heavily modified habitats' (cities and 'extremely intense modern agricultural landscapes'). Cultural landscapes are those used by people but 'still rich in biodiversity, beautiful to look at' ('landscapes of fairy tales, without dragons') [18]. This paradigm is also used by the British scholar O. Rackham [19]. Unlike their Russian counterparts, British researchers view modern landscapes as a result of interaction between different components of the environment without distinguishing between natural and cultural ones [20].

The 'cultural' paradigm is used in numerous works focusing on the conservation of natural and cultural heritage in different regions of the world. In the USA, the cultural geography paradigm dates back to 1962. It is associated with the works of J. Jackson and the *Landscape* journal [21]. As to modern scholars, it is worth mentioning the Polish researcher J. Bogdanovski, the German scholar M. Dietrich, Australians L. Leader-Elliot and R. Maltby, their US colleague R. Longstreth, the Canadians G. Swinnerton and S. Buggey [22–26]. A special – standard-setting – aspect of cultural landscapes was popularised by the famous US scholar, author of *Land Ethics*, Aldo Leopold [27]. Naturally, the cultural paradigm in landscape studies is actively supported by the UNESCO World Heritage Committee [28].

The name of H. Küster [29–31] is associated with a special paradigm that does not have exact counterparts in Russian geography. It can be called the 'comprehensive' paradigm. Küster suggests calling all natural and cultural landscape simply landscapes, since changes in landscapes by human occupation takes place almost everywhere. Küster's works address a broad range of problems – from studying landscape-forming process of different origins to research on the age of cultural landscapes and their secondary succession. Of special interest are his arguments about the advantages of traditional over intensive land use (observed in the long-cultivated districts of Europe) and the triad of statements: 'if managed in a traditional way, agricultural areas could look very different... They look very similar, if intensive agriculture is applied. ...if the landscapes all look similar, regional identity will lose its basic foundation and ultimately vanish'.

The authors of this study define the modern landscape as a certain image of a territory developed under the influence of various natural and anthropogenic factors.

3 Methodology

The natural complexes of the Kaliningrad Region were first studied at the level of individual components. In the post-war period, the Presidium of the Academy of Sciences of the USSR organised a major expedition to the region to study local geology, soils, and vegetation [32]. I. G. Vazhenin and V. I. Belyakova studied the mineralogical composition of regional Quaternary deposits and analysed the agrochemical properties of soils [33]. On the territory of today's Kaliningrad Region, soils have been studied since the mid-nineteenth century. However, such studies were rather sporadic. The soil study unit headed by Prof A. A. Zavalishin focused on the current condition of soils, local features of source formation, and patterns of soil distribution [34]. The studies resulted in the preparation of a regional soil map. A special study addressed the polders of the Kaliningrad Region [35]. The expedition also focused on the regional vegetation [36, 37].

Despite the in-depth study into natural components, a detailed comprehensive research on landscapes was not carried out. Researchers focused on the landscape zoning of territories, and exhaustive landscape mapping was not performed. Detailed research on the seacoast of the Kaliningrad Region was carried out over several decades [38]. A landscape map of the region at 1:500,000 prepared by A. A. Sukhov and I. I. Kozlovich and based on the hypsometric approach to landscape identification was published in the *Geographical Atlas of the Kaliningrad Region* [39].

In 2006, based on the European practices of landscape planning, a research group supervised by V. P. Dedkov and G. M. Fedorov prepared a landscape programme and a framework landscape plan for the territory of the Kaliningrad Region [40].

In 2003–2011, landscape-planning initiatives were launched in the Kaliningrad Region, focusing on both the region in general and its individual municipalities [41–43]. These efforts emphasised the urgent need for a landscape map at a scale of at least 1:200,000, which has not been prepared so far. Efforts aimed to create such a map began in 2003. In 2003–2005, we carried out a landscape survey of the southern part of the river Neman's ancient delta (Slavsk district) at 1:50,000 and the towns of Svetlogorsk and Zelenogradsk at 1:10,000. A landscape map of the Sambia Peninsula at 1:50,000 was produced in 2006–2008 [44, 45]. In 2009–2010, the landscape survey was extended to the whole Kaliningrad Region. It was performed at a scale of 1:200,000. An original map of regional landscapes and physiographic zoning was published in 2011 [46, 47]. A landscape survey was conducted at the level of genetic types classified by the specific features of the sedimentary base, soils, vegetation, and other landscape components.

4 Research Results and Discussion

The region's landscape structure is a result of interaction between both natural and artificial components. Jointly, they shape the vertical and horizontal landscape configuration and determine the complex structure of landscape cover within a small territory.

4.1 Characteristics of Landscapes' Major Natural Components: Relief, Climate, Soils, and Vegetation

4.1.1 The Region's Geomorphological Features and Quaternary Deposits

The key differentiation factor at the level of the landscape type is the sedimentary base – the relief type and deposit genesis. Rolling plains of the ground moraine account for a significant part of the region's area. The Warmian and Vištytis Uplands, as well as the Sambia moraine node, are terminal moraine formations. Glaciolacustrine lowlands account for a significant part of the territory. The region's north-west is a flat lowland of the river Neman's ancient delta. The northern coast of the Vistula Lagoon and the area between the Neman and the Šešupe are home to ancient alluvial plains, which were exposed to aeolian processes in the postglacial period. There are several ancient valleys shaped by the flows of glacial melt waters. The Curonian and Vistula Spits are coastal plains with dunes and the *Palve* relief on aeolian sands. The seacoasts of the Sambia Peninsula are a combination of abrasive, accumulative, and transitional abrasive-accumulative forms. All relief forms have been changed by human occupation, since the region's territory was populated a very long time ago. The relief transformations date back to the Bronze Age, when the peninsula's many tumuli were created.

A comparison of our field observations and an analysis of archive and literary sources show that there are a number of contentious issues in identifying the genesis of certain landscape areas of the Kaliningrad Region. Some authors identify the Sambia moraine plateau on the Sambia Peninsula [48–51]. Others unite the Sambia moraine plateau with the Instruch ridge – a narrow strip of terminal moraine formations on the right bank of the Pregolya. Schlicht [52] identifies a complex of terminal moraine, moraine, and acqueo-glacial formations. We suggest identifying the Sambia terminal moraine node combining horseshoe terminal moraine ridges, areas of irregular fluvioglacial plains, ground moraine plains, and individual kames.

Field studies made it possible to create a geomorphological map of the Kaliningrad Region at 1:540,000 [53]. The region's territory demonstrates the following relief types: irregular and rolling moraine plains, terminal moraine ridges and uplands, fluvioglacial plains with kames, flat limnoglacial plains, relict hummocky alluvial plains, relict delta lowlands, modern river valleys, ridge and aeolian dune plains and fan sand massifs, and sea and lagoon coasts (abrasive, accumulative, and transitional forms).

Different geneses and physical and chemical properties of the Quaternary deposits account for a rich variety of soils and a motely landscape structure of the territory. A wide variety of edaphic types of habitats shape favourable conditions for natural vegetation communities and agriculture in the region. Oligotrophic mesoxerophilic conditions are created by marine sand deposits and Neogene sands. Meso-oligotrophic and mesotrophic habitats with humidification types ranging from mesophilic to hygrophilic correspond to lagoon deposits of varying granulometric structure, ancient-alluvial sands and loamy sands, light boulder and boulderless deposits of the ground and terminal moraine, and ancient delta and acqueo-glacial sands and loamy sands. Richer deposits of the same humidification type are represented by boulderless or boulder – often carbonate – loams of the ground moraine and the glaciolacustrine heavy loams, silt, and laminated clays. Hygrophilic habitats are characterised by peat deposits of upland (oligotrophic), transitional (mesotrophic), and lowland (eutrophic) mires.

Alongside the Quaternary deposits and local relief, the uniqueness of regional landscape is accounted for by climatic features.

4.1.2 Climate

According to B. P. Alisov's classification, the climate of the Kaliningrad Region is part of the Atlantic continental area of the middle latitude zone, the South Baltic subarea close to Western Europe in terms of circulation processes [54].

The key factors behind the climatic features of the Kaliningrad Region are as follows:

- The influence of the Atlantic Ocean and Eurasia
- Its location in the central zone of middle latitudes in the coastal area of the Baltic Sea
- Flat relief
- The territory's openness to the eastern atmospheric transport
- Winter location of the polar front over the territory of the Kaliningrad Region

All these factors, alongside the region's small area, account for the slight differences in the climate conditions in different parts of the region manifested in the following phenomena:

- An increase in the precipitation amount during west winds in the region's southwest and west.
- The frequency of winds of 2–20 m/s decreases eastwards – from 314 days in Baltiysk to 251 days in Chernyakhovsk.
- High frequency of gale-force and hurricane west winds.

- Total average long-term precipitation ranges from 650 mm in the south (village of Zheleznodorozhny) to 798 mm in the centre (Kaliningrad) and 777 mm and 797 mm in the west (Svetlogorsk and Zelenogradsk, respectively).

In the region, liquid precipitation accounts for 76% of all precipitation, freezing for 15%, and frozen to 9%.

Air temperature is affected by both the region's latitude and the influence of the Atlantic Ocean. The uniqueness of the temperature regime is most evident when considering long-term average highs ranging from $-0.5°C$ in the West to $-4.6°C$ in the East. On the coast (Baltiysk, Svetlogorsk), the average December temperature is $0°C$. The average temperature of the warmest summer month is $+17.1$ to $+17.7°C$. The climatic features manifested in a high degree of humidification, warm winters, and a long vegetation period shape the characteristics of regional vegetation and soils.

4.1.3 Vegetation

The territory of the Kaliningrad Region belongs to the landscape subzone of coniferous-deciduous forests with oaks and lime trees. Within the floristic zoning system, the region belongs to the Baltic-Belarusian subprovince of the North-European taiga province of the forest zone, as well as to the Kaliningrad geobotanical district characterised by the presence of beeches, hornbeams, ashes, maples, and elms in the forest stand.

Before the intense economic occupation, the territory was fully covered in forests. After World War II, forests accounted for 12–13% of the region's territory. Today, they occupy 24% of the territory. One third of the forest stand is planted forests.

The territory of the region is divided into four forest subdistricts:

1. The black alder forests of the Neman lowland
2. The planted pine forest and spruce-deciduous forests of the region's east
3. Planted spruce birch forests and pine and black alder forests covering most of the territory
4. Planted – including mountain – pine and black alder forests of the Curonian and Vistula Spits and the Pregolya's ancient delta.

Today's vegetation cover of the Kaliningrad Region is represented by dry and flood-meadows. Their species composition is affected by the position on the floodplain, humidification, and soils.

The region's vegetation has been radically transformed by human occupation: almost all forests and meadows are either planted or improved (including amelioration and the planting of grasses and valuable tree species). Today, forests account for 20.4% of the region's territory. Only 20% of meadows comprise flood-meadows; the others are originally artificial dry meadows. The vegetation cover of some planted forests – for instance, spruce and pine ones – contains species characteristic of deciduous forests (ferns, anemones, etc.). Part of planted climax forests is hardly distinguishable from natural forest communities, since it has already developed the typical structure and composition of undergrowth and grass canopy of a local forest.

4.1.4 Soils

The Kaliningrad Region belongs to the Eastern European soil area; it is situated at the border of two soil provinces – the Baltic characterised by soddy-podzolic soils and the Western European province characterised by brown earths. The border between these provinces is blurred and rather broad; therefore, both types of soils are found in both western and eastern districts of the region. The spread of these soils is affected by soil forming species and the humidification of habitats.

Brown earths developed on the light ground and terminal moraine deposits under normal humidification conditions on positive landforms under deciduous forests. The boulder and boulderless light and medium loams of the ground moraine in the conditions of normal humidification developed sod-low and meso-podzolic soils. Muck- and sod-high-podzolic soils are associated with the areas of fluvioglacial heavy loams and clays. Due to the aquiclude properties of soil forming species, these soils are often waterlogged, and they show traces of gley processes. Muck-podzolic, humus-illuvial, and podzolic soils developed on the light terminal moraine, ancient delta, and ancient alluvial deposits. The many-century process of the territory's agricultural cultivation contributed to the transformation of soils into artificial analogues, which interrupted the podzolic and gley processes as a result of fertilisation, liming, and drainage.

4.2 Metachronicity of Regional Landscapes

The formation of the territory's natural landscapes is an evolutionary process affected by both external factors and internal processes. In a mature landscape, all natural components are interconnected. Their 'adjustment' is a long-term processes affected by the evolution of each component. Therefore, complex landscapes have complex spatial and temporal structures.

A specific feature of regional natural landscapes is the metachronicity of their components. The age of landscapes is associated with the moment of the emergence of the sedimentary base. Superficial deposits and major landforms developed in the Pleistocene account for a large part of the territory. Part of younger landscape complexes (mire landscapes, river terrace complexes) emerged in the Holocene. The youngest dynamic landscapes are modern river floodplains and seacoasts undergoing the process of landscape genesis – alluvial and maritime accumulation and abrasion. The soils and vegetation have gone through several development stages following the climate trends. The Palaeogene and Neogene flora was mostly subtropical with wide-leaved elements – magnolias, tulip trees, palms, elms, willows, maples, lime trees, pines, larches, spruces, hemlocks, and ginkgoes.

In the Late Pleistocene, the edge of the melting glacier was covered in the periglacial tundra-/forest-steppe vegetation – the dwarf birch, avens, moss, lichen, common mugwort, and xerophilic true grasses. Later, it was replaced by sombre birch-pine and spruce forests. Approximately 8,000 years ago, the warmed climate initiated the spread of wide-leaved species – the oak, lime tree, maple, beech, and elm. The current composition of local vegetation developed 2,500 years ago [55]. Today's flora of the Kaliningrad Region is composed of species that migrated from Eastern Alps and the Carpathian Mountains. However, there are also preglacial and mid-Holocene relicts – the northern firmoss, martagon lily, snowdrop anemone, big-flowered foxglove, and cloudberry. Apparently, the region's soils are at a transitional stage of evolutionary succession from typical sod-podzolic to brown earth soils, which accounts for the wide variety of transitional forms with evidence of podzolic processes and lessivage.

Different ages of the landscape components, including the sedimentary basis, are one of the factors behind the complexity of the region's landscape structure.

4.3 Genetic Typology of the Landscape's Natural Basis: The Territory's Landscape Structure

The landscape structure of the Kaliningrad Region is characterised by the following genetic types of landscapes – landscapes of glacial and fluvial origin and sea and marine landscapes (Fig. 1). In their turn, they are divided into the following types of natural landscapes:

- Ground moraine plains
- Terminal moraine uplands
- Glaciolacustrine plains
- Coastal landscapes
- Ancient delta landscapes
- Valley landscapes
- Aeolian ancient alluvial plains

Glacial landscapes account for at least 80% of the region's territory, except for its northern part and the sea and lagoon coasts. Glacial landscapes are divided into the following types: rolling plains of the ground moraine, terminal moraine uplands and ridges, flat glaciolacustrine plains, hummocky fluvioglacial plains, and glacial meltwater flow valleys.

Moraine plains – predominantly irregular and rolling – found in the centre of the region stretch from the Sambia Peninsula in the West to its eastern borders. The Kaliningrad Region is dominated by a system of horseshoe terminal moraine ridges. Terminal moraine ridges form a chain of hills (of a maximum height of 110 m)

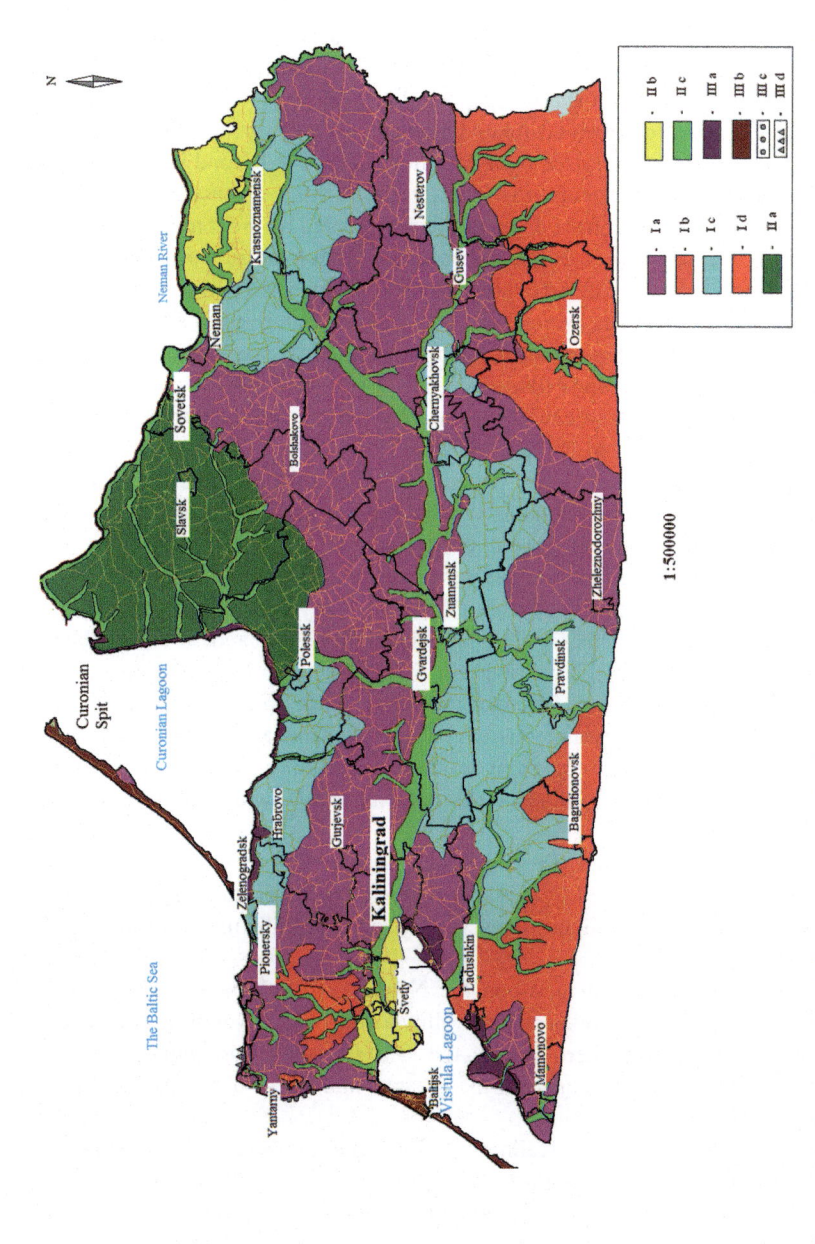

Fig. 1 Types of landscapes in the Kaliningrad Region. *I*. Glacial landscapes: (*a*) rolling plains of the ground moraine; (*b*) terminal moraine uplands and ridges; (*c*) flat glaciolacustrine plains; (*d*) hummocky and fluvioglacial plains. *II*. Fluvial landscapes: (*a*) ancient delta lowlands; (*b*) aeolian ancient alluvial plains; (*c*) valley complexes. *III*. Marine and lagoon landscapes: (*a*) coastal lagoon lowlands; (*b*) coastal aeolian formations; (*c*) accumulative seacoasts; (*d*) abrasive seacoasts

covered in mixed forests. The peninsula has retained various fluvioglacial landforms – kames. Another glacial landscape massif is found in the south of the region. Two spurs of the Warmian-Vištytis Upland, whose central part is located in Poland (the Suwałki Upland), run through the region's territory – in the south-west and south-east. The maximum height (242 m) is observed near Lake Vištytis – a lake of glacial origin. Terminal moraine uplands have a broken relief, mostly covered with forest massifs on brown earth and cryptopodzolic soils.

A vast glaciolacustrine plain is situated between the two spurs of the Warmian-Vištytis Upland. It is characterised by a flat tilted relief. The second largest area of glaciolacustrine deposit distribution is found in the region's north-west. The Kaliningrad Region's glaciolacustrine deposits are represented by predominantly brown and red plastic clays, which – despite their heavy granulometric structure – form fertile soils. These plains are occupied by agricultural lands and partly waterlogged. An exception is the glaciolacustrine plain of the heavily waterlogged southern coast of the Curonian Lagoon. This plain is mostly occupied by wet deciduous forests partly by agricultural lands on polders.

Coastal landscapes are represented by abrasive and accumulative-abrasive coast of the Kaliningrad Peninsula and the accumulative coast of the Vistula and Curonian Spits. The accumulative bodies of spits are composed of aeolian complexes, including moving and stabilised dunes. The coasts of the Vistula and Curonian Lagoons are covered in coastal meadows and coastal lowland mires.

Aqueous landscapes are found in the north-west and north-east of the region, as well as the southern part of the Sambia Peninsula. They also include all modern river valleys. Fluvial landscapes include an ancient delta lowland, aeolian ancient alluvial plains, and modern valley complexes.

The ancient delta of the river Neman is situated in the region's north-western borderlands. It is a vast lowland plain featuring lowland and upland mires and polders, characterised by a complex hydrographic network. There are also inland dunes and fan sand massifs. Most of the ancient delta lowland is occupied by agricultural lands.

The valleys of the Neman, Pregolya, and Šešupe and their tributaries have distinctive, partly waterlogged floodplains with oxbow lakes and numerous arms in the lower reaches. The floodplains are covered in hygrophilic meadows and black alder forests.

In the region's north-west, between the Neman and Šešupe and southwards, there is a large massif of aeolian ancient alluvial deposits covered in deciduous-coniferous forests. The southern part of the sand massif is underlain by glaciolacustrine clays; this area accounts for the greatest number of peat bog massifs on this territory.

Almost all the territory of the Kaliningrad Region is a coastal area; therefore, its western part is home to marine and lagoon landscapes – coastal lagoon lowlands, coastal aeolian formations (the Curonian and Vistula Spits), and accumulative and abrasive seacoasts.

4.4 Characteristics of Modern Landscape's Anthropogenic Components: The Impact of Agricultural Cultivation of Regional Territories

The mosaic landscape structure of the Kaliningrad Region is reinforced by the impact of long-term cultivation of the territory. It has been cultivated for seven centuries. In particular, vast massifs of polders have been created over this period alongside numerous systems of transportation and drinking water canals. By the beginning of the twentieth century, the region's territory was densely populated. Many areas had a population density of over 100 people per km^2. The territory was fully cultivated; agricultural lands accounted for 60% of the area, over a half being croplands. By 1939, a dense motorway and railway network ran across the territory. In 1939, the density of paved roads reached 40 km per 100 km^2, that of railways exceeded 13.8 km per 100 km^2. Therefore, the territory's long-term intensive occupation has almost completely transformed the natural environment, primarily, its soils and vegetation.

Undisturbed landscapes are represented by coastal meadow, partially by lowland mires, and by the central part of the terminal marine ridges. Anthropogenic agricultural lands and forestry landscapes account for most of the region's territory. Residential and transport landscapes and polders have been transformed most radically. The surviving polders are a rare example of artificial agricultural landscapes with a regulated hydrological and chemical regime. Part of the anthropogenic landscapes (mineral deposits, mounds erected for non-existent railways, abandoned farms, quarries, and peat fields) was withdrawn from economic use. Today, they are transforming into secondary natural-anthropogenic landscapes. These landscapes are entering the stage of secondary succession.

In 2003–2012, we studied landscape successions in key areas and landscape profiles using the method of landscape analogues. Soils and vegetation were considered as landscape development indicators [56].

The general pattern of modern landscape development in the Kaliningrad Region is cyclic succession. The initial transformation stage is landscapes covered in native forests (mixed spruce-deciduous and deciduous-spruce forests and deciduous forests in the west) on brown earth soils (sod-low and meso-podzolic and deeply podzolised and muck-podzolic humus-illuvial soils).

As a result of long-term cultivation of the territory, most natural landscapes turned into productive anthropogenic modifications, primarily agricultural ones. Earlier vegetation was replaced by its artificial counterparts, including planted dry meadows. Planted monodominant spruces, pines, English oaks, and introduced species account for 40% of the forest area. Soils assumed artificial features, being represented by brown cultivated unsaturated and sod-meadow varieties. Agricultural soils were treated with clay and sand to improve their physical properties. Most of the region's territory was treated by closed and open drainage. The general pattern of secondary succession cycle can be divided into historical and genetic sequences: forest – meadow – cropland and forest – cropland. These sequences repeated more than once, which resulted in the development of cultivated agricultural landscapes. As the land use system changes, part of landscapes is involved in the process of natural reclamation, i.e. secondary post-agrarian succession. A full secondary cycle occurs on abandoned croplands. Land recovery consists of several stages – each associated with a certain condition of soils and vegetation.

The general pattern of secondary successions on the agricultural landscapes of the Kaliningrad Region is as follows: pioneer vegetation community → bunchgrass meadow → firm bunchgrass meadow in the case of increased humidity – waterlogged meadow with willows → the 'sapling pole' stage – rich soils of normal humidification grow aspens and humid soils black alders (a quasi-primary community). If blocking types are introduced (goldenrods, the blue lupine, cow parsnip), succession becomes serial, i.e. it does not reach the primary (quasi-primary) stage. The final stage of secondary succession should be the return to native vegetation and soils; however, this is not observed in the Kaliningrad Region. Moreover, some authors stress that this process takes approximately 70–100 years in southern taiga and mixed forests.

Fallow lands create a complex system of recovering landscapes at different stages of natural reclamation. Today, recovering landscapes account for over 50% of the region's territory. Agricultural lands, primarily pastures, comprise 25%. Forest massifs and mire complexes account for another 19%. The remaining 6% is occupied by towns and villages, transport infrastructure, and water objects.

Therefore, the modern Kaliningrad landscapes are complex metachronous complexes, whose landscape structure has a complicated mosaic structure, which is accounted for by their glacial origin and the following natural evolution of the soil and vegetation. The economic cultivation of the region's territory has radically transformed natural landscapes. They also contribute to the differentiation of landscapes. Recovering and natural landscapes result in the rich ecological diversity and sustainability of the region's territory.

4.5 Landscape-Dependent Settlement in the Kaliningrad Region

Landscape-dependent settlement means that the location of towns and villages depends on the natural features of the environment – river valleys (terraces, edges), south-facing slopes, seacoasts, and plots of easily cultivated soils. An analysis of the landscape pattern of the Kaliningrad Region and actual population density in 1939 and 2009 makes it possible to identify the following natural districts with different degrees of dependence on landscape in the past and the present:

1. Neman's ancient delta lowland district
2. Modern valleys of large rivers
3. Flat plains of glaciolacustrine origin
4. Terminal moraine uplands with a rugged relief

5. Rolling plains of the ground moraine
6. Ancient alluvial sand massifs
7. Coastal area of the Baltic Sea and the lagoons

4.5.1 The Neman's Ancient Delta District

Most of the Neman's ancient delta is below sea level (reaching −1.4 m). The long cultivation history created an intricate system of lands with a regulated hydrological regime – polders.

The 1939 population density map shows a high level of cultivation with a rural population density of above 50 people per km^2. Settlement was dispersed – alongside large villages with an average population of 245 people, there were over a 100 isolated farms.

The district's current population is concentrated in several areas (with a density of above 50 people per km^2). The settlement pattern is strongly dependent on landscape complexes. The western part of the district is almost unpopulated, being occupied by impassable fens. Several villages situated on the lagoon coast in the pre-war period disappeared as the operation of adjacent pumping units was discontinued. That is why the territory is waterlogged more heavily than before. Today, there are only two villages – Mysovka and Prichaly – on the coast of the Curonian Lagoon. The most populated area is the central part of the district. Settlements are situated in the uplands. The least populated area is the 'trans-canal zone' (the northern part of the district separated from the main territory by the canalised Matrosovka River). Most of the territory of the Neman's ancient delta has a population density of below ten people per km^2.

4.5.2 Modern Valleys of Large Rivers

The largest towns and villages of the Kaliningrad Region are located in river valleys. As a rule, all these settlements have a centuries-long history. Today, just as in the pre-war period, the Pregolya's valley forms a natural axis of population concentration in the region. Another band of high population density is associated with the Neman Valley. Here, a change in the absolute population density was accompanied by a reduction in the area of territories with a population density of above 100 people per km^2.

4.5.3 Flat Plains of Glaciolacustrine Origin

The flat relief and soil features shaped the district's settlement structure. Both modern and pre-war settlement structures can be identified as either narrow or wide ranged based on the features of agricultural production and the size of agricultural units. An average population density exceeded 50 people per km^2;

today it is approximately 20 people per km². In the pre-war period, the least populated districts were associated with large mire massifs (the Zalau mire, fens of the Curonian Lagoon coast, and the bogs of region's north-east). The area of unpopulated territories has increased almost fourfold. The cause is not always the territory's landscape features. Today, there are two large military training areas in the areas. When they were founded in the 1950s, the local population was relocated.

4.5.4 Terminal Moraine Uplands with a Rugged Relief

Most terminal moraine uplands are covered in deciduous and deciduous-coniferous forests. The only exception is the northern spur of the Vištytis Upland called Lysye Gory. In the pre-war period, the settlement pattern of this district was narrow ranged, since settlements emerged in the areas with the most favourable environmental conditions – on the shores of lakes, at forest edges, and in the valleys of small rivers. Settlement areas are relatively small and have a high population density of above 100 people per km². Today, the terminal moraine uplands are the least populated area of the Kaliningrad Region. Population is concentrated in several isolated areas, whereas a high population density is observed only in towns.

4.5.5 Rolling Plains of the Ground Moraine

This district incorporates most of the region's territory. In the pre-war period, the average rural population density in the district exceeded 50 people per km². Settlement concentrations accounted for large areas with a population density of above 100 people per km². The settlement structure was uniform. Today, the average rural population density is below ten people per km². The settlement pattern suggests a strong association with natural complexes – large plots of land remain unpopulated.

4.5.6 Ancient Alluvial Sand Massifs

Not unlike the pre-war period, this district is covered in forests. The difference between the 1939 and 2009 population density is striking. Although the environmental conditions remained unchanged, the current population density is far below a one-tenth of the pre-war level. Today, most of the district's territory has a population density of below ten people per km².

4.5.7 Coastal Area of the Baltic Sea and the Lagoons

Across the world, coastal zones of seas and lagoons are the most attractive area for settlement. As a rule, they are densely populated. The coast of the Baltic Sea in the

Kaliningrad Region is an exception. Only the northern coast of the Sambia Peninsula is densely populated. Unlike their counterparts in the neighbouring European States, vast territories of the coastal zone (western coast) have a population density of below ten people per km^2. Partly, this is explained by the presence of a large military training area. In the pre-war period, the population density of this district was above 50 people per km^2. Not unlike the pre-war period, the coast of the Curonian Lagoon is less cultivated than that of the Vistula Lagoon. The Curonian Lagoon is surrounded by lands with a population density of below ten people per km^2, which is largely explained by the high proportion of mires. The coast of the Vistula Lagoon is as densely populated as in the pre-war period. The district's beneficial economic and geographical position – partly accounted for by transport accessibility and the vicinity of the regional centre – is inherited. Over 70 years, this district has been a 'growth pole' being home to numerous industrial facilities and fishing ports. The district still has a high development potential.

Therefore, the settlement system of the identified districts of the Kaliningrad Region shows a varying degree of landscape dependence, which is not always inherited historically. In many cases, it relates to other factors – primarily, political, socioeconomic, and demographic ones – creating the opportunities for cultivating and transforming environmental complexes. The territory of the Kaliningrad Region consists of natural districts showing a varying degree of landscape dependence. In some of them, the settlement pattern is very different from the pre-war one. Three natural districts – the Neman's ancient delta, the ancient alluvial sand massif between the Neman and Šešupe, and terminal moraine uplands – have a steep settlement gradient.

4.6 Modern Landscapes as Polygenetic Systems of Areal and Network Environmental Components

The region's modern landscapes developed on the basis of natural complexes created by the last Quaternary glaciation and affected by the processes shaping the natural environment in the postglacial period. Therefore, there is a need to study the structure of natural landscapes, in particular, their genetic types. Since the region's soils and vegetation radically transformed over the study period, the key research objects are the parent rock and relief.

Further, it is important to estimate the degree of natural landscapes' transformation by human occupation. This requires identifying such components of a landscape's material layer that will make it possible to assess the human impact on landscapes and identify representative time samples. The region's territory has been cultivated for many centuries. However, this transformation was gradual. The traces of Prussian cultivation are found only at archaeological sites (primarily, tumuli); they do not comprise a significant proportion of the regional landscapes. At the same time, the modern settlement pattern is largely inherited from the pre-war period, being an important component of all cultural landscapes. Moreover, studying regional population distribution will make it possible to estimate the human impact on the landscape [55].

An important factor behind the transformation of a natural landscape into a cultural (modern) one is agriculture, primarily, crop farming. Therefore, one of the major objectives of research on cultural landscapes should be land use analysis. Other components of the material layer of a cultural landscape – transport infrastructure, industrial, military, and recreation facilities linked with the settlement system – should be taken into account.

A cultural landscape is much more strongly affected by both spatial and temporal changes than a natural one. Therefore, research on modern landscapes is impossible without studying the degree of their transformation over a short period. The most convenient time samples are those of 1939 (a pre-war condition and the maximum cultivation degree) and the present.

Estimating the degree of modern landscapes' transformation over a certain period requires studying their dynamics and comparing past and present landscapes. There are no representative data on the region's pre-war cultural landscape. Therefore, there is seemingly nothing against which to compare current landscapes. However, there is a solution. The key to solving this problem is Yu. G. Saushkin's definition of the cultural landscape. People create cultural landscapes based on natural ones changing their components. However, the new landscape still lives by the laws of nature. If human impact on a landscape is discontinued, the landscape will develop according to the same laws but without human participation. Here, one can recall the work of V. P. Semenov-Tyan-Shansky who distinguished among primeval, semiwild, cultural, abandoned, and reclaimed landscapes [57]. In this classification, cultural landscapes are the highest rung on the ladder (maximum human participation), those preceding it are steps up (human participation increases), and those following it are steps down (human participation decreases).

Therefore, the algorithm of studying modern cultural landscapes cannot be reduced to analysing only the natural components of the environment. It should focus on factors and patterns of the modern development of regional territorial complexes forming the socioeconomic components of a landscape environment. As well as their natural counterparts, these components can be network or areal. Areal components include settlement and land use systems, network components isolated industrial and economic facilities, transport systems, and borders. Areal network components are residential territories. On the one hand, settlement systems are network elements of the basic territorial framework. As the mapping scale is reduced, even large residential areas turn into isolated dots. On the other hand, at a closer look, any residential territory is an areal form, since it has a certain area and it can be divided into zones, which is one of the major features of an area. Therefore, residential zones can be classified as a transitional type of spatial units.

5 Conclusions

The modern appearance of Kaliningrad landscapes shows similarities with European Plain landscapes of the last Quaternary glaciation's accumulation zone in the temperate climate zone. The key components of landscapes indicative of their genesis are the sedimentary base (Quaternary deposits) and the relief. One can distinguish the following genetic types – landscapes of glacial, fluvial, marine, and lagoon origins. In turn, they are divided into ground moraine plains, terminal moraine uplands, glaciolacustrine plains, coastal landscapes, ancient delta lowlands, valley landscapes, and aeolian ancient alluvial plains. Soils and vegetation, which evolved in the postglacial period from tundra to forest vegetation, have undergone radical transformation by human occupation over the past millennium. A specific feature of regional landscapes is the metachronicity of their components. The sedimentary basis of most of the territory developed in the Pleistocene. Part of younger landscape complexes (mire landscapes, river terrace complexes) emerged in the Holocene. The youngest, most rapidly changing landscapes are the modern river floodplains and seacoasts undergoing the process of landscape genesis – channel and marine accumulation and abrasion. Soils and vegetation went through several development stages governed by climate trends. In the Late Pleistocene, the glacier's edge was covered in periglacial tundra and forest-steppe vegetation. Later, it was replaced by open birch-pine and spruce forests having supplanted deciduous forests as the climate grew warmer. The final composition of local vegetation was attained approximately 2,500 years ago. However, it has been radically transformed over the past millennium. Today, forests account for 20% of the region's territory. Autochthonous forests are almost absent, with the exception of large massif's cores. Most forests were planted at different times. Most of the region's territory is occupied by agricultural lands. Its appearance is shaped by land use rather than natural processes.

The modern landscape environment of the Kaliningrad Region consists of areal and network (linear) elements of both natural and anthropogenic origin. A combination of these elements in space shaped modern landscapes. Natural areal elements of the modern landscape environments include sedimentary complexes and river basins, anthropogenic ones consist of land use areas, which affect the transformation and composition of soils and vegetation, and artificial elements (buildings and structures). Natural network elements are river networks and isolated water objects, coastal areas of seas and lagoons, gullies, and dry channels. Anthropogenic ones are transportation objects (roads, bridges, etc.) and isolated industrial and communications facilities. Depending on the scale, these objects can assume either an areal or network character. At a smaller scale, settlements are classified as network components, since they can be reduced to point features; at a larger scale, they are areal. The landscape structure of the region and its parts also contains natural-anthropogenic areal and network elements, namely, the passages and nodes of the territory's natural ecological framework.

The key factors behind the landscape genesis on the territory of the Kaliningrad Region are the land use and settlement systems. In turn, the modern settlement system is partly affected by the landscape conditions. The landscape dependence of settlement is not always inherited historically. In many cases, it is affected by different factors – primarily, political, socioeconomic, and demographic – which create opportunities for the cultivation and transformation of territorial complexes. In the Kaliningrad Region, there are districts with a varying degree of landscape dependence of the modern settlement pattern, the latter often different from the pre-war one. The most radical changes in the settlement structure – as compare to the pre-war data – are observed in three natural regions: the Neman's ancient delate, the ancient alluvial sand massif between the Neman and Šešupe, and the terminal moraine uplands, where uninterrupted but narrow-ranged settlement was replaced by wide-ranged settlement.

References

1. Kagansky VL (2009) Kulturnyy landshaft: osnovnye kontseptsii v Rossiyskoy geografii (Cultural landscape: key concepts in Russian geography). Observatoriya kultury (Observatory of Culture) 1b:62–70
2. Koshurov BI (2000) Razvitie geoekologisheskikh terminov i ponyatiy (Development of geoecological terms and concepts). Problemy regionalnoy geoekologii (Problems of Regional Geoecology) 3:144–157
3. Reymers NF (1990) Prirodopolzovanie: slovar-spravochnik (Nature management: a reference book). Mysl, Moscow, 638 pp
4. Romanova EP (1997) Sovremennye landshafty Evropy (bez stran Vostochnoy Evropy), uch. posobie (Modern landscapes of Europe (excluding Eastern Europe). A textbook). Moscow State University Press, Moscow, 312 pp
5. Kalutskov VN (2007) Issledovaniya kulturnogo landshafta v Rossii: metodologiya, istoriya voprosa, sovremennaya situatsiya (Studies of cultural landscapes in Russia: methodology, history, the present). In: Proceedings of the cultural landscapes as a heritage future: a comparative study of the UK and Russia workshop, Moscow, e-publication
6. Kalutskov VN, Ivanova YA, et al. (1998) Kulturnyy landshaft Russkogo severa: Pinezhye i Pomorye (Cultural landscape of Russia's North: the Pinega area and the White Sea coast). FBMK, Moscow, 136 pp
7. Kalutskov VN, Krasovskaya TM (2000) Predstavleniya o kulturnom landshafte: ot professionalnogo do mirovozzrensheskogo (The idea of cultural landscape: from professional to philosophical views). MSU Vestnik Geogr Ser 4:3–6
8. Nizovtsev VA (2009) Struktura, ierarkhiya i klassifikatsiya kulturno-istoricheskikh landshaftov (Structure, hierarchy, and classification of cultural and historical landscapes). In: Materialy Vserossiyskoy nauchnoy konferentsii «Seliverstovskie chteniya»: Geografiya i Geoekologiya na sovremennom etape vzaimodeystviya prirody i obshchestva (Proceedings of the 'Selivestrov readings: geography and geoecology at the modern stage of interaction between nature and society' all-Russian conference). Saint Petersburg State University Press, Saint Petersburg, pp 693–698
9. Vedenin YA, Kuleshova ME (2001) Kulturnyi landshaft kak obyekt kulturnogo i prirodnogo naslediya (Cultural landscape as an object of cultural and natural heritage). Izvestiya Russ Acad Sci Geogr Ser 1:7–14

10. Vedenin YA, Kuleshova ME (2004) Kulturnye landshafty kak kategoriya naslediya (Cultural landscapes as a heritage category). www.heritage.UNESCO.ru
11. Kuleshova E (2000) Kulturnye landshafty: obshchie predstavleniya, ponyatiya, podkhody k otsenke (Cultural landscapes: general ideas, concepts, and approaches to assessment). Ekologisheskie problemy sokhraneniya istoricheskogo i kulturnogo naslediya (Ecological problems of historical and cultural heritage conservation). Rossiyskiy nauchno-issledovatelskiy institut kulturnogo i prirodnogo naslediya (Russian Research Institute of Cultural and Natural Heritage), Moscow, pp 37–52
12. Etnokulturnye issledovaniya (Ethnocultural studies) (2004) Staryy sad, Moscow, 464 pp
13. Yamskov AN (2003) Etnoekologicheskie issledovaniya kultury i kontseptsiya kulturnogo landshafta (Ethnoecological cultural studies and the concept of cultural landscape). Kulturnyy landshaft: teoriya i regionalnye issledovaniya (Cultural landscape: Theory and regional studies). Moscow State University Press, Moscow, pp 62–77
14. Kagansky VL (2001) Kulturnyy landshaft i sovetskoe obitaemoe prostranstvo; Sbornik statey (Cultural landscape and the Soviet inhabited space: a collection of papers). Novoe literaturnoe obozrenie, Moscow, pp 45–59
15. Rodoman BB (1999) Territorialnye arealy i seti. Ocherki teoretisheskoy geografii (Territorial areas and network. Works on theoretical geography). Oykumena, Smolensk, 256 pp
16. Sauer CO (1963) The morphology of landscape. In: Leighly J (ed) Land and Life: A Selection from the writings of Carl Ortwin Sauer. University of California Press, Berkley, pp 315–350. First published in Geography 2.2 (1925)
17. Hartshorne R (1939) The nature of geography. A critical survey of currents thought in the light of the past. Association of American Geographers, Lancaster
18. Lowton J (2004) Foreword. In: Dietrich M, Straaden JV (eds) Cultural landscapes and land use. The nature conservation – society interface. Kluwer Academic, Boston, pp xi–xiii
19. Rackham O (1988) Trees and woodland in a crowded landscape – the cultural landscape of the British Isles. In: Birks H (ed) The cultural landscapes: past, present and future. Cambridge University Press, Cambridge, pp 53–78
20. Landscape character assessment: guidance for England and Scotland (2000) Land Use Consultants and Department of Landscapes, University of Sheffield, Sheffield, 63 pp
21. Jackson J (1984) Discovering the vernacular landscape. Yale University Press, New Haven, p 156
22. Bogdanovski J (1993) Landscape architecture: preservation and modelling. In: The tradition of 777 training in landscape architecture in Poland. Osrodek Ochrony Zabytkowego Krajobrazu, Narodowa Institucia Kultury, Warsaw, pp 1–85
23. Dietrich M (2004) Reflections on the intelligence of natural systems. In: Dietrich M, Straaden JV (eds) Cultural landscapes and land use. The nature conservation – society interface. Kluwer Academic, Boston
24. Leader-Elliot L, Maltby R, Burke H (2004) Understanding cultural landscapes – definition. www.fhrc.flinders.edu.au
25. Longstreth R (2008) Balancing nature and heritage. In: Longstreth R (ed) Preservation practice. Cultural landscapes. University of Minnesota, Minneapolis, pp 1–15
26. Swinnerton G, Buggey S (2004) Protected landscapes in Canada: current practice and future significance. George Wright Forum 21(2):78–92
27. List P (2004) Aldo Leopold on the ethical foundation for conserving biodiversity. In: Dietrich M, Straaden JV (eds) Cultural landscapes and land use. The nature conservation – society interface. Kluwer Academic, Boston, pp 13–24
28. UNESCO World Heritage Center (2003) 140 pp. whc.unesco.org
29. Küster H (1995/1999) Geschihte der Landschaft in Mitteleuropa. Von der Eiszeit bis zur Gegenwart. Verlag C.H. Beck, Munich
30. Küster H (1995) Postglaziale Vegetationsgeschichte Sudbayerns: Geobotanische Studien zur Praehistorischen Landschaftskunde, German edn. Akademie Verlag, Berlin, 373 pp

31. Küster H (2004) Cultural landscapes – an introduction. In: Dietrich M, Straaden JV (eds) Cultural landscapes and land use. The nature conservation – society interface. Kluwer Academic, Boston, pp 1–12
32. Grebenshchikova AA (1952) Torfyanoy fond Kaliningradskoy oblasti (Peat resources of the Kaliningrad Region). Izdatelstvo Upravleniya torfyanogo fonda pri Sovmine RSFSR (USSR Council of Ministers's Peat Department Press), Moscow, 68 pp
33. Vazhenin IG, Belyakova VI (1959) Agrokhimisheskaya kharakteristika pochv (Agrochemical characteristics of soils). Agrokhimisheskie raboty Kaliningradskoy oblasti (Agrochemical studies in the Kaliningrad Region). Izdatelstvo Akademii nauk SSSR (Academy of Science of the USSR Publishing House), Moscow, pp 28–66
34. Zavalishin AA, Nadezhdin BV (1961) Pochvennyy pokrov Kaliningradskoy oblasti (The soil cover of the Kaliningrad Region). Pochvy Kaliningradskoy oblasti (Soils of the Kaliningrad Region). Izdatelstvo Akademii nauk SSSR (Academy of Science of the USSR Publishing House), Moscow, pp 6–130
35. Panasin VI, Bormatenkov OA (1969) Poldernye zemli Kaliningradskoy oblasti (Polders of the 811 Kaliningrad Region). Kaliningradskoe knijnoe izdatelstvo, Kaliningrad, 48 pp
36. Leontyev VL (1955) Nekotorye osobennosti lesov Kaliningradskoy oblasti (Some features of the Kaliningrad Region's forests). Tr. Botanicheskogo in-ta im. V.L. Komarova Akademii nauk SSSR (Publications of the V.L. Komarov Institute of Botany of the Academy of Sciences of the USSR), Geobotany Series 10:16–28
37. Pobedimova EG (1955) Sostav, rasprostranenie po rayonam i khozyaystvennoe znashenie flory Kaliningradskoy oblasti (Composition, district distribution, and economic significance of the Kaliningrad Region's flora). Tr. Botanicheskogo in-ta im. V.L. Komarova Akademii nauk SSSR (Publications of the V.L. Komarov Institute of Botany of the Academy of Sciences of the USSR), Series 3, Geobotany 10:26–38
38. Orlenok VV (2004) Geografiya Yantarnogo kraya (Geography of the Amber region). Yantarnyy skaz, Kaliningrad, 416 pp
39. Sukhov AA and Kozlovich II (2002) Landshaftnaya karta Kaliningradskoy oblasti M 1: 500 000 (Landscape map of the Kaliningrad Region of scale 1:500 000). In: Orlenok VV (ed) Atlas Kaliningradskoy oblasti (Atlas of the Kaliningrad Region). Kaliningrad State University Press, Kaliningrad, pp 120–121
40. Dedkov VP, Fedorov GM (2006). In: Dedkov VP (ed) Prostranstvennoe, territorialnoe i landshaftnoe planirovanie v Kaliningradskoy oblasti (Spatial, territorial, and landscape planning in the Kaliningrad Region). Immanuel Kant Russian State University Press, Kaliningrad, 185 pp
41. Romanova EA (2009) Landshaftnaya struktura i sistema rasseleniya Kaliningradskoy oblasti (Landscape structure and settlement pattern in the Kaliningrad Region). In: Materialy Vserossiyskoy nauchnoy konferentsii «Seliverstovskie chteniya»: Geografiya i Geoekologiya na sovremennom etape vzaimodeystviya prirody i obshchestva (Proceedings of the 'Selivestrov readings: geography and geoecology at the modern stage of interaction between nature and society' all-Russian conference). Saint Petersburg State University Press, Saint Petersburg, pp 261–266
42. Romanova EA, Vinogradova OL (2006) Landshaftnoe planirovanie territorialnykh edinits Kaliningradskoy oblasti (Landscape planning of territorial units of the Kaliningrad Region). Materialy XI Mezhdunarodnoy landshaftnoy konferentsii, 22-25 avgusta 2006 g (Proceedings of the 11th international landscape conference, August 22-25, 2006). Moscow State University Press, Moscow, pp 689–692
43. Romanova EA, Vinogradova OL (2006) O vidakh planirovaniya territorii (na primere territorialnykh edinits Kaliningradskoy oblasti) (On the types of territory planning: The case of the Kaliningrad Region). Vestnik of the Immanuel Kant Russian State University 7:17–28
44. Orlenok VV, Romanova EA, Vinogradova OL, Stanchenko LY (2004) Landshaftno-ekologisheskoe planirovanie Svetlogorskoy rekreatsionnoy zony (Landscape and ecological planning of the Svetlogorsk recreation zone). «Pribrezhnaya zona morya: morfolitodinamika i

geoekologiya» (Coastal marine area: morpholithodynamics and geoecology). Kaliningrad State University Press, Kaliningrad, pp 224–227

45. Orlenok VV, Romanova EA, Vinogradova OL (2005) Landshaftno-ekologisheskoe planirovanie munitsipalnykh obrazovaniy Kaliningradskoy oblasti (Landscape and ecological planning of municipalities in the Kaliningrad Region). In: Trudy XII sezda Russkogo geograficheskogo obshchestva, t.4 (Proceedings of the 12th congress of the Russian Geographical Society, Volume 4), Saint Petersburg, pp 185–189
46. Romanova EA, Vinogradova OL (2011) Landshafty (Landscapes). In: Orlenok V, Fedorov G (eds) Kaliningradskaya oblast. Atlas (The Kaliningrad Region. An atlas). Kollektsiya, Kaliningrad, pp 26–27
47. Romanova EA, Vinogradova OL (2011) Fiziko-geograficheskoe rayonirovanie (Physiographic zoning). In: Orlenok V, Fedorov G (eds) Kaliningradskaya oblast. Atlas (The Kaliningrad Region. An atlas). Kollektsiya, Kaliningrad, pp 30–31
48. Bukreeva LM, Vinogradov SS, Starikov NN (1949) Resursy mineralnykh stroitelnykh 862 materialov Kaliningradskoy oblasti i perspektivy ikh ispolzovaniya (Mineral construction 863 material resources in the Kaliningrad Region and the prospects of their use), vol 1. Izdatelstvo Akademii Nauk, Moscow, 864 118 pp
49. Vereyskiy NG (1946) Poyasnitelnaya zapiska k geologisheskoy karte chetvertishnykh otlozheniy Kaliningradskoy oblasti RSFSR i prilegayushchikh rayonov Polshi (An explanatory note to the geological map of quaternary deposits in the Kaliningrad Region of the RSFSR and the neighbouring Polish districts). Rukopis v 4. Geol. upr. (Manuscript), Moscow, 126 pp
50. Zagorodnykh VA, Kunaeva TA (2005) Geologiya i poleznye iskopaemye Kaliningradskogo regiona (Geology and minerals of the Kaliningrad Region). Yantarnyi skaz, Kaliningrad, 208 pp
51. Kurkov AA, Sukhova AA (1972) K landshaftnoy kharakteristike Kaliningradskogo 873 poluostrova (On the landscape characteristics of the Kaliningrad peninsula). In: Borisov AA 874 (ed) Izushennost prirodnykh resursov Kaliningradskoy oblasti (Studies into natural resources 875 of the Kaliningrad Region), Nauka, Leningrad, Nauka, pp 120–121
52. Schlicht O (2002) Das westliche Samland. Bd. 2. Verlag von Frankfurt am Main, Frankfurt, p 456
53. Romanova EA, Vinogradova OL (2011) Geomorfologiya (Geomorphology). In: Orlenok V, Fedorov G (eds) Kaliningradskaya oblast. Atlas (The Kaliningrad Region. An atlas). Kollektsiya, Kaliningrad, pp 18–19
54. Barinova GM (2002) Kaliningradskaya oblast. Klimat (The Kaliningrad Region. Climate). Yantarnyi kray, Kaliningrad, 413 pp
55. Orlenok VV, Barinova GM, Kucheryavyy PP, Ulyashev GL (2001) Vishtynetskoe ozero: priroda, istoriya, ekologiya (Lake Vištytis: nature, history, ecology). Kaliningrad State University Press, Kaliningrad, 212 pp
56. Vinogradova OL (2012) Nekotorye aspekty antropogennoy evolyutsii landshaftov Kaliningradskoy oblasti (v period s 1950 po 2010 gg. (Some aspects of anthropogenic evolution of landscapes in the Kaliningrad Region in 1950-2010). In: Issledovanie territorialnykh sistem: teoriya, metodologiya i prikladnye aspekty: Materialy Vserossiyskoy nauchnoy konferentsii s mezhdunarodnym ushastiem, 4 – 6 oktyabrya 2012 g., gorod Kirov (Studies of territorial systems: theory, methodology, and applied aspects: Proceedings of the all-Russian research conference with international participation, October 4 – 6, Kirov). Loban, Kirov, pp 153–159
57. Semenov-Tyan-Shansky VP (1928) Rayon i strana (District and country). GIZ, Moscow and Leningrad, 312 pp

The History and Pattern of Forest and Peatland Formation in the Kaliningrad Region During the Holocene

T.V. Napreenko-Dorokhova and M.G. Napreenko

Abstract Investigations of mire peat deposits in the Kaliningrad Region such as analysis of the botanical composition of peat, pollen analysis and radiocarbon dating provide an opportunity to work out some important palaeogeographical issues related to zonal (climax) and azonal vegetation formation, climatic changes as well as human impact to the environment.

Based on the outcomes of authors' peat investigations on a number of mires, the main pattern of forest and peatland formation during the Holocene were defined in the Kaliningrad Region in two largest landscape areas: glaciolacustrine plain in the central part of the region and coastal lowland which comprises a vast territory along Curonian Lagoon and in the Neman Delta.

It was stated that the territory of the Kaliningrad Province is to be divided, in palaeoenvironmental respect, onto two different parts each of those could be united with the neighbouring regions of Poland and Lithuania. Both parts are similar, in palynological respect, in the Early and the Middle Holocene but had been obtaining distinguishes in the Late Holocene when conifers became dominating in the northeastern part of the region while broad-leaved were common in the southern parts, where *Carpinus* and *Fagus* became essential components in forest vegetation.

A mire formation was mainly caused by the paludification processes on dryland. Nevertheless, mires in the coastal area along the Curonian Lagoon are peat bodies of a complex genesis their different parts developed in various ways at the initial stage of formation. Large raised bogs are rather recent geological bodies in the region, their transition into *Sphagnum* bog stage and major formation occurred only

---

T.V. Napreenko-Dorokhova (✉)
Shirshov Institute of Oceanology, Russian Academy of Sciences, Moscow, Russia
e-mail: tnapdor@gmail.com

M.G. Napreenko
Immanuel Kant Baltic Federal University, Kaliningrad, Russia
e-mail: maxnapr@gmail.com

in the end of the Middle Holocene and the Late Holocene which is later than in the most part of the forest belt in European Russia.

Keywords Botanical composition of peat, Holocene, Palaeogeography, Palynology, Stratigraphy

Contents

1	Introduction	122
2	Study Sites and Methods	123
	2.1 Geographical Location	123
	2.2 Analysis of the Botanical Composition of Peat and Peat Bed Structure	125
	2.3 Pollen Analysis	126
	2.4 Radiocarbon Dating	126
3	Results and Discussion	126
	3.1 The History of Forest and Peatland Formation on Pregolya Glaciolacustrine Plain During the Holocene	126
	3.2 The History of Forest and Wetland Formation on Lower Neman Lowland During the Holocene	133
	3.3 Comparison of the Palaeoreconstructions in the Study Areas of the Kaliningrad Region with the Environmental Development in Adjacent Regions During the Holocene	139
4	Conclusions	141
References		144

1 Introduction

A period of recent environment formation started on the territory of the Kaliningrad Region in the postglacial time during which present-day features of climate, soils and vegetation developed together with ecosystem alterations caused by human impact.

Since peatlands are considered as a natural record of vegetation dynamics, they play a great role for investigations of ecosystem formation which can be transcribed by means of palaeobotanical researches (incl. pollen analysis and study of the botanical composition of peat) and radiocarbon dating.

In this respect, the detailed palaeobotanical survey on key natural objects in a certain landscape is of great importance for environmental reconstruction providing an opportunity to reveal local vegetation patterns and to set up probable succession factors in the area. There are a number of landscape types in South-Eastern Baltic Region that having formed here after the ice sheet ablation, the largest among them are glaciolacustrine plains in the central part of the region and coastal lowlands which comprise a vast area along Curonian Lagoon and in the Neman Delta (Fig. 1).

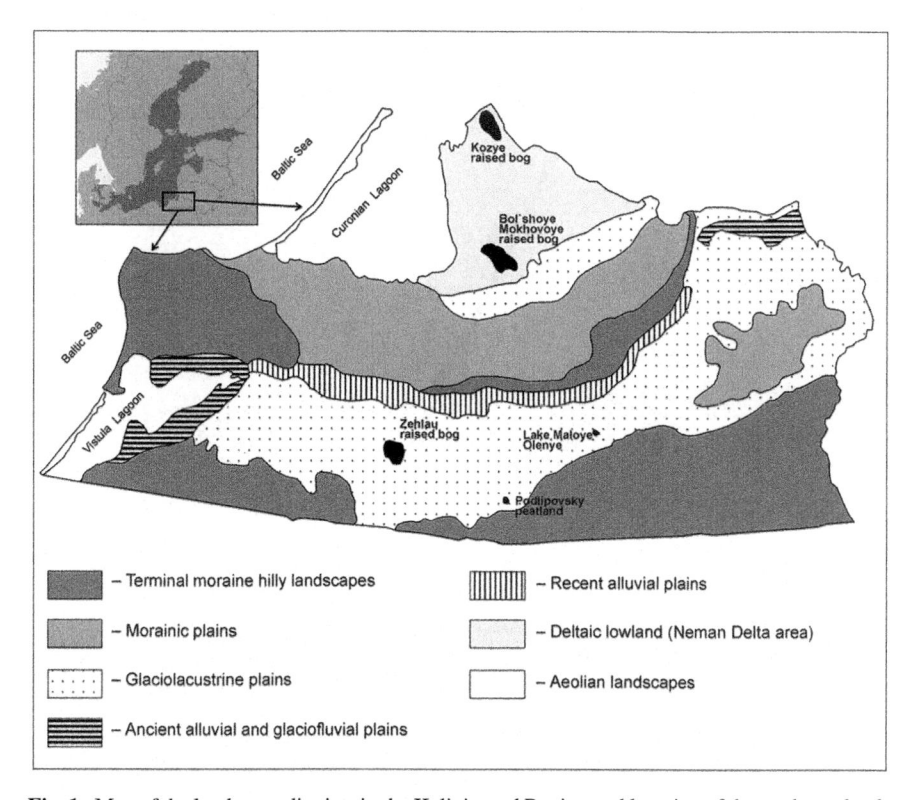

Fig. 1 Map of the landscape districts in the Kaliningrad Region and location of the study wetlands

2 Study Sites and Methods

2.1 Geographical Location

Five wetland ecosystems were chosen as study objects on the territory of the Kaliningrad Region, three of them are on Pregolya Glaciolacustrine Plain (Zehlau raised bog, Podlipovsky peatland and terrestrialized lake Maloye Olenye with a riparian mire) and two ones are on Lower Neman coastal lowland (raised bogs Bol'shoye Mokhovoye and Kozye) (Fig. 1).

All the investigations on large bogs were carried out along the transect lines which stretched across the whole peatland area (Figs. 2 and 3). The survey included the following operations: levelling procedure to determine the bog surface topography, manual peat bed probing to determine a floor relief of the bog depression, selective coring of the peat bed and collection of peat deposits samples.

The survey is based on comprehensive approach which implies a combination of various palaeoenvironmental methods, such as pollen analysis, plant macrofossil determination and radiocarbon dating.

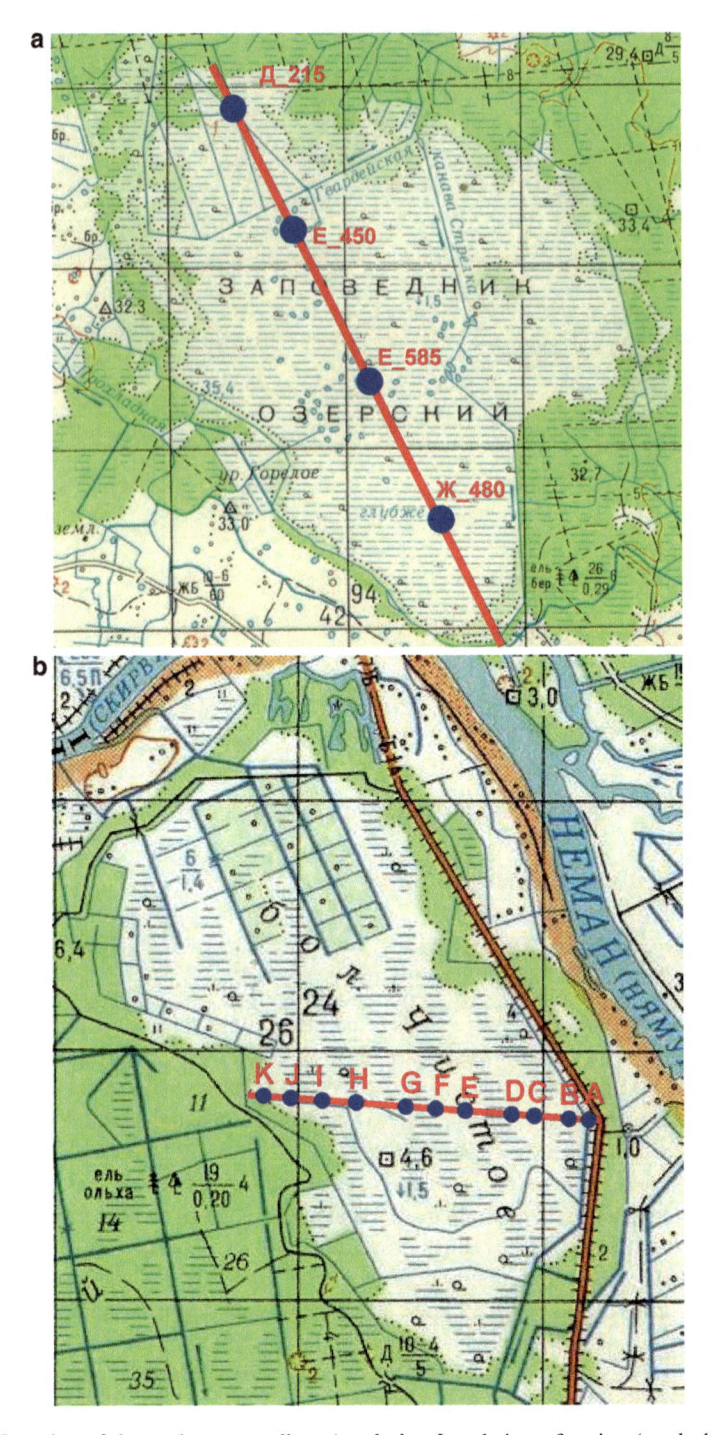

Fig. 2 Location of the study transect lines (*marked red*) and sites of coring (*marked blue*) on Zehlau raised bog (**a**) and Kozye raised bog (**b**) (basemap: Kaliningrad Region 1: 100,000). See Fig. 1 for location

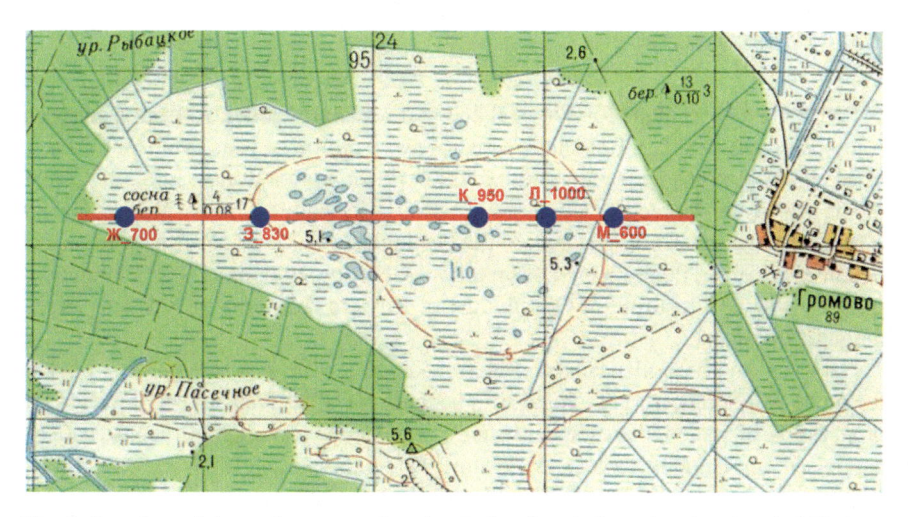

Fig. 3 Location of the study transect line (*marked red*) and sites of coring (*marked blue*) on Bol'shoye Mokhovoye raised bog (basemap: Kaliningrad Region 1: 100,000). See Fig. 1 for location

Coring of peat layers and retrieving peat samples for further analysis were carried out by means of the Russian D-corer with the semicylindrical sample chamber (model TBG-66).

2.2 Analysis of the Botanical Composition of Peat and Peat Bed Structure

On Zehlau raised bog, four core samples (Fig. 2a) were retrieved from the peat layers and subsampled onto 134 samples of peat which had been afterwards microscopically analysed.

On Kozye raised bog, a transect line was established in the southern part of the mire that is not disturbed by peat extraction (Fig. 2b) and where 346 peat samples were collected from 11 cores.

There have been sampled and analysed 265 peat samples from 5 cores on Bol'shoye Mokhovoye raised bog along the transect line in the north-western part of the mire that is least influenced by human impact (Fig. 3).

On riparian mire of Maloye Olenye Lake, 33 peat and gyttja samples were collected in the monoliths which have been retrieved from the terrestrializing buoyant mat and near-bottom gyttja layers.

Processing and macrofossil analysis of 778 peat samples were performed in the Laboratory of Geoecology of the Atlantic Branch of the P.P. Shirshov Institute of Oceanology of Russian Academy of Sciences (Kaliningrad, Russia).

2.3 Pollen Analysis

Pollen analysis was made for 476 peat and gyttja samples collected from 4 cores on study peatlands (Zehlau, Bol'shoye Mokhovoye, Podlipovsky and Maloye Olenye). These investigations were carried out in the following laboratories: Ecology Centre of the Christian Albrecht's University in Kiel (Germany), Greifswald University (Germany), Marine Geology Dept. of the Geological Institute of Polish Academy of Sciences (Gdansk, Poland) and Laboratory of Geoecology of the Atlantic Branch of the P.P. Shirshov Institute of Oceanology of Russian Academy of Sciences (Kaliningrad, Russia).

Samples for the pollen analysis were taken from the peat monoliths in every 5 cm and were processed in laboratories according Faegri-Iversen technique [1] for preparation of pollen/spores specimens. Our palaeoenvironmental reconstructions are based on Gudelis Holocene Scheme [2] elaborated for the Eastern Baltic Region.

2.4 Radiocarbon Dating

Radiocarbon dates have been obtained for each study peatland (31 samples from 5 cores were sampled for ^{14}C dating). Absolute age was determined by Dr. E. Zazovskaya using radiocarbon method (^{14}C isotopes content) in the Radiocarbon Laboratory of the Institute of Geography of Russian Academy of Sciences (Moscow, Russia), laboratory index "IGAN."

Obtained radiocarbon dates were calibrated using the programme CALIB (version 7.1.0 ^{14}ChronoCentre, Queen's University Belfast) by means of calibration curve IntCal13 [3].

3 Results and Discussion

3.1 The History of Forest and Peatland Formation on Pregolya Glaciolacustrine Plain During the Holocene

As it shown on the Fig. 1, three various peatlands were chosen as key study sites for interpreting the history of forest and peatland formation on Pregolya Glaciolacustrine Plain, these are large raised bog Zehlau in the centre of the Kaliningrad Region and two wetlands in the District of Tchernyakhovsk: Podlipovsky peatland (under peat extraction) and Lake Maloye Olenye which is being at the moment terrestrialized with the buoyant mats (Fig. 1).

Zehlau raised bog. Peat bed is mostly composed of continuous peat layers which replace each other across the whole peat bed (Fig. 4), and herewith swamp and fen

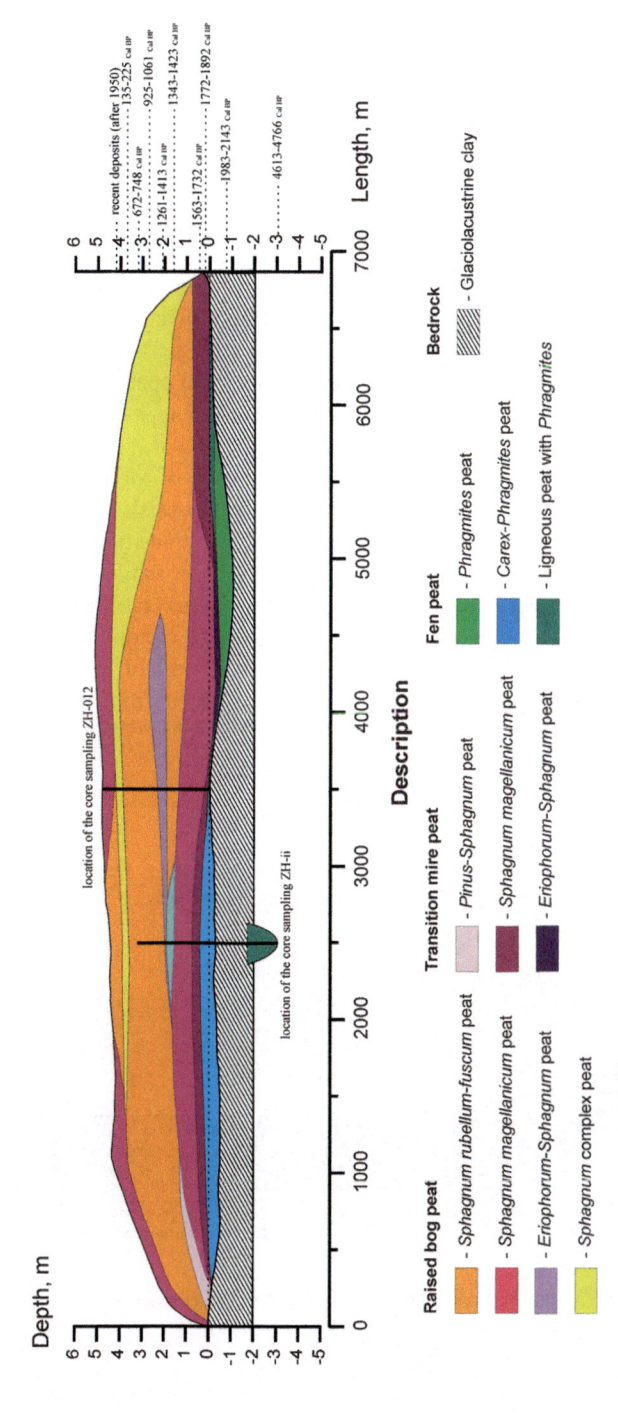

Fig. 4 Litho-stratigraphic cross-section for Zehlau raised bog

peat layers are very thin (15–80 cm) being abruptly replaced upwards by a thick layer (4–5 m) of *Sphagnum* moss peat with various decomposition grades. These peat layers have in the early going the character of transition mire deposits but pass rapidly into a raised bog peat. Thus, the whole Zehlau peat bed is to be defined as a thick (5–6 m) raised bog bed with the dominance of various kinds of *Sphagnum* moss peat.

Peat bed on a study part of Maloye Olenye Lake is represented by a thick floating mat which forms a broad belt of bog vegetation around the lake having originated as a result of terrestrialization process (Fig. 5). Being 3.5–4 m thick, the mat reaches its widest size – up to 100 m width – in the south-western part of the lake. The peat bed of the mat is separated from a thick near-bottom layer of gyttja deposits by a 3–5 m layer of water body with a free-flowing peat (Fig. 5).

Closer to the lake centre, the mat passes steeply to an open water surface. There is also a peat bed below, on a depth 80–150 cm, which appears to be a submerged central part of the mat.

A main peculiarity of the peat bed on riparian mire of Maloye Olenye Lake is complete absence of fen and swamp peat. The bed consists of two parts with a distinct boundary: the raised bog peat layer and the transition mire layer. Both layers contain *Sphagnum* mosses as dominant remnants, while their species composition in these horizons is different. It's also a remarkable fact that raised bog deposits build up lower two-thirds of the mat (120–350 cm on the coring site) and they are covered upwards by a thin layer (100–120 cm) of transition mire peat (Fig. 5). Such kind of peat bed structure is notably different from a general sequence of peat layers in the bed.

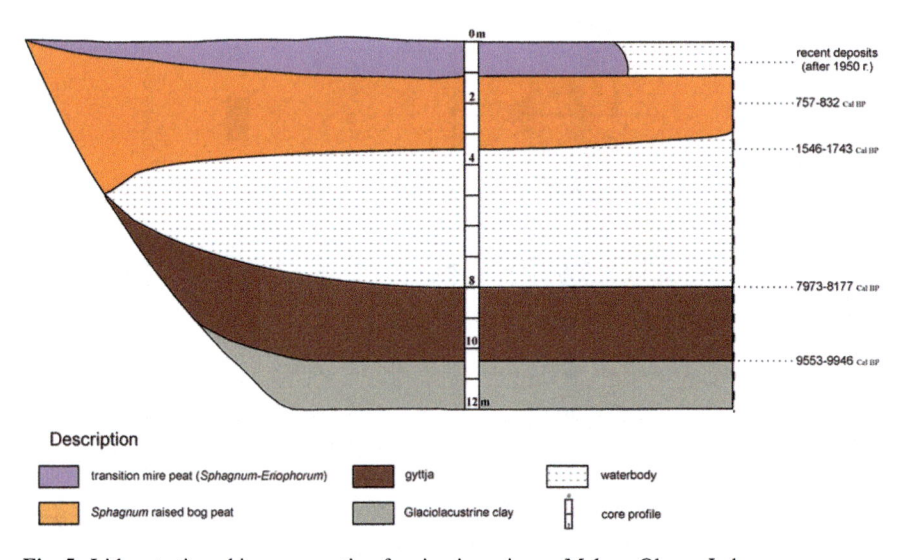

Fig. 5 Litho-stratigraphic cross-section for riparian mire on Maloye Olenye Lake

Thus, taking the peat bed structure on study sites as well as ^{14}C datings into account, the general pattern of mire ecosystem development on Pregolya Glaciolacustrine Plain can be considered as follows:

1. The early Holocene. Peat accumulation was absent in this area or had been in some places under extremely slow way. Wide-scale mire formation set up here only at the beginning of the Sub-Boreal.
2. Mire vegetation development originated here on a flat glaciolacustrine plain surface with sparse depressions, mostly from one "genetic centre."
3. Paludification processes were a prevailing mode of mire formation in that area having had alder swamps with a tall-grass ground layer of *Phragmites* and sedges as a primary peat accumulating mire community. Peat accumulation rate was low (0.4–0.5 mm per year).
4. Stage of swamp was very short on a flat surface of the glaciolacustrine plain, arboreal layer vanished rapidly from the primary mire communities having formed rather thin seams of ligneous peat.
5. Reed and sedge fens were the next mire stage that expanded here on vast areas after filling all the depressions with peat and levelling the surface. These communities formed 1–2 m thick fen peat layers which are in turn replaced with a transition mire peat.
6. Transition of mires into a raised bog stage occurred on Pregolya Glaciolacustrine Plain in the first half of the Sub-Atlantic. Raised bogs developed at a very quick rate in the area and formed thick (several metres in thickness) *Sphagnum* peat deposits which caused an essential degree of convexity of the bog surface. A peat accumulation rate had been increasingly growing up to 2–2.5 mm per year at the average.
7. During the last 200 years, in response to drainage ditching across the bogs in the area, their peat bed surface has sunk in 2–3 m having reduced bog domes convexity. In this way, raised bog Zehlau, which is found beyond the coastal area, obtained some features of "oceanicity" (in particular, flat central plateau) which are specific only to mires of coastal lowlands.

Pollen diagrams, generated for the study mire ecosystems (Figs. 6, 7, and 8), provide a foundation for the palaeoreconstruction of vegetation development on the territory of Pregolya Glaciolacustrine Plain:

1. According to the pollen spectra on the diagrams and radiocarbon data, it could be stated that cryophilic vegetation of the steppe-tundra was common on the territory of Pregolya Glaciolacustrine Plain at the end of Last Glacial and the Older Holocene (11,500–10,200 cal BP); at the same time, first communities with *Pinus* and *Betula* had been gradually spreading in the area.
2. In the Early Holocene (the Boreal period, 10,200–8,600 cal BP), *Pinus* and *Betula* forests prevailed here, whereas some thermophilic species were an essential part in their species composition.
3. Since the Middle Holocene (8,600 cal BP), deciduous broad-leaved forests with a dominance of *Quercus, Ulmus, Tilia* and *Corylus* had been expanding here as a

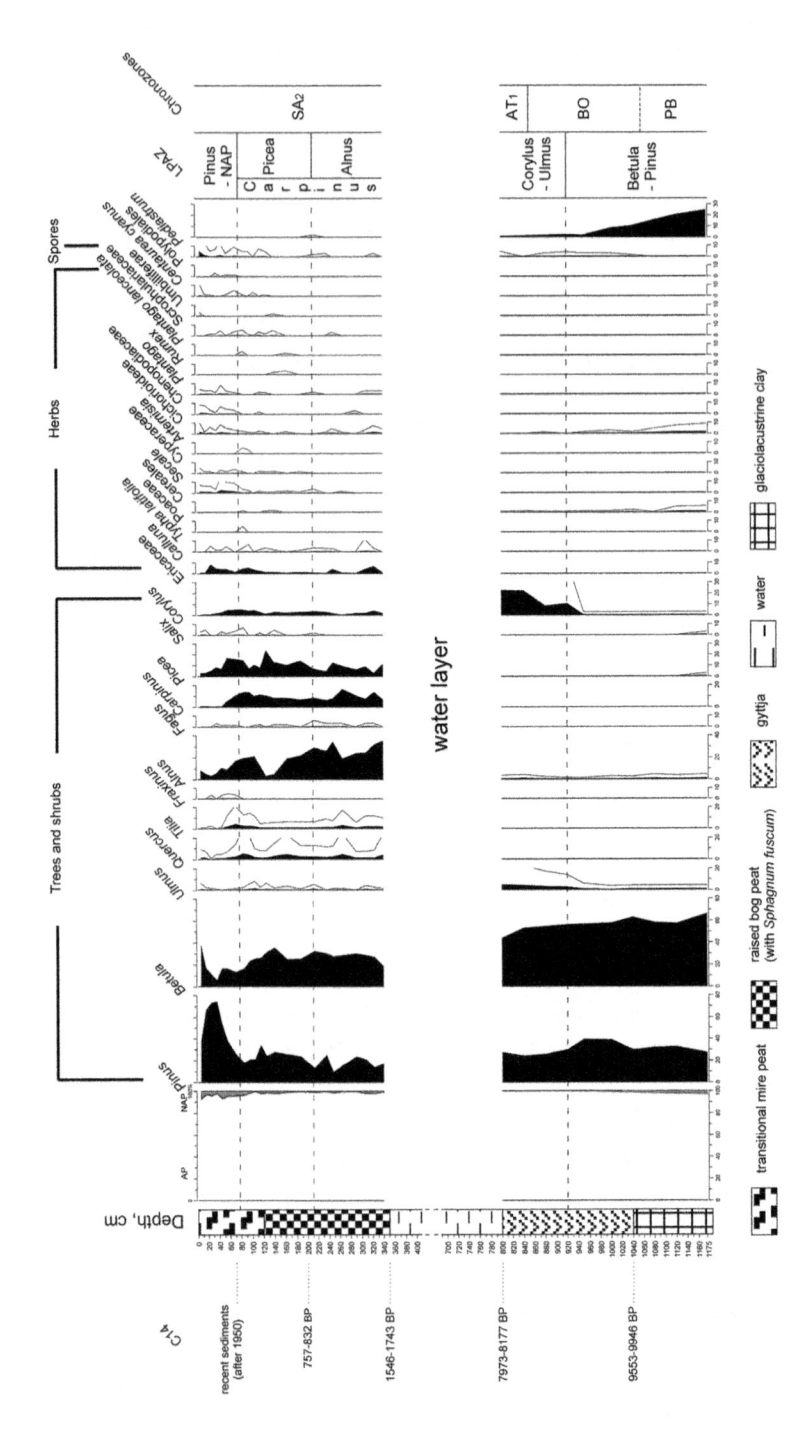

Fig. 6 Pollen diagram for Maloye Olenye Lake

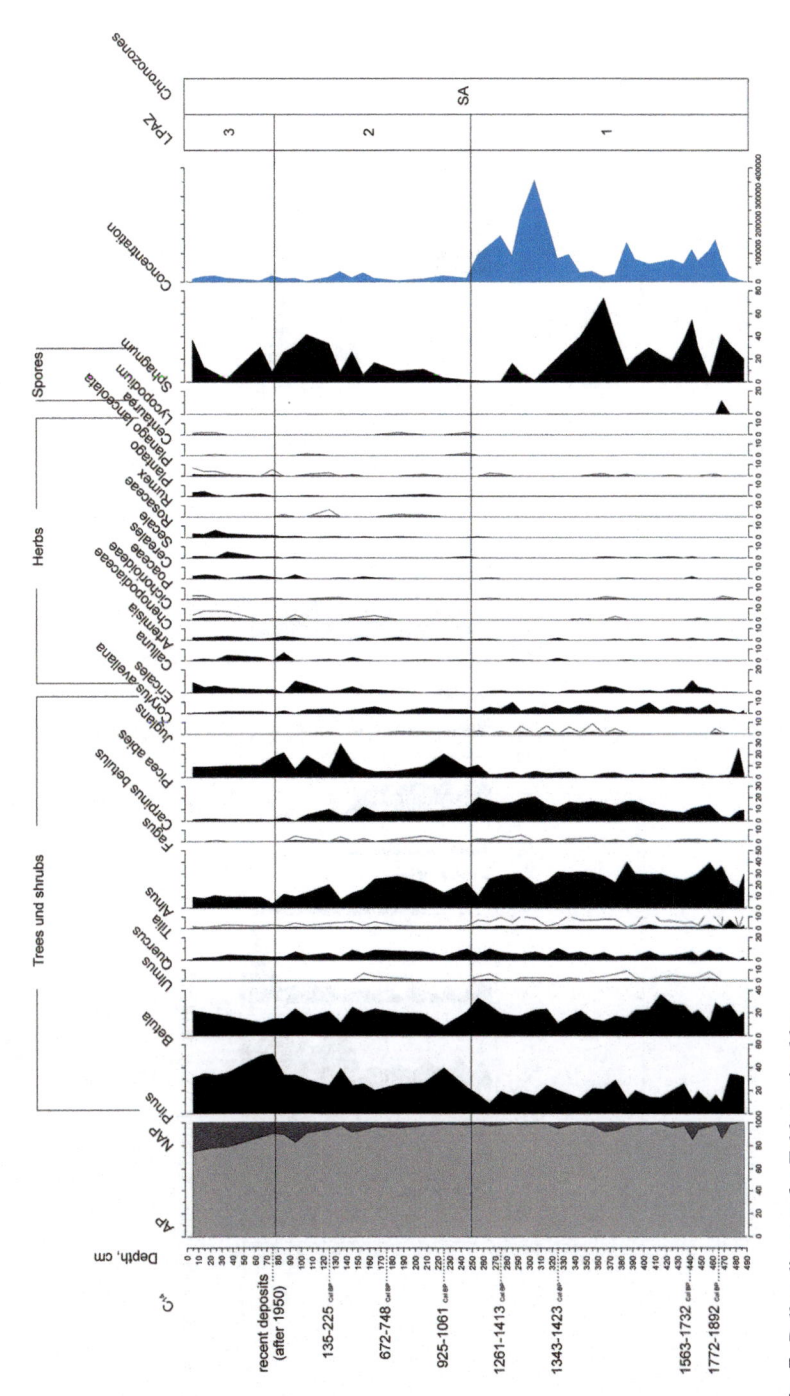

Fig. 7 Pollen diagram for Zehlau raised bog

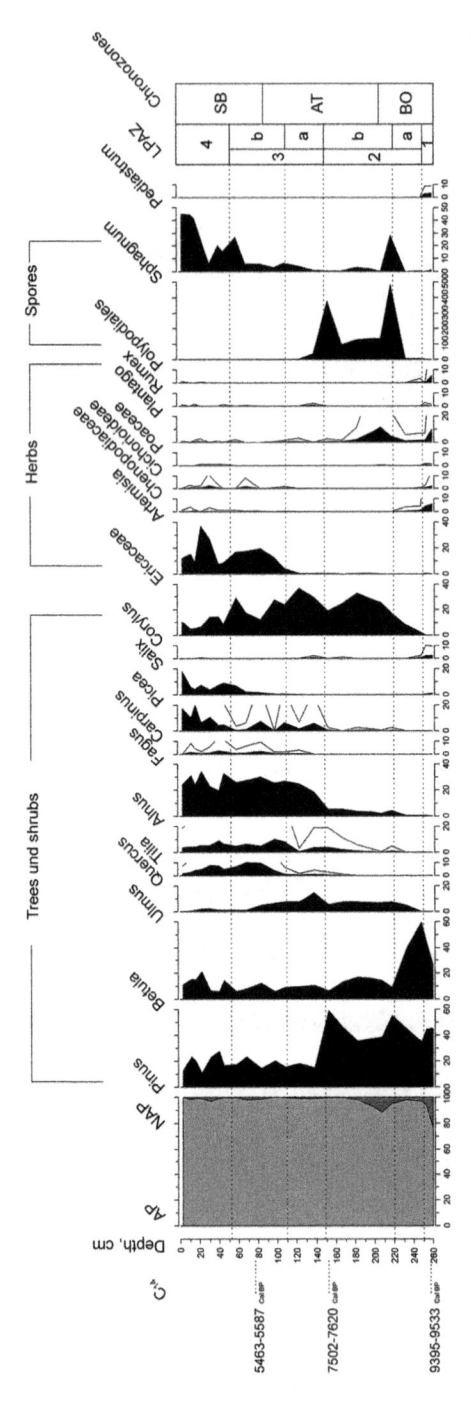

Fig. 8 Pollen diagram for Podlipovsky peatland

main vegetation type. These communities (so-called nemoral complex) became an overwhelming ecosystem in this landscape to the middle of the Atlantic.

4. Dominance of the "nemoral" broad-leaved forests kept on here as well in the Sub-Boreal (5,700–2,600 cal BP) indicating the fact of an adequate humidity in the area during this period.
5. Some cooling and humidification occurred in the Sub-Atlantic, though the first half of this period (2,600–1,200 cal BP) was, likely, warmer in the area. Nemoral broad-leaved forests kept their dominance but underwent certain alterations; in particular, the hornbeam (*Carpinus betulus*) and the beech (*Fagus sylvatica*) became significant components in their species composition. The latter species, according to the pollen diagrams, is thought to play an important role in the forest community structure on Pregolya Glaciolacustrine Plain up to the beginning of intensive land reclamation here in the Middle Ages.

   In general, species composition of arboreals was, apparently, richer at that time inasmuch as, it included some species which have now more southern distribution range (*Juglans sp.*, *Tilia platyphyllos*).
6. More essential cold snap occurred in the second half of the Sub-Atlantic (1,200–400 cal BP) which caused a reduction of broad-leaved and increase of coniferous as well as formation of special type of forest communities – the spruce-broad-leaved forests.
7. Since the seventeenth century (400 cal BP), strongly enhanced human activity effected harsh alterations in the forest cover on Pregolya Glaciolacustrine Plain which resulted in a large reduction of broad-leaved and spruce-broad-leaved forests in the area and, simultaneously, caused an increase of agricultural areas and synanthropic habitats as well as secondary pine and birch stands.

3.2 The History of Forest and Wetland Formation on Lower Neman Lowland During the Holocene

Peat bed structure of two large raised bogs – Bol'shoye Mokhovoye and Kozye which are located in the coastal zone of the Curonian Lagoon, in the southern and the northern part of the Neman Delta, respectively – gives an insight about wetland development in Lower Neman Lowland (Fig. 1).

Bol'shoye Mokhovoye raised bog has a very rugged floor topography in the mire depression which is separated by the ancient morainic humps onto several minor kettles with 4–9 m depth range. The peat bed is thick (4–11 m) and underlaid by morainic sands and adjacent silt underneath (Fig. 9).

Fen peat deposits are mostly 3 m thick increasing in the deepest kettles up to 5 m. Transition mire peat is performed with a thin layer (0.25–1 m) and overlaid by a thick raised bog *Sphagnum* peat bed (5–6 m mean thickness) with a dominance of *fuscum* peat.

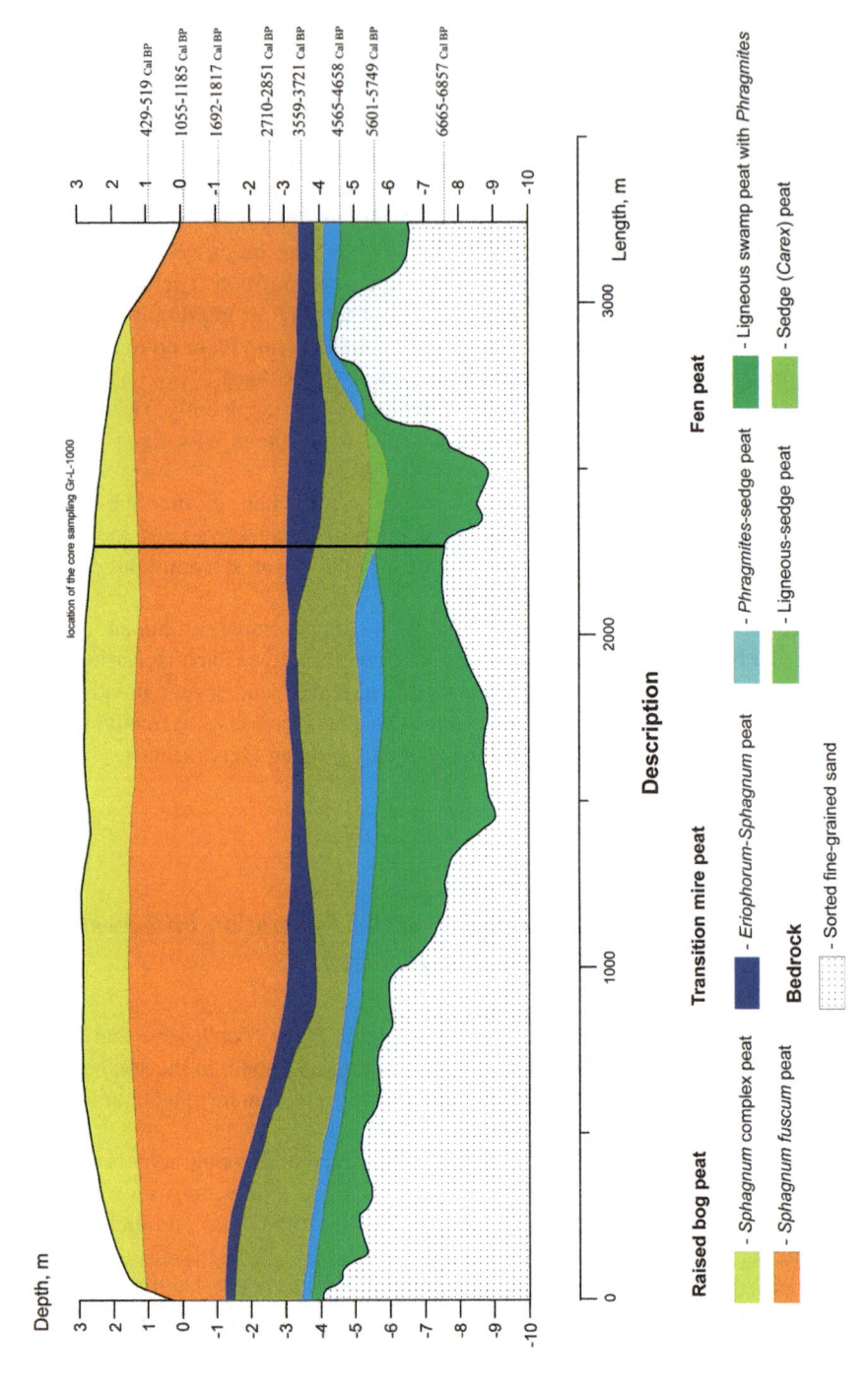

Fig. 9 Litho-stratigraphic cross-section for Bol'shoye Mokhovoye raised bog

Mire depression on Kozye raised bog is also separated onto two big kettles with a rugged floor as it is conspicuously seen on the cross-section (Fig. 10). The mire floor raises obviously in its centre forming a sandy ridge composed of the elder morainic terrace material.

Both kettles are filled with the fen and swamp peats in their lower parts (see litho-stratigraphic profile on Fig. 10), while in a centre of the bed, this kind of peat is being almost vanished and replaced with the thin interlayer of transition mire peat or directly with raised bog peat.

The raised bog peat layers compose a thick homogenous complex which covers all the kettles levelling relief topography of the mire depression and makes Kozye bog an entire mire area. The most part of the raised bog peat bed is also performed by *fuscum* peat.

Basing on above described peat bed structure along with the ^{14}C datings, the following main stages of Lower Neman Lowland development were defined:

1. In postglacial environment in Lower Neman Lowland with very rugged topography and large amount of ridges and kettles with different size and depth, the mire formation in the Neman Delta initiated in numerous inter-mound depressions of the ancient moraine. Their peat beds had been merging afterwards into entire mire bodies.
2. Peat accumulation took place in this area in the Early Holocene only along the coastal zone in the deep kettles which remained here after water level sinking in Curonian Lagoon at the end of Ancylus stage. This process went slowly on and did not occur in the other part of the area.
3. An intensive mire formation commenced in the Neman Delta only at the end of the Atlantic (5,800 cal BP).
4. The paludification processes prevailed though initial stages of mire formation in the deep kettles were connected with the terrestrialization of small water bodies. In the latter case, hydrophilic *Hypnum* moss communities became the first peat-forming associations, while in the paludificated areas these were wet alder swamps with tall reed and sedge stands in the ground layer.
5. The stage of swamp was rather long-term having provided thick layers of ligneous peat.
6. As far as thickness of ligneous peat layers increased alder swamps had been replacing into vast reed and sedge fens, many primarily separate mire "genetic" centres merged, having formed the an entire peat bed.
7. Mire transition into a raised bog stage occurred in Lower Neman Lowland only at the end of the Sub-Boreal. Raised bogs developed here very rapidly due to very fast increase of peat accumulation rates which resulted into a formation of thick *Sphagnum* peat beds with a distinct convexity.
8. During last 200 years, since implementation of plans for wide-scale drainage in the area, numerous melioration ditches were established around the mires having affected their peat bed dome convexity. Peat body surface sank in 2–3 m and it was recorded in Pregolya Glaciolacustrine Plain.

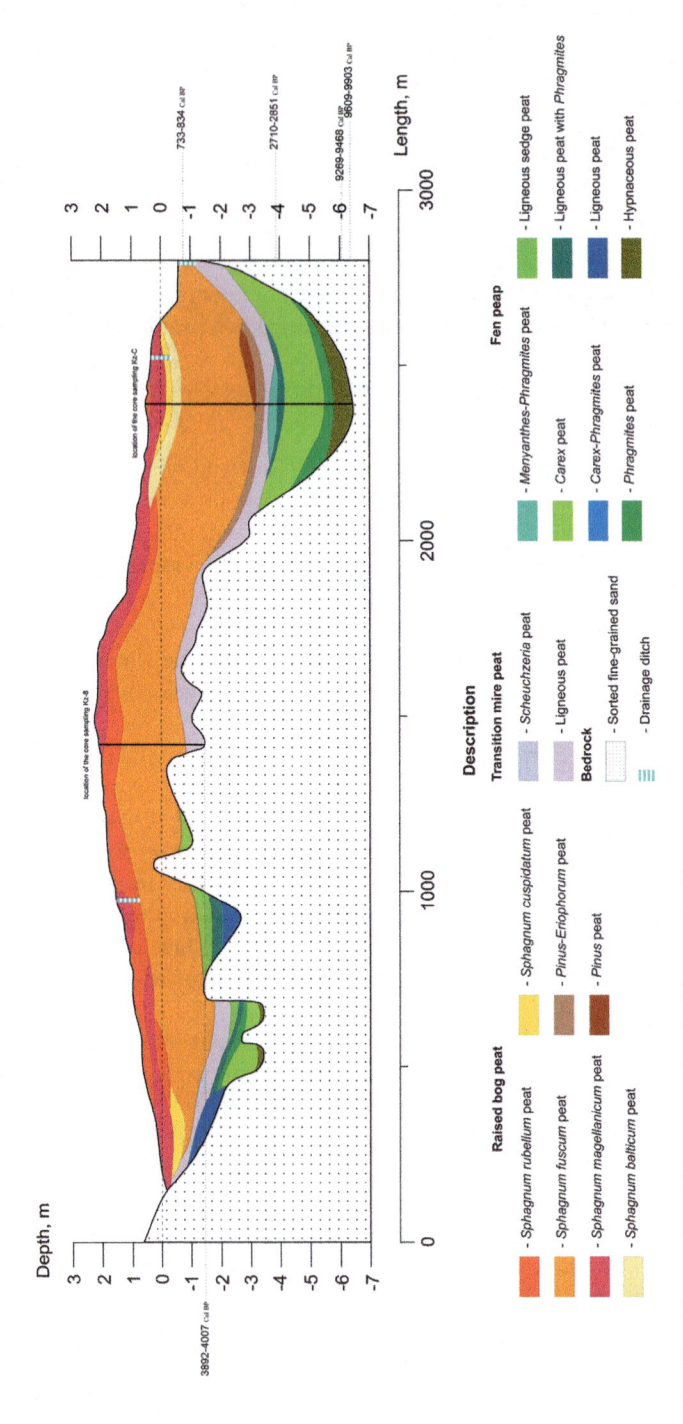

Fig. 10 Litho-stratigraphic cross-section for Kozye raised bog

9. Peat accumulation rates varied substantially during different mire development stages in this region. It was approximately 2 mm/year during alder swamp stage in the Atlantic, subsided up to 1 mm/year in the Sub-Boreal, while it rose up to 2–2.7 mm/year at *Sphagnum* raised bog period in the Sub-Atlantic.

Pattern of the forest vegetation development in the Lower Neman Lowland was defined basing upon a pollen spectra analysis carried out for Bol'shoye Mokhovoye mire (Fig. 11) which is considered as a reference ecosystem in the area:

1. The Early Holocene events and first half of the Atlantic are absent in the peat deposits of Bol'shoye Mokhovoye mire.
2. The second part of the Atlantic (AT_2, 7,400–5,700 cal BP) is usually considered as a climatic optimum of the Holocene in the area. There has been recorded a distribution maximum for the broad-leaved of the "nemoral complex" with a rich species composition (*Quercus, Ulmus, Tilia, Corylus*) as well as maximum for *Alnus* which corresponds to the active successions of the alder swamps on the litho-stratigraphic cross-section (Fig. 9).
3. The environment at the end of the Atlantic was similar to those during the Sub-Boreal and at the beginning of the Sub-Atlantic (AT2-SB-SA_1, 5,700–2,600 cal BP). The climatic conditions in the Lower Neman Lowland were still mild and humid with a strong influence of adjacent Curonian Lagoon so broad-leaved forests of the "nemoral complex" (*Quercus, Ulmus, Tilia, Corylus*) kept their predominant distribution and rich species composition. Alder and pine forests reduced their abundance, while *Picea abies* and *Carpinus betulus* were permanently spreading to the area.
4. From the beginning of the Sub-Atlantic till the middle of the eighteenth century (SA_1, 2,600–250 cal BP), some cold snap and humidification are thought to occur in the area; meanwhile the climatic environment in the first half of the Sub-Atlantic was, apparently, warmer. The broad-leaved forests were still most common vegetation type but they changed their structure due to integration of *Picea* and *Carpinus* which have increased abundance in that region. A new type of deciduous forests, the spruce-broad-leaved, became common in the study landscape.

    Abundance of alder swamps increased also during this period. That fact is, apparently, connected with a shift of the Neman River mouth into study area. This changed hydrological environment and caused a raise of paludification. In general, hard approachability of Lower Neman Lowland makes forest use here weakly possible and prevented clearing of pristine deciduous forests as well as their replacement with the secondary pine and birch forest stands. That fact explains a long-term distribution of the broad-leaved and spruce-broad-leaved primary forests in this landscape in the Late Holocene.
5. The Latest Holocene (from the middle of the eighteenth century till the present time) is clearly bound up with an intensive human activity that caused distinct alterations within the vegetation structure in the area, in particular, a substantial reduction of primary broad-leaved and spruce-broad-leaved forests, drainage on

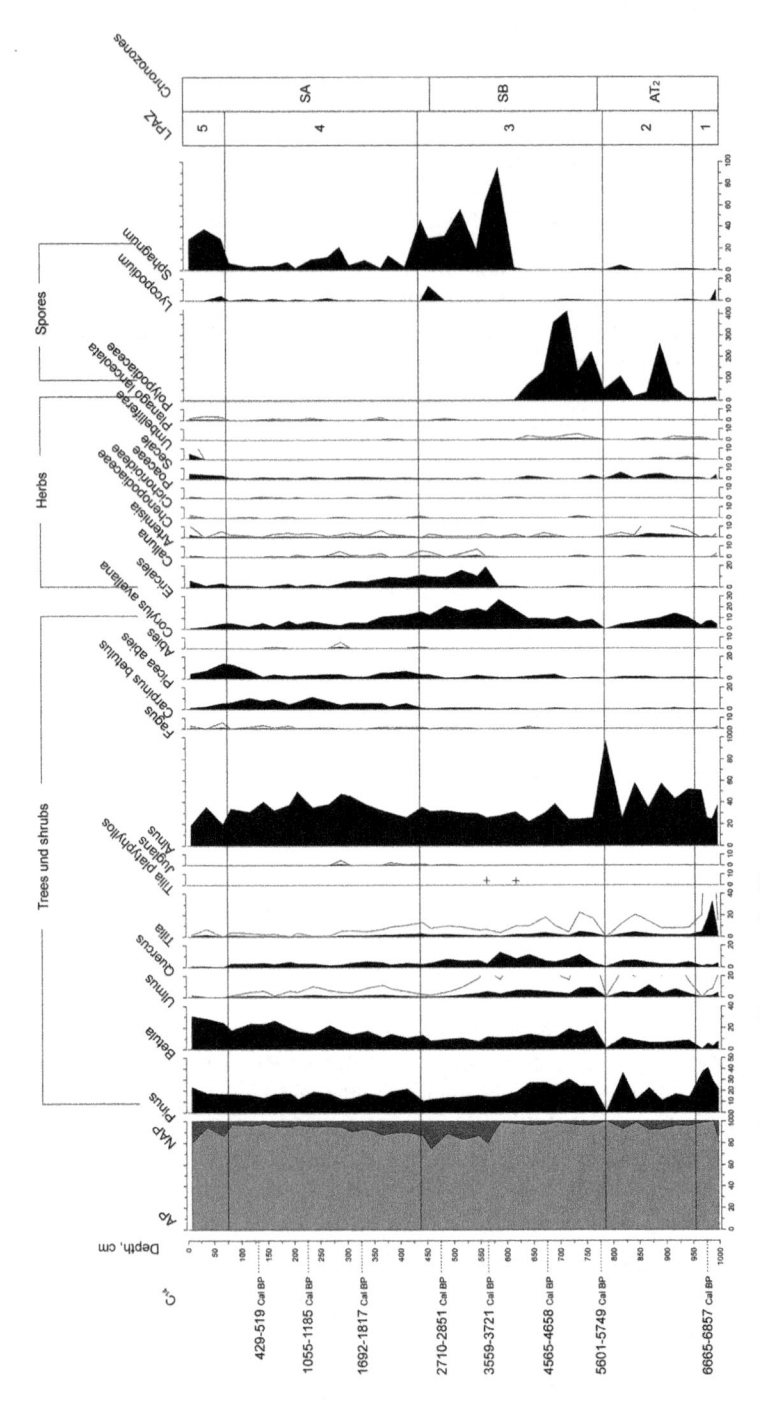

Fig. 11 Pollen diagram for Bol'shoye Mokhovoye raised bog

vast areas and their reclamation for the agricultural use. Secondary vegetation – birch and pine forests – became widespread.
6. Set up pattern of vegetation development in Lower Neman Lowland during last 5,500 years, based at first on course of pollen spectra and sequence of peat deposits, allow to state that this territory, at least its southern part, despite of its minor elevation above the sea level, often storm events and Littorina Sea transgression, had not been flooded by coastal water of the Curonian Lagoon in that period.

3.3 Comparison of the Palaeoreconstructions in the Study Areas of the Kaliningrad Region with the Environmental Development in Adjacent Regions During the Holocene

Outcomes of the researches provide a foundation for comparison of the environment development in the study areas with the data on landscape evolution in the neighbouring countries as well as for contribution to general insights on environment formation on the territory of the Kaliningrad Region.

Pollen diagrams generated for the mires in the southern part of the Kaliningrad Region, i.e. Pregolya Glaciolacustrine Plain, can be united with the diagrams from Northern Poland though the latter contain to some extent more *Carpinus* and *Fagus*. Nevertheless, these species are performed on the diagrams by continuous curves having certain quantitative value on the spectra in the Late Holocene. These species are thought to be an important component of the broad-leaved and spruce-broad-leaved forests in the southern part of the Kaliningrad Region.

Pollen diagrams from the northern area of the Neman Delta are rather to be united with the Lithuanian diagrams, inasmuch *Carpinus* and *Fagus* are not so distinct components in a species composition there. Consideration of pollen diagrams for the other northern areas in the region [4, 5] allows to extend this conclusion onto a whole northern part of the Kaliningrad Region of Russia.

Thus, the Kaliningrad Region, from the palaeoenvironmental point of view, could be divided onto two different parts, each of those is to be united with the adjacent regions of Poland and Lithuania. A boundary between these parts stretches nearly along the rivers Deima, Pregolya and Pissa. It is, apparently, to be considered as a boundary between the zone of Hemiboreal forests (Sub-Taiga) and the zone of Middle-European and East European deciduous forests within the phytosociological (geobotanical) regional subdivision system (Fig. 12).

This conclusion concerns also a question on zonal status of the alder swamps in the coastal area of the Neman Delta along the Curonian Lagoon. These ecosystems are usually considered as belonging to the nemoral vegetation (i.e. a part of the zone of deciduous forests) [6].

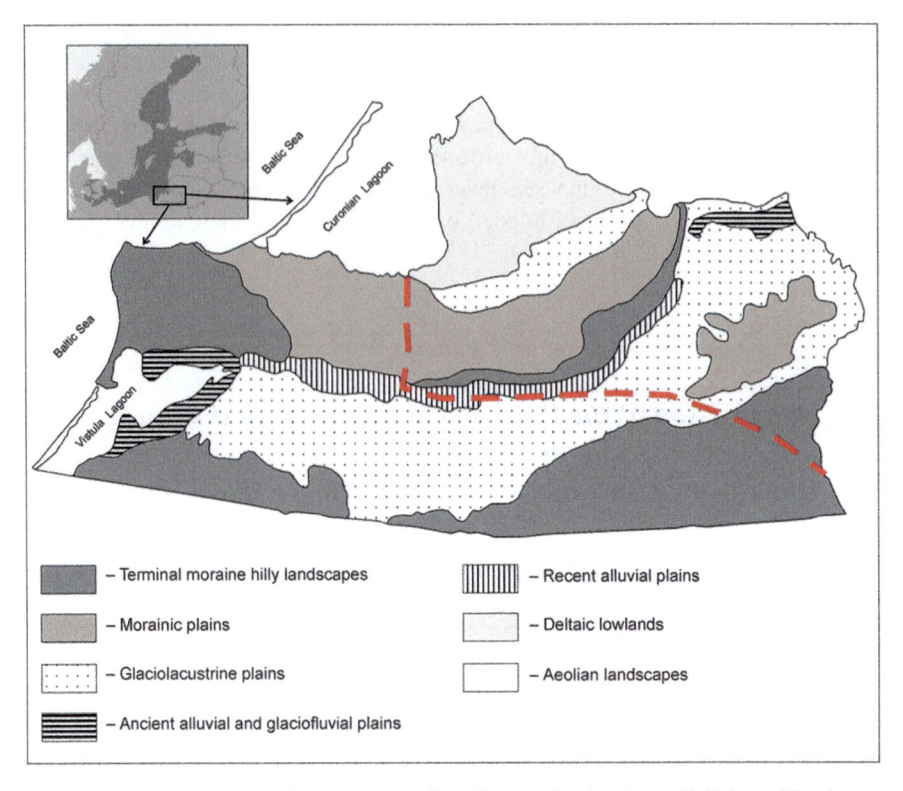

Fig. 12 Suggested boundary between two pollen diagram districts in the Kaliningrad Region

At the same time, according to the obtained pollen spectra, this territory within the Lower Neman Lowland area ought to be referred to the Hemiboreal zone.

Taking into account a formation history of this area as well as a habitat character of the alder swamps which is being flooded with the lagoon water during storm surges, these ecosystems are, likely, to be considered as extrazonal nemoral vegetation within the Hemiboreal (Sub-Taiga) zone.

Wide spread of the alder swamps in the Neman Delta area is thought to be bound with a rise of the lagoon water level during the Littorina Sea period. This fact coincides with the pollen amount increase of *Alnus* on the diagram for Bol'shoye Mokhovoye raised bog (Fig. 11). The same increase of *Alnus* pollen amount is also recorded in the adjacent regions of Lithuania [2] and Poland [7, 8] which reflects an intensive spreading of alder swamp communities, apparently, caused by flooding of the coastal areas in Lower Neman Lowland.

These processes changed, probably, the erosion base level of rivers in the whole region which led to rewetting of many topography depressions and caused rapid *Alnus* growth in these habitats. This fact is obviously confirmed by our pollen diagrams (Figs. 8 and 11) as well as by data on botanical composition of peat in the lower layers on Zehlau raised bog that are mainly performed by strongly decomposed swamp peat with numerous ligneous remnants of *Alnus*, i.e. before peat accumulation launch at 4,800–4,600 cal BP in the lowest depressions, this area were already covered with communities.

High ground water level and often storm run-up resulted in a substantial rewetting in the area of the Neman Delta, having forced a mire formation process on that territory and, consequently, caused a wide extrazonal expansion of alder swamps which are one of the distinctive landscape features in Lower Neman Lowland.

Another question under discussion in recent times is a reason of distribution range disjunction the deciduous and conifer forests. According to S. Turubanova [9], the spruce (*Picea*) was an inherent component of the broad-leaved forests during all the Holocene; therefore spruce-broad-leaved forests were, likely, the main biome type in a whole region of Northern Eurasia in the Holocene. It is also confirmed in our investigation, in particular, such kind of ecosystem became, apparently, a widespread in the region since the Sub-Boreal, just in this time pollen diagrams record an expansion of *Picea* on the territory of South-Eastern Baltic from the northern and eastern regions. But herein, the "nemoral complex" species were also common and abundant in these ecosystems up until the fifteenth to sixteenth centuries both in southern and northern parts of the region (Figs. 6, 7, 8, and 11, see also [4]).

Disjunction of deciduous and conifer forests occurred during second half of the Late Holocene as a result of strongly enhanced human activity. Regular burning out aggravated growth conditions either for the broad-leaved tree species and for the whole "nemoral" flora. These affects coincided in time with some cold snap in the Late Holocene (Little Ice Age) that in turn changed northern distribution range of most "nemoral" species and caused its shift southwards. According to our data, this range differentiation came in South-Eastern Baltic Region along with the elimination of beech (*Fagus*) from the forest ecosystems.

Thus, the outcomes of our investigation enable to draw more exactly the boundaries of zonal vegetation units which were naturally formed in the area in the Late Holocene, at the same time, they were affected essentially under intensive human impact in the latest phase of the Sub-Atlantic. Recent retreat of these boundaries is connected with a pressure on forest resources during the last 200 years.

4 Conclusions

Based on the outcomes of our investigations and taking into account their comparative analysis with some key literature references [4, 5, 10, 11], the main pattern of forest and peatland formation in the two largest landscape areas in the Kaliningrad Region during Holocene can be stated in the following lines:

1. In the Pre-Boreal (11,500–10,200 cal BP), the territory of the Pregolya Glaciolacustrine Plain was occupied by cryophilic vegetation of steppe-tundra, while spreading of pine and birch communities went gradually on.

In the Boreal (10,200–8,600 cal BP), pine and birch forests were already predominant in the area, but their structure included an essential part of thermophilic species.

For the territory of the Neman Delta, events of these time periods are absent in peat deposits on Bol'shoye Mokhovoye raised bog.

2. In the Atlantic (8,600–5,700 cal BP), the deciduous broad-leaved forests with a rich species composition had been expanding in both study areas as a main vegetation type. The core element in these ecosystems was so-called nemoral complex of species (*Quercus, Ulmus, Tilia, Corylus*). Such kind of forests covered the most part of an even territory in Pregolya Glaciolacustrine Plain, while in Lower Neman Lowland, they occupied only uplands being replaced in depressions by the alder swamps (*Alnus glutinosa*) that shared a dominance with the "nemoral complex" forests in this landscape.

The deciduous broad-leaved forests continued to prevail in both territories still in the Sub-Boreal (5,700–2,600 cal BP), inasmuch as a climate drought in that period, likely, not so obviously existed here as it is assumed for the other regions, and a moistening was more essential.

3. Some cooling and humidification occurred in the Sub-Atlantic; herein its first half (2,600–1,200 cal BP) was characterized by warmer environment. Deciduous forest kept their dominance but underwent some structural alteration – the hornbeam (*Carpinus betulus*) became an important element in the species composition in both landscapes, while in the southern part of the region, in Pregolya Glaciolacustrine Plain, the beech (*Fagus sylvatica*) became also one more key element of the broad-leaved communities. This species is not considered here as a remote strange, quite opposite – a distinct component of the deciduous forests in that territory before its intensive reclamation since mediaeval time. The common species composition of the broad-leaved was still richer and included some species that have now more southern range (*Juglans sp., Tilia platyphyllos*).

4. In the second half of the Sub-Atlantic (1,200–400 cal BP), more essential cold snap caused a reduction of broad-leaved and increase of coniferous in both study areas which resulted in a formation of a new forest type – the spruce-broad-leaved forests. The pure broad-leaved of the "nemoral complex" altered to a specific deciduous forest subtype – the oak-hornbeam-lime forests which are considered as a zonal forest vegetation peculiar for the whole Eastern European Region within the belt of deciduous forests.

In this respect, vast areas of alder swamps in Lower Neman Lowland, being a peculiar feature in this landscape, could be considered as an extrazonal broad-leaved vegetation within the Hemiboreal (Sub-Taiga) zone.

5. Initial stages of mire formation were performed in two study areas in different way, since wetland development started in each landscape in a very different geomorphologic environment.

Peat accumulation began in Lower Neman Lowland in separate kettles ("genetic centres") which were abundant in rugged relief of the ancient moraine but isolated from each other by numerous ridges and mounds with different sizes

(Figs. 9 and 10), [10]. The bottom of most kettles is situated lower the lagoon level and the area is, therefore, weakly naturally drained that determined a long-term duration of the swamp and fen stages with the dominance of alder and reed communities. Only having filled with peat a whole kettle and having formed distinct *Sphagnum* peat layers, separate peat beds merged into the entire mire system (Figs. 9 and 10).

In glaciolacustrine landscapes in the central part of the region, mire formation initiated on a flat plain surface with sparse depressions, mostly from one "genetic centre" (Fig. 4). A stage of swamp lasted for a very short period; herein, arboreal layer was rapidly extinct from the primary mire communities, having been replaced with vast *Phragmites* fens, and then in a short period with *Sphagnum* communities of the raised bog (Fig. 4) [4].

6. Large raised bogs developed on the territory of the Kaliningrad Region mainly via paludification processes, nevertheless, mires in the coastal area along Curonian Lagoon formed in various ways during their initial stages, and these bogs are, therefore, the ecosystems of complex genesis.

Paludification origin for mires in glaciolacustrine plains is proved by flat bottom relief of mire kettles, indistinct mire depressions in landscape topography, common mire location on watersheds, lack of gyttja deposits and species composition of plant remnants in peat (Fig. 4), [4].

Paludification is also confirmed for many mire kettles in Lower Neman Lowland but a number of them, usually the deepest ones, are proved to be of a lacustrine genesis as having deposits of gyttja and hydrophilic hypnaceous moss peat. Having formed in separate kettles, peat beds of various structures merged into the entire mire body only at the stage of the raised bog in the Late Holocene (Figs. 9 and 10) after extremely rapid *Sphagnum* peat growth.

7. Mires in the coastal zone were affected by transgressions of the Baltic and its lagoons which resulted into formation of alternating layers of gyttja, mineral alluvium, shell deposits and water veins in the lower part of peat bed [10]. Mire formation is thought to occur here via a terrestrialization of the shallow lagoon basin, a process which is considered now [12, 13] as a peculiar type of mire development, so-called primary mire formation or Küsten-Überflutungsmoore, that has been still taking place in the coastal zone of South and Eastern Baltic.

Thus, high ground water table and storm run-up effect caused a significant inundation in the area and acceleration of mire formation process in this landscape while the most territory of the Lower Neman Lowland was not flooded with the lagoon water during last 5,500 years.

8. Peat accumulation rates varied substantially on different stages of mire development in study landscapes in the coastal area and glaciolacustrine plains.

In general, peat accumulation in coastal landscapes during the Early and the Middle Holocene was higher than in central part of the Kaliningrad Region (2 mm per year in the Atlantic and 1 mm per year in the Sub-Boreal for alder swamps in Lower Neman coastal lowland while 0.4–0.5 mm per year in the same ecosystems in Pregolya Glaciolacustrine Plain during the Sub-Boreal).

Nevertheless, peat accumulation rate increased up to 2–2.7 mm per year during raised bog peat formation in the Sub-Atlantic in both landscapes.
9. Having been affected by human activity, vegetation structure suffered an essential change on the territory of Pregolya Glaciolacustrine Plain since the beginning of the seventeenth century (400 cal BP), while in Lower Neman Lowland, such alterations are recorded only since the middle of the eighteenth century (250 cal BP), apparently, due to its hard approachability that restricted human impact onto pristine forests in this area. Intensive land use and clear cutting resulted into a large reduction of broad-leaved and spruce-broad-leaved forests in the area and, simultaneously, caused an increase of agricultural areas and synanthropic habitats as well as secondary pine and birch stands in both study landscapes.

Since establishing the melioration ditches across the most mires in the region, the bog peat beds sank in 2–3 m, having caused a reduction of the mire dome convexity. Primarily convex large raised bogs in Pregolya Glaciolacustrine Plain, in particular Zehlau, became more flat in the centre (bog plateau) having obtained some "oceanicity" features that are proper only to the coastal mires in the Lower Neman Lowland.
10. According to studied pollen spectra, the territory of the Kaliningrad Region doesn't form an entire palaeoenvironmental district, and is to be divided, in this respect, onto two different parts each of those could be united with the neighbouring regions of Poland and Lithuania. A boundary between these two identified palynological districts is stretched along the rivers Deima, Pregolya and Pissa. They are palynologically similar in the Early and the Middle Holocene, while they had been obtaining distinctions in the Late Holocene when conifers became dominating in the north-eastern part of the region having gradually replaced communities of the "nemoral complex," but the latter remained common in the south and especially in the south-western parts of the Kaliningrad Region where they were key components of forest vegetation in the Late Holocene together with *Carpinus* and *Fagus*.

References

1. Faegri K, Iversen J (1989) Textbook of pollen analysis, The Blackburn Press, 328 p
2. Gudelis VK (1975) Palaeogeographical scheme for the late glacial and postglacial period in the Eastern Baltic Region [Skhema paleogeograficheskogo razvitiya Pribaltiki v pozdne-poslelednikovoye vremya]. In: Gudelis VK, Emelianov EM (eds) Geology of the Baltic Sea, Mokslas, Vilnius, 380 p. (in Russian)
3. Reimer PJ, Bard E, Bayliss A, Beck JW, Blackwell PG, Bronk RC, Buck CE, Cheng H, Edwards RL, Friedrich M, Grootes PM, Guilderson TP, Haflidason H, Hajdas I, Hatté C, Heaton TJ, Hoffmann DL, Hogg AG, Hughen KA, Kaiser KF, Kromer B, Manning SW, Niu M, Reimer RW, Richards DA, Scott EM, Southon JR, Staff RA, Turney CSM, van der Plicht J (2013) IntCal13 and Marine13 radiocarbon age calibration curves 0–50,000 years cal BP. Radiocarbon 55(4):1869–1887

4. Arslanov KA, Druzhinina O, Savelieva L, Subetto D, Skhodnov I, Dolukhanov PM, Kuzmin G, Chernov S, Maksimov F, Kovalenkov S (2011) Geochronology of vegetation and paleoclimatic stages of South-East Baltic coast (Kaliningrad region) during middle and late Holocene. Geochronometria 38(2):172–181
5. Bitinas A, Druzhinina O, Damušytė A, Napreenko-Dorokhova T, Guobytė R, Mažeika J (2017) The lower reaches of the Nemunas River at the end of the Last (Weichselian) Glacial and beginning of the Holocene. Geol Q 61(1):156–165
6. Yurkievich ID, Geltman VS, Lovchiy NF (1968) Types and associations of the black alder forests (according investigations in Byelorussia). "Nauka I tekhnika," Minsk, 376 p. (in Russian)
7. Miotk-Szpiganowicz G, Zachowicz J, Uścinowicz SZ (2008) Review reinterpretation of the pollen and diatom data from the deposits of the southern Baltic lagoons, vol 23. Polish Geological Institute Special Papers. Gdańsk, pp 45–70
8. Ralska-Jasiewiczowa M, Latałowa M, Wasylikowa K, Tobolski K, Madeyska E, Wright H, Turner CH (eds) (2004) Late Glacial and Holocene history of vegetation in Poland based on isopollen maps. Polish Academy of Sciences, Kraków, 444 p
9. Turubanova SA (2000) European forests in the Holocene – what have there been? Forest Bull 4 (16):C.17–19. (in Russian)
10. Gams H (1932) Schriften der Phys.-ökon. Gesellschaft zu Königsberg i. Pr. B. 67. H. 3/4: S.74–88. (in German)
11. Gams H, Ruoff S (1929) Schriften der Phys.-ökon. Gesellschaft zu Königsberg i. Pr. B. 66. H 1:S.1–193. (in German)
12. Sjörs H (1983) Mires: swamp, bog, fen and moor. In: Gore AJP (ed) Regional studies. Elsevier Scientific, Amsterdam, pp 69–94
13. Succow M, Joosten H (2001) Landschaftsökologische Moorkunde, Stuttgart, 622S. (in German)

Eolian Coastal-Marine Natural Systems in the Kaliningrad Region

I.I. Volkova, T.V. Shaplygina, N.S. Belov, and A.R. Danchenkov

Abstract The landscape structure of natural dune complexes is given on the example of the Curonian and the Vistula Spits. The main natural factors determining the transformation of the Eolian coastal-marine natural complexes (NC) were considered. Original techniques are presented with a view to assessing the digression, the potential and integrated (modified) resistance of eolian coastal-marine natural systems considering their natural features, the current state resulting from a long-term influence of natural and anthropogenic factors, determining the direction and progress of the evolutionary transformation of the Spits' landscapes, as well as the peculiarities of their use. Spatial differentiation of the Eolian coastal-marine natural complexes was fulfilled by indicators of digression and sustainability. The morpholithodynamic changes in the coastal sea zone were assessed at key sites of the Curonian and the Vistula Spits, using the technology of surface laser scanning. The historical preconditions were considered, and the present-day nature management at the Curonian and the Vistula Spits was analyzed as well as areas of its optimization.

Keywords Assessing, Digression, Eolian coastal-marine natural complexes (NC), Resistance, Surface laser scanning (SLS)

I.I. Volkova (✉), T.V. Shaplygina, and N.S. Belov
Department of Geography, Nature Management and Spatial Development, Immanuel Kant Baltic Federal University, Kaliningrad, Russia
e-mail: volkova.bfu@yandex.ru; tshaplygina@gmail.com; belovns@gmail.com

A.R. Danchenkov
Immanuel Kant Baltic Federal University, Kaliningrad, Russia

Laboratory of Geoecology, The Atlantic Branch of Shirshov Institute of Oceanology, RAS, Kaliningrad, Russia
e-mail: swdartvader@mail.ru

Contents

1 Introduction 148
2 Research Methods 150
3 Results 160
4 Conclusions 174
References 176

1 Introduction

The Curonian and the Vistula Spits are large accumulative forms of the shore of the South-East Baltic (Fig. 1). The Curonian Spit with the length of 98 km lies within the Kaliningrad Region (49 km) and Lithuania; its width ranges from 0.4 km north of the village of Lesnoy to 3.8 km near the village of Nida (Fig. 2) [1–5]. Out of 55 km of the Vistula Spit, 25 km is part of the territory of the Kaliningrad Region (the Baltic Spit), 30 km belongs to Poland; its width ranges from 0.3 to 1.8 km (Fig. 3) [2, 6, 7].

The Curonian and the Vistula Spits are made up of Quaternary deposits of modern (Q_{IV}) and upper (Q_{III}) sections [8, 9].

In accordance with the general trends of formation of dune complexes in the terrain of the Spits in the direction of the sea-lagoon, the following geomorphological zones are identified: a seabeach, a coastal dune ridge complex (a foredune and a coastal dune ridge), a pre-dune flat land (palve), dune bodies, a lagoon platform (lagoon palve), and a beach adjacent to the lagoon [10, 11].

The terrains of the Curonian and the Vistula Spits in their present-day outlook have been shaping up during the past several 100 years under the influence of natural and anthropogenic factors, and they are of eolian coastal-marine kind [12]. The structural organization of the terrains of the Spits features a similarity in general terms at the level of large morphologic units in the eolian coastal-marine terrain (areas), displaying a spatial replaceability in the direction of the sea-lagoon: a coastal stripe (beach area), coastal dunes (a foredune, a coastal dune ridge), a pre-dune flat land (palve), dune ridges, and a lagoon platform [7, 12]. The main differences were identified between smaller terrain structures – natural boundaries and clusters of natural boundaries: on the Vistula Spit – the lack of wandering dune massifs located in the near-lagoon area; a cavernous nature of the palve as a result of transversely oriented dune ridges; on the Curonian Spit – a weak manifestation of the coastal dune ridge; the presence of natural complexes developing on moraine loams; of a high moor; lakes, etc. The differences in smaller morphologic units of the Spits' terrain primarily concern their morphometric, biocenotic, and soil characteristics [13].

The current outlook of the dune terrains of the Spits is largely determined by morphologic structures being stabilized by the vegetative ground cover.

The main features of the Spits' soils are their youth and the presence of buried horizons developed as a result of sand movements [7]. The soils on the Spits are of four types: podzolic, moory-podzolic, sod-gleyed, and high peaty moory, of which

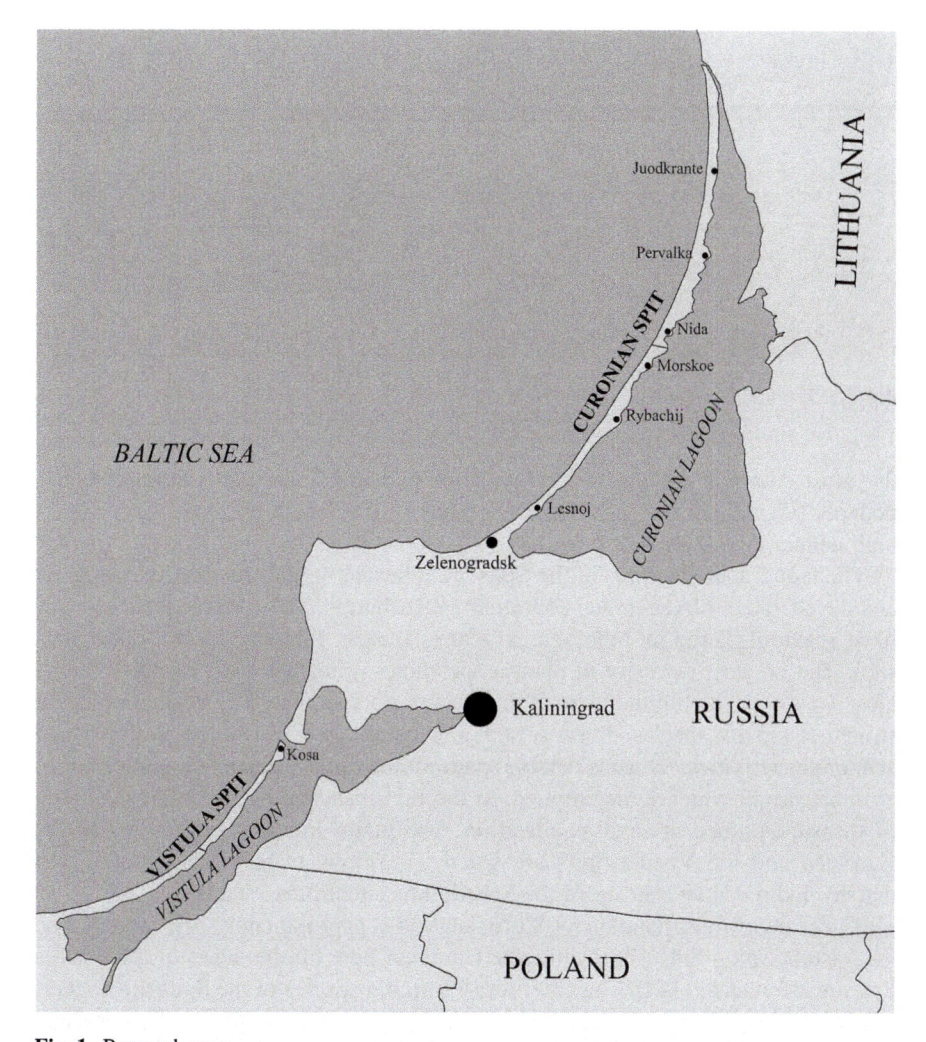

Eolian Coastal-Marine Natural Systems in the Kaliningrad Region

Fig. 1 Research area

Fig. 2 Curonian Spit

Fig. 3 Vistula Spit

the most common are the two subtypes of podzolic soils (podzolic and soddy-podzolic) developing on positive topographic forms (dune hilly surfaces, ridges, high undulating surfaces of palve).

The topographic features of the Spits are reflected in the structure, the diversity and the character of vegetation community distribution. The zonal sequence shows itself regarding large morphologic structures (beach, foredune, palve, dune massifs). The greatest diversity in plant associations is characteristic of palve and is closely associated with the level of groundwater occurrence. The vegetation cover structure and its species diversity are greatly affected by the recreational and economic activities. A considerable part of the Spits' plant association is of artificial origin which is determined, in the first instance, by the implementation of measures aimed at sand stabilization. The major kinds of vegetation on the Curonian and the Vistula Spits are woody, meadow, marshy, and littoral. The marshy and meadow cenoses of the Spits hold a subordinate status.

The share of forest land on the Curonian Spit is approximately 70%, and that on the Vistula Spit – 80% [7]. Unlike the Curonian Spit, all the dunes of the Vistula Spit are afforestated [14]. The chief forest forming species of the Spits are – *Pinus sylvestris*, *Picea abies, Betula pendula* and *B. pubescens*, *Alnus glutinosa* and *A. incala*, *Populus tremula*; with *Pinus sylvestris* as prevalent. A separate specimen of cenoses on the Spits is developing on wandering and semi-fixed dune sands where psammophytic vegetation prevails. The coastal cenoses of the Curonian and the Vistula lagoons are represented by belts of reed, cane, cow lilies and water lilies, large-leaved pondweeds, micro- and macrophytes characteristic of open parts of basins [4].

2 Research Methods

With a view to studying current trends of development of eolian coastal-marine natural systems in the Kaliningrad Region, original techniques were elaborated and adjusted to the peculiarities of this particular type of terrain.

The sustainability of natural territorial complexes can be defined as their ability to retain the initial properties and functions, as well as historically developed interconnection of their elements and components. Each natural complex features its own span of variability in response to the effect of the factor (of a set of factors) within which a possibility is preserved to change while retaining its structural and functional properties at a particular period in its development. Thus, it is possible to talk of an oscillation about a certain ecosystem homeostasis. Such a state can be considered conceivable ecosystem stability. That notwithstanding, the intensity of the effect of natural and anthropogenous factors is not disastrous. Otherwise, changes become irreversible and are capable of leading to the destruction of the ecosystem. The indicator of potential stability of the natural complex is based on the principle of grade componentwise estimation of five indicators identified taking into consideration the peculiarities of the coastal-marine type of terrain: the relief, the degree of manifestation of exogenetic processes, ground waters, soils, and vegetation (Table 1).

The most persistent forms of the Spits' terrain are plain, plain-billowy, and hummocky-billowy surfaces corresponding on the Spits to a disinflationary-depositional plain of low, medium, and high level. The ridge relief features a great diversity of forms and is actually a combination of dune ridges particularly, dune massifs and isolated dome-shaped dunes with different gradient slopes: from slightly sloping of 3–5° to abrupt – 30–40°.

Exogenic processes under the conditions of the Spits are capable of not only heavily transforming natural complexes, but in some cases can lead to a sharp decrease in their stability which is especially important for the coastal dune ridge system. Under the conditions of the coastal-marine terrain, three groups of exogenic processes of various localization were identified: destructive, landwaste transfer, and accumulative. Destructive processes are represented by deflation (dune ridge systems), washout (beach area), abrasion (in the area of moraine ingressing onto the Curonian Spit); transfer (transiting) of landwaste (alongshore load transport, sand material from damaged areas of the shore coming deep into the Spits, etc.); accumulative processes, like deflationary ones, are characteristic of dunes and dune ridges without a vegetational cover or those poorly stabilized by the vegetation. In order to assess stability, an indicator was introduced that reflects the size of the natural site affected by exogenic processes and the character of their manifestation.

In accordance with the moisture regime on the Spits, there were identified areas with normal humidification (medium and high palve, the ground waters level (GWL) is 1–2 m); with constant insufficient (ridge systems, isolated dunes, GWL is constantly over 3 m); with constant excessive (low palve, GWL – <1 m); occasionally excessive (medium and partly low palve, GWL in individual years reaching the ground surface); occasionally insufficient (hummocky-billowy palve, GWL in some periods – 3–4 m, under hillocks – 3–6 m). In certain dry-weather periods of the year as a result of a GWL drop, some parts of low and medium palve experience the lack of moisturization.

Table 1 Matrix for calculation of potential stability of eolian coastal-marine natural complexes

Indicators Indicator value/grade	Relief	Manifestation of exogenetic processes	Ground waters (character of humidification)	Soils	Vegetation
	Ridge with slopes of over 15°	Manifested throughout, affect more than 50% of the area	Insufficient humidification	Beachy and slightly humic sands	Failed areas with psammophytic wild grasses and mossy-lichenous aggregations; open-growing artificial stands of pine tree; dead covering and lichenous pine forests
	1	1	1	1	1
	Ridge with slopes of 6–15°	Manifested throughout, affect 26–50% of the area	Occasional insufficient humidification	Podzolic (surface and micropodzolic)	Mossy pine forests; filical spruce forests, bare lichenous-mossy areas with thinned psammophytic wild grasses
	2	2	2	2	2
	Ridge with slopes of <6°	Manifested locally, affect 11–25% of the area	Normal humidification	Turfy-podzolic	Mossy-gramineous pine and spruce forests, grassy lime woods, psammophytic-motley grass-grasses bare areas
	3	3	3	3	3
	Hummocky-billowy	Manifested locally, affect 5–10% of the area	Occasionally excessive	Peaty- and humous-podzolic	Wood-sorrel, birch forests, forb and bog-grass alder forests, wet and bog-grass non-forested areas
	4	4	4	4	4
	Plain, plain-billowy	Absent or affect <5% of the area	Excessive	Turfy	Gramineous birch and aspen forests, helad willow woods, shrubby motley grass-grasses bare areas
	5	5	5	5	5

The main soil-forming material on the Spits is eolian sands, in some areas – moraine loams which can be slightly spanned by sands (the Curonian Spit, near the village of Rybachij) and eolian slime ground (the Vistula Spit, approximately 21st kilometer). For sands stabilization, of great significance is the degree of soils reinforcement with root systems, the thickness and character of the underlayment, a high humus content in soils, etc. Unfixed and semi-fixed sands feature lowest stability. The substrate dynamism determined by the activity of eolian processes, a constant income of sand material, peculiar vegetation that does not favor the development of humic layers prevent soils from shaping up. Surface- and micropodzolic soils shape up on more quieted sands. Their formation is connected with tree and shrubbery vegetation and developed ground cover. These soils feature thin soil horizons, an immature underlayer which, if damaged, fails to constrain deflation. Soddy-podzolic soils are spread within podzolic and marshy-podzolic kinds of soils (soddy- micropodzolic, soddy-shallowpodzolic, soddy-podzolic earthen-gleyed). They are distinguished by a relatively high degree of sodding, with the underlayer thickness of 3–4 cm. Peaty-podzolic and humous-podzolic soils have similar conditions and shape up on low levels of terrain under the influence of excessive humidification. Peaty-podzolic earthen-gleyed soils have, as a rule, a well-defined peaty floor whose thickness ranges from 8 to 16 cm and which is quite firmly reinforced by plant roots. Humous-podzolic soils feature an unmarked or feebly marked (0–1.5 cm) horizon A_0, while the humous horizon A_1 is quite thick (6–18 cm).

The vegetative component of natural sites under the conditions of the Spits is extremely fast moving and the response of its diffcrent tiers varies over a wide range. The most endurant component of the phytocoenosis is the timber stand. It should be noted that the majority of forests on the Spits are of artificial origin. The vegetational cover grading is based on such characteristics as self-restoring capacity, the concordance of forest plantation to the forest site type, capacity class, and others. Under the conditions of dune coastal complexes, the distortion of the grass-moss layer (especially at positive topographic forms) furthers sand substrate movement and emergence of deflation focuses.

All the indicators were estimated in points from 1 to 5. For each natural complex, the point total was calculated. According to the calculations obtained, the indicator of conceivable stability of natural complexes (NCs) was distributed into four categories: strongly unstable NCs (5–10 points), unstable NCs (11–15 points), weak-stable NCs (16–20 points), and stable NCs (20–25 points).

Thus, the conceivable stability of natural complexes is determined by historically developed peculiarities of their components and does not consider their current changes resulting from the effect of forces of various origins which makes it difficult to assess the profundity of anthropogenic transformation of natural complexes of the Spits, its territorial differentiation, and interconnection with natural processes. It is proposed to use the digression rate as an indicator adjusting the prospective stability of an individual natural complex. The latter considers the distinguishing features of the site and is based on a set of natural and anthropogenic markers which in this case are viewed as indicators of the state of the coastal-marine terrain.

The relief, being against this background quite a dynamic component, is one of the priority assessment characteristics in determining digression. The share of the area with failures in the relief is used as a marker of its state. As the research conducted showed, this indicator demonstrates the highest variability for the beach, foredune, scour terraces, unfixed and poorly fixed dune ridge and massifs. In this case, the degree of relief displacement can vary from <10% (for example, areas of relatively stable foredune and the beach) to 100% (areas of destroyed foredune with the lacking beach). Fixed dune ridges and massifs occupy an intermediate position. The most statical forms of relief of the dune terrain are plain, plain-billowy, and hummocky-billowy surfaces corresponding on the Spits to the deflationary-depositional plain (palve) of low (the groundwater level is <2 m), medium (2–4 m), and high (over 4 m) level.

The methodology of assessment of natural complexes digression is founded on an integrated index based on grade componentwise assessment of 15 indicators of natural and anthropogenic genesis, identified taking into consideration natural peculiarities of the eolian coastal-marine terrain and prevalent forms of human impact (Table 2). Among the chief indicators are as follows: the share of the area with relief failures; the share of saturation and bogginess area; the share of the area with disturbed structure of the underlayer; the share of the area lacking underlayer; lowering of the projective cover of soil continuum; the share of ruderal species; foliage cover of ruderal species; the share of tree and shrubbery vegetation damaged by phytophagans and (or) stricken with diseases; the share of tree and shrubbery vegetation with mechanical defects; the character of distribution of vegetative strap; among the additional indicators are the following: the amount of solid domestic waste and (or) industrial waste; the share of the area contaminated by petroleum products; the share of the area affected by the fire; the share of the area exposed to haying and cattle grazing; the share of conducted melioration measures.

All the indicators were assessed in points from 0 to 5, where 0 point characterizes the lack of indicator manifestation, 1 point – a very low degree of its manifestation, 2 points – a low degree, 3 points – medium, 4 points – high, and 5 points – a very high degree. The point total was calculated for each natural complex. Different significance and completeness of the effect of indicators in various NCs was surmounted by introducing weighted coefficients. The integrated index of natural complexes' digression is calculated by the following formula:

$$D = \sum_{i=1}^{n} \alpha_i p_i k_i,$$

where D – the integrated index of digression; α_i – a component of incidence vector (1 or 0, if the attribute "works" or does not); k_i – a weighted coefficient (from 0 to 1; $\sum k_i = 1$); p_i – an indicator (in points).

The results of the calculation of the integrated index of digression for each natural complex were normalized regarding maximum possible values and represented in the form of a five-stage grid of digression: stage I – <0.15 (very

Table 2 Matrix for calculation of natural complexes digression

	Main										Additional					
Indicators	Share of the area with relief failures (%)	Share of saturation and bogginess area (%)	Share of the area with disturbed structure of the underlayer (%)	Share of the area lacking underlayer (%)	Lowering of the projective cover of soil continuum (%)	Share of ruderal species (%)	Foliage cover of ruderal species (%)	Share of tree and shrubbery vegetation damaged by phytophagans and (or) stricken with diseases (% of plants)	Share of tree and shrubbery vegetation with mechanical defects (% of plants)	Character of distribution of vegetative strap	Amount of solid domestic waste and (or) industrial waste (units/m^2)	Share of the area contaminated by petroleum products (%)	Share of the area affected by fire (%)	Share of the area exposed to haying and cattle grazing (%)	Share of conducted melioration measures (% of the necessary volume)
Indicator value/point	–	–	–	–	–	–	–	–	–	–	–	–	–	–	>90 or there is no necessity
															0
	0	0	0	0	0	0	0	0	0	0	0	0	0	0	60-90
	<3	<3	<5	<3	<5	<5	<5	<5	<5	Sporadic	<0.1	<0.1	<0.1	<5	1
	1	1	1	1	1	1	1	1	1	1	1	1	1	1	30-60
	3-10	3-10	5-10	3-10	5-10	5-10	5-10	5-10	5-10	Scattered	0.1-0.5	0.1-2	0.1-5	5-10	2
	2	2	2	2	2	2	2	2	2	2	2	2	2	2	10-30
	10-20	10-20	10-25	10-20	10-25	10-25	10-25	10-25	10-25	Scattered in groups	0.5-1.5	2-5	5-15	10-25	3
	3	3	3	3	3	3	3	3	3	3	3	3	3	3	<10
	20-50	20-50	25-60	20-50	25-60	25-50	25-60	25-50	25-50	Dense in groups	1.5-3	5-10	15-40	25-40	4
	4	4	4	4	4	4	4	4	4	4	4	4	4	4	Not conducted
	>50	>50	>60	>50	>60	>50	>60	>50	>50	Abundant throughout	>3	>10	>40	>40	5
	5	5	5	5	5	5	5	5	5	5	5	5	5	5	

weak); stage II – 0.15–0.30 (weak); stage III – 0.31–0.45 (medium); stage IV – 0.46–0.60 (strong); and stage V – >0.60 (very strong).

The dependence of integral (modified) stability on digression is not linear and can be described by the following formula: Sm = Sp × (1 − D), where: Sm – integral (modified) stability, Sp – prospective stability, and D – digression.

The general trends of prospective stability decreasing depending on digression stages look as follows: the stability of NCs that are at the first to second stages of digression decreases insignificantly, it is of steady nature and time-expanded. Stage III sees (under relatively fixed conditions) a certain latent period of changes accumulation and weakening interelement ties which is expressed in a certain stabilization of the process. The fourth stage is marked by sharp destructive changes of all the components of the natural complex, the influence on the neighboring natural systems is local; stage V marks a deep disruption of natural complexes and can carry a threat for the neighboring NCs.

The matrix for the calculation of the integral (modified) stability of natural complexes is given in Table 3. The point total was calculated for each natural complex. Based on the obtained calculations, the distribution of the index of the integral (modified) stability of natural complexes within five categories was conducted: those NCs that lost their stability potential (0–5 points), highly unstable NCs (5.1–10 points), unstable NCs (10.1–15 points), weak-stable NCs (15.1–20 points), and stable NCs (20.1–25 points).

The application of surface laser scanning (SLS) in studying processes shaping up the coastal zone has not, unfortunately, been sufficiently addressed in research literature both in Russia and abroad. As a rule, when mentioning laser scanning of the coastal area, examples are given from studies conducted with the help of LIDAR – Laser Induced Direction and Ranging or Light Detection and Ranging. Today, laser scanning is provisionally divided into three large groups – air (LIDAR), surface (SLS), and mobile (MLS). Surface laser scanners are divided, according to the principle of range finder functioning, into two groups – impulse (a long range, a relatively low rate of functioning), phase (a short range, an extremely high rate of functioning). The application of LIDAR survey is justified when the matter concerns large in extent territories or relatively stable large morphologic forms. The structure of the coastal zone of the South-East Baltic makes its use inexpedient on the Kaliningrad coast. The application of mobile scanning systems is impossible due to technical features of the system. Therefore, an optimal way of obtaining accurate and reliable information about the morphometric properties of the above-water coastal zone is surface laser scanning.

In order to monitor the above-water coastal zone of the Baltic Sea within the boundaries of the Kaliningrad Region on the accretion coasts, key areas were chosen with the length of up to 200 m, which were examined with the use of technology of surface laser scanning in the period from 2011 until 2015. Two pulse scanners TOPCON GLS-1500 and a positional system Topcon GR-5 were used in the research.

The experience of the application of the technology of surface laser scanning on natural sites showed its high efficiency, especially on the areas with high dynamical

Table 3 Matrix for calculation of integral (modified) stability of eolian coastal-marine natural complexes

Digression	Integral (modified) stability																								
	5	6	7	8	9	10	11	12	13	14	15	16	17	18	19	20	21	22	23	24	25
0.00	5	6	7	8	9	10	11	12	13	14	15	16	17	18	19	20	21	22	23	24	25
0.05	4.75	5.7	6.65	7.6	8.55	9.5	10.45	11.4	12.35	13.3	14.25	15.2	16.15	17.1	18.05	19	19.95	20.9	21.85	22.8	23.75
0.10	4.5	5.4	6.3	7.2	8.1	9	9.9	10.8	11.7	12.6	13.5	14.4	15.3	16.2	17.1	18	18.9	19.8	20.7	21.6	22.5
0.15	4.25	5.1	5.95	6.8	7.65	8.5	9.35	10.2	11.05	11.9	12.75	13.6	14.45	15.3	16.15	17	17.85	18.7	19.55	20.4	21.25
0.20	4	4.8	5.6	6.4	7.2	8	8.8	9.6	10.4	11.2	12	12.8	13.6	14.4	15.2	16	16.8	17.6	18.4	19.2	20
0.25	3.75	4.5	5.25	6	6.75	7.5	8.25	9	9.75	10.5	11.25	12	12.75	13.5	14.25	15	15.75	16.5	17.25	18	18.75
0.30	3.5	4.2	4.9	5.6	6.3	7	7.7	8.4	9.1	9.8	10.5	11.2	11.9	12.6	13.3	14	14.7	15.4	16.1	16.8	17.5
0.35	3.25	3.9	4.55	5.2	5.85	6.5	7.15	7.8	8.45	9.1	9.75	10.4	11.05	11.7	12.35	13	13.65	14.3	14.95	15.6	16.25
0.40	3	3.6	4.2	4.8	5.4	6	6.6	7.2	7.8	8.4	9	9.6	10.2	10.8	11.4	12	12.6	13.2	13.8	14.4	15
0.45	2.75	3.3	3.85	4.4	4.95	5.5	6.05	6.6	7.15	7.7	8.25	8.8	9.35	9.9	10.45	11	11.55	12.1	12.65	13.2	13.75
0.50	2.5	3	3.5	4	4.5	5	5.5	6	6.5	7	7.5	8	8.5	9	9.5	10	10.5	11	11.5	12	12.5
0.55	2.25	2.7	3.15	3.6	4.05	4.5	4.95	5.4	5.85	6.3	6.75	7.2	7.65	8.1	8.55	9	9.45	9.9	10.35	10.8	11.25
0.60	2	2.4	2.8	3.2	3.6	4	4.4	4.8	5.2	5.6	6	6.4	6.8	7.2	7.6	8	8.4	8.8	9.2	9.6	10
0.65	1.75	2.1	2.45	2.8	3.15	3.5	3.85	4.2	4.55	4.9	5.25	5.6	5.95	6.3	6.65	7	7.35	7.7	8.05	8.4	8.75
0.70	1.5	1.8	2.1	2.4	2.7	3	3.3	3.6	3.9	4.2	4.5	4.8	5.1	5.4	5.7	6	6.3	6.6	6.9	7.2	7.5
0.75	1.25	1.5	1.75	2	2.25	2.5	2.75	3	3.25	3.5	3.75	4	4.25	4.5	4.75	5	5.25	5.5	5.75	6	6.25
0.80	1	1.2	1.4	1.6	1.8	2	2.2	2.4	2.6	2.8	3	3.2	3.4	3.6	3.8	4	4.2	4.4	4.6	4.8	5
0.85	0.75	0.9	1.05	1.2	1.35	1.5	1.65	1.8	1.95	2.1	2.25	2.4	2.55	2.7	2.85	3	3.15	3.3	3.45	3.6	3.75
0.90	0.5	0.6	0.7	0.8	0.9	1	1.1	1.2	1.3	1.4	1.5	1.6	1.7	1.8	1.9	2	2.1	2.2	2.3	2.4	2.5
0.95	0.25	0.3	0.35	0.4	0.45	0.5	0.55	0.6	0.65	0.7	0.75	0.8	0.85	0.9	0.95	1	1.05	1.1	1.15	1.2	1.25
1.00	0	0	0	0	0	0	0	0	0	0	0	0	0	0	0	0	0	0	0	0	0

activity [15]. The key areas are located in the zone of the beach – the coastal dune ridge complex. The calculations of morphometric indices were made on the basis of triangulation models developed in conformity with the "point cloud." The conducted surface laser scanning and processing of the obtained data were followed by the development of 3-D models of the key areas of the coastal zone of the Curonian Spit. In developing graphic models, the software systems ScanMaster 2.7 and ArcGis 9.3 were used. Also, it is possible to work with the "raw point cloud" in the software package AutoCAD, TopoCAD, Quick Terrain Modeler. This enabled us to obtain graphic detailed geomorphologic data, as well as 3-D characteristics of the research target.

Scanning of natural sites in the coastal zone with a complex geomorphologic structure enabled us to identify a number of methodical specifics:

- the more complex the morphometry of the area is, the more scanner observation points should be laid. This is accompanied by a decrease in scanning density and predominance of narrow-angle scanning, which results in increasing the time of conducting works.

  The operation speed depends on scanning parameters. The choice of scanning parameters should be determined by the tasks realized at key areas, for the purposes of an optimum compromise between the speed and the quality of the data obtained. Figure 4 shows schematically options of the quality of measurements on a complex surface. The most appropriate for the coast of the Kaliningrad Region is setting the scanning angle at 360° with the resolution of $30 \times 30 \times 10$, which provides an opportunity to attain the operation time of scanning system at the observation point of 20–25 min, in conjunction with a panoramic photo.

- when selecting areas, the following approach is appropriate:
  - for long-term observations, priority should be given to morpholithodynamically uniform areas with apparent destructive processes and with the length of up to 0.5 km;
  - for short-time observations, priority should be given to very dynamic areas with the length of not more than 0.25 km;

- for the observation over damaged areas, the length is determined by the extent of the damaged area itself.
- the TLS technology is most effective for the areas which are inaccessible for traditional measurement techniques.
- the stage-by-stage approach of processing the materials obtained as a result of scanning is directly connected with the objectives of scanning: in obtaining current morphometric indicators, it is sufficient to make use of triangulate model built on the basis of the "point cloud"; for the comparative analysis of the long-term data – the application of raster models involving the use of the relevant programmes, for example, ArcGIS, QGIS, SAGA etc.

It should be noted that the speed and the quality of scanning are in direct relationship to the weather conditions and methodological approach to scanning. The presence of even negligible precipitation makes scanning difficult.

After scanning, the results are processed in ScanMaster (Fig. 5).

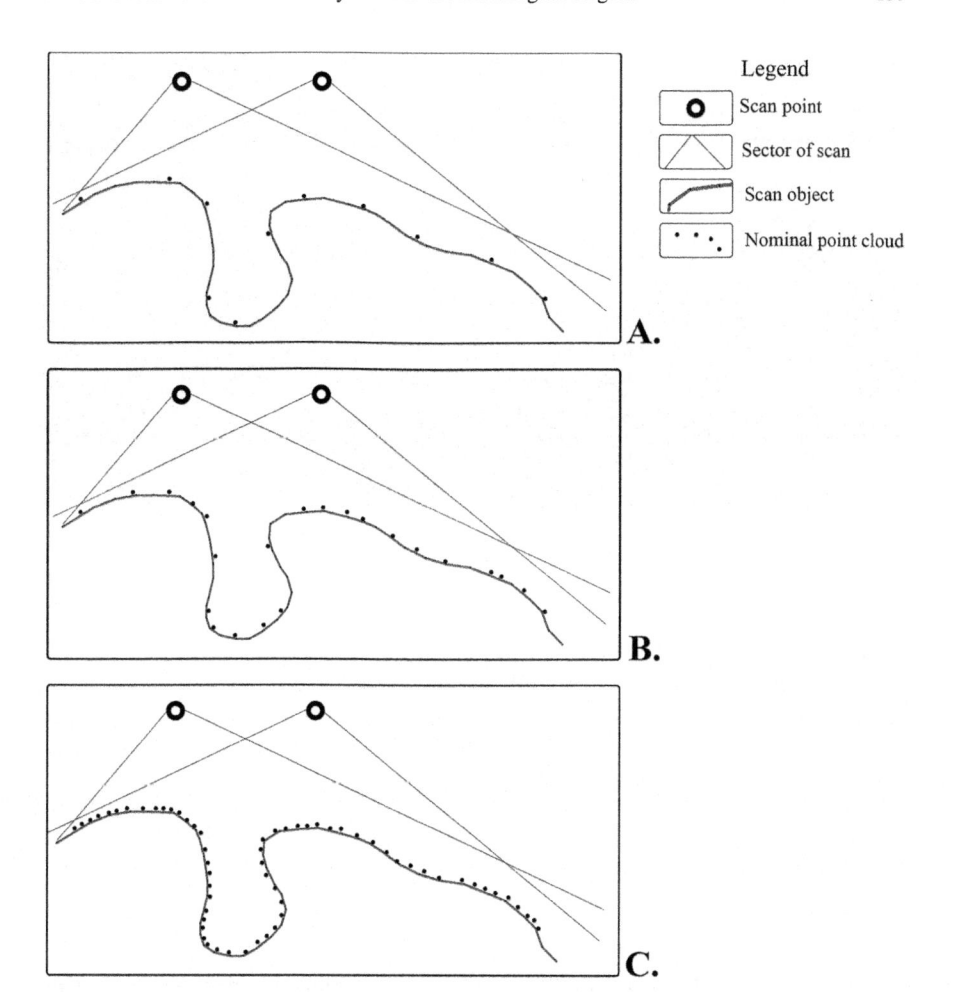

Fig. 4 Options of degree of measurements on a complex surface. Option (**a**) High operation speed, low quality of the point cloud (suitable only for the reconnaissance investigation). Option (**b**) Medium operation speed, medium quality of the point cloud (optimum option). Option (**c**) Extremely low operation speed, the highest quality of the point cloud (suitable only in case of extremely complex site or with geological tracking)

In ScanMaster it is possible not only to contract scans but also to produce preliminary calculations, first of all, such as estimation of cubage, computation of linear characteristics, and cross section preparation. The processing of the "point clouds" with resulting reference surfaces makes it possible to further obtain planimetry making use of the software package Golden Software Surfer (Fig. 6).

After processing the results and obtaining linear and volumetric characteristics, the data are transferred to the software package ArcGis where the subsequent calculations are conducted. Two methods are used in modelling surfaces – that of natural neighborhood and Kriging (Table 4).

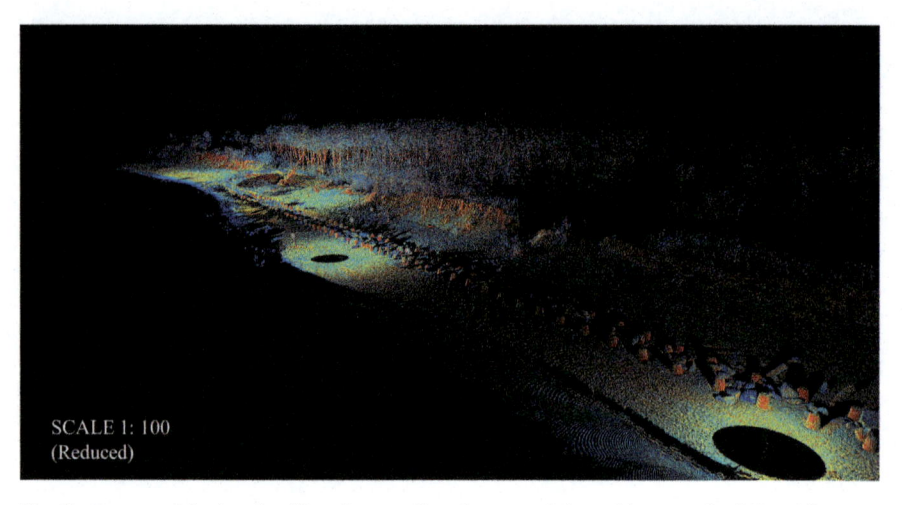

Fig. 5 Contracted "point cloud" on the area "beach – coastal dune ridge complex" (first kilometer of the Curonian Spit)

The application of these methods of modelling reference surfaces pursues two tasks: one involves the calculation of the deficiency and proficit of fragmentary material; the other lies in the necessity of proper visualization of the processes. The application of the ArcGIS satisfies both tasks, since its tools set is the most complete of all the competitive packages.

3 Results

The principal natural factor determining the evolvement of eolian coastal-marine natural complexes of the Kaliningrad Region is wind and wave activity (Fig. 7). According to the morpho- and lithodynamic attributes in the nearshore zone, the following areas are distinguished: those of stream-bank erosion, of sediments transit and of their aggradation.

The length of the wash-out site on the Curonian Spit (the town of Zelenogradsk – the village of Rybachij) is 34 km with the speed of coast retrogression of 1–1.8 m/year [16]. The coast site at the root area of the Spit is exposed to the heaviest wash-out with the foredune having been almost completely washed out, and the dune rampart being intensively washed out which is accompanied by the development of a wash terrace. Between the villages of Rybachij and Morskoe, a transit coast area is distinguished with the length of 22 km with the beach width of up to 20–40 m. The site of aggradation within the boundaries of the Curonian Spit (from the village of Morskoe as far as the Russian-Lithuanian border) is characterized by the widest beach – 50 m and more [16].

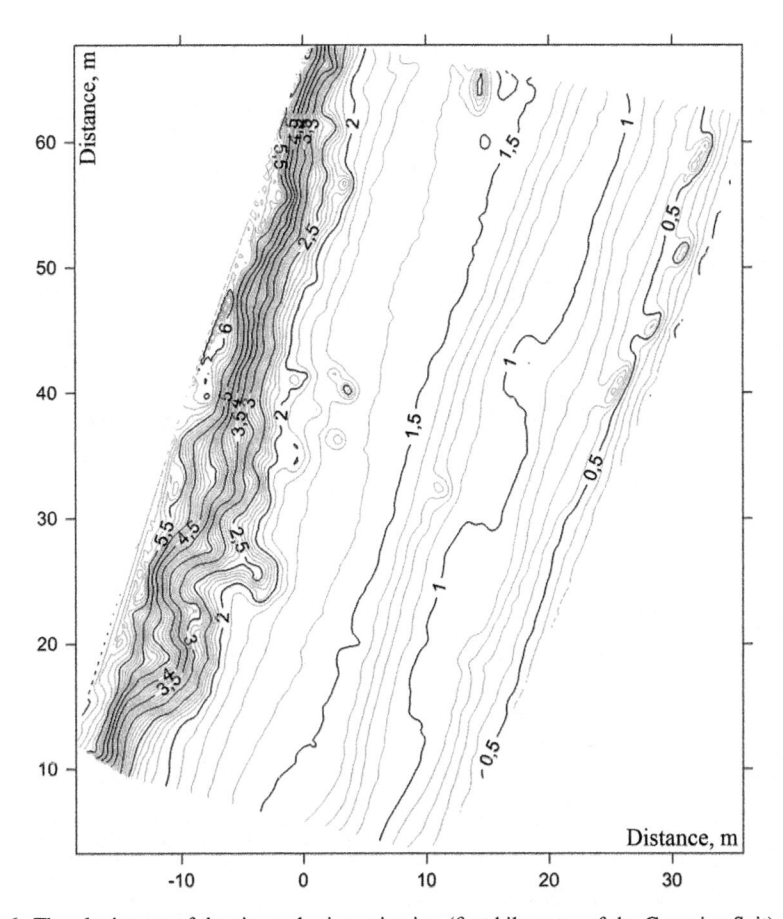

Fig. 6 The planimetry of the site under investigation (first kilometer of the Curonian Spit)

Table 4 The comparison of two interpolation methods

	Natural neighborhood	Kriging
Advantage	High operation speed Accessibility of calculation process	Possibility of assessment of the quality of modelling conducted Control of the process parameters
Disadvantage	Limitation of the distance covered by the dependence between the data	Complexity of conducting consistent kriging
Compensation	High density of the data measured in the course of scanning	A more precise surface model

The geoenvironmental monitoring on the Curonian Spit involving the technology of surface laser scanning was conducted from 2011 till 2015 on two key sites: in the southern part of the Spit, in the immediate vicinity of the town of Zelenogradsk, and the village of Morskoe.

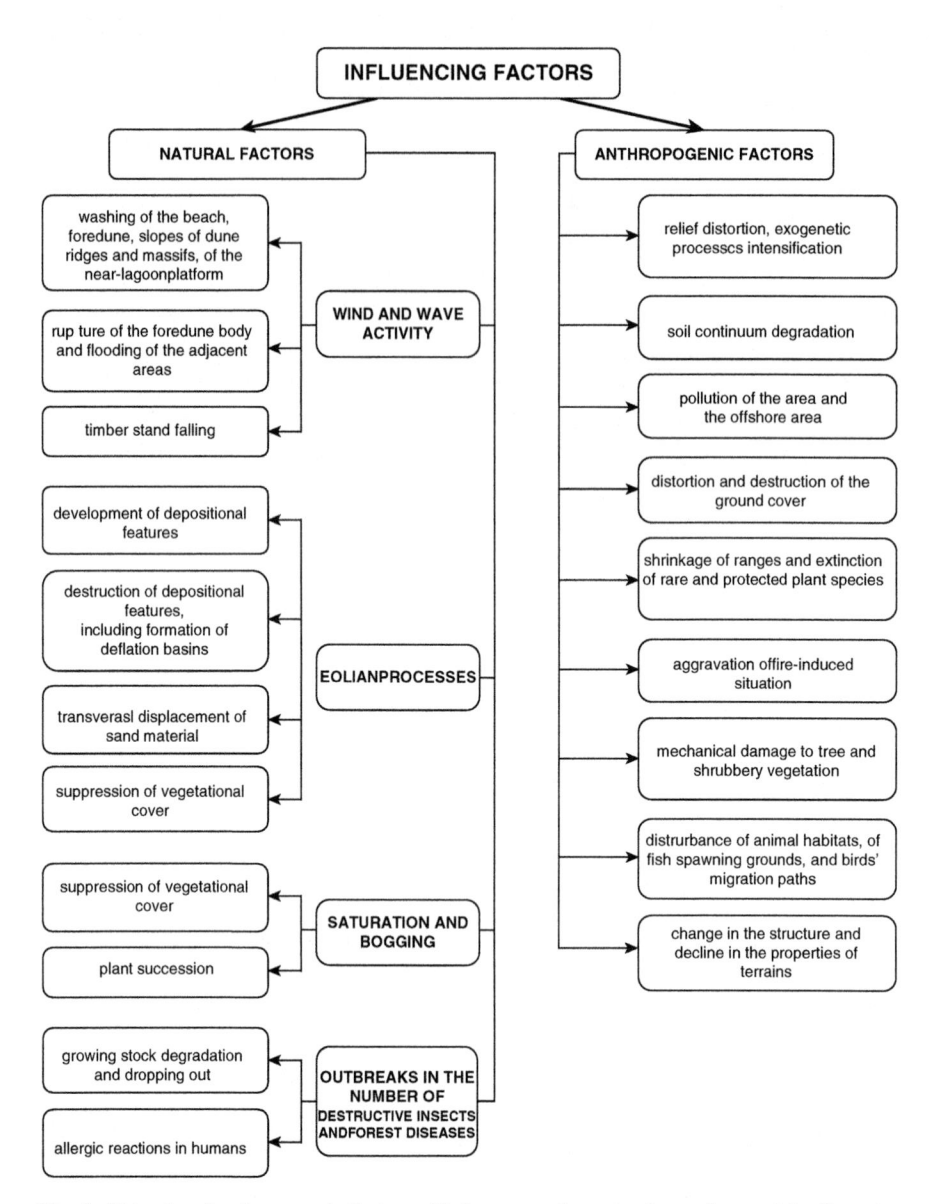

Fig. 7 Natural and anthropogenic factors of influence on the natural complexes of the Curonian and Vistula Spits

One of the key sites is located in the area of the first kilometer of the Curonian Spit in the wash zone and is characterized by a high dynamism of the processes going on. In connection with the lack of sediments on the submersed beach approach, the income of fragmentary material is obstructed here, which results in reducing the width of the beach and worsening of the condition of the coastal dune ridge complex (Figs. 8 and 10).

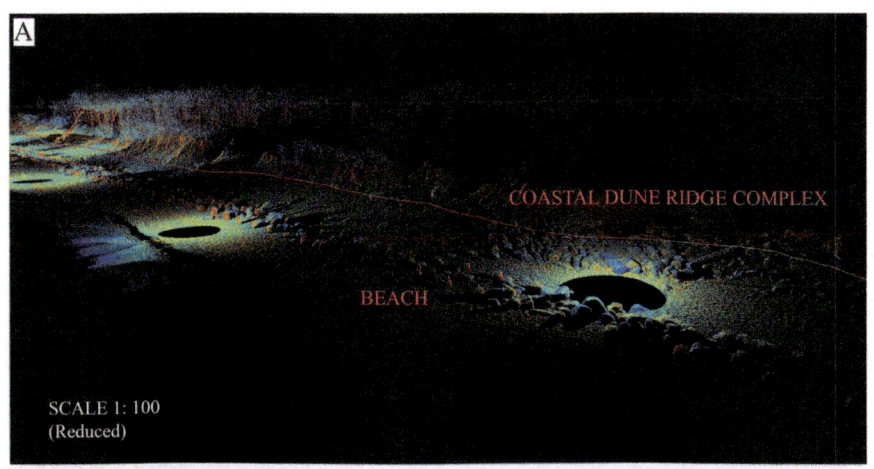

Fig. 8 The key site of the "beach – coastal dune ridge complex," first kilometer of the Curonian Spit (**a** "point cloud," **b** photo)

The other key site is located in the area of the village of Morskoe in the aggradation zone. The investigation carried out in the years 2011–2015 shows its relative stability in the beach zone – the coastal dune ridge complex (Figs. 9 and 11).

Once the point clouds were processed, simulation models of the area under investigation were developed in ArcGIS (Figs. 10 and 11).

The main parametrical indices are shown in Table 5 and in Figs. 12 and 13.

Fig. 9 The key site of the "beach – coastal dune ridge complex" in the area of the village of Morskoe (**a** "point cloud", **b** photo)

The analysis of the results of scanning of the first key site showed a multidirectional year to year trend data of morphometric indices of the beach and coastal dune ridge complex against the background of the general trend – the predominance of the washout process. Within the period of 2013–2014, there was observed an increase in its area by 1,265.3 m^2, of its volume – by 2,700 m^3; the beach width grew by 2 m on average. A change in the condition of the site under

Fig. 10 A triangulate model of the key site of the "beach – coastal dune ridge complex" in the area of the first kilometer of the Curonian Spit according to the data of surface laser scanning (**a** 2013, **b** 2014, **c** 2015)

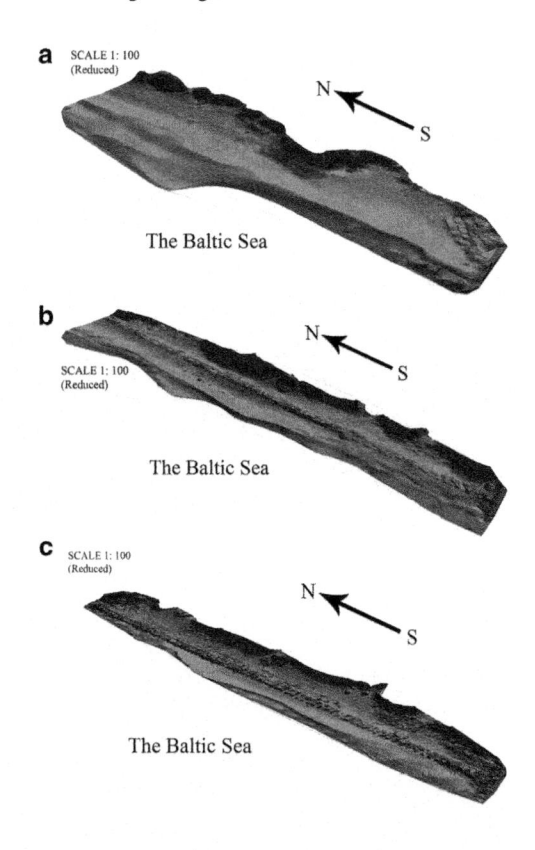

investigation in 2015 was caused by the storm *Felix* which resulted in a drastic reduction of the amount of sand material in the area "beach – coastal dune ridge complex" by 4,200 m^3. On the whole, estimating the trend from 2013 till 2015, one can confirm a decrease in the ultimate width of the beach within this period by 7 m (from 32 to 27 m), in the minimal width – by 1 m, with a drop in the average width by 3 m (Fig. 12). There was recorded a piecemeal stoping of the upwind slope and, locally, a decrease in the height of the coastal dune ridge complex up to 1.2 m; on separate sites there was observed a formation of deflation basins, and in spots there were seen ruptures of the ridge body. Against the background of the general negative change of the beach zone, the role of the coast protection works should be noted, in particular, benefits of the installation of tetrapods which promoted the development of a new generation of the coastal dune ridge complex.

The site in the area of the village of Morskoe shows an example of a relatively sound sea coast. Thus, the maximum width of the beach within the period of 2013–2015 increased by 4 m (from 55 to 61 m); the minimal width features stable values (46–48 m); the average width did not virtually change, and the height of the coastal dune ridge was constant (Fig. 13). These morphometric characteristics demonstrate the prevalence of accumulative processes on the site in question. The amount of sand material within the period of 2013–2015 increased by 991 m^3, and

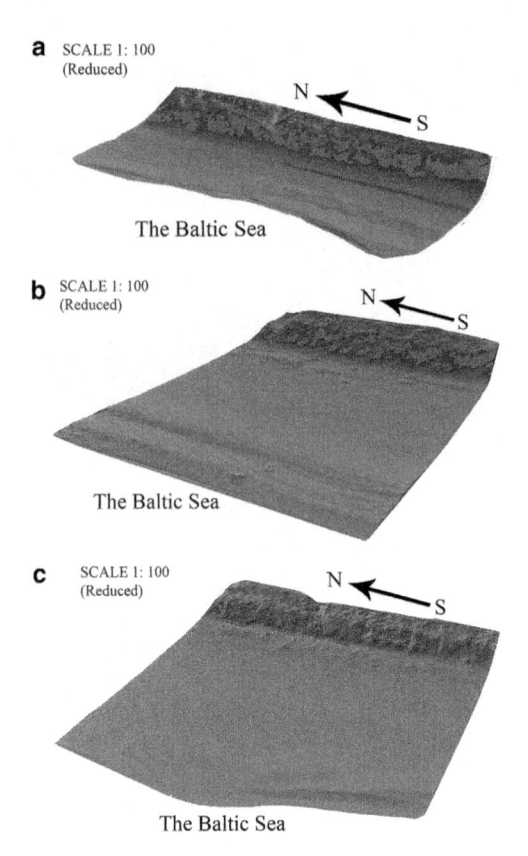

Fig. 11 A triangulate model of the key site of the "beach – coastal dune ridge complex" in the area of the village of Morskoe according to the data of surface laser scanning (**a** 2013, **b** 2014, **c** 2015)

Table 5 The parametrical indices of the key sites of the Curonian Spit

	Site location/years					
	Area of the first kilometer of the Spit			Area of the village of Morskoe		
Parameters	2013	2014	2015	2013	2014	2015
Length of the site under investigation (m)	200	200	200	70	70	70
Area of the site under investigation (m^2)	5,696.3	6,961.6	5,253.1	5,240	5,220	6,080
Amount of material on the site under investigation (m^3)	7,100	9,800	5,600	12,150	12,385	13,141

the area – by 840 m^2. The after-storm examination showed that the wind and wave activity on the site did not have a considerable negative impact. At the same time, shortly after the storm *Felix* there was a short-term reduction of the beach width at

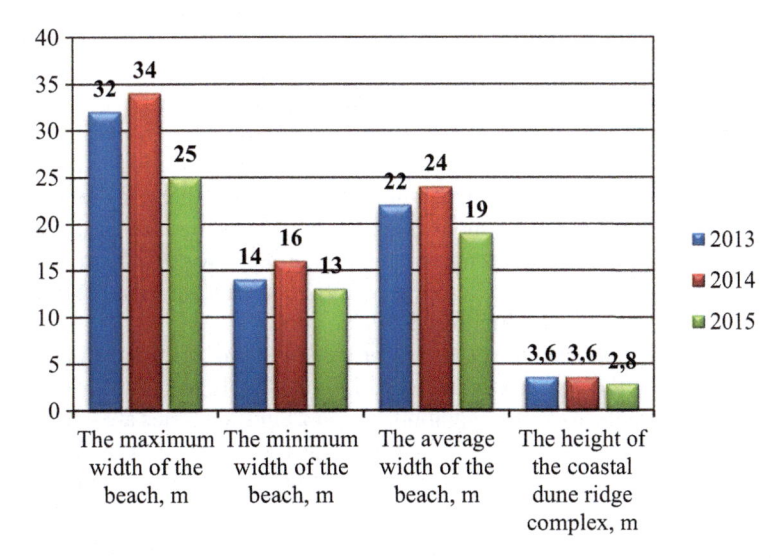

Fig. 12 Year to year trend data of morphometric indices on the site "beach – coastal dune ridge complex" (first kilometer of the Curonian Spit)

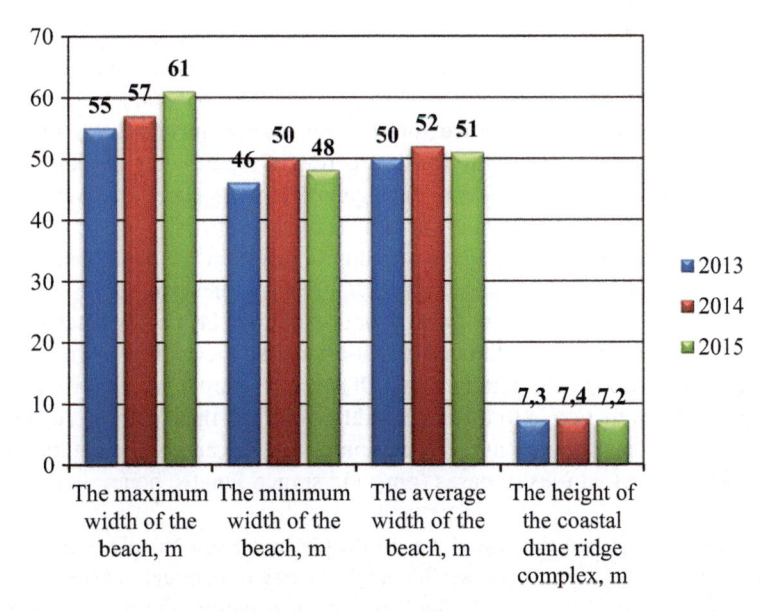

Fig. 13 Year to year trend data of morphometric indices on the site "beach – coastal dune ridge complex" (area of the village of Morskoe, the Curonian Spit)

the root site of the Curonian Spit up to 1–4 m, and at the site in the area of the village of Morskoe – up to 25–30 m. However, the regeneration of the former site took place much later, and it was not as significant as that of the latter site which is determined by the predominance of different degree of intensity of the processes affecting their formation.

In the northern part of the Vistula Spit under the influence of the breakwater piers of the entrance channel, a zone of the seacoast washout shaped up with the length of 5 km with the wash-out rate reaching 2.8 m/year [17]. At the initial 0.5–0.6 km, the shore is in a critical condition which poses a threat to the villages on the Spit. To the south of the washout zone, there are transit zones (with the beach width of 30–40 m) and sediments accumulation (the beach width is 50 m and more).

In the near-lagoon area of the Curonian and Vistula Spits, the erodible type of the shore prevails (22.2 and 12.4 km, respectively) [17].

The degree of manifestation of eolian processes, being one of the main relief-forming factors of the coastal-marine terrain, immediately depends on the stabilization of dune complexes with vegetation. The most unstable sites are the foredune, unconsolidated dune massifs, taluses of dune ridges and massifs exposed to the sea and the lagoon. The disruption of the structure of the foredune and the deflation of the dune ridge are accompanied by the withdrawal of sand to the adjacent sites of palve, which in some cases leads to serious disruptions of the vegetation cover up to complete dieback of the forest sites. The activization of eolian processes in the recent years has been enhanced by a registered increase in the number of visitors to the Spits [18, 19].

An important factor of influence on the palve is saturation and formation of marshes characteristic of predominantly lower areas of Spits and near-lagoon territories located on low levels covered primarily by pure alder forests or of mixed species composition, spruce forests, and marshy meadows. A rise in the level of ground waters can lead to the destruction of the growing stock on considerable areas. A run-up rising water in the lagoons is accompanied by swamping and saturation of the low-lying areas of the near-lagoon shore.

A phyto-pathological factor plays an important role in the development of the ecological state of wooded natural sites. Outbreaks in the number of destructive insects (gypsy moths and processionary moths, alder leaf beetle, eight-toothed bark beetle, etc.) are connected with both climatic factors, and with an increase in the amount of food resources in certain years. Of special hazard regarding worsening of forest pathology situation in forests are windthrown trees, thick planting, dead trees, and others. The deterioration in the condition of forests can also be determined by the development of various diseases (pine and spruce fungus, honey fungus, pitch cancer).

On the whole, the natural complexes of the Curonian and Vistula Spits feature a high dynamism of natural processes having seasonal and annual variations.

The influence of anthropogenic factors is of dual nature. On the one hand, the growth in the number of holiday-makers, the development of the Spits areas (hay harvesting, grazing lands, household plots, construction, housing and utilities infrastructure, etc.) are accompanied by the disruption of both separate components, and natural complexes at large, by a decline in their robustness and a loss of their recreational value. On the other hand, an integral part of the ecosystem exploitation on the Curonian and Vistula Spits is conducting environmental measures (building of coast protection works, fixation of the foredune, establishment of new and rehabilitation of old planted vegetation, environmental harvesting and sanitary felling, insect pest control and control of forest diseases, fire-prevention and fire-fighting, etc.) furthering both maintenance of the functionality of the Spits' natural complexes, and leading in some cases to their considerable transformation [18, 19].

The differences in the structure of natural complexes of the Curonian and Vistula Spits (the absence of unfixed dune massifs on the Curonian Spit, a cavernous nature of palve, an extensive dune ridge complex – on the Vistula Spit and the others) determine the inhomogeneity of their spatial differentiation in the rate of their potential stability. 29% and 10% of the areas of the Curonian and Vistula Spits are regarded as stable, weak-stable areas – 33% and 35%, unstable – 7% and 35%, and strongly unstable – 31% and 20%, respectively.

The similarity of natural and anthropogenic factors leading to the transformation of natural complexes of the Spits determines common patterns of distribution of their territories by digression stages (Fig. 14). Stage I on the Spits is predominant and reflects the states of the majority of natural complexes of the deflationary and depositional plain (palve); part of the latter, being within the influence of recreational and economic zones, shows the state corresponding to stages II–III of digression. Stage V features an extremely subordinate status and is characteristic of primarily greatly degraded sites of the coastal dune ridge complex. The prevalence of stages I–II is connected with the environment-oriented status of the Curonian Spit, and on the Vistula Spit – with a limited transport accessibility for holiday-makers [13, 18].

The first and second digression stages in conditions of the coastal-marine type of the terrain are characterized by the changes that do not constitute a threat to the natural complexes. In this case, the following should be taken into consideration: whether it is a long-term stage of the NC existence or a short-term stage of development when the trends of its transfer to higher stages of digression are evident. As a rule, in the former case such a period constitutes up to 10 years and more, while in the latter case the period is 3–5 years with an apparent destructurization of the components-indicators (soil, vegetation) and a severe disruption of the relief. A third stage of digression in the conditions of the Spits is transitional and is supported mainly due to a recreational factor. Unlike the previous ones, this digression stage considerably reduces the natural stability of the chief components of the NC and is starting for its serious changes. Along with this, if the character and the intensity of factors action do not change significantly for a long period of time, the natural complex can remain at this stage for quite a long time and weakening of interelement links will feature a sluggish character.

The most hazardous are the fourth and fifth digression stages. Their localization is primarily in the contact areas sea-land, lagoon-land and primarily involves sites of the foredune, slopes of the dune ridge exposed to the lagoon. The factors leading NCs to these digression stages do not only considerably lower the potential stability of the natural complexes, but in certain cases further their destruction. Under the Spits conditions, these stages are either a corollary of the storm activity, or a result

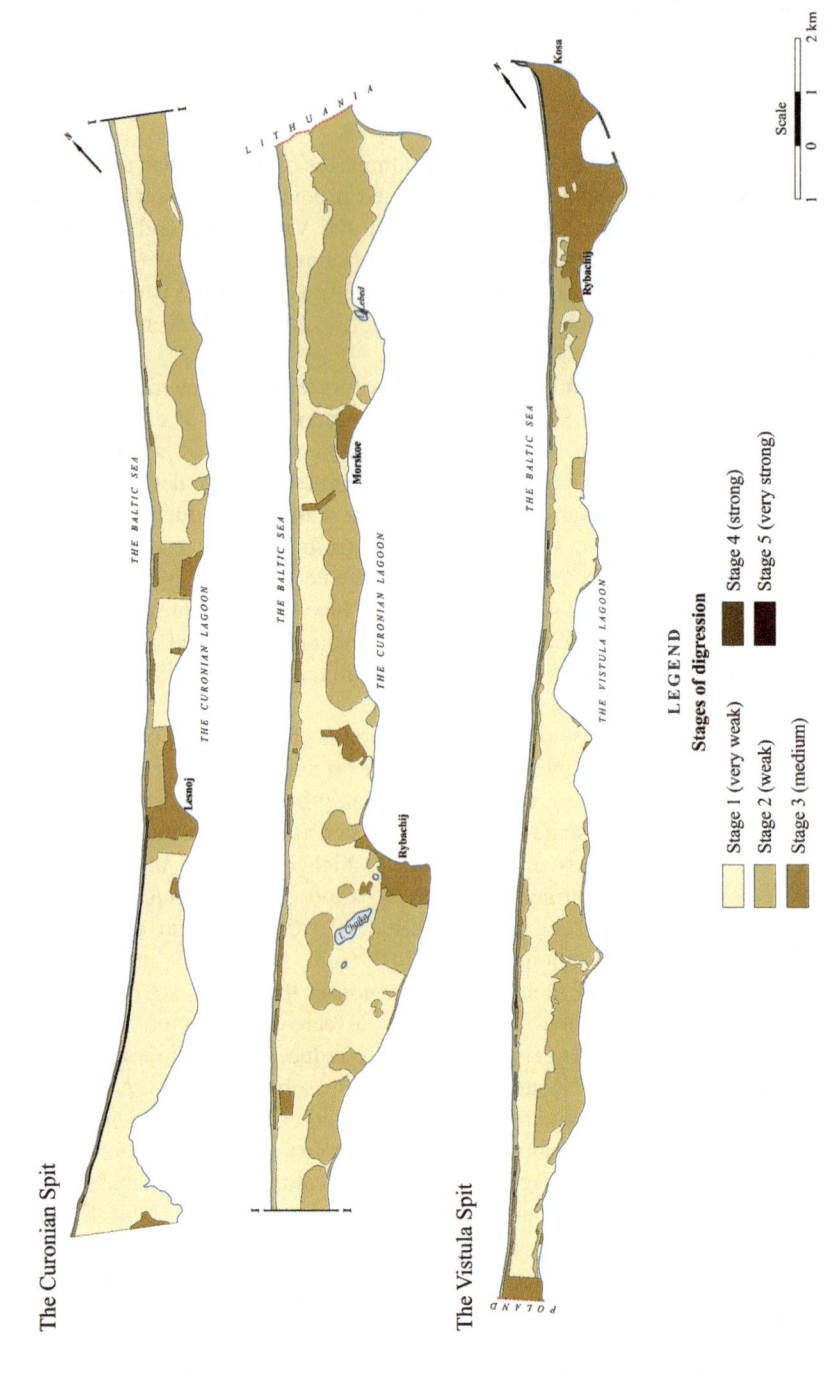

Fig. 14 Digression of natural complexes of the Curonian and Vistula Spits

of the combination of the wind-wave and recreational factors in the coastal zone. The activization of eolian processes results not only in changing the morphometric indices of the site, but also frequently in the disturbance of the adjacent areas.

The assessment of the integrated (modified) stability of the natural complexes of the Curonian and Vistula Spits (Fig. 15) showed that among stable are 29% and 10% of the Spits' territories, among weak-stable – 33% and 35%, unstable – 7% and 34.5%, and strongly unstable – 30.5% and 20%, respectively. The areas that have virtually lost the potential of stability do not represent integrated natural complexes on the Spits, they feature fragmentary position primarily within the coastal dune ridge and occupy not more than 0.5% of the area.

Among strongly unstable are primarily natural complexes of the coastal dune ridge complex and, first of all, recent generation dune ridges. There are occasional sites in this area with extremely low potential of stability. Among unstable are the sea beach and natural complexes of the dune ridge of the middle part of the Spit stabilized by tree vegetation and dune massifs featuring active eolian processes. Weak-stable are the natural complexes of the deflationary and depositional plain (palve) primarily of medium and high level with respective structure of the forest cover. Among stable are the natural complexes of the low and partially medium palve with forest vegetation and, more rarely, meadow vegetation. The breakout of modified stability shows a certain inhomogeneity which is connected with the differences in the structure of the Spits' natural complexes.

On balance, the integrated (modified) stability mirrors general trends of the natural stability of the natural complexes. Chief changes mainly concern the areas exposed to the influence of the factors of biological and human impact, in particular, the fire-induced factor, the effect of phytophagous organisms (first of all, on forest cenoses and others), and the recreational factor. Separate sites of the Curonian Spit near the dune massifs feature abnormal conditions of humidification which leads to the development of extreme conditions for the vegetation cover, and, as a consequence of this, hinders the development of plant communities optimal of such areas and frequently results in allogenic successions.

The current state of the natural complexes of the Curonian and Vistula Spits alongside with natural factors is to a considerable degree determined by the pattern of their use. Fishery was for a long time the major kind of activity on the Spits. In the late nineteenth to early twentieth century, the natural resource management acquired a recreational trend. At that time, quite varied and well developed for the period recreational infrastructure shaped up on the Spits. The agricultural exploitation of the areas was of a subordinate status. A large scale forest clearance determined the necessity of conducting a series of actions aimed at fixation of shifting sands which had for a long time posed a grave threat to the settlements. The construction of the coastal protection embankment (a foredune) started in the nineteenth century with a view to protecting the Spits' natural complexes.

After WWII, the development of the Curonian and Vistula Spits took different routes. This was mainly determined by the construction in the pre-war period on the territory of the Vistula Spit of a military airfield with the relevant infrastructure. In the aftermath of the War, the chief kind of the environmental management on the

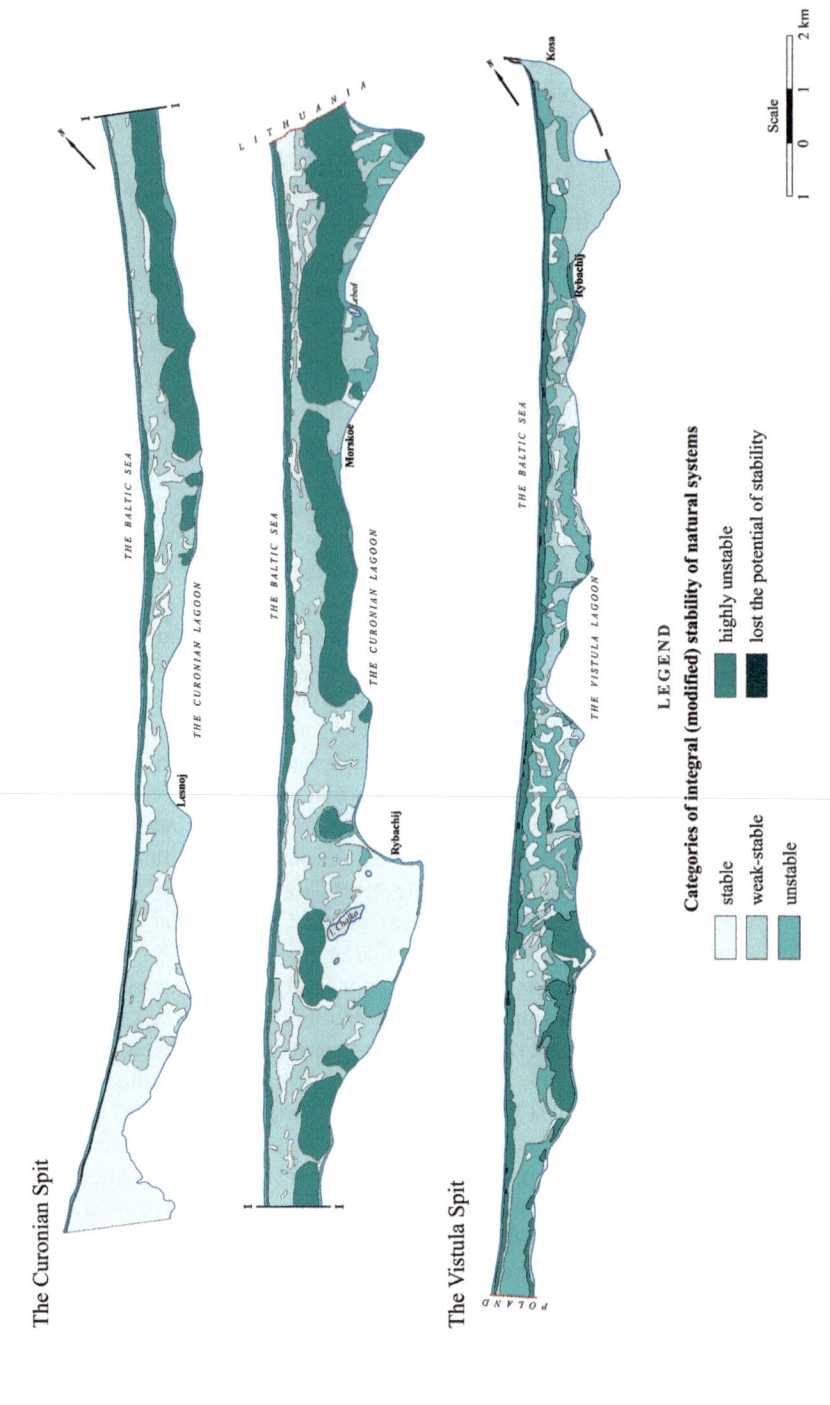

Fig. 15 Integrated (modified) stability of the natural complexes of the Curonian and Vistula Spits

Spit was its use for military purposes. The establishment on its territory in 1963 of a faunal area of regional significance "Vistula Spit," which functioned until 2004, did not change the structure of nature management at large. As a result, such an essential leisure activity site as the Vistula Spit dropped out of the recreational sector for several decades. The main kinds of natural resource management on the Curonian Spit in the post-war period were recreational, attended by an active development of the relevant infrastructure, forestry and fisheries. In 1967, with a view to retaining and increasing the population of the hunting fauna and studying bird migrations on the Curonian Spit, a wildlife reserve of local significance was established, and in 1987 the national park "Curonian Spit" was set up.

The chief kind of today's environmental management on the Curonian and Vistula Spits is recreational activities which on the Curonian Spit is controlled within the national park "Curonian Spit" and is carried out within the boundaries of the specially allotted recreational area, while on the Vistula Spit it is of uncontrollable character with the trend towards an increase in the recreational presence. The economic use or the use for military purposes are of secondary nature and are limited in the former case to the areas of settlements, and in the latter – to the sites of military facilities. A compulsory part of the environmental management on the Spits is conducting environmental protection measures aimed at supporting the functionality of the natural complexes. Such measures are currently carried out extensively on the Curonian Spit (involving the foredune fixing, planting of new and rehabilitation of old forests, lowering fire danger to forests, their protection from destructive pests and diseases, etc.), while being kept to a minimum on the Vistula Spit.

The peculiarities of the present-day natural-anthropogenic transformations of the eolian coastal-marine natural complexes of the Curonian and Vistula Spits, determined by their characteristics and usage pattern, condition a set of measures aimed at the regeneration, stabilization, and prevention of unfavorable natural and anthropogenic impact on the terrains, at ensuring sustainable utilization of resource potential and overcoming a contradiction between the conservation and economic-recreational interests.

An important component in the optimization of the conservational ecosystem exploitation on the Spits is conducting regular geoenvironmental monitoring; attributing a conservation status (that of a national park) to the Vistula Spit with consecutive development of the cross-border cooperation, in particular within the establishment of the Russian-Polish trans-border specially protected natural reservation; environmental and educational activities (organization and conducting environmental tours, ecological events, working with schoolchildren, etc.) and research activities.

With relation to the optimization of the recreational ecosystem exploitation on the Curonian Spit, it is currently important to modernize the recreational areas of the national park "Curonian Spit" involving an enhancement of the level of amenities, including the construction of new accommodation facilities, provision of state-of-the-art service, and reducing the load on the adjacent areas. As concerns the Vistula Spit, it is necessary to conduct land-use planning of the territory within the national park identifying and developing recreational zones which will allow to lower the loads and prevent the destruction of the Spit's natural complexes. An integral part of the optimization of the recreational ecosystem exploitation is conducting regular environmental and recreational monitoring of the recreational zone of the national park "Curonian Spit," of the scheduled national park "Vistula Spit" and the areas adjacent to them, and also controlling recreational flows.

Within the improvement of the residential ecosystem exploitation on the Curonian and Vistula Spits it is advisable to modernize the settlements, involving the construction of new and reconstruction of the current facilities of transport and engineering (water supply, wastewater disposal) infrastructure.

A high percentage of forest land on the Curonian and Vistula Spits determines the necessity of conducting forest improvement measures aimed at retaining the sustainability of forest ecosystems (planting new and rehabilitation of old forests, measures aimed at protecting forests from fires, destructive insects and diseases, improvement of sanitation conditions, etc.).

With a view to regenerating and preserving the functionality of the maritime sequence of the Spits, measures are conducted involving the construction of special coast protection facilities, physical fixation of the coastal dune ridge complex and poorly fixed dune massifs, and phyto-reclamation works.

Since farming operations on the Spits are of secondary character and are mainly concentrated in the area of the village of Rybachij (the Curonian Spit), the village of Kosa and at the 21st kilometer (the Vistula Spit) (haying, cattle grazing), the chief measure aimed at lowering the impact of this kind of environmental management on the Spits' natural complexes is setting rates for loads.

4 Conclusions

The longitudinal studies of the deposition coasts of the Baltic Sea within the Kaliningrad Region showed the necessity of an individualized approach to the natural sites with high dynamic properties, which was reflected in the application of original methodological approaches to the assessment of their digression, potential and integrated (modified) stability of eolian coastal-marine natural complexes, developed through the example of the largest depositional features of the Baltic – the Curonian and Vistula Spits. The procedures are based on the integrated index relying on the graded componentwise assessment of indicators of natural and anthropogenic genesis, identified taking into consideration natural peculiarities of the eolian coastal-marine terrain, the established system of environmental management and the intensity of human impact at the present stage.

While selecting the assessment rates, a number of provisions were taken into consideration: the Spits' terrains sustain an integrated effect of natural and anthropogenic factors, which in certain periods can cause irreversible alterations; the impact of factors on different NCs shows a strong variability of their manifestations and consequences; the predominance and the combination of factors influencing different NCs are complex; the key natural factor determining the development of the eolian coastal-marine natural complexes in the Kaliningrad Region is the wind-wave activity, a number of factors show a recurrence in their action; a principal anthropogenic factor exerting influence on all the NCs of the Spits and, in a number of cases, adjusting the course of natural processes, is recreational; soils, vegetation and ground waters play a key steadying role in the Spits' terrains.

The analysis of the natural stability of the natural complexes of the Curonian and Vistula Spits showed a similarity in the breakdown of stages of potential stability regarding the category of "weak-stable" – approximately 1/3 in terms of the area. The rest of the categories on the Spits shows a certain variability, which is, first of all, connected with the structural peculiarities of the Spits' natural complexes: the number of natural complexes falling into the category of "stable" is 19% higher on the Curonian Spit, which is first and foremost associated with the structure of the deflationary-depositional plain. The amount of unstable natural complexes is 18% higher on the Vistula Spit, which might primarily be connected with a large number of intricately orientated dune ridges giving a cavernous character to the deflationary-depositional plain; the number of strongly unstable natural complexes on the Curonian Spit is 11% higher than that on the Vistula Spit, which is first and foremost associated with the presence of moving dune massifs or dunes poorly fixed with vegetation.

The integrated (modified) stability mirrors the general trends of the natural stability of the natural complexes. The major changes primarily concern the areas exposed to the influence of factors of biological and human impact, in particular, an igneous factor, the impact of phytophagans (first of all, on forest cenoses, etc.), and a recreational factor.

As regards the breakdown of the natural complexes of the Curonian and Vistula Spits by digression stages, common patterns were identified: predominant is Stage I (was recorded for 55% and 52% of the Spits' territories respectively); highly subordinate status is taken by Stages IV and V (approximately 1% of the territory respectively). A high level of digression is characteristic of the natural complexes of the coastal zone (beach, coastal dune ridge complex).

The monitoring of the coastal zone with the use of the technology of surface laser scanning, conducted within the period of 2011–2015 on the deposition coasts, showed a high efficiency of the assessment of the morphometric indicators taking into consideration both linear and volumetric characteristics of the components of the surface coastal ecosystems, and the application of the software packages ArcGis 9.3., AutoCAD enabled us to visually represent the morpholithodynamic processes and to subsequently simulate them.

Acknowledgements The work was supported by the RFBR grant No. 12-05-00530-a.

References

1. Boldyrev VL (1998) Kurshskaya kosa: sostoyanie beregovoy zony i voprosy beregozashchity (The Curonian Spit: the state of the coastal zone and issues of coast protection). In: Slobodyanik VM, Manukyan AR (eds) Problemy izucheniya i ohrany prirody Kurshskoy kosy: collection of papers. SE "KGT", Kaliningrad, pp 87–99
2. Krasnov EV, Blazhchishin AI, Shkitsky VA (1999) Ekologiya Kaliningradskoy oblasti (Ecology of the Kaliningrad oblast). Yantarny Skaz Press, Kaliningrad, p 188
3. Kulakov VI et al. (2008) Kurshskaya kosa. Kulturny landshaft (The Curonian Spit. A cultural landscape). Yantarny Skaz Press, Kaliningrad, p 432
4. Mares K (1959) In: Yankyavichus K (ed) Itogi kompleksnogo issledovaniya (Kurshyu Mares. Results of comprehensive studies). Vilnius, p 547
5. Teplyakov GN, Boldyrev VL (2003) Formirovanie, sostoyanie i problemy sohraneniya landshaftov Kurshskoy kosy (The development, state and issues of preservation of the Curonian Spit terrains). Problemy izucheniya i ohrany prirodnogo i kulturnogo naslediya. NIA-Priroda Press, Moscow, pp 20–40
6. Boldyrev VL, Bobykina VP (2001) Obshchie cherty morfologii i dinamiki Vislinskoy kosy (Common features of morphology and dynamics of the Vistula Spit). Ekologicheskie problemy Kaliningradskoy oblasti i Baltiiskogo regiona: collected research papers. Kaliningrad University Press, Kaliningrad, pp 88–92
7. Volkova II (1995) Dyunnye kompleksy yugo-vostochnoy Baltiki i puti ih ratsionalnogo osvoeniya (Dune complexes of the south-east Baltic and the ways of their rational utilization). PhD thesis, Kaliningrad, p 185
8. Korneevets LV (1998) Gidrogeologicheskie usloviya i osnovnye cherty rezhima gruntovyh vod Kurshskoy kosy (Hydrogeological conditions and the main features of the ground-water conditions of the Curonian Spit). In: Slobodyanik VM, Manukyan AR (eds) Problemy izucheniya i ohrany prirody Kurshskoy kosy: collection of papers. SE "KGT" Press, Kaliningrad, pp 230–240
9. Korneevets LV (1990) Prichiny podtopleniya Kurshskoy kosy i rekomendatsii po ohrane ee prirody (Reasons of saturation of the Curonian Spit and recommendations on its nature conservation). PhD thesis, 11.00.11, Leningrad, p 137
10. Badyukova EN, Zhindarev LA, Lukianova SA, Solovieva GD (2004) Geomorfologia Kurshskoy kosy (The geomorphology of the Curonian Spit). In: Orlenok VV (ed) Pribrezhnaya zona morya: morfolitodinamika i geoekologiya: conference materials. Kaliningrad State University Press, Kaliningrad, pp 65–70
11. Solovieva GD, Badyukova EN (1997) Geomorfologicheskaya harakteristika Vislinskoy kosy Baltiiskogo morya (The geomorphologic characteristic of the Vistula Spit of the Baltic Sea). Geomorfologiya 2:82–89
12. Basalikas AB (1977) Landshafty Litvy (Terrains of Lithuania). Vilnius, p 240
13. Shaplygina TV, Volkova II (2013) Prirodnaya i antropogennaya transformatsiya landshaftov i reliefa Kurshskoy i Vislinskoy kos (The natural and anthropogenic transformation of terrains and relief of the Curonian and Vistula Spits). Geomorfologiya 1:95–103
14. Volkova II, Korneevets VS, Fedorov GM (2002) In: Fedorov GM (ed) Vislinskaya (Baltiiskaya) kosa. Potentsial vozmozhnostey (The Vistula (Baltic) Spit. The potential of possibilities). KSU Press, Kaliningrad, p 72
15. Belov NS, Volkova II, Shaplygina TV, Danchenkov AR (2014) Spetsifika primeneniya tehnologii nazemnogo lazernogo skanirovaniya pri monitoringe beregovoy zony (The characteristics of the application of the technology of surface laser scanning in the monitoring of the coastal zone). Mine Surv Bull 5:49–51
16. Zhamoida VA, Ryabchuk YP, Kropachev YP, Boldyrev VL, Sivkov VV (2008) Proyavlenie sovremennyh litodinamicheskih protsessov v beregovoy zone Kurshskoy kosy (The manifestation of modern lithodynamic processes in the coastal area of the Curonian Spit). Problemy izucheniya i ohrany prirodnogo i kulturnogo naslediya natsionalnogo parka "Kurshskaya

kosa": collected research papers, vol 6. Immanuel Kant State University of Russia Press, Kaliningrad, pp 149–166
17. Bobykina VP (2008) Morfodinamika beregov Baltiiskoy kosy (The morphodynamics of the Baltic Spit shores). In: Dinamika pribrezhnoy zony besprilivnyh morey: materials of the international conference (school – workshop), Terra Baltika, Kaliningrad, pp 37–38
18. Shaplygina TV (2010) Geoekologicheskaya otsenka sostoyaniya prirodnyh kompleksov Kurshskoy i Vislinskoy kos (The geoenvironmental assessment of the conditions of natural complexes of the Curonian and Vistula Spits). PhD thesis, Kaliningrad, p 198
19. Shaplygina TV, Volkova II (2013) Sovremennye prirodno-antropogennye predposylki transformatsii eolovyh pribrezhno-morskih prirodnyh kompleksov (Modern natural and anthropogenic preconditions of transformation of the eolian coastal-marine natural complexes). Vestnik Baltiiskogo Federalnogo Universiteta im. I. Kanta 1:39–46

Factors and Patterns of Current Development of Territorial Units in the Kaliningrad Region

Elena A. Romanova, Olga V. Vinogradova, and Darja V. Sergeeva

Abstract The article examines social and economic factors and laws shaping patterns of development of the regional area over the past 100 years – settlement and land-use systems (retrospective analysis of their variation and mechanisms impacting landscape environment), influence of geopolitical factors (borders: their position, functional types and mechanisms of impacting landscape environment), and principal trends and scenarios in the future development of the Kaliningrad Region. The current settlement system of the Kaliningrad Region is defined by a settlement network of East Prussia established by 1945, which largely followed the pattern of hydrographic network and landscape structure of the territory, by the postwar system of administrative-territorial division and management, by transformation of the regional transport system, by specifics of socioeconomic development of the area shared by all former Soviet Republics, and by current demographic processes. Using the Kaliningrad Region as a case study, some notions of landscape environment as a zone of intersection of areal and network components and their borders were introduced; a number of new directions in the development of geographic lymology were validated.

Keywords Current landscape, Land use, Settlement systems, State and administrative borders

Contents

1 Introduction 180
2 Theoretical Background 180
3 Methodology 181
4 Research Results and Discussion 183

E.A. Romanova (✉), O.V. Vinogradova, and D.V. Sergeeva
Institute of Environmental Management, Territorial Development and Urban Planning,
Immanuel Kant Baltic Federal University, A.Nevskogo Str. 14, Kaliningrad 236041, Russia
e-mail: alberta63@mail.ru

4.1	Retrospective Analysis of Settlement System Development over the Past 100 Years and Its Results: Evidence from Landscapes	183
4.2	Factors Causing Changes in the Settlement System in the Territory of the Kaliningrad Region	184
4.3	Spatial Features of the Current Settlement System in the Territory of the Kaliningrad Region	185
4.4	Retrospective Analysis of Land-Use System Over the Past 100 Years and Its Result: Evidence from Landscapes	186
4.5	Current State of the Land-Use System of the Kaliningrad Region	188
4.6	Current State of Areal-Networking Components of Landscape Environment	189
4.7	Current State of Network Components of Landscape Environment	193
4.8	The Influence of Borders on Other Components of Landscape Environment (Historical, Functional, and Spatial Aspects)	196
5	Conclusions	197
	References	198

1 Introduction

At present, the landscapes of the Kaliningrad Region area represent a complex system of territorial units with various degrees of their natural base transformation as well as at various stages of their development. At the foci of growth, an intense land use and planning continues with further construction of a cultivated landscape. In distressed regions, landscapes growing feral as well as feral landscapes predominate (using terminology proposed by Tyan-Shanskyi [1]). Landscapes growing feral and feral landscapes correspond to different stages of secondary succession. Without denying the fact that present-day landscapes in any territory are formed on the basis of natural landscapes, we believe that it is not a viable option to limit oneself to the study (and considering the duration of their anthropogenic transformation, also restoration) of only natural basis [2]. Over the past several centuries, the landscape environment of the Kaliningrad Region has been affected primarily by social and economic factors, which shaped the present-day layout and state of extant landscapes. Using these landscapes as a basis for analysis, we have suggested a new methodology for studying landscape areas subject to a long-term reclamation. The suggested methodology claims that studies of extant landscapes should comprise not only investigating structure of natural landscapes in a given area but should also investigate a spatial aspect of settlement systems as well as land-use dynamics over the period under investigation.

2 Theoretical Background

Up to date several research paradigms for investigating extant landscapes can be distinguished in international and Russian geographical science. Specific features of the Russian school of landscape study are related to its being rooted in physical geography, while Western school of landscape studies has been developing within the framework of social geography following K. Sauer's work [3]. The authors of the present study draw upon classical works in Russian geography, which represents a complex geographical approach without division of geography into physical and socioeconomic branches, Rodoman [4] and Solntsev [5], as well as ideas of contemporary scientists Kolbovskyi [6], Gutnov and Glazychev [7].

The area of the Kaliningrad Region does not have any landscape borders; however, it has many other kinds of borders: administrative, state, and ethnic which have been repeatedly transformed during contemporary history. Thus, the region appears to be a suitable object of inquiry for specifying the principal factors and patterns shaping contemporary territorial units – landscapes.

3 Methodology

The source material for investigating current landscapes of the Kaliningrad Region forms several strata (blocks) of evidence. The first (base) stratum of information comprises the structure of natural landscapes and their genetic appearance as identified based on relief and quaternary deposits constituting the area. This evidence is provided by the map of genetic types of landscapes using the data of landscape survey of the area carried out in 2003–2011 [8].

The second stratum of evidence comprises a present-day system of settlement in spatial relation. The third stratum is analogous to the second one and deals with a spatial concept of the previous settlement system (for this we have selected a time sample – 1939). E.A. Romanova compiled maps of the actual population density for two-time samples – 1939 and 2009 [9]. The population distribution over the area shows not only the dynamics of settlement landscapes but also the intensity of landscape load. Private subsidiary farms – dachas and vegetable plots – occupy mostly the areas nearest to inhabited localities. Recreational load is higher around the towns; the closer one gets to major populated areas, the higher is the road network density, etc. Calculation of real population density was carried out both within settlement borders and the areas located within walking distance from the inhabited locality taking into account the specifics of their landscape and land use.

Superimposing the second and third base strata using GIS enabled us to compile a map reflecting the degree of transformation of current landscapes, which shows areas with variable degrees of secondary succession of landscapes, which, in fact, helps to identify the areals where transformation of landscapes by anthropogenic activity continues or, on the contrary, has practically discontinued. Nevertheless, this kind of evidence cannot be absolutely reliable, since thanks to advanced technologies humans have learned to alter relief, soil landscape, and vegetation cover even on uninhabited terrain. For instance, there are several military training areas in the Kaliningrad Oblast (Region), such as artillery practice and bombing grounds, both operating and abandoned ones. The areas have already been uninhabited for several decades. In order to eliminate such errors, the compilation of the map of areals of anthropogenic relief was complemented by a ground observation carried out in the last 5 years by participants of a landscape expedition.

Moreover, a cross-spectrum analysis of the first, second, and third strata makes it possible to establish a link between population settlement and natural environment. In 1939, the areals with high population density (over 100 people/km^2) covered almost the entire area of the region forming an uninterrupted band along the Sambia coastline, along the Pregel River, and concentrating around the towns of Königsberg (Kaliningrad), Insterburg (Chernyakhovsk), Gumbinnen (Gusev), and Tilsit (Sovetsk). The high population density areals in the western and eastern parts of the region were counterbalanced. Low population density areals (below 10 people/km^2) were located in the northwest of the region, where there are still massifs of lowland and raised bogs, as well as in individual remote areas covered with forests. According to 2009 data, the areals with population density over 100 people/km^2 had become dissociated, while their area had considerably decreased. The southmost tip and eastern and southeastern areas of the region are almost unpopulated. Most of the area of the Kaliningrad Region has population density below 10 people/km^2. The western seacoast of Sambia Peninsula as well as its central areas is scantily populated. The majority of the population of the region is localized in the regional capital and around it. Compared to the 1939 settlement system, the current settlement system shows stronger association with natural landscapes than the prewar one. There is also a strong correlation between low population density in the areas and their specific landscape features creating conditions unsuitable for land use.

The fourth stratum of evidence – the current land-use system – has been compiled in the course of a more detailed investigation of the area at the level of municipality and lower. The fifth stratum of the information is confined not to areals but to a network. Depending on the scale, it comprises information on the existing point (fixed) and linear facilities (settlements, roads, dams, etc.). Combining areal and network approaches in studies of cultural landscapes not only increases research possibilities but also serves as a basis for forecasting their further transformation patterns. Moreover, comparing and contrasting various areal strata enable us to obtain additional strata that facilitate understanding of the nature of the phenomenon. In this respect, settlement zones play a special role since depending on the research scale they can be regarded as either areal or as network ones.

The application of the methodology described above allowed to identify two principle factors which form the current landscape environment of the Kaliningrad Region and influence the main developmental patterns of its territorial units both conventionally natural units (forests and specially protected nature conservation areas) and conventionally anthropogenic ones (settlement zones agricultural land).

4 Research Results and Discussion

4.1 Retrospective Analysis of Settlement System Development over the Past 100 Years and Its Results: Evidence from Landscapes

Settlement system of any territory is formed under the influence of a range of factors – the level of socioeconomic development of society, the distribution of economic units, demographic composition of its population, administrative and territorial system, as well as natural features of the area. The settlement system appears to be more dynamic than the extant network of settlements and reflects primarily socioeconomic and administrative-managerial state of the region since it depends on a network of administrative-territorial division, which predefines population service system and geography of the labor market [10, 11]. Obviously, a landscape structure of the area indirectly affects the settlement system via established network of settlements, which is directly related to relief features, hydrographic network, as well as soil and vegetation cover. Besides, the natural conditions of the area affect the existing current settlement system because they determine the efficiency of capital investment in infrastructure development and construction, as well as the positioning of industrial facilities utilizing natural resources. In order to investigate current landscapes of the Kaliningrad Region and determine the degree of their transformation, it is necessary to consider not only the current settlement system in the region but also to compare it to the previous one – the settlement system which existed in the prewar period.

The Kaliningrad Region is an old-cultivated area whose settlement system has been developing over many centuries. One of the specific features of the region is a complete substitution of the local population, which took place after World War II due to the accession of part of the former Eastern Prussia to the Soviet Union as part of the Russian Federation. Repatriation of the German population completed in 1948 and population of the area with immigrants from Central Russia, Belorussia, and the Ukraine, establishing a new political and economic system, had triggered dramatic changes in the management system and determined special aspects in redevelopment of residential places and trends in infrastructure development. The extant settlement system of the region demonstrates, on the one hand, some similarities with settlement systems characteristic of other subjects of the Russian Federation in the non-black earth zone (non-chernozem zone) of the Russian Plain, and on the other hand, it is the system inherited from the settlement system in the northern part of Eastern Prussia.

Before the termination of World War II, the area presently occupied by the Kaliningrad Region belonged to Eastern Prussia and was densely populated. The total population that lived within the borders of the present-day region in question as of 17.05.1939 was 1,107,197 people. The total population of the Kaliningrad Region as of 1.01.2015 is 968,944 people. In 1939, the average population density was 83.5 people/km², while as of January 1, 2009, the average population density was only 70.7 people/km², and as of January 1, 2015, it was 64.06 people/km².

In 1939, the north of Eastern Prussia was largely rural. Industry concentrated in the regional capital and major towns, while locally only small businesses operated whose production was based on local raw materials. In 1939, the rural population was 479,777 people (43.3% of the total population); at least 60% of the economically active population was involved in agricultural production. In individual territorial subdivisions, average rural population density was high – 30 to 63 people/km². Average population size of a rural settlement varied from 138 people (Schlossberg region) up to 525 people (Samland Region) [12]. There were some regional variations in the distribution of the rural population. In lowlands of the Neman River and other major river valleys, linear settlements predominated; the population also centered along the canals. In upland areas small and medium-sized isolated farmyards adjoining forest edges typically occurred. Circular and star-shaped settlements around Königsberg extended as far as the seacoast [13].

In 2009 the rural population of the Kaliningrad Region was 219,935 people (23.5% of the total population of the region). Average population density for individual municipalities varied from 6.7 people/km² (Krasnoznamenskyi District) up to 38 people/km² (Zelenogradskyi and Guryevskyi Districts). Average population size of a rural settlement varied from 109 people (Ozyorskyi District) up to 284 people (Slavskyi District). The maximum in rural population for the Kaliningrad Region was observed in 1960 (219,400 people), since then it has been steadily declining. The minimum rural population size in the region was recorded in 1983 (179,200 people). Since 1994 until 2009 there was a trend toward an increase in not only absolute size of the population of the region but also an increase in the share of the rural population in relation to the residential population of the region – from 21.6% (1994) up to 23.5% (2009) and 22.3% (2015). An increase in the size of rural population in the 1990s is generally linked to the outflow of urban population into rural areas due to the systemic crisis of that period [14]. An increase in the size of rural population in the past 5 years can be explained exclusively by migration processes since natural population growth in many districts of the region still remains negative.

4.2 Factors Causing Changes in the Settlement System in the Territory of the Kaliningrad Region

The current settlement system of the Kaliningrad Region has been shaped by the following factors:

- A settlement network of former Eastern Prussia established by 1945 which was largely associated with patterns of hydrographic network and landscape structure of the territory

- A postwar system of administrative-territorial division and territory management
- Changes in the transportation network of the region which manifested themselves in downsizing of the regional railroad network and interrupted routes of communication with the southern neighbor for decades
- Specifics of social and economic development of the area shared by all post-Soviet republics which determined specialization of production units of the districts and were consequential for population settlement
- Demographic processes

Postwar changes affected the transportation system of the territory. Until 1990, the Kaliningrad Region was a closed territory. The roads connecting it with its southern neighbor were blocked (at the moment, there are four operating border-crossing points on the southern border of the region). On the contrary, northern and eastern borders of the region were connecting it to the mainland of the country in the Soviet times. Road communications there did not stop until the Baltic Republics regained their state independence (at present, four border-crossing points are in operation there). Eastern Prussia had an extended railroad network providing connection to all major and smaller towns of the region as well as to many rural settlements. After the war, most railroad tracks were removed and at present only the main lines are in operation. Unlike railroad tracks, road network has been almost entirely preserved.

The settlement system in the region was also affected by the demographic factors, which, in general, held true of many other regions of the Russian Federation, mainly, a negative natural population increase. Another characteristic feature of the region is a constant migration inflow of people.

Among the factors outlined above, natural characteristics of the territory play a special role. On the one hand, it was a landscape structure of the territory which had shaped the structure of the prewar settlement system. On the other hand, the settlement system itself has been dramatically transformed over the past decades and has acquired some features similar to other regions of non-chernozem belt of Russia, namely, it has become more polarized.

4.3 Spatial Features of the Current Settlement System in the Territory of the Kaliningrad Region

Comparison of settlement systems of the Kaliningrad Region in 1939 and 2009 reveals that at present the territory of the trigon is less densely populated than it was before the war, while the majority of the population is currently concentrated in the regional capital. Within the region, one can distinguish several zones whose settlements have become distinct in terms of their economic role and geographical position [15]: the west of the region (including Kaliningrad and its residential neighborhoods, seaside resort subzone, coastal defensive-industrial subzone, access to the Russian-Polish border and agricultural inside of the peninsular), a farther commuter zone (the west of Gvardeiskyi, Polesskyi, and Pravdinskyi Districts), the north or the northern periphery of the region (Prinemanye, i.e., the Neman River area), the inside periphery (including Chernyakhovskyi, Gusevskyi Districts, and eastern Gvardeiskyi and Polesskyi Districts), and southeastern cross-border periphery (including Ozerskyi, Nesterovskyi Districts, as well as eastern Pravdinskyi District). The residential neighborhoods are the most populated areas with population density 186 people/km^2, which is approaching European indicators, while the near-border areas are the least densely populated and low urbanized [14, 16].

Changes in the settlement system are manifest in landscape changes. At the first stage of succession, derelict and neglected farmyards are overgrown with *Petrophytum* plants and later on, alongside primary soil formation on top of construction waste, the weed stage follows. Further direction in the evolution of the land plot depends on the degree of humidification of the area.

4.4 Retrospective Analysis of Land-Use System Over the Past 100 Years and Its Result: Evidence from Landscapes

In any area, a land-use system defines a current state of its landscapes. A land-use system is dynamic in terms of both time and space. In landscape structure of individual territories, alongside anthropogenic modifications of geographic systems, there are landscapes withdrawn from economic use and being at various stages of renaturalization. At the same time, some previously fallow lands are reintroduced into economic use, and some of the lands change their category – some lands are allotted for residential or production uses, while others become involved in agricultural use.

Timewise, such variations in land use are generally synchronous for all parts of the country and the region since they are related to certain events on a global and national scale.

Spatially, however, variations in land use are not synchronous since they depend on a multitude of local factors: geographical position of a regional unit in question, its natural conditions and resources, quality of human capital, regional policies in industry and agriculture, as well as accessibility of advanced agricultural technologies to the producer.

Russia has experienced two major crises in land use over the past 70 years: the former was related to World War II, while the latter was caused by a transition from socialist state-controlled type of economy to a market-driven one. During the first crisis, vast areas of land in the European part of Russia were not only abandoned but also altered in the course of military operations. During the second crisis (in the 1990s), much of the agricultural lands became derelict, and livestock population considerably decreased. Thus, arable land, hayfields, and pasture grounds became overgrown. In the past 20 years, post-agricultural progressive (secondary) successions have become a typical process in most of the rural regions of Russia [16, 17]. At the same time, the process of recovery of previously derelict lands to agricultural use has been observed in some Russian regions.

Likewise, the Kaliningrad Region did not manage to evade the land-use crises. However, the process of current landscape formation in the region possesses some specific features, which are conditioned both by its natural peculiarities, and the entire history of land assimilation and development [18].

Before World War II, the area of the Kaliningrad Region formed the northern part of Eastern Prussia. In 1939, the forests covered 14% and bogs covered 12% of its territory, while the rest of the land was occupied by settlements, road network, and agricultural lands. Most of the forests were reclaimed, and replanting of fir trees and oaks was carried out. Forest areas reclaimed by planting common alder were subsequently replanted with more valuable species of trees. According to the Central Statistical Office data, in 1943 the share of land involvement in agricultural production in the part of Eastern Prussia that later on formed the Kaliningrad Region was very high – up to 76% (the same indicator in the Leningrad Region for the same period was 11–12%) [19]. In the prewar period, among agricultural land the share of arable land was 68%, the share of hayfields was 12%, and pasture grounds was 18%. Another 2% was accounted for by orchards and other land types. Agriculture in Eastern Prussia specialized in dairy cattle husbandry and swine rearing but also had a considerable share of crop growing represented mainly by forage production. The techniques used in agriculture were considered to be the most advanced at that time.

A special feature of the territory is a vast area of polder land, which is the largest in Russia at the moment. Most of the polder land is located in the ancient estuary of the River Neman. Over 50% of the old alluvial low-lying area is covered with polder land. Before the war, up to 80% of the region's area was drained. Polder land occupied about 78 thousand hectares of land comprising 58 polders and 96 pumping stations. At present polder system in the old estuary of the River Neman covers the area of 65.7 thousand hectares comprising 32 polders with 65 pumping stations (49 of which are currently functioning). Polders are areas of land where controlled irrigation-drainage land reclamation is carried out using field or subfield drainage. The system includes a drainage network, constructions for water discharge, and border (protective) dikes. Polders are used in intensive agriculture, mainly for growing forage grasses, potatoes, and vegetable crops. Moreover, many rural settlements are located in the polder areas. In the years after the war, the features of land use in the newly founded Kaliningrad Region were shaped by the following factors: changes in the number and quality of human resources (the German population that left the region was never sufficiently replaced by migrants from other regions of the USSR and demobilized servicemen), vast areas with disturbed lands and most recent belligerent landscapes, poor state of the reclamation system and extensive littering of the upper surfaces, and soil with metal and explosive materials (bomb shells, land mines, and other kinds of weapons). Such conditions made for a very slow regeneration of agriculture in the region. By 1965, the area of agricultural lands had increased up to 54%; among them, 46% was arable land, 22% was hayfields, and 32% was pasture grounds. Crop yields were low and so were milk yields.

By the end of 1980s and early 1990s, a certain increase in agricultural production was marked: by 1985, cereal crop yields had reached 3 60 kg per hectare (36 dt/ha); yield of potatoes was 110–120 dt/ha. The area of reclaimed land had reached a postwar maximum: in total, 1,036 thousand hectares of land had been drained (about 65% of the region's area), out of which 730 thousand hectares was agricultural lands. Specialization of agriculture had not changed focusing primarily on cattle breeding and forage production.

The crisis of the 1990s affected the economy of the Kaliningrad Region in the same way as it affected other non-chernozem regions of Russia. The level of production in agriculture dropped: from 1990 to 1999 the areas under crop decreased by 36%, cattle population decreased by 65%, and among them cows decreased by 56%. The proportion of livestock production in gross production had gone from 70% in 1990 down to 51% in 2000 and 45% in 2004. According to official figures, the share of agricultural land was reduced to 48% in 2006 [20]. A similar trend was observed in most Russian regions.

4.5 Current State of the Land-Use System of the Kaliningrad Region

At present, a sustainable growth of agricultural production in the Kaliningrad Region is observed. In June 2014, the area under crops made up 222 thousand hectares (in 2009 and 2011 the area under crops was 165.2 and 143.6 thousand hectares correspondingly) [21–23]. Involvement of unutilized lands in agricultural turnover is one of the priorities in the agricultural sector of the Kaliningrad Region. In order to address the challenge, the Government of the Kaliningrad Region adopted a target program for the Kaliningrad Region "Involvement of agricultural land unused for its intended purpose in agricultural production in the Kaliningrad Region for the period of 2011–2016." Within the framework of the program, over 100 thousand hectares of cultivated land was reintroduced in economic turnover during 2011–2014, and now the proportion of utilized agricultural land makes up on average 65% in the region. The areas under crops had increased by half, namely, by 78.4 thousand hectares. In 2014, agricultural producers planned to reintroduce into turnover over 20.0 thousand hectares of unutilized land. As of July 21, 2014, 18.3 thousand hectares had been reintroduced. The other 22 thousand hectares of land reintroduced into agricultural production are accounted for by hayfields and pasture grounds. It has been planned to increase the proportion of reintroduced agricultural land up to 75% by the year 2016. Spatially – in terms of individual municipalities in the region – this process is nonsynchronous. For example, in 2013 the proportion of areas under crops in different districts was on average 23.4% of the total cultivated land with maximum values marked in Gusevskyi and Nesterovskyi Districts (40.5 and 52.0% correspondingly), while minimum values were marked in Gvardeiskyi, Bagrationovskyi, Zelenogradskyi, and Slavskyi Districts (11.2–15.0%).

Grain crop yield is an indicator of intensive agriculture. In 2014, due to favorable weather conditions, mean yield of grain legumes in the region reached 410 kg per hectare (41 dt/ha). This figure placed Kaliningrad Region within the top ten regions of Russia. The region took the first place for rape yield (25 ht/ha). Croppage has increased recently. As of November 2014, the harvest of grain crops and grain legumes was 438 thousand tons and including rape made up 530 thousand tons. In terms of regional areas, these indicators were distributed unevenly.

In recent years, alongside traditional dairy cattle husbandry, a new direction in agriculture has emerged – beef husbandry. In Nesterovskyi District, the number of livestock had increased from 9.2 up to 15.4 thousand heads, while in Ozyorskyi District it had increased from 3.2 up to 27.6 thousand heads from 2009 till 2013. Consequently, this trend resulted in increase in areas covered with forage crops as well as hayfields and pasture grounds. The largest areas with forage crops were planted in Bagrationovskyi and Nesterovskyi Districts (42.7 and 42.9% correspondingly of the total area under crops) in 2013. Dairy cattle husbandry continues to develop: average milk yield per cow per year was 5,486 kg in 2013 (which was only 4,285 kg in 2009).

Interestingly, land areas under crops and the number of workers involved in agriculture have decreased compared to 1990, while the croppage has increased due to application of intensive methods in agriculture. At present, all stages of transformation of the landscape environment are represented in the territory of the Kaliningrad Region including those where a progressive succession continues. Over decades, many of these territorial complexes have turned into a kind of "nature reserves." They have had a beneficial influence on biodiversity of the territory and become nodes for emerging ecological framework of the region. A mosaic pattern of current landscapes in the Kaliningrad Region makes it necessity to study thoroughly the local conditions when developing spatial and strategic plans. Only using this approach the peculiar feature of landscape environment can be turned into an advantageous feature ensuring a sustainable development of the region in the future.

4.6 Current State of Areal-Networking Components of Landscape Environment

Settlement zones are referred to as areal-networking components of a landscape environment. On the one hand, patterns of settlements represent a networking element of areal settlement systems, and reduction in a plotting scale makes even large settlement zones appear as dots on the map. On the other hand, on closer inspection, any settlement zone can be regarded as an areal formation because it occupies a certain area and, therefore, allows for a further subdivision into districts within this area, which is one of the main properties of an areal. On these grounds, settlement zones can be classified as an intermediate type of spatial formation.

Kaliningrad There is only one truly large town in the Kaliningrad Region just like in the former northern Eastern Prussia – Kaliningrad, former Königsberg. According to V.P. Semenov-Tyan-Shanskyi's terminology, it can be referred to as the town "proper." Indeed, the population size of Kaliningrad is 419.2 thousand people (01.01.2014), which is over 40% of the population in the region (currently the Kaliningrad Region). The town was founded due to its convenient geographical position (river estuary, close access to the sea, intersection of trading routes) and has been developing under the influence of economic and geographical factors. Kaliningrad (Königsberg) is by definition the capital city of the region.

However, it is not only its geographical position, which makes a peculiar feature of Königsberg but also the fact that it was a university town. Beginning with 1,545, it was Königsberg University that singled out Königsberg from the list of provincial German towns. Gradually, natural sciences and studies became prevailing there. University buildings (institutes) were located all over the town area. The faculty of some of them included hundreds of researchers and teachers.

Many of the world's greatest towns fall by the wayside at certain moments of their history. The same destiny befell Königsberg. The devastating World War II did not bypass that once flourishing town. Uncertainty in the future of the town during the first postwar decades, mentality of the new inhabitants of Kaliningrad and general trends in the Soviet policy and economy management common for the entire country were largely responsible for the present-day image of the town which has been formed for over half a century.

Despite the diversity of spatial organization, the town has preserved its major directions of traffic and radial-ring structure characteristic of ancient European towns. Besides, there are many historical buildings in the town (built in the nineteenth century and earlier). In fact, considering recent infill construction, over half of the buildings in the residential area were built before the war. Buildings of the Soviet period account for about 25% of all buildings, while most recent construction accounts for about the same level – 23%. The calculation was carried out based on the area covered by buildings of a corresponding time period (age). The calculation of the correlation in question carried out in relation to square meters presents a distorted picture: low-story houses and private detached houses prevailed among the prewar buildings (in square meters, prewar housing constitutes only a quarter of the entire housing stock of the town).

Among special features of the current landscape structure in Kaliningrad is a traffic network, which was inherited from the prewar period and ensures continuity of spatial structure of the town.

The inherited traffic network is a principle source of current problems in the town. Traffic capacity of the main streets does not meet the demands of growing traffic flows. The complicated situation is further exacerbated by another fact: there are fewer bridges across the river than there used to be before the war (there are only five bridges in operation including Trestle bridge). All of these factors have led to a significant congestion of the traffic streams at several points in the town center, which enhances contrasts of Kaliningrad urban environment.

Sovetsk This is the second largest town in Kaliningrad Region. Its population size is 42.6 thousand people. This is a historical town, whose name was Tilsit before the war. Significant landmarks in European and Russian history are associated with this town. In terms of landscape, specific features of the town are related to its riverine and cross-border location. It is located on the River Neman, which is not the largest river in the region but serves as a natural historical frontier. Sovetsk suffered less significantly in the last war and thus retained entire blocks with densely situated multistory buildings of the prewar period (in Königsberg, by contrast, such blocks were turned into ruins and did not survive until the present time). Spatial structure of the town is semi-radial because unlike the regional capital, it is situated on one slope of the river valley. Road network gets more crowded at the main river crossing point, the famous Queen Louise Bridge. Industrial enterprises also tend to be located closer to the river. There is a river port in the town. Besides, Sovetsk is an important railroad junction in the region. Unlike Kaliningrad where railroad lines divide the town into distinct zones ("islands"), a spatial structure in Sovetsk is less complex because the railroad cuts off only the westmost section of Sovetsk without entering the center. A special feature of Sovetsk is vast green areas such as parks, squares, and forested areas which occupy over 30% of the entire town area. Residential areas make up about 45% of the town territory, while about 20% is accounted for by industrial and transportation zones. Residential outskirts of the town are more rural rather than town-like in appearance: low-story detached houses are situated on large plots of land.

Chernyakhovsk The town is situated at the junction of major highways connecting the region through neighboring states with Central Russia and Moscow. The population of Chernyakhovsk is 39.4 thousand people. The town is located at the confluence point of the two rivers, Angrapa and Instruch, and was named Insterburg before the war. In terms of landscape, a specific feature of the town is a large area with prewar buildings including castles of the Teutonic Order Insterburg and Georgenburg. Another feature is a large number of garrison towns scattered over the area. This is a totally inherited feature because military units are housed in the prewar barracks. The relief of the town was formed in river valleys which makes it more complex compared with the abovementioned towns. Just like in Sovetsk, a railroad junction is located away from the town center and cuts off southern outskirts from the main area of the town. Residential areas occupy up to 60% of the town area (including almost rural outskirts and garrison towns); about 20% of the town area is covered with green zones including flood meadows of Instruch floodplain. Industrial and transportation zones occupy about 15% of the town area.

Other towns There are over 18 smaller towns in Kaliningrad region whose population size does not reach the officially adopted in Russia "urban requirement" of 12 thousand people. The largest among smaller towns is Baltyisk. In terms of population, it is followed by Gusev and Svetlyi, which fall into the category of "semi-medium" towns. Krasnoznamensk is the smallest town among them (3,751 people). All of these towns apart from Primorsk function as administrative centers which is their primary function because the industrial potential of these towns is quite insignificant.

Rural settlements There are 1,801 rural settlements in the Kaliningrad Region (2009). Their distribution over the area of the region is uneven. The maximum settlement density (over 10 settlements per 100 km^2) is found on Kaliningrad Peninsula as well as in Guryevskyi and Polesskyi Districts, that is, within the commuter belt around Kaliningrad.

Overwhelming majority of the present-day rural settlements (98%) have a prewar history. In general, these settlements have retained their prewar nucleus of buildings and houses. There is usually an old church ruined to a varying degree in the center of the settlement. Afterwar construction occupies on average from 30 to 60% of the residential area in the settlement. However, the spread in values over the regional area is quite significant: many small settlements and isolated farmyards comprise entirely prewar buildings, while larger settlements, previously former collective farm (kolkhoz and sovkhoz) premises, are made up almost entirely by buildings dating to 1960s–1980s. Landscape appearance of the settlement depends totally on the period of its construction.

The prewar residential areas consist, as a rule, of spacious redbrick dwelling houses or stuccoed houses on a foundation made from crude stone. Household outbuildings, mostly well wrought, roomy, and high, are also made from boulders or red brick. These are horse stables, cattle sheds, and thrashing barns. The buildings of the Soviet period are standardized and thus typical of all districts in the region (and almost of all non-chernozem regions in Russia). Dwelling houses are designed for one family and built from sand-lime brick. Nearby are low sheds or pre-engineered two-story apartment buildings. Here and there, farms can be found. They are low concrete constructions. Post-Soviet buildings are, generally, represented by detached houses varied in their architectural design, which quite often have a garage.

Other residential areas A special type of a residential area, which does not fall into any settlement category, is represented by dacha settlements (or gardening communities). The phenomenon of *dacha* (a small cottage in the countryside with a vegetable plot and a garden), as a unique type of "a town dweller's second house" having agricultural purposes (typical of the Moscow region and of Central Russia, in general), has been studied in detail by Russian researchers [24]. In Kaliningrad Region, dacha settlements frequently either occupy partially the territory of a town (as, e.g., in the regional capital) or are situated in the suburban areas of larger settlements. The following types of dacha settlements (referred to as dachas for short) can be distinguished based on their location: dachas within the area of the regional capital; dachas of Kaliningrad dwellers located outside the town; and dachas of dwellers from other towns in the region. The dachas in the region differ from each other not only in their location but also in their building type, size of a land plot, and land-use pattern.

4.7 Current State of Network Components of Landscape Environment

Network components of landscape environment comprise primarily objects that have linear or point-type character, in other words, elements of transport and industrial infrastructure.

Motorways and their infrastructure Kaliningrad Region has a dense network of motorways, most of which have hard covering. The density of motorway network varies from one regional district to another from 16 km/100 km^2 in the central region to 89 km/100 km^2 in the south. Average density of motorway network with hard covering constitutes 41.7 km/100 km^2 (cf. 0.37 km/100 km^2 on average in Russia and 36 km/km^2 in the vicinity of Moscow). Many of the motorways in the region have over a 100-year-long history and specific appearance in terms of landscape. Because of trees planted on either side of the roads, they looked more like alleys with Berlin highway being an exception among prewar motorways. Along the motorways, not only various species of linden (occurring most often along the roads) were planted but also valuable species of tree such as oaks, ash trees, maple trees, and even fruit trees in the south. Another special landscape feature of old motorways is their large sett or cobblestone paving that has been preserved in some road sections (e.g., a sett pavement near the settlement Mezhdurechye on route to Moscow).

Infrastructure of the motorways comprises road interchanges, crossovers, and bridges. Road interchanges are associated with recent roads, but a few road interchanges were retained on Berlin Highway. Most of the crossovers are located in the regional capital, but there are also crossovers in Sovetsk and Chernyakhovsk. They were built before the war and pass over (or under) railway lines. The region has an extensive river system; therefore, there are many motorway bridges in the region built both recently and before the war. Some bridges are of historical significance, for example, an old bridge across the River Pregolya in the settlement Znamensk, a suspension "Dutch-style" bridge across the River Lugovaya in Slavskyi area, and a railroad bridge across the River Krasnaya near the settlement Dmitriyevka.

Railways and their infrastructure Before the war, the territory of the region used to be covered with a dense network of railways. In 1939, their total mileage was 1,823 km (including 442 km of narrow-gauge track with track gauge 750 mm). Besides 184 stations and 240 roadside stations were in operation. At present only major railway lines are retained with the total mileage about 730 km. Most of the railroads in the region are single-track and non-electrified railways (only 14% of the railroads are electrified, mainly railways leading to the seacoast). The gauge width of the railways complies with Russian standards (1,520 mm). An exception to this rule is a spur track going from South Railway Station to Poland (and farther to Berlin) as well as direction from Zheleznodorozhnyi to Chernyakhovsk, which has a standard European gauge width (1,435 mm). European track also enters the territory of the region in Bagrationovsk. Transshipment of cargo from European to Russian gauge width is carried out at stations Chernyakhovsk and Dzerzhinskaya Novaya in Kaliningrad.

Railways that disappeared after the war (which means about two thirds of their prewar mileage) form a special type of landscape, which, despite "natural" character of vegetation, still retains some features of a cultural landscape:

- Firstly, the forms of relief least prone to damage have been preserved (bodies of railroads, excavations). Sixty years later, after the railway line was removed, they can be clearly seen on the surface of the ground. The vegetation of the former bodies of railroads (which can be over 4 m high) is similar to that of earth dams (since railway bodies were made from boulder-pebbled material).
- Secondly, in some places the infrastructure of a removed railway line has been preserved; generally, these are crossovers. For example, on Kaliningrad Peninsula there used to be a narrow-gauge track connecting Marienhof station (present-day Pereslavskoye-Zapadnoye) and Gaffken station (present-day Parusnoye). A large viaduct of that railway built from crude stone still exists today. The viaduct goes across the valley of the River Nelma and is still used as a country road. Besides there remained some deep excavations (up to 15 m deep) in places where the railway crossed western spur of terminal moraine upland.
- Thirdly, preserved platforms and station buildings have become special components of a cultural landscape. Because many railways disappeared after the war, these constructions survived only in settlements. Station buildings are used as dwelling houses, while storehouses are used as sheds. Besides, in some places, platforms remained intact.

Airfields and their infrastructure There are several airfields in the region; however, not all of them are used for their intended purpose, and some of them have been suspended.

Water routes The position of the Kaliningrad Region predetermines the presence of harbors and port facilities. The port complex of the region comprises commercial port and fishing port in Kaliningrad, a port in Svetlyi, a port in Pionerskyi, terminals for transshipment of oil products in settlement Izhevskoye and Kaliningrad, as well as a ferry terminal in Baltyisk. The base of the Baltic Naval Fleet is located in Baltyisk. Port infrastructure includes not only mooring areas but also Kaliningrad Sea Canal, river boat yards, and bulkheads, which come under the authority of Maritime Administration of the Port of Kaliningrad. Many rivers and canals in Kaliningrad Region suitable for navigation have hardly been used for this purpose since 1994. The system of inland waterways comprises rivers Neman, Pregolya, Deima, Matrosovka, Nemonin, and Lugovaya as well as canals Primorskyi, Polesskyi, and Ozerkovskyi.

The presence of a river or sea port largely shapes spatial structure of the town, because it is waterways which have been the most convenient means of transportation since medieval times. In coastal and riverside towns, circular streets usually follow the contour of former defensive walls, while radial streets converge either in the town center (which in most European and Russian towns is located at the point of a river bend) or in the sea harbor. Thus, spatial structure of Kaliningrad, Sovetsk, and other ports of the region is not exceptional in this respect.

Industrial zones and enterprises Most of industrial enterprises in the region are located in Kaliningrad and its suburban area. In fact, all industrial centers in the west of the Kaliningrad Region (to the west of Polessk–Pravdinsk line) belong to Kaliningrad industrial hub. The other two smaller hubs are located in the north (the hub Sovetsk–Neman) and in the east (the hub Gusev–Chernyakhovsk) of the region.

In terms of space, the Kaliningrad industrial hub is heterogeneous: the concentration of industrial zones increases in the direction of the regional capital getting more crowded at its borders and then stretches in tongue-shaped areas in western, north-eastern, and southern directions from it, following largely the contours of a hydrographic network and railway lines.

Similar spatial patterns are inherent to industrial hubs of a lower order in the region: older enterprises tend to gravitate to rivers and railway lines, while more recent ones tend to be located in the suburbs of towns and quite often oriented toward motor routes.

Oil-recovery facilities and quarry-dumping complexes represent a special type of industrial landscape. Over 25 oil deposits have been discovered in Kaliningrad Region in recent time (two of them located in the Baltic Sea area). Commercial oil production has been carried out since 1975. The company OOO "LUKoil-Kaliningradmorneft" is developing 18 land oil deposits [25], which are located in Bagrationovskyi, Pravdinskyi, Gvardeiskyi, and Slavskyi Districts. Twenty-two kilometers from the coast, an offshore oil deposit Kravtsovskoye (D-6) is being developed. From there, oil is transferred to an oil-gathering facility "Romanovo" via a subsea pipeline.

There are numerous quarry-dumping complexes over Kaliningrad Region, which differ in their size. They are developed mostly for construction materials such as sand and sand-gravel aggregate (less often, red glacio-lacustrine clay). The largest quarries are confined to old alluvial deposits of the River Pregolya (settlements Ozerki and Pushkarevo), fluvio-glacial formations or uplands of terminal moraines.

The quarries are an example of a recent landscape, which has completely lost its natural foundation. Drastic alterations impact not only vegetation and soil cover (which become completely destroyed and never recuperate to their original condition) but also a relief and geological foundation because a multimeter stratum of quaternary deposits is extracted.

Power engineering facilities The following types of electric power plants are located in the Kaliningrad Region: thermal power plants, small hydro power plants, and alternative sources of energy using wind power.

An essential component of the power system is power transmission lines. These are linear objects, which cross all landscapes and condense at the points of large populated areas. In the forests, they pass along cutover patches.

The other components of landscape environment in the Kaliningrad Region comprise belligerent landscapes and point-type objects, recreational facilities, hydro-engineering objects (of ameliorative and transportation significance), as well as objects of communications and navigation.

4.8 The Influence of Borders on Other Components of Landscape Environment (Historical, Functional, and Spatial Aspects)

A significant role in landscape development is played by borderlines. A border is a real or conceived line separating territories with distinct quantitative or qualitative spatial properties: natural, social, economic, and political. Borders perform a variety of functions as barriers, points of contact, reflection, connecting points, etc. The functions of borders can change over time. To a greater extent, this concerns human-made borders. The borders can be subdivided into natural and anthropogenic ones. Both types of borders can be either real, objective ones or constructed, defined solely by human volition (they are usually shown on the maps but cannot be seen afield).

Natural borders in the Kaliningrad Region Natural borders are always real ones, they are different in their width, they can be more or less clearly defined, and they have a different degree of fixedness. Running a boundary line correctly depends entirely on the accuracy of the method of their definition. Hydrographic and orographic boundaries as well as boundaries of areals of quaternary deposits distribution are clear and narrow since they visually reflect changes in natural habitats: heights, roughness of relief, grain-size composition of deposits, and coastline of a water body. These boundaries influence other landscape components and determine soil moisture and distribution of plant associations. They are the most stable boundaries since their evolution progresses very slowly. On the other hand, this border type can be quite dynamic. The most dynamic is a coastal zone border, which is quite mobile and changes not only due to disastrous natural phenomena but also due to wave-built processes and longshore currents.

Anthropogenic borders in the Kaliningrad Region Boundaries of land-use and functional zones are most clearly defined in the region: boundaries of industrial, settlement, agricultural zones, etc. The boundary of the sea resort zone of the Kaliningrad Region is also well delimited. This is a narrow band of settlements with organized leisure/holiday activities (Baltyisk, Yantarnyi, Primorye, Otradnoye, Svetlogorsk, Pionerskyi, Kulikovo, Zelenogradsk) and some areals of unofficial leisure/holiday activities. These boundaries are a result of a purposeful influence of human activity on the landscape. However, under certain circumstances (e.g., change in the land use and development rules or change of owner) they can disappear over time. Another feature of these boundaries is their natural and historical pre-determinacy.

Administrative and state borders are entirely different by nature. They are defined and delimited by humans, sometimes without taking into consideration any natural features of the territory, though quite often natural divides are used as a basis for drawing an administrative or state border.

For example, the northern border of the Kaliningrad Region passes along the River Neman, while its southern border was drawn arbitrarily (in terms of landscape). Another feature of these borders is their crucial influence on the land-use and settlement systems, the changes in which immediately affect all processes of recent landscape genesis. In this respect, the leading role is played by state borders. Transformation of a landscape environment of the Kaliningrad Region, which was manifest in change of direction of development (some territories growing feral and others being developed), took place as a result of changes in state borders position after the war, which caused dramatic changes in the systems of settlement and land use. Before World War II East Prussia was divided into three government districts: Königsberg, Gumbinnen, and Allenstein. The area of the present-day Kaliningrad Region occupies part of the government districts of Königsberg and Gumbinnen.

Current borders of administrative units in the Kaliningrad Region are not easily identifiable on-site because they are meaningful only from the point of view of territorial management. They determine a land-use type of landscapes, but since land-use types in adjacent districts are practically identical, they do not affect landscape appearance to a large extent.

5 Conclusions

At present, the landscapes of the Kaliningrad Region represent an elaborate system of territorial complexes manifesting various degrees of their natural base transformation and being at various stages of their development. In the past centuries, the landscape environment of the Kaliningrad Region territory has been formed under the influence of socioeconomic factors, which shape current appearance and condition of extant landscapes. Based on their analysis, a methodology for studying landscape areas subjected to long-term reclamation has been suggested. According to the methodology, investigating recent landscapes should comprise not only studies of natural landscape structure of the area, but it is also necessary to carry out the analysis of the present-day and previous settlement systems from a spatial perspective and to consider land-use dynamics of the area over a time period in question.

The current settlement system of the Kaliningrad Region is defined by a settlement network of East Prussia established by 1945, which largely followed the pattern of hydrographic network and landscape structure of the territory; by the postwar system of administrative-territorial division and management; by transformation of the regional transport system, which was manifest in reduction of the railway routes and discontinued communication with the southern neighbor of the region; by specifics of socioeconomic development of the area shared by all former Soviet Republics, which determined specialization of district economy and had effect on patterns of settlement; and by current demographic processes. Comparison of settlement systems over the Kaliningrad Region area in 1939 and 2009 showed that, at present, the area is less densely populated than before the war, and the majority of the population now is concentrated in the regional capital and around it, which impacted landscape appearance in the northeast of the region.

Having overcome several crises related to World War II and transformation of the economic system in the 1990s, the present-day land use in the Kaliningrad Region is characterized by the growth in agricultural production, which is manifest in involvement of previously derelict land in agricultural turnover. At the same time, all stages of landscape environment transformation are represented in the regional area including those with secondary succession. Many of those territorial complexes have turned into unique nature reserves, which affect biodiversity level of the territory. The main feature of the regional area is a mosaic pattern of recent landscapes since natural frontiers are enhanced by the specifics of land-use and settlement systems.

Of special significance for current landscape genesis are state borders of the Kaliningrad Region, whose recent age determined the time span of landscape transformations caused by their change in the postwar period.

References

1. Semyonov-Tyan-Shanskyi VP (1928) Raion i strana [Region and country]. -M.- L.: GIZ, 312 p
2. Romanova EA, Vinogradova ОЛ, Gagiyeva VE (2012) Sovremennoye sostoyaniye landshaftnoy sredy Kaliningradskoy oblasti (metodologiya i osnovniye rezultaty) [Current state of landscape environment of the Kaliningrad region (methodology and main findings)]. Vestnik BFU im. I. Kanta, issue 7, Natural Sciences. BFU im. I. Kanta Publishing, Kaliningrad, pp 149–156
3. Sauer CO (1963) The morphology of landscape. In: Leighly J (ed) "Land and Life": a selection from the writings of Carl Ortwin Sauer. University of California Press, Berkley, pp 315–350, first publ. in Geography 2.2 (1925)
4. Rodoman BB (2002) Polyarizovannaya biosphera: sbornik statei [Polarized biosphere: a selection of articles]. Oykumena, Smolensk, 336 p
5. Solntsev NA (2001) The doctrine of the landscape: selected works. M.V. Lomonosov Moscow State University, Geogr. Fact. Publishing House of Moscow State University, Moscow, 383 p
6. Kolbovskyi EU (2008) Landscape planning. Academy, Moscow, 336 p

7. Gutnov AE, Glazychev VL (1990) Mir arkhitektury: litso goroda [The world of architecture: the face of the town]. M.: Molodaya Gvardiya, 350 p
8. Romanova EA, Vinogradova OL (2011) Landshafty [Landscapes]. In: Orlyonok VV, Fedorov GM (eds) Kaliningradskaya oblast. Atlas [The Kaliningrad region. Atlas]. Masterskaya "Kollektsiya," Kaliningrad, pp 26–27
9. Romanova EA, Gagiyeva VE (2011) Stepen transformatsii landshaftov: Kaliningradskaya oblast [The extent of landscape transformation: the Kaliningrad region]. In: Orlyonok VV, Fedorov GM (eds) Kaliningradskaya oblast. Atlas [The Kaliningrad region. Atlas]. Masterskaya "Kollektsiya," Kaliningrad, pp 28–29
10. (2004) Generalnyi plan Kaliningrada. Osnovniye polozheniya do 2015 goda. [General Urban Plan of Kaliningrad. Principle provisions until 2015]. In: Kondakova TL (ed) Otv. za vypusk. Izd-vo Tenaks Media, Kaliningrad, 40 p
11. Antipova EA (2008) Geodemograficheskiye problemy i territorialnaya struktura selskogo rasseleniya Belarusi [Geo-demographic issues and territorial structure of rural settlement in Belarus]. BSU, Minsk, 327 p
12. Barran FR (1994) Stadte-atlas, Ostpreussen
13. Levchenkov AV (2004) Vzaimosvyazi sistemy selskogo rasseleniya i landshaftnykh faktorov Kaliningradskoy oblasti [Interrelation between the rural settlement system and landscape factors in the Kaliningrad region]. Vestnik KSU, issue 5. Series: Ekologiya regiona Baltyiskogo morya [Ecology of the Baltic Sea region]. pp 33–39
14. Fedorov GM (2001) Naseleniye Kaliningradskoy oblasti. Demograficheskiye usloviya obosnovaniya Territorialnoy kompleksnoy skhemy gradostroitelnogo planirovaniya razvitiya territorii Kaliningradskoy oblasti i yeyo chastei. Monografiya [Population of the Kaliningrad region]. Demographic conditions for substantiation of the Territorial scheme for town planning development of the Kaliningrad region and its individual areas. Monograph]. KSU Publishing, Kaliningrad, 111 p
15. Fedorov GM (1985) In: Agafonov NT (ed) Geodemograficheskaya tipologiya [Geodemographic typology]. LSU Publishing, Baton Rouge, LA, 152 p
16. Lyuri DI, Goryachkin SV, Karavayeva NA, Denisenko EA, Nefedova TG (2010) Dinamika selskokhozyaistvennykh zemel Rossii v XX veke i postagrarnoye vosstanovleniye rastitelnosti i pochv [Dynamics of agricultural lands in Russia in the XX century and post-agrarian soil remediation and vegetation restoration]. M.: GEOS, 416 p
17. Tishkov AA (2006) Rol suktsessiy v formirovanii raznoobraziya sostoyanyi landshafta na regionalnom urovne [Role of successions in the formation of diversity in landscape states at a regional level]. In: Dyakonov CN, Kasimov NS et al (eds) Proceedings of XI international conference "Landhsftovedeniye: teoriya, metody, regionalniye issledovaniya, praktika" [Landscape studies: theory, methods, regional studies, practice]. Faculty of Geography, MSU, East Lansing, MI, pp 288–289
18. Vladimirov VV, Naimark NI (2002) Problemy razvitiya teorii rasseleniya v Rossii [Issues of development of settlement theory in Russia]. Editorial URSS, Moscow, 376 p
19. Gusev AL, Maslov VN (eds) (2006) Istoriya selskogo khozyaistva Kaliningradskoy oblasti: 1945–2006 gg [History of agriculture in the Kaliningrad region: 1945–2006]. IP Mishutkina I. V., Kaliningrad, 464 p
20. Bazy dannykh munitsypalnykh obrazovanyi Rossyiskoy Federatsii. www.gks.ru/dbscripts/munst/. Accessed 15 Nov 2014

21. Khovanskyi MA. O sostoyanii selskogo khozyaistva v Kaliningradskoy oblasti v 2014 godu. Doklad na zaklyuchitelnom forume po mezhdunarodnomy proyektu "Podderzhka i razvitiye selskogo predprinimatelstva: ot mestnogo opyta k transgranichnomy sotrudnichestvy" [On the state of agriculture in the Kaliningrad region in 2014. Report presented at the final forum of the international project "Support and development of agricultural entrepreneurship: from local practices to cross-border cooperation"]. Kaliningrad. Accessed 14 Nov 2014
22. Tselevaya Programma Kaliningradskoy oblasti "Vovlecheniye v selskokhozyaistvennoye proizvodstvo ne ispolzuyemykh po tselevomu naznacheniyu zemel selskokhozyaistvennogo naznacheniya Kaliningradskoy oblasti na period 2011–2016 godov" [Target Program of the Kaliningrad region "Involvement of agricultural lands unutilized according to their intended purpose into agricultural production of the Kaliningrad region during the period of 2011–2016"]. Official Web-site of the Ministry of Agriculture of the Kaliningrad Region. http://mcx39.ru. Accessed 18 Nov 2014
23. Territorial Authority of the Federal State Statistics Service of the Kaliningrad Region. http://Kaliningrad.gks.ru/. Accessed 18 Nov 2014
24. Nefedova TG (2004) Gorodskaya selskaya Rossiya [Urban rural Russia], www.polit.ru. Accessed 13 Oct 2013
25. Orlyonok VV, Fedorov GM (2005) Regionalnaya geografiya Rossii. Kaliningradskaya oblast: uchebnoye posobiye [Regional geography of Russia. The Kaliningrad region: a textbook]. I. Kant KSU Publishing, Kaliningrad, 259 p

Specific Features of Urban Geosystems in the Kaliningrad Region

Elena A. Romanova, Olga L. Vinogradova, Vladislav V. Danishevskij, and Irina V. Frizina

Abstract The article describes the evolution of the urban settlement of the Kaliningrad Region, examined the structure and characteristics of the landscape environment of the cities of the region, analyzed their natural and anthropogenic components, and proposed a methodology for assessing the quality of the urban environment according to the degree of their construction, using satellite information, and the degree of transport connectivity of Kaliningrad. The system of urban settlement of the Kaliningrad Region in general has been inherited since the time of East Prussia but has undergone significant changes related to changes in state and administrative borders, the change in the ethnic composition of the population, and the cardinal restructuring of the management system. In modern times, urban settlement is formed against the backdrop of modern socioeconomic and geopolitical processes associated with the exclave position of the region in the Russian Federation and the difficult economic situation in the country and in the world as a whole. Currently, there is one big city in the Kaliningrad Region – the regional center – around which the city agglomeration is formed. The remaining cities of the region are welterweight and small. The stability of the modern geodemographic situation is ensured by migration growth, and the small size of the region and the presence of good quality intra-regional transport infrastructure (roads and railways) ensure the stability of the population of cities remote from the center due to pendulum migrations.

Keywords Building percentage, Landscape, Transport connectivity, Urban environment, Urban settlement

---

E.A. Romanova (✉), O.L. Vinogradova, V.V. Danishevskij, and I.V. Frizina
Institute of Environmental Management, Territorial Development and Urban Planning, Immanuel Kant Baltic Federal University, A.Nevskogo str. 14, Kaliningrad 236041, Russia
e-mail: alberta63@mail.ru

Contents

1 Urban System: Types and Specific Features 202
   1.1 Structure of Urban Geosystems 202
   1.2 Functioning of Urban Geosystems 204
   1.3 Functional Zoning of the Area 204
   1.4 Urban Frames 205
2 Retrospective Analysis of Urban Settlement Evolution in the Area Over the Past 100 Years 206
3 Review of Components in Urban Environment 207
   3.1 Natural Host Landscapes 207
   3.2 Relief, Grounds, and Soils 208
   3.3 Water Bodies and Air Masses 210
   3.4 Vegetation 211
4 Methodology of Urban Environment Quality Assessment Using Satellite Data 213
5 Evaluation of Degrees in Development of Urban Transportation Systems (Transport Accessibility Chart of Kaliningrad) 216
References 221

1 Urban System: Types and Specific Features

At the present stage of societal evolution, its distinctive feature is a quick growth of towns and an increasing number of town dwellers. Within urban settlements a specific environment for human life is being formed which is generally referred to as an urban (or urbanized) environment. The purpose of the extant urban environment is not only to meet the needs of society and individuals for subsistence and to provide protection from natural disasters and social calamities but also to develop cultural and physical potential of humans.

Towns are characterized by a high concentration of population, objects, and processes within a limited area. Another property of the towns is their multi-functionality and, consequently, a complex structure as well as the rhythm and rate of processes and cycles, which are occurring faster in urban environment than in analogous natural one.

1.1 Structure of Urban Geosystems

Urban geosystems possess a more complex structure than natural ones, which is caused by inclusion into their composition flows of substances, energy and information arising spontaneously and controlled by humans, as well as by inclusion of new morphological elements.

Urban systems comprise several subsystems [1]:

1. *Natural subsystem* comprising a host landscape where a town is developing. The features of a natural landscape help to locate the place where a town is to arise;

they influence spatial structure of towns, possibilities of their growth and development, as well as certain functional properties of towns and their specialization. If the process proceeds in an evolutionary manner, the founding of a human settlement gravitates to the most convenient and protected locations: the mouths of rivers, sea harbors, uplands, etc. Thus, natural obstacles largely predetermine a town planning pattern, for instance, the growth of a town along a river or a seacoast. Likewise, health resort towns are developing in areas with high medicinal and health-improving potential, while growth of tourist centers can be accounted for by certain natural environment suitable for sports and recreation. In some cases, disappearance of towns can be triggered by natural factors involving not only natural disasters but also soil impoverishment and depletion of mineral resources and water sources. Natural factors (such as climate, peculiarities of grounds, dangerous natural phenomena) shape conditions of living of a town population as well as conditions of a town construction and its functioning. In large towns, natural environment is deteriorating very fast, and in order to provide comfortable habitat conditions, the society is creating an artificial one (or quasinatural environment). This new environment (landscaped green areas, parks, squares, etc.) requires constant maintenance to ensure processes of its functioning and renewal, otherwise it tends to deteriorate very quickly.
2. *Technogenic subsystem* comprising an artificial habitat for humans, namely, a complex of constructions, buildings, objects, machines, and mechanisms. The creation of a technogenic subsystem is accompanied by accumulation of both natural and man-made components and materials within a relatively small area: stone, concrete, asphalt, metals, plastic materials, etc. Within the framework of this artificial environment, systems of water cycles (sewage systems, water supply system), matter cycles (technological flows, waste transportation, etc.) and information cycles (introduction of new genetic material) function.
3. *Social subsystem* comprising society and social institutions. This subsystem plays a steering role in urban geosystems. It is precisely spiritual, social, political, economic, and ecological needs and demands of the population that specify the development of towns and intensity of the use of natural landscape potential.

Changes in landscapes include the following:

– Transformation of one or several components
– Restructuring of a horizontal or vertical landscape structure
– Emergence of additional energy sources
– Emergence of anthropogenic components in landscape structure (constructions, mechanisms, substances)
– Decrease in diversity and area occupied by structural elements of natural landscapes

1.2 Functioning of Urban Geosystems

In urbanized conditions, both external and internal connections between the components and elements of a landscape are disturbed. These connections are realized via functional links, namely, an aggregate of natural and anthropogenic processes involving transportation, exchange, and transformation of matter and energy. A considerable effort is invested in deceleration of natural processes undesirable in urban conditions (such as underfloods and floods, karst processes, landslides, etc.).

In urban conditions, the rates and volume of water and mineral turnover change.

In urban geosystems, a change in patterns of a natural hydrographic network occurs, and new patterns are formed. Changes are also triggered in the intensity of land and groundwater runoff, hydrological regime in the area due to regulation, and dissection of natural water reservoirs due to formation of new watershed divides and talwegs as well as in dewatering and underflooding of grounds.

In urban geosystems, a geochemical turnover has both natural and anthropogenic components: reinforcement of erosion processes, the use of artificial soils, surface planning, concentration of substances, and formation of reliefoids.

Biological turnover of substances in towns also undergoes significant changes: there is a pervasive replacement of a natural biocenosis with an artificial one, uncharacteristic to latitudinal conditions. Besides, there is a change in biological productivity and biomass removal.

The functioning of the technogenic subsystem is supported by new energy sources; among them are mechanical, thermal, and chemical types of energy. A specific feature of urban geosystems is a high concentration of energy within a landscape.

In the towns, not only the composition and vertical structure of a landscape are changed (increase or decrease in stratification, removal of natural bodies, and introduction of artificial ones), but also horizontal structure and vertical connections in all landscape elements are disrupted.

A natural morphological structure (a landscape base) of the town is simplified due to disappearance of facies and individual land plots as a result of planning of streams and rivers as well as surface levelling. At the same time, neolandscapes are created in towns. They have a status of land plots and include railroad embankments, artificial water bodies, underground constructions, bulwarks, and excavations. Building construction, creation of green areas also makes urban structure more complex. Thus, urban geosystems represent a complicated combination of natural, altered, and artificial elements in the landscape structure.

1.3 Functional Zoning of the Area

The town area is generally quite heterogeneous with particular areas designed for specific functions: residential areas, areas for recreation, industrial production, etc.

A totality of urban areas (or zones) having their specific purposes constitutes a functional structure of the town.

Functional zoning is a part of an area planning scheme which is to be approved [2]. Functional (area) zoning in urban areas is conducted in order to identify either one or several functions in already existing zones.

A functional zone is a portion of the territory, which is characterized by a special legal regime for utilization of land plots.

There are various types of functional zones, among them residential, industrial, agricultural, transportation, nature protection, recreational (which often overlaps with residential zones just like industrial ones), belligerent zones, as well as zones intended for special purpose. For each zone type, a distinct regime of environmental management is established.

Implementation of zoning involves identification of areas performing one, two, and even three functions, for example, a residential-natural zone or a public-production-residential zone.

Depending on which types of functional zone predominate, the following types of towns can be distinguished:

1. Towns with large industrial enterprises
2. Towns with smaller industrial enterprises
3. Centers of cooperative industry
4. Transportation centers
5. Towns as scientific and experimental centers
6. Towns as holiday resorts

1.4 Urban Frames

A current concept of a town landscape is based on the notions concerning natural, natural and ecological, historical and cultural, as well as socioeconomic frames of the town area [3].

1. *Natural frame* comprises basic elements of a landscape structure in the area and natural components of a geosystem. It predetermines the conditions for construction and life-sustaining activities.
2. *Natural and ecological frame* is a system of interrelated natural and quasinatural town areas. It ensures environmental balance in the town.
3. *Socioeconomic frame* of the area comprises economy-related and residential functional zones in the town.
4. *Historical and cultural frame* comprises monuments of historical and cultural heritage in the town area.

2 Retrospective Analysis of Urban Settlement Evolution in the Area Over the Past 100 Years

There were 22 towns in the Kaliningrad Region in 1939, and 13 out of them played a role of administrative centers. East Prussia was a separate region, which had a unified settlement system and comprised several administrative districts, whose network of settlements, in turn, formed settlement systems at a lower level: districts of Königsberg, Gumbinnen, West Preussen, Allenstein, and Memel. By the same token, those districts were divided into smaller administrative units. When the area of East Prussia was divided, the Russian Federation obtained about half of the district of Königsberg and a major part of the district of Gumbinnen; other districts were transferred to Poland, while Memel was transferred to Lithuania. Moreover, six lower-level administrative units turned out to be separated by a state border. Therefore, a unified settlement system disintegrated, and the formation of new separate systems in Poland and the Kaliningrad Region began.

In 1939, the largest town in the area was Königsberg (with a population size of 372,164 residents) being an administrative center of the district of the same name. The second large town was Tilsit (59,105 residents). Large industrial enterprises were located in Königsberg, Tilsit, and Insterburg. Gumbinnen functioned mostly as an administrative center of the district of the same name. The leadership index of the regional capital was 1.9 before 1939. Small towns had small-scale enterprises mainly for processing local products of agriculture (creameries and cheese dairies) or for using local construction materials (brick factories) [4].

An administrative-territorial division of the Kaliningrad Region underwent multiple alterations during the postwar period. At present, in compliance with Federal Act № 131 "On common principles in organization of local governance in the Russian Federation" [5], 7 urban districts (*okrug*) and 15 municipal districts (*raion*) were established in the Kaliningrad Region. There are 21 towns in the Kaliningrad Region at the moment. In 2009, the regional capital still remained the largest town in the region (with population 419,180 people, which constitutes 44.8% of the total population in the region and 58.6% of the total urban population). The second largest town is Sovetsk (Tilsit) with population of 42,619 people. Compared to 1939, the leadership index of the regional capital has gone up to 2.5. The Kaliningrad Region is a territory of smaller towns. Most of the municipal districts have a population below 10,000 people and low industrial potential. They function as public service centers. According to classification adopted in Russia [6], most of the towns in the region belong to the category of small towns with population between 5,000 and 20,000 people (42.8%), while 28.6% of the towns are considered to be the smallest ones since their population does not exceed 5,000 people. Medium-sized towns make up 23.8% of the urban settlements. Only the regional capital is qualified as a large town (with population between 100,000 and 500,000 people).

Due to fundamental changes in the territorial administration system over the past 70 years, the hierarchy of towns has also undergone profound transformations.

The calculation of the ratio of values "town status–town size" shows that the curve of the actual population size for the top ten towns in the region is significantly detached from the curves calculated for both 1939 and 2009. However, while the population size of smaller towns in East Prussia practically coincides with the estimated values, the population size of the present-day towns in the Kaliningrad Region does not correspond to them at all. This fact serves as evidence for a consolidated role of the regional capital and insufficient rates of development of the other towns in the region.

The level of urbanization in the region can be identified on the basis of the proportion of urban population in the total population size [7, 8]. In the light of interpretation adopted currently in economic geography, between 1979 and 1999, the settlement in the Kaliningrad Region corresponded to the fourth phase in settlement evolution (urbanization). During the aforementioned period, urban population was increasing dramatically, while rural population and population of smaller towns were decreasing. The period between 1939 and 1979 shows departure of the settlement values from standard evolutionary phases, which is accounted for by profound transformations in the life of the regional community: World War II, accession of the region to the Soviet Union, repatriation of the native population during 1945–1948, and its substitution with settlers from the other Russian regions. The period between 1999 and 2009 indicates a cessation in urban population increase and a slight increase in rural population, which could provide evidence to the beginning deconcentration of population, a characteristic of the following evolutionary phase in settlement. On the whole, based on the urbanization rate data (76.4% of urban population), the Kaliningrad Region occupies a middle position among other subjects of the Russian Federation in the Northwestern Federal District. In this respect, the maximum urbanization rate is marked in Murmanskaya region (91.2%), while the minimal one was marked in Nenetsky Autonomous District (64.2%) [9].

3 Review of Components in Urban Environment

3.1 Natural Host Landscapes

The towns of the Kaliningrad Region, just like other towns in the world, are typically located at the interface of two or more kinds or even types of natural (host) landscapes. For example, coastal towns of Svetlogorsk, Zelenogradsk, and Pionersky are located at the interface of landscapes of marine and glacial origin. Zelenogradsk, for instance, is situated within the following kinds of landscape comprising marine accretion and abrasion shores, coastal lagoonal lowlands (a new construction area in the southeast part of the town), maritime eolian formations (western edge of the town and the root of the Curonian Spit), and flat-undulating morainic plain (the main part of the town). The town of Baltiysk is situated at the interface of marine accretion and abrasion shores, maritime eolian formations, and coastal lagoonal lowlands.

The second group of towns – the towns on the shores of lagoons – comprises Svetlyi and Ladushkin. Ladushkin covers an area of the coastal lagoonal lowlands and morainic plain, while the valley side of Svetlyi is occupied by an old deltoid lowland of the river Pregolya.

Most towns in the Kaliningrad Region are located on the rivers: Kaliningrad, on the banks of the river Pregolya and its tributaries; Guryevsk, on the river Guryevka; Gvardeisk, at the confluence of two rivers, Pregolya and Deima; Chernyakhovsk, at the confluence of the rivers Pissa and Instruch; Gusev, on the river Pissa; Ozyorsk, on the river Angrapa; Pravdinsk, on the river Lava, the towns of Sovetsk and Neman are on the river Neman; Krasnoznamensk, on the river Sheshupa; Mamonovo, on the river Mamonovka; and Polessk, on the river Deima. The valley sides on which the towns are located differ in their origin: they are terminal morainic ridges, morainic flat and undulating glaciolacustrine plains, and accretion shores of lagoons.

Specific features of the host landscapes predetermine not only peculiarities in town planning but also give rise to certain negative processes.

Since some of the towns are located in coastal areas, their expansion is naturally constrained by the sea. They also experience a threat of carving and erosion of the shores. This problem is particularly urgent for Zelenogradsk and Svetlogorsk lying on steep valley sides. The rate of shore erosion of northern shores of the Kaliningrad Peninsula reaches 0.5–0.7 m per year. Due to coast protection works, the rate of erosion is reduced by about 12 cm per year. However, coastal erosion poses a threat to buildings in the west area of Zelenogradsk and the central area of Svetlogorsk.

The towns located along the lower course of the river Pregolya (as far as Gvardeisk) experience underfloodings due to water surges caused by west winds.

3.2 Relief, Grounds, and Soils

In urban conditions, even components of the most substantial lithogenous group of geosystems are thoroughly transformed to the extent of removal of natural bodies and their replacement with artificial ones. For instance, in the course of construction of buildings and other facilities, surface planning as well as soil filling or removal is carried out. In Kaliningrad, technogenic deposits are the thickest in the central, oldest part of the town reaching 0.2–0.5 m on the right bank and 2.5–3.4 m on the left bank of the river [10]. This difference is obviously due to greater roughness of natural relief on the right bank. The composition of technogenic deposits includes construction debris, household waste, or a mixture of both with natural grounds as well as sand, gravel, or pebble filling. The age of the grounds varies being in some cases 500 years old. For instance, there is evidence of ground filling on the island dating back to the construction works of the cathedral and ground filling dating back to the construction of a defensive earthwork (along the present-day Litovskyi

Val Street and in place of Chernyshevskogo Street). Among the prevailing types of substrates are those formed during carpet bombings (especially in the town center) as well as current substrates filled in during vertical surface planning for building construction. According to drilling data, the town of Gumbinnen (Gusev) is built almost entirely on filled grounds in the middle of valley plain of the river Pissa. The thickness of grounds under the old buildings reaches almost 2 m [11]. In Sovetsk, in the course of construction of a kindergarten, a sand cushion 1 ha in area and up to 4 m thick was made in place of a lowland bog. In other towns of the region, a filling with grounds was used in construction of dikes, earth mounds, dams, etc. Earth-fill dams in Tapiau (Gvardeisk), Tilsit (Sovetsk), Rauschen (Svetlogorsk), Ragnit (Neman), and Insterburg (Chernyakhovsk) were erected in the construction of mill ponds and castle ponds as long ago as thirteenth to sixteenth century.

Urban soils formed on artificial and altered grounds possess specific chemical and physical properties, which are different from the typical properties of zonal soils. Soils of this kind are referred to as *urbozem* (meaning "urban soil," the word formed on the analogy with *chernozem* meaning "black soil") [12–14]. Their profile is formed as a result of technogenesis-related processes. Characteristic of urbozem is the presence of buried soils, ground fills, and inclusions, among them artificial ones.

Soil profile № 3, laid in the square between Universitetskaya Street and Generala Sommera Street, revealed the upper filling soddy horizon up to 8 cm thick, dumped on top of the layer of construction debris (red brickbats), and mixed up with sand (about 15 cm thick). Below were buried soils: primitive soddy urbozem with a profile form A'(5–7 cm), AC'(8–10 cm), C'. The topmost section of the profile was classified as *technozem*, that is, artificially created type of soil.

Soil profile № 12 was laid down on the I.Kant Island, 50 m from the cathedral. The analysis of morphological composition revealed alternating layers of sod and sand fillings (without any evidence of soil formation) up to 70 cm in depth. Below 50 cm was a layer of brick rubble under which was a thin soddy horizon on a laminated fluvial filling.

Studies of soil profiles № 16 and № 23 (Arsenalnaya Street, Kaliningrad, and Sadovaya Street, Zelenogradsk) showed that soils in old gardens and vegetable plots in the towns are characterized by high soil fertility, increased humus horizon, satisfactory physical qualities, and appropriate degree of aeration. Judging by their physical and chemical as well as morphological properties, those soils can be classified as soddy arable soils.

Due to multiple fillings, grain-sized composition of urban soils became lighter with considerable increase in fraction over 1 cm [15]. The density of a soddy horizon in a background (soddy-podzolic) soil is 0.7–0.9 g/cm^3, while the density of the upper soddy horizon in town is about 1.8–1.9 g/cm^3, and in areas with trampled vegetation, it is up to 1.9–2.3 g/cm^3.

Changes in chemical composition of urban soils are caused by a number of factors: by a supply of substances uncharacteristic to natural soils, including pollutants, and attenuation of a soil formation process.

An introduction of substances of alkali composition and termination of podzolization process changes not only their acidity level and buffer capacity but also cation exchange value, cation exchange capacity, and composition of soil absorption complex.

The level of exchangeable soil acidity in Kaliningrad and Zelenogradsk is equal to 7.0–8.5, while for background natural soils, the value is 4.0–5.0 pH units.

The level of urban soil buffering in terms of alkalization is 1.9–2.1 times lower compared to background soils, while in terms of acidification, it is 1.6–1.7 times higher. In most soil samples from the profiles in Kaliningrad and Zelenogradsk, the value of cation exchange capacity varies within the range of 3–19 mg/eqv. per 100 g of soil [16]. In background soils, cation exchange capacity values are usually within the range of 8–12 mg/eqv. per 100 g of soil.

Most urban soils are superposed with artificial covering (asphalt, concrete, paving slabs, etc.). In the center of Kaliningrad, the covering of soils reaches 30–65%, while on the town edges, it is 12–17%. In smaller towns with a great deal of detached houses, such as Zelenogradsk, the degree of soil covering does not exceed 11–13%.

A characteristic feature of all urban soils is their contamination with oil, heavy metals, and other substances [17, 18].

3.3 Water Bodies and Air Masses

Water bodies serve as sources of fresh and utility water, receivers of various types of waste (household and industrial waste), and places of recreation for town dwellers. In towns, characteristic processes concerning water bodies are the following: altering of an existing water system and creating new artificial water bodies (canals, water supply reservoirs); alteration of pattern of the river basins due to construction of dikes, dams, and other hydro-technical utilities; alteration of water regime in the area due to regulation of surface and subsurface runoff, artificial drying out or watering of grounds; and alteration of chemical properties of natural water and its pollution.

An active transformation of a natural water system in the towns of the Kaliningrad Region began with the arrival of Teutonic Order to the area. The building of fortresses and other defensive fortifications was accompanied by construction of dams and ponds (the Lower and the Upper Lakes in Kaliningrad, the Tikhoye Lake in Svetlogorsk, etc.). Some streams were straightened and turned into canals (stream Zooparkovyi and Goluboy in Kaliningrad, river Pissa in Gusev, river Svetlogorka in Svetlogorsk, etc.). In order to provide water supply to Königsberg residents, the canal Landgraben (Pityevoy) was built, which supplied water from a cascade of freshwater lakes on Sambia Peninsula. The system still functions effectively today.

Part of Zelenogradsk (Kranz before 1946) is located within a boggy lagoonal lowland. Before the war, in order to drain the town area, polders with pumping stations were built. The stations pumped water into the canal Kranzbeek and then into the lagoon. The polders were surrounded with earthworks and dams up to 1.5–2.0 m high. At present, the pumping stations are out of operation, and the town area is inundated by bog waters.

Surface runoff is regulated by a complex system of water pipes and sewage, including storm water drainage system, as well as by water extraction for utility and industrial purposes. The regime of water bodies also depends on collection of snow and change in vegetation cover in towns.

Water leakage from the municipal sewage and piping system causes processes of soil erosion, ground subsidences, and landslides (as, e.g., near the Lower Lake).

The pollution of water bodies in the towns is caused both by discharge of contaminated waters and by supply of substances from the atmosphere. In Kaliningrad, 7 out of 13 lakes have been declared unsuitable for swimming over a period of several years on the basis of their microbiological and chemical parameters [19].

The influence of anthropogenic processes on climate in towns is manifest primarily in changes of temperature regime of air masses, in the content of solid matter and gases including aero-pollutants, in changes of wind pattern and self-purification capacity of the atmosphere [20]. All of these changes are distinctly manifest in large towns. However, similar phenomena have been observed in Kaliningrad and other towns in the region. For example, from 2009 to 2012, the number of air samples with increased maximum allowable concentrations (MAC) content made up between 0.5 and 0.8% of the total number of air samples in towns. Those were samples collected mainly close to the motor roads and industrial enterprises of Kaliningrad, Guryevsk, and Sovetsk [21]. The increase in MACs was registered by state monitoring agencies for contents of suspended matter, nitrogen dioxide, formaldehyde, and benzopyrene.

A more favorable situation with regard to atmospheric air is observed in the coastal towns of the Kaliningrad Region where atmosphere purification takes place due to their adjacency to a large water body and a high number of windy days per year. Air masses in smaller towns of the Kaliningrad Region, such as Nesterov, Ozyorsk, Krasnoznamensk, and others, do not experience any significant anthropogenic pressure due to the absence of industrial enterprises.

3.4 Vegetation

Urban vegetation is formed as a result of gradual ousting of indigenous plant species from the town area and deliberate introduction of cultivated plant species and plants used for landscaping as well as due to self-sustained distribution of synanthropic species. A purposeful, related to town planning planting of green landscaped areas in the towns of the Kaliningrad Region started at the end of nineteenth century – beginning of the twentieth century. In every town a park was laid out. By 1928, a total area of public gardens and parks in Königsberg is made up of about 630 ha, which means 24 m^2 per one resident [21]. By 2003 these figures went down to 470 ha and 6.8 m^2 correspondingly. The area of trees and shrubs, including suburban forests, reaches 100 m^2 per one Kaliningrad resident [22].

Wood species from the North and South America, South and Central Europe, Central Asia and Asia Minor, China, and Japan were used for planting ginkgo, northern catalpa, blood-red hawthorn, horse chestnut, blue spruce, and Amur cork tree as well as decorative shrubs and trees. Thus alien plant species make up to 80% of the overall species composition. Many species of decorative shrubs were used in planting, for example, various species of lilac, mock orange, etc. Later on, planting of green areas was carried out randomly. Among the trees planted were mostly poplar, Norway maple, and drooping birch. At present, in Kaliningrad and other towns, old trees predominate, over 70% of the total number of trees. About 40% of trees and shrubs in Kaliningrad (including lawns) are in a satisfactory condition, while the rest are degrading.

On the whole, the ecological balance of the town areas is supported by domestic plants in private gardens. In smaller towns of the region and even in Kaliningrad, the area of private garden spaces is quite large.

In towns, large areas are covered with synanthropic species such as common dandelion, common burdock, common plantain, slender plantain, buckthorn plantain, blind weed, hop trefoil, mountain clover, bulbous bluegrass, black elder, and other species. Survey of vegetation on the grounds surrounding buildings in Kaliningrad showed that synanthropic index of those plant communities reaches 50–70% (discount areas in Universitetskaya Street, Krasnaya Street, Yemelyanova Street, and Proletarskaya Street). Urban weedy communities are characterized by depauperization of species composition, decrease in projective cover, and stratification in comparison to floodplain of natural territories.

Increased environmental pressure (trampling down, overcrowding, as well as soil and air pollution) affects the state of plants in the town: the rate of plant disease is going up, while plant productivity is going down, etc. Studies of test plants (common dandelion) from discount areas with various degrees of mechanical stress have indicated that the height of plants decreases by 30–70%, flower size by 12–15%, and germination capacity of seeds by 12–18%.

Urban green spaces should play both aesthetic and protective role under conditions of a powerful anthropogenic pressure. However, because of their vulnerability, trees and shrubs in Kaliningrad perform their function only partially, especially in the town center. In the other towns of the Kaliningrad region, the greenery copes with its role quite successfully.

Apart from the components altered by human activity, urban geosystems comprise artificial elements such as building and facilities as well as vehicles and machinery. In towns, due to housing differentiation in terms of building height, a unique profile is formed with positive and negative elements of neorelief. A profile with two peaks of building height is typical of Kaliningrad and smaller towns of the Kaliningrad Region: the town center has three- to four-storey buildings, while the town periphery has recent multi-storey buildings. A classic situation came about in terms of historical differentiation of the buildings: the age of the buildings decreases from the town center toward its periphery. The same trend holds true of smaller towns in the region. For instance, the following pattern of zoning of the town depending on the age of buildings is typical of Zelenogradsk and Svetlogorsk: a historical core of the town with prewar three- to four-storey buildings, a belt of prewar detached houses (one- to two-storey buildings with a half storey), generally four-storey houses of the 1960–1970s, housing estates with standard five-storey buildings of 1970–1980s, housing estates with recent private detached houses (two to three storeys), and multi-storey apartment buildings (12–20 storeys). As a result of intense bombing raids, Kaliningrad practically lost its historical center, and now the town center presents an area with a melange of buildings: prewar individual buildings, blocks of houses of 1960–1970s, and recent in-fill constructions.

4 Methodology of Urban Environment Quality Assessment Using Satellite Data

Apart from natural components, urban environment contains artificial, man-made components referred to as *technosphere*. Its constituents comprise production and its results, architectural complex of the town, and transportation.

The final and probably the most important component of an urban system is its population. It acts both as a consumer of results of production activity and at the same time as a bearer of various nonmaterial needs. People's socially oriented interests include a wide range of needs pertaining to culture, environment, aesthetics, national identity, economy, and politics. Town infrastructure is aimed at meeting these diverse needs of the population as a whole and individuals as subjects of relationships with other components of urban system.

Various components of urban environment are closely interrelated. In the course of their interaction, contradictions between individual elements tend to aggravate. As a result of an active and transformational human agency, a new ecological environment with a high concentration of anthropogenic factors has emerged. Some factors such as air pollution, high level of noise pollution, and electromagnetic radiation are an immediate result of industrialization, while others such as concentration of enterprises within a limited area, high population density, migration processes, etc. are consequences of urbanization as a form of settlement [23].

Urban environment for habitation (living) is a totality of concrete and fundamental conditions created by humans and nature within the borders of a populated locality, which affect the level and quality of human livelihood.

This concept is basic and relevant to another two concepts:

- Quality of urban environment for habitation (living) is a capacity of an urban environment to meet objective needs and requirements of town dwellers in compliance with norms and standards of living generally accepted at a certain point in time.

- Quality assessment of urban environment for habitation (living) is identification of quantitative values for qualitative indicators for urban environment for habitation, which enables us to provide comparison with established criteria.

The aforementioned definitions were formulated and introduced into scientific discourse by the Ministry of Regional Development of the Russian Federation, all-Russian Public Organization "Russian Union of Engineers" and Federal Agency for Construction, and Housing Maintenance and Utilities of the Russian Federation.

Environmental status can be improved by various measures:

- Technological measures (application of advanced technologies)
- Technical measures (upgrading of waste treatment facilities)
- Structural measures (moving enterprises pollutants outside town limits and developing environmentally conscious enterprises)
- Architectural and planning measures (creating protective zones)

International experience shows that it is possible to assess the quality of urban environment in a variety of ways using a variety of methodologies based on different approaches to the interpretation of the concept *urban environment*. At the same time, a shared feature of most methodologies is the use of a more or less constant set of factors (indicators) (both objectively measured and subjectively assessed) characterizing such aspects of urban life as connectedness of transportation, housing conditions, service availability, environmental situation, public amenities, safety and security, and a general feeling of comfort experienced by people in the urban environment [24].

This list may include the intensity of land use (intensity of development), which is determined by the area location within the entire structure of the town in accordance with plans of street and road network and a system of town area centers designed within the framework of the General Urban Plan. The intensity of town area use is characterized by indicators of development density, proportion of area development and density of street, and road network expressed by a standard size of a town block [25].

One of the ways of obtaining those indicators and, consequently, of assessing the quality of urban environment is to use satellite data. Special software is used to interpret satellite imagery. Then data regarding construction areas, roads, town blocks, green areas, and other objects are computed. Using the latest satellite images, one can obtain much more information than from official town archives. Besides, this is up-to-date information. Being guided by town planning documentation, construction standards, and regulations, as well as by other statutory technical, economic, and legal instruments regulating town planning activities and using satellite data and criteria, one can effectively assess the quality of urban environment.

Taking into account statutory documents for residential construction [26] as well as using the data obtained after interpreting satellite images of Kaliningrad and after processing of their results, a map can be compiled which allows to assess effectively the intensity of land use (intensity of development) and draws certain conclusions concerning the quality of urban environment in Kaliningrad.

At the present stage, we are investigating town areas, which are qualified as residential areas in compliance with the General Urban Plan of Kaliningrad.

Zoning of the town area is one of the fundamental instruments for regulating town planning activity. It establishes a framework of conditions for urban area use. These conditions are obligatory for all participants of town planning activity with respect to function, density, and type of development as well as landscape arrangement of the area.

Zoning concept described in the General Urban Plan of Kaliningrad is based on conclusions of a complex town planning analysis. It takes into account historical and cultural as well as urban planning features of the town, the existing specifics of land use, characteristic features of cultural heritage, and information of a land cadaster. While specifying zones in the town, due consideration was given to provisions of town planning and Land Code of the Russian Federation, to the requirements of town planning norms and rules concerning zones with restricted regime of town planning activity [27].

Below is presented a list of zones specified in the principal drawing of the General Urban Plan of Kaliningrad:

- Residential zones – comprise building of various types in accordance with building height and density, multi-storey housing estates; two- to four-storey houses (including terraced houses), low-rise private detached houses with private garden spaces, horticultural communities (which are considered to be perspective zones for residential construction)
- Public and business zones – comprise healthcare facilities and social welfare offices, institutions of higher education and secondary-level vocational training, sports complexes and facilities, other public and business zones (administrative, business, cultural and entertainment, shopping areas, and other objects)
- Multifunctional zones – comprise public and business zone in the town center, public and residential zones, multifunctional production and business, as well as shopping areas
- Industrial zones – comprise production and utility areas as well as areas for utility services of residential areas
- Natural and recreational zones – comprise parks, squares, gardens, boulevards; forest-parks, municipal forests, recreational zones
- Specially protected natural areas
- Zones of engineering and transport infrastructure – comprise railroad area, ports and port facilities, sewage treatment facilities of other objects of engineering and transportation
- Zones of agricultural use
- Other zones – comprise cemeteries, green areas for special purposes, other green areas, other town areas
- Zones for perspective town development

A colored scale of the charts presented below specifies land-to-building ratio (land-use ratio) in residential zones of Kaliningrad:

- Light-gray color – low (below 5%)
- Gray color – medium (5–20%)
- Dark-gray color – high (20–35%)
- Black color – very high (over 35%)

According to normative standards, the limit is established at 40% [27].

In Alexander Kosmodemyanskyi settlement located in the west part of Kaliningrad, houses with one to two apartments predominate. These are primarily German houses built before the war, and consequently, land-use ratio in the settlement is low and medium (Fig. 1).

If the area closer to historical town center is considered, for instance, the sector delimited by Prospekt Mira, Sovetskiy Prospekt, Marshala Borzova Street, Sportivnaya Street (Fig. 2), which comes under Tsentralnyi (Central) town district, one can see that this ratio is high and very high. This area is built mainly with German apartment buildings and five-storey apartment buildings constructed after 1946. The area with high land-to-building ratio (over 35%) in Fig. 2 is a historical area of prewar buildings characterized by high density. Having analyzed aerial photographs taken in 1944 [28], one can say that land-use ratio in this part of Königsberg was about 70–75% (Fig. 3).

Based on the evidence provided by the maps, one can conclude that despite a rather developed infrastructure in Kaliningrad, land-to-building ratio, being one of quality indicators of urban environment, lies within normative standards.

5 Evaluation of Degrees in Development of Urban Transportation Systems (Transport Accessibility Chart of Kaliningrad)

Transportation system of the town is a major factor specifying functionality of its current spatial structure.

The transportation system of the present-day Kaliningrad is largely determined by the General Urban Plan. Measures aimed at improving and developing of transportation in Kaliningrad were proposed in the General Urban Plan taking into consideration international, federal, and regional programs. The draft of the General Urban Plan offers the best option for transport infrastructure development taking into account maximum possibilities under present conditions and preserving historical environment of the town. The draft is based on the most rational changes and additions to the existing transportation network. In many cases, the changes are fundamental; however, they do not lead to deep contradictions within the existing town infrastructure. Resolving transportation problems in the town is possible only

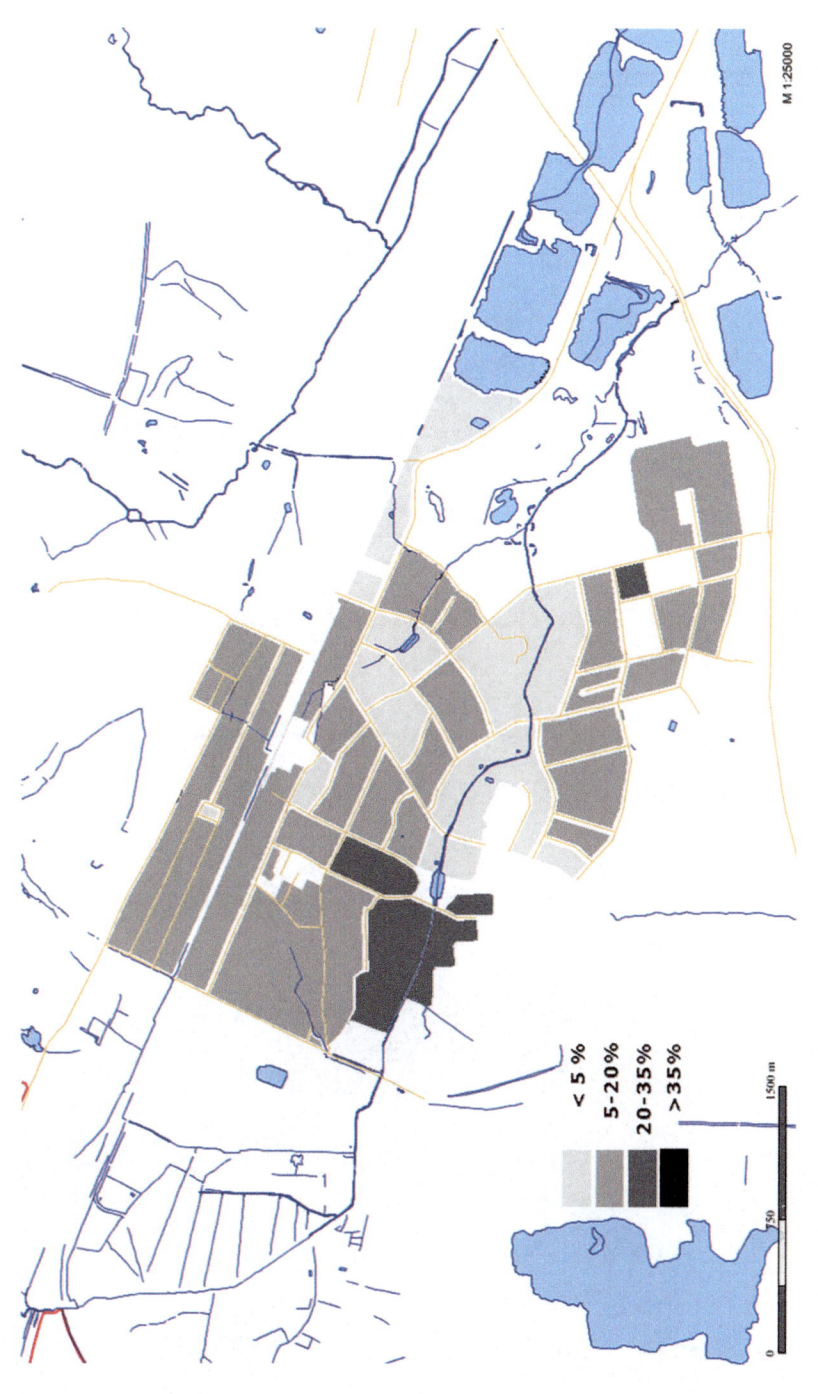

Fig. 1 Degree of area development in Alexander Kosmodemyanskyi settlement

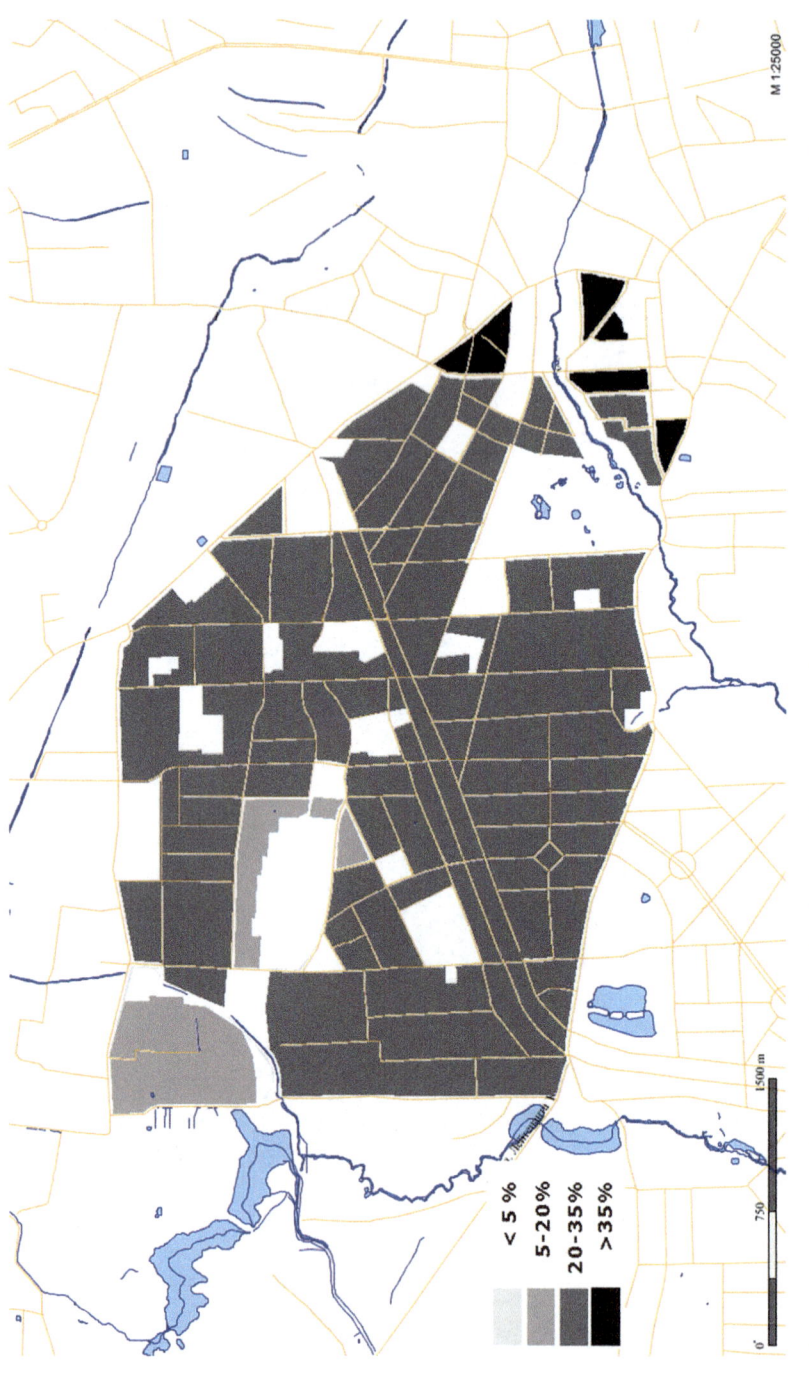

Fig. 2 Degree of the area development delimited by Prospekt Mira, Sovetskiy Prospekt, Marshala Borzova Street, Sportivnaya Street (Kaliningrad)

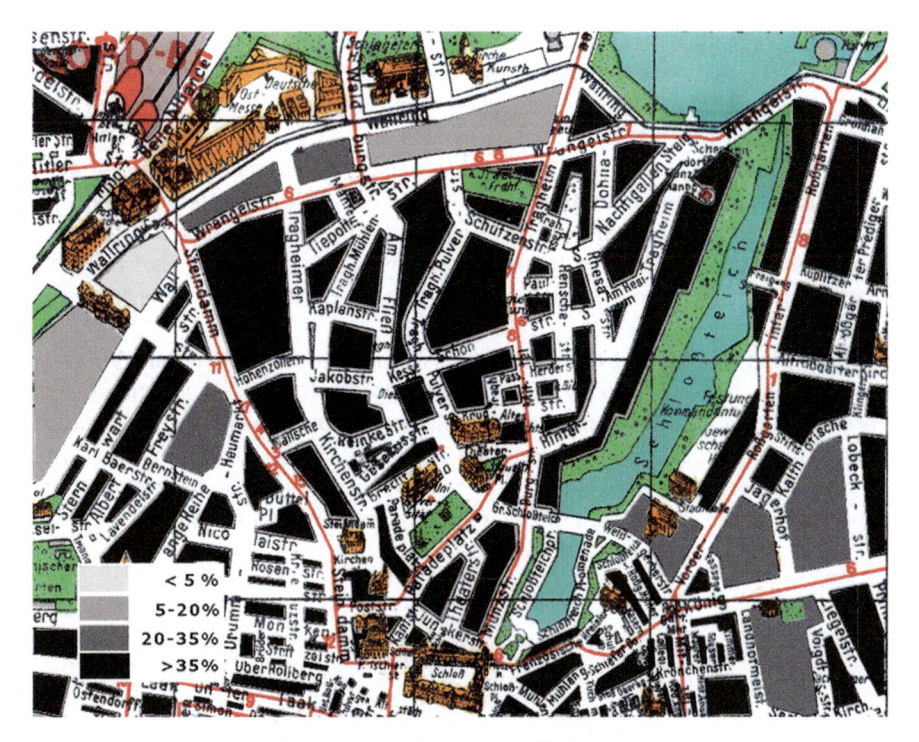

Fig. 3 Degree of the area development in the center of Königsberg (1944)

using a complex approach combining a tactful rebuilding of the network in historical areas with dramatic improvements at individual nodes of the network.

Despite availability of various means of transportation and routes in Kaliningrad, it is not always easy to choose the quickest and most convenient route to travel from one part of the town to another. This is accounted for by a number of problems. The degree of connectedness of road network in Kaliningrad is very low. Leninsky Prospekt can provide a relevant example. The street does not have an alternative, that is why Leninsky Prospekt is overloaded with traffic. It seems neither reasonable nor convenient to redirect transportation via the double-deck bridge or the island (there are fewer passengers traveling by those routes than via Leninsky Prospekt). Moreover, there are much more traffic jams on all the other bridges across the river than on the Trestle bridge. Parked cars are another factor, which contributes to traffic overloading in Leninsky Prospekt. A special lane for public transport is used by parked cars along the street section between the intersection point of Epronovskaya Street and Leninsky Prospekt and as far as the Trestle bridge. Thus, busses are forced to maneuver around parked cars and cannot gain necessary speed to drive faster. The lack of bridges across the river is another cause of transportation problems in the town. Among bridges in operation at the moment are the Trestle bridge, Vysokyi (Oktyabrsky) bridge, Berlinskyi (Berlin) bridge, Derevyannyi bridge, and double-deck bridge. The situation was somewhat improved when the second Trestle bridge was put in operation: it crosses both water courses of the river Pregolya, passes over the Island Oktyabrsky connecting 9 April Street on the right bank and Dzerzhinskogo Street on the left bank. Due to all these factors, access of public transportation to such densely populated town districts as Baltyiskiy and Selma is severely limited, especially during the rush hour.

Maps of public transportation access to various town districts in Kaliningrad were compiled for the year 2013. Means of transportation included in the maps were busses, public taxi vans, trams, and trolley busses. The maps showed public transport access both from the town center to its edges and from the outskirts back to the center (Fig. 4).

Since the town's public transportation is a crucial factor for a sustainable urban development, it is urgent to develop a program for the future transportation system ensuring a balanced development of public and private means of transportation, optimization of route network based on actual passenger traffic flows, ensuring increase in the number of vehicles for all means of transportation, increase in traffic capacity of road network, and upgrading it in compliance with modern standards.

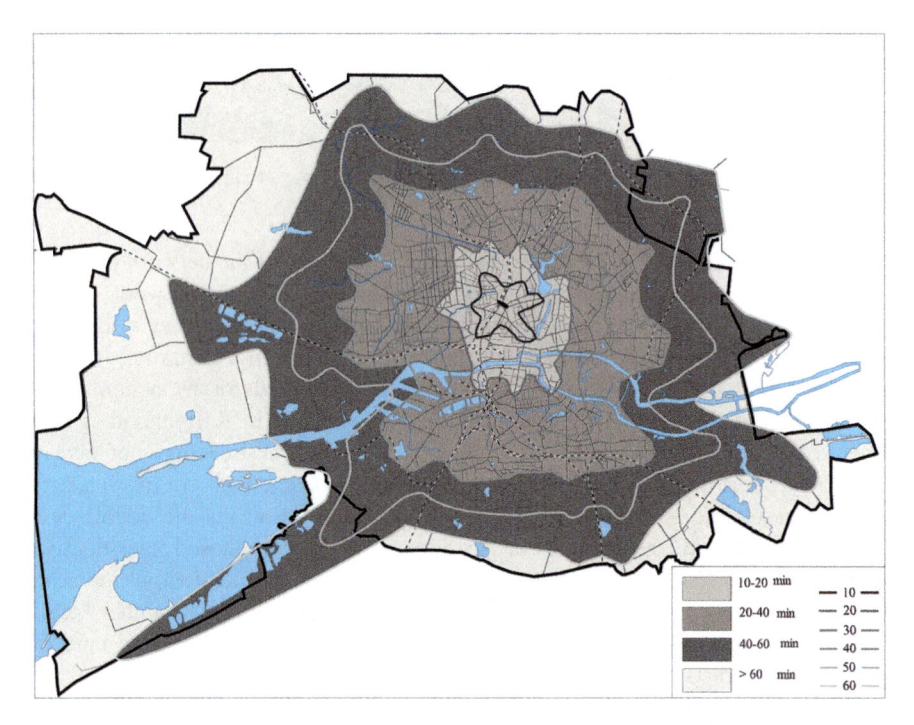

Fig. 4 Schematic map of public transportation access to various districts in Kaliningrad: from the center to the fringe of the town by bus during the rush hour. The four areals are delimited on the map: the central light-gray areal with public transport access of 10–20 min, which is surrounded by the areal in a darker shade of gray indicating transport access of 20–10 min, then follows the dark-gray areal indicating transport access of 40–60 min and, finally, comes the light-gray areal closer to the town edges indicating public transport access over 60 min

In order to create conditions for development of transportation infrastructure, it is necessary:

- To ensure fast development of street and road network as well as rebuilding of existing roads in order to upgrade them to current standards and requirements of road traffic
- To link street and road network of the town and its suburban zone by developing already available network of main roads, by constructing duplicating highways and upgrading them
- To create a system of effective cooperation between municipal and private public transportation
- To reduce road accident rate involving ground transportation

References

1. Romanova EP (1997) Sovremenniye landshafty Yevropy (bez stran Vostochnoy Yevropy) [Recent European landscapes (excluding countries of East Europe)]. M.: MSU Publishing, Toruń, 312 p
2. Gradostroitelnyi Kodeks Rossiyskoy Federatsii (2005) [Town-planning Code of the Russian Federation]. M.: Prospekt, 103 p
3. Kazakov LK (2011) Landshavtovedeniye [Landscape studies]. M.: Publishing Center "Akademiya", 336 p
4. Barran FR (1994) "Stadte-atlas, Ostpreussen"
5. Federal Act № 131 "Ob obshchikh printsypakh organizatsii mestnogo samoupravleniya v Rossiyskoy Federatsii" ["On Common Principles in Organization of Local Governance of the Russian Federation"]
6. Semagin Yu A (2006) Territorialnaya organizatsiya naseleniya [Territorial organization of the population]. M.: Izdatelsko-torgovaya korporatsiya "Dashkov & Co", 244 c
7. Mironenko NS (2001) Stranovedeniye: teoriya i metody [Country-study: theory and methods]. M.: Aspekt-Press, 268 p
8. Gibbs J (1963) The evolution of population. Econ Geogr 2
9. Mezhdunarodnaya ekonomicheskaya statistika [International statistics in econom-ics]. http://statinfo.biz/HTML/M61F290A3821L1.aspx. Accessed 22 Mar 2015
10. Galai AV, Savostina OA, Yu SL (2012) Antropogennaya transformatsiya relyefa g. Kaliningrada. [Anthropogenic transformation of relief in Kaliningrad]. Antropogennaya geomorfologiya: nauka i praktika. [Anthropogenic geomorphology: science and practice]. Proceedings of XXXII Plenary Meeting of Geomorph. Commission, Russian Academy of Sciences, Belgorod. M.-Belgorod: ID "Belgorod", pp 189–192. Accessed 25–29 Sept 2012
11. Orlyonok VV, Barinova GM, Kucheryavyi PP, Ulyashev GL (2001) Vyshtynetskoye ozero: priroda, istoriya, ekologiya [The Vyshtynyetskoye Lake: nature, history, environment]. KSU Press, Kaliningrad, 212 p
12. Dobrovolskyi GV (ed) (2003) Antropogenniye pochvy (genesis, geografiya, rekultivatsiya) [Anthropogenic soils (genesis, geography, reclamation)], M.: Oykumena, Smolensk, 267 p
13. Dobrovolskyi GV (1997) Pochva, gorod, ekologiya. [Soil, town, environment]. M.: Fond "Za ekonomicheskuyu gramotnost", 310 p. http://soils.narod.ru/appendices/tpo.html. Accessed 12 Feb 2015
14. Kurbatova AS, Bashkin VN, Myagkova AD, Reshetina TV, Savelyeva VA, Toshcheva GP, Yakovlev AS (2003) Metodicheskiye ukazaniya po otsenke gorodskikh pochv pri razrabotke gradostroitelnoy i arkhitekturno-stroitelnoy dokumentatsii [Methodological guidelines on

assessment of urban soils in development of town-planning and architectural and construction documentation]. GlavAPU MOSKOMARKHITEKTURY, AO "MOSPROEKT", Scientific-Research and Design and Survey Institute of Urban Environment. Moscow. http://www.fileswat.com/file. Accessed 24 Jan 2015

15. Salikhova EV, Savostina OA, Vinogradova OL (2003) Transformatsiya osnovnykh svoistv urbanozyomov Kaliningrada [Transformation of the main features of urbanozyom in Kaliningrad]. Vestnik KSU. Ekologiya regiona Baltiyskogo morya [Environment of the Baltic Sea region], vol 1, pp 98–102
16. Vinogradova SS (2013) Bufernaya yomkost pochv kak ikh osobennost k podshchelachiv-aniyu [Buffer capacity of soils as their property for alkalization] Vestnik BFU im. I.Kanta, vol 1, pp 102–109
17. Salikhova EV, Savostina OA (2001) Ekologo-geokhimicheskiye aspekty sostoyaniya gorodskoy sredy Kaliningrada [Environmental and chemical aspects of urban environment in Kaliningrad]. Geografiya, obshchestvo, okruzhayushchaya sreda: razvitiye geografii v stranakh Tsentralnoy i Vostochnoy Yevropy. [Geography, society, environment: development of geography in the countries of Central and East Europe. Abstracts. Part 1. KSU Publishing, Kaliningrad, pp 133–148
18. Yu SL (2009) Raspredeleniye tyazholykh metallov v pochvakh i rastitelnosti gorosdkilh ekosistem Kaliningradskoy oblasti [Distribution of heavy metals in soils and vegetation of urban ecosystems in the Kaliningrad region]. Vestnik RSU im. I.Kanta. Series – Natural Sciences, vol 1, pp 81–85
19. Doklad: Ob ekologicheskoy obstanovke v Kaliningradskoy oblasti v 2012 godu (2013) Report: on environmental state of the Kaliningrad region in 2012. The Government of the Kaliningrad Region, Kaliningrad, 164 p
20. Kurbatova AS, Bashkin VN, Kasimov NS et al (2004) Ekologiya goroda [Ecology of the town]. M.: Nauchnyi Mir, 620 p
21. Salikhova EV (2006) Retrospektiva formirovaniya ekologicheskogo karkasa Kaliningrada (sistema ozeleneniya) [Retrospective studies of formation of environmental frame in Kaliningrad (a system of vegetation planting)]. Vestnik BFU im. I.Kanta, №7. Kaliningrad, BFU Publishing, pp 34–39
22. Sostoyaniye okruzhayushchei prirodnoy sredy Kaliningradskoy oblasti v 2003 godu (2003) The state of environment in the Kaliningrad region in 2003. Committee for Natural Resources of the Kaliningrad Region, Kaliningrad, 160 p
23. Sitarov VA, Pustovoitov VV (2000) Sotsialnaya ekologiya [Social ecology]. M.: Publishing Center "Akademiya", 280 p
24. Decree of the Ministry of Regional Development of Russia of 09.09.2013 N 371 "Ob utverzhdenii metodiki otsenki kachestva gorodskoy sredy prozhivaniya" ["On approval of methodology for quality assessment of urban environment for habitation"]
25. Government Decree of the City of Moscow of 25.01.2000, N 49 "Ob utverzhdenii norm i pravil proyektirovaniya planirovki i zastroyki Moskvy MGSN 1.01-99" ["On approval of norms and rules for planning design and development in Moscow MGSN 1.01-99"]
26. SP 42.13330.2011 Gradostroitelstvo. Planirovka i zastroika gorodskikh i selskikh poseleniy. Aktualizirovannaya redaktsiya SNiP 2.07.01-89*, prilozheniye G (obyazatelnoye). Normativniye pokazateli plotnosti zastroiki territorialnikh zon [Town-planning. Designing and real-estate developing of urban and rural settlements. Revised edition SNiP 2.07.01-89*, annex G (obligatory). Regulatory indicators of housing density of territorial zones
27. Generalnyi plan Kaliningrada. Osnovniye polozheniya do 2015 goda (2004) [General Urban Plan of Kaliningrad. Principle provisions until 2015]. Otv. za vypusk T.L. Kondakova. Kaliningrad. Kaliningrad: Izd-vo TENAKS MEDIA, 40 p
28. Town Plan of Konigsberg, published by War Office, 1944, URL: http://www.etomesto.ru/map-kaliningrad_1944kenigsberg/

Environmental Features of Watercourses in the Kaliningrad Region

T.A. Bernikova, N.N. Nagornova, N.A. Tsoupikova, and S.V. Shibaev

Abstract This paper describes environmental features of the watercourses in the Kaliningrad Region. Hydrological characteristics of small watercourses were analysed within the catchment areas of the Curonian and Vistula (Kaliningrad) lagoons and separately for the rivers flowing directly into the Baltic Sea. The Pregolya and the Neman river basins (within the region boundaries) were considered particularly. Data on morphometry, hydrometry, hydrology and hydrochemistry are represented regarding all watercourses.

Keywords Chlorophyll *a*, Dissolved oxygen, Morphometric characteristics, Organic substances, River basin

Contents

1	Introduction	223
2	Materials and Methods	224
3	Results and Discussion	225
4	Conclusions	261
References		263

1 Introduction

Watercourses in the Kaliningrad Region are important throughways and the key element of water supply in many settlements. They are used for relaxation and recreational fishing, serve as spawning and nursery areas for valuable fish species

---

T.A. Bernikova (✉), N.N. Nagornova, N.A. Tsoupikova, and S.V. Shibaev
Department of Ichthyology and Ecology, Faculty of Bioresources and Nature Usage, Kaliningrad State Technical University, 1, Sovietsky prospect., Kaliningrad 236000, Russia
e-mail: bernikovy@gmail.com; nagornova@klgtu.ru; tsoupikova@klgtu.ru; shibaev@klgtu.ru; shibaev.s@gmail.com

and have importance for the commercial fishery. However, they make a definite contribution to the pollution of the Baltic Sea and its lagoons. Over 95% of the watercourses in the Kaliningrad Region are classified as "small." These streams are an initial segment for the formation of water resources in the territory. Water quality and ecological status of larger streams and water bodies depend largely on the environmental situation in the catchment areas of such watercourses. Each river is an aquatic ecosystem with a cyclical exchange of mineral and organic substances and energy. The key role in structure and functioning of this ecosystem belongs to water providing a habitat for organisms. Obviously, environmental research of watercourse ecosystems now is one of the most up-to-date areas of environmental protection activities. Despite the exceptional importance of small rivers, long-term data on their hydrological regime is absent, and observations did not embrace a significant part of small watercourses until very recently as the integrated environmental monitoring system in the Kaliningrad Region does not include permanent lookout stations on small rivers.

Increasing deviation of natural-anthropogenic geosystems in river basins from their natural condition caused an "issue of small rivers" and conditioned the necessity to address this issue which appears mainly in degradation of river geosystems due to intensive economic use of rivers and their catchments, discharge of polluted wastewaters to the rivers and transboundary anthropogenic load.

In this context, the purpose of this paper is to describe environmental features of watercourses and their catchment areas including the analysis of hydrological and hydrochemical peculiarities of the watercourses.

2 Materials and Methods

This description of watercourses is based on the material of observations made in 1993–2000 [1–5], 1999–2006 [6–13] and 2007–2014 [12, 14–27] using also open sources of literature.

Terms of research were planned to capture all the water regime phases of the rivers. Within the period from 2007 to 2009, the watercourses were examined mainly during autumn floods (September–November) and summer low streamflow periods (May–August) interrupted with short-term river floods. The most detailed data was collected in 2010, when the observations covered all four water regime phases [16–18, 28, 29] and for the first time were carried out in winter, and also for the first time, all transboundary watercourses along the southern border of the region were studied. Study of chlorophyll a in the Neman river began in 2007–2014 [30].

Location of cross sections was determined by the size of the river, the presence of potential contamination sources and the possibility of approaching the projected points. There were assigned estuarine cross sections (outfalls) to estimate the nutrient load to the coastal area of the Baltic Sea; water quality at such cross sections sums up the environmental situation of the river basin as a whole. To assess the quality of water coming from the neighbouring countries, the border stations were chosen at the transboundary watercourses as close to the state frontier as possible. To characterize the major rivers, the cross sections were located along the rivers up- and downstream of the main large settlements (potential point sources of discharge of pollutants), in areas receiving drainage waters from agriculturally used land and wetlands (to monitor the diffuse runoff from the catchment area). The points possibly closest to the headwaters or to the state border (for transboundary rivers) were selected as baseline monitoring stations (Fig. 1, Table 1).

Observations were based on publicly available standard methods [31–33]. Water temperature and dissolved oxygen level were measured with WTW-3751, electrical conductivity and pH with WTW Multiline P3. Water samples for hydrochemical testing were taken with a bucket; plastic bottles were used as containers. Hydrochemical examination and determination of chlorophyll a was carried out in the hydrochemical laboratory of the Department of Ichthyology and Ecology (by using a spectrophotometer UNICO 2100). The volume of chlorophyll a samples varies depending on the season (0.25 dm^3 in May–July, 0.5 dm^3 in August–November). Chlorophyll a concentration was calculated pursuant to the formula recommended in [34].

3 Results and Discussion

1. Geological and geomorphological factors and climatic conditions support the high level of water resources security in the region. However, specific relief features result in that drainage divides are well defined only within upland districts. The maximum height of drainage divides is 161 m (Ozerskaya ridge), the height of the other two ridges (Instruchskaya and Dobrovolskaya) and of the Sambian upland does not exceed 100–60 m, and the height of other water divides is 60–40 m or less. Divides on the plains are often swamped and ill-defined.

The river network is well developed and very dense (up to 1.35 km/km^2). Watercourses in the region are classified as plain streams and belong to the Baltic Sea catchment area. Under the watershed principle, the watercourses were divided into the following groups (Table 2, Fig. 2).

Almost all the rivers in the Neman delta and at the Curonian lagoon coast are connected by channels and form an integrated drainage system, which makes it impossible to define the drainage divides: all of them are attributed to the catchment area of the Curonian lagoon. Description of natural conditions of the landscape and climatic areas is given in Table 3.

Regarding a range of natural and anthropogenic factors, there are the most favourable conditions for small watercourses draining the delta landscapes and

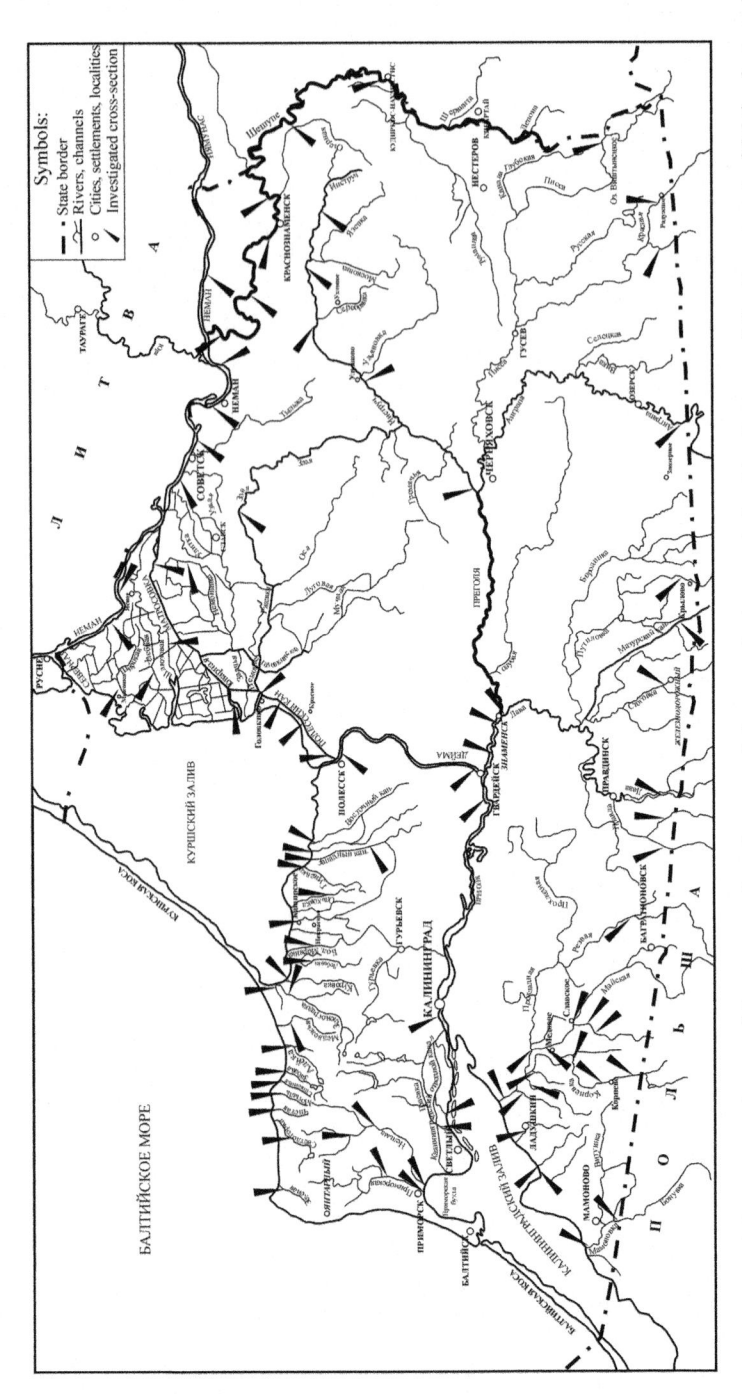

Fig. 1 Location of the cross sections used for integrated environmental river monitoring, carried out by the team of the Department of Ichthyology and Ecology, KSTU (English equivalents of geographical names are given in Table 1)

Table 1 List of geographical names on the maps of the Kaliningrad Region

Hydrographical network			
English	Russian	English	Russian
Aleyka	Алейка	Ovrazhka	Овражка
Angrapa (Węgorapa)	Анграпа (Венгорапа)	Ozernaya	Озерная
Baltic Sea	Балтийское море	Pissa	Писса
Bolshaya Moryanka	Большая Морянка	Pokosnaya	Покосная
Borodinka	Бородинка	Polessky canal	Полесский канал
Chernaya	Черная	Pravda	Правда
Chistaya	Чистая	Pregolya	Преголя
Curonian lagoon	Куршский залив	Primorskaya	Приморская
Deima	Дейма	Primorsky canal	Приморский канал
Dunayka	Дунайка	Prokhladnaya	Прохладная
Glubokaya kanava (ditch)	Глубокая канава	Putilovka	Путиловка
Golovkinsky canal	Головкинский канал	Rezvaya	Резвая
Golubaya	Голубая	Rzhevka	Ржевка
Grayevka	Граевка	Russkaya	Русская
Gremyachya	Гремячья	Seletskaya	Селецкая
Guryevka	Гурьевка	Serebryanka	Серебрянка
Instruch	Инструч	Severnaya	Северная
Kaliningrad (Vistula) lagoon	Калининградский (Вислинский) залив	Sheshupe (Šešupė)	Шешупе
Kaliningrad bypass channel	Калининградский отводной канал	Shirokaya	Широкая
Khlebnaya	Хлебная	Shlyuzovaya river	Шлюзовая
Kornevka	Корневка	Skirvite	Скирвите
Krasnaya	Красная	Spokoynaya	Спокойная
Kurovka	Куровка	Stogovka	Стоговка
Lava (Łyna)	Лава (Лына)	Svetlogorka	Светлогорка
Lobovka (Kraynyaya)	Лобовка (Крайняя)	Trostyanka	Тростянка
Lugovaya	Луговая	Tumannaya	Туманная
Malaya Moryanka	Малая Морянка	Tylzha	Тыльжа
Mamonovka	Мамоновка	Ulitka	Улитка
Matrosovka canal	Матросовка	Ulyanovka	Ульяновка
Mayskaya	Майская	Uzkaya	Узкая
Medvezhya	Медвежья	Vika	Вика
Moskovka	Московка	Vitushka	Витушка
Motyl	Мотыль	Vostochny canal	Восточный канал
Muchnaya	Мучная	Yazevka	Язевка
Nelma	Нельма	Zabava	Забава
Neman	Неман	Zapadny canal	Западный канал

(continued)

Table 1 (continued)

Nemonin	Немонин	Zayachya	Заячья
Nemoninka	Немонинка	Zelenogradka	Зеленоградка
Nemoninsky canal	Немонинский канал	Zelyonaya	Зеленая
Olkhovka	Ольховка	Zlaya	Злая
Osa	Оса		
Cities, settlements, localities			
English	Russian	English	Russian
Bagrationovsk	Багратионовск	Nesterov	Нестеров
Baltiysk	Балтийск	Nemanskoe	Неманское
Bolshoe Selo	Большое Село	Polessk	Полесск
Chernyakhovsk	Черняховск	Pionersk	Пионерск
Gvardeysk	Гвардейск	Pravdinsk	Правдинск
Guryevsk	Гурьевск	Primorsk	Приморск
Gusev	Гусев	Slavsk	Славск
Kaliningrad	Калининград	Sovetsk	Советск
Kashirskoe	Каширское	Svetly	Светлый
Khrabrovo	Храброво	Yantarny	Янтарный
Krasnoznamensk	Краснознаменск	Yasnoe	Ясное
Ladushkino	Ладушкино	Zheleznodorozhny	Железнодорожный
Levoberezhnoe	Левобережное	Znamensk	Знаменск
Mamonovo	Мамоново		

landscapes of the Vishtynets upland, rather favourable conditions – for the rivers of Sambia Peninsula.

Watershed of the Pregolya river and subwatersheds of the Angrapa and the Instruch rivers forming the Pregolya as a result of confluence and subwatersheds of some major Pregolya's tributaries (the Lava, the Golubaya, the distributary Deima) occupy significant part of the Kaliningrad Region's territory (Table 4).

Watersheds of the Prokhladnaya, the Mamonovka, the Nelma and the Primorskaya rivers, subwatersheds of the Neman's tributaries (the Tylzha, the Sheshupe) and of the small rivers flowing into the Baltic Sea and the Vistula and the Curonian lagoons occupy the rest part of the region. Over 95% of the total length of the rivers in the region falls to the share of the rivers in which basin areas are less than 2,000 km^2 (Table 5).

The forest coverage of the watersheds is low, mostly about 13–14% [35]. Watershed slopes (Table 6) are also very small (the mean value is below 0.001).

Due to the small value of slope, the estuarine and adjacent parts of the rivers are in the backwater of the receiving water body and very dependent on the wind-driven surges. Most small rivers (especially within the Neman delta, the Pregolya river valley, at the coast of the Curonian and the Vistula lagoons) are channelized, and the water pumped out from the polder lands is dumped therein.

Table 2 Groups of watercourses within the Kaliningrad Region [23]

Watershed number shown in figure	Watershed name	Subwatershed number shown in figure	Subwatershed name
10.	Pregolya river basin	10.1	Deima river (distributary) basin
		10.2	Lava river basin
		10.2.1	Pravda river basin
		10.2.2	Stogovka river basin
		10.2.3	Putilovka river basin
		10.3	Golubaya river basin
		10.4	Angrapa river basin
		10.4.1	Pissa river basin
		10.5	Instruch river basin
11.	Neman (Nemunas) river basin	11.1	Drainage area tributary to the eastern part of the Curonian lagoon (the Matrosovka canal, the Golovkinsky canal and such river as the Shlyuzovaya, the Khlebnaya, the Uzkaya, the Shirokaya, the Zayachya, the Ulitka, the Zlaya)
		11.2	Sheshupe (Šešupė) river basin
1.	Watersheds of small and very small rivers flowing northwards and discharging into the Baltic Sea (the Zelyonaya, the Svetlogorka, the Chistaya, the Motyl, the Spokoynaya, the Zabava, the Aleyka, the Medvezhya)		
2.	Primorskaya river basin		
3.	Nelma river basin		
4.	Grayevka river basin		
5.	Drainage area tributary to the eastern part of the Vistula (Kaliningrad) lagoon		
6.	Ditches and land drainage canals in the area of settlement Ladushkino		
7.	Mamonovka river basin		
8.	Prokhladnaya river basin		
9.	Watersheds of small and very small rivers flowing northwards and discharging into the southern part of the Curonian lagoon (the Zelenogradka, the Bolshaya Moryanka, the Malaya Moryanka, the Kurovka, the Lobovka, the Olkhovka, the Guryevka, the Zapadny canal, the Vostochny canal)		

Many rivers flowing into the Baltic Sea have shifting beds near their mouths, and displacement movement can reach 200 m and more [6, 7, 11, 12, 25–27, 42].

Almost the entire territory of the region is dewatered one way or another. There are a large number of outlet drainage canals on floodplains, in low-lying areas of the region; many rivers are transformed into drainage canals. Estuarine parts of the most rivers flowing into the Curonian and the Vistula lagoons are regulated by polder systems, some of which are currently damaged to one extent or another.

Runoff coefficient varies between 0.19 (the Lava river) and 0.31 (the Instruch river). Mean specific runoff of the most rivers is in the range of 5–7 $l/s/km^2$. Only

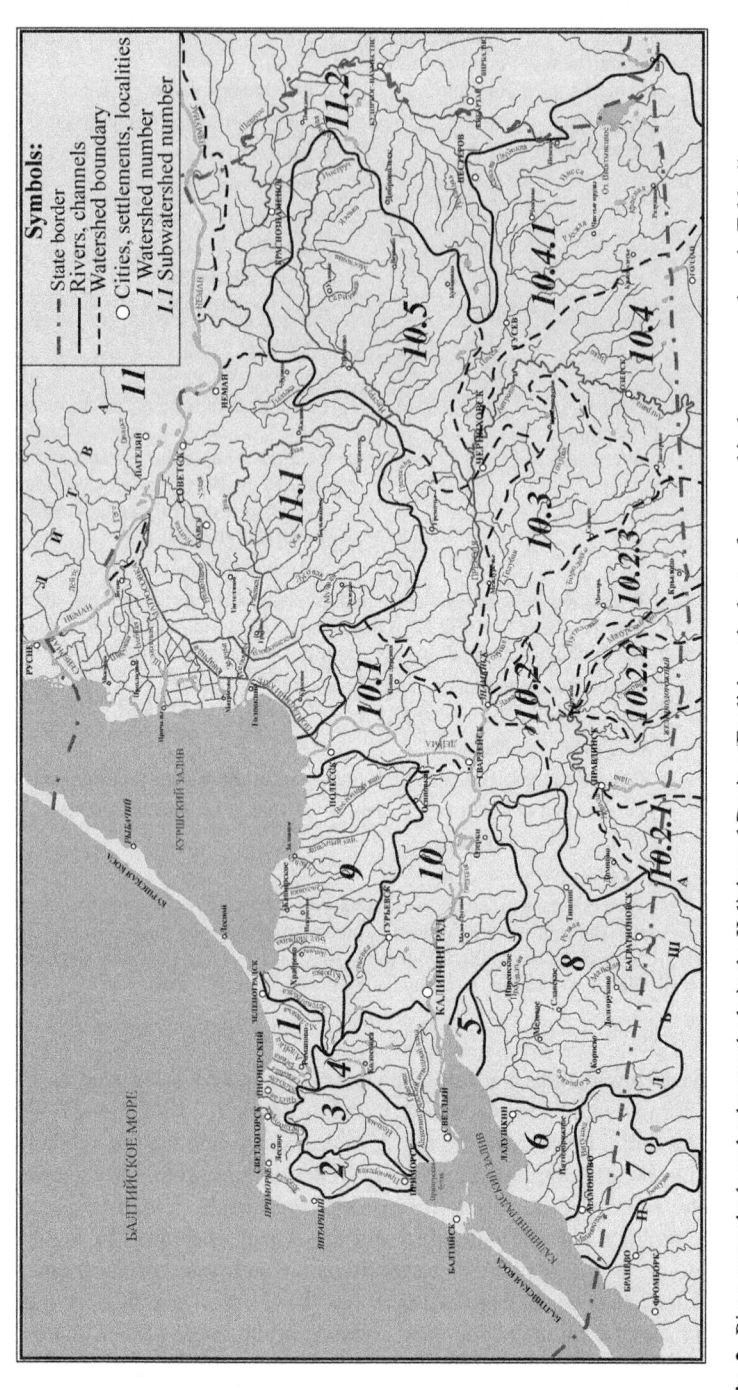

Fig. 2 River watersheds and subwatersheds in the Kaliningrad Region (English equivalents of geographical names are given in Table 1)

Table 3 Description of natural conditions of the landscape and climatic areas (compiled according to the data [35, 36])

Climatic area number	Landscape areas	Predominant soil type	Soil texture	Forest coverage, %	Swamp area, %	Groundwater level, m	Rivers
I	Sambian upland, estuarine sandy lowlands	Sod-low podzolic and mesopodzol	Sandy, sandy loam	13	3	0–3, up to 5	Nelma, Primorskaya, Grayevka, Zelyonaya, Svetlogorka, Chistaya, Spokoynaya, Medvezhya, Motyl, Aleyka, Zabava
II	1. Warmian upland, western part of the Pregolskaya lowland	Soddy gley	Light loamy	14	3	0–3	Mamonovka, Prokhladnaya
	2. Polesskaya lowland, Severo-Vostochnaya plain	Soddy gley	Medium loam	32	15	0–3	Zelenogradka, Bolshaya Moryanka, Kurovka, Lobovka, Olkhovka, Guryevka, Zapadny canal, Vostochny canal, Deima
	3. Deltaic lowland of the Neman river	Peat humus, alluvial boggy	Medium loam	21	36	0–1	Rivers and channels of the Neman river delta
III	1. Pregolskaya lowland	Sod mesopodzol	Heavy loam	10	5	0–3	Lava, Golubaya
	2. Neman-Sheshupe plain, Sheshupe lowland, Angrapa-Sheshupe plain, Instruch ridge	Sod mesopodzol, soddy gley	Medium and heavy loam	9–13	3–10	0–3, up to 5	Instruch, Angrapa
	3. Vishtynets upland	Sod-low podzolic	Light loamy, sandy loam	14	3	0–5	Pissa, Krasnaya

Table 4 The major rivers in the Kaliningrad Region [37–39]

River	Where the river flows into	Length, km Total	Length, km Within the Kaliningrad Region	Catchment area, km^2	Annual runoff, km^3
Neman	Curonian lagoon	937	115	101,000	25.1
Sheshupe (Šešupė)	Neman	308	102	6,120	0.52
Pregolya	Vistula lagoon	123	123	13,600	1.39
Instruch	Pregolya	101	101	1,250	0.27
Angrapa (Węgorapa)	Pregolya	169	95	3,960	0.72
Lava (Łyna)	Pregolya	289	56	7,130	1.31
Deima	Curonian lagoon	37	37	463	0.962
Pissa	Angrapa	98	98	1,360	0.273

Table 5 Natural features of the river basins belonging to the different groups of watercourses

Group of watercourses[a]	Length, km	Basin area, km^2	Watercourses
Streams	<10	Below 50	The Zelyonaya, the Chistaya, the Motyl, the Spokoynaya, the Pokosnaya
Very small	10–25	50–2,000	The Aleyka, the Zabava, the Medvezhya, the Svetlogorka, the Guryevka, the Trostyanka, the Kurovka, the Lobovka, the Olkhovka, the Nemoninka
Small	26–100		The Prokhladnaya, the Mamonovka, the Nelma, the Primorskaya, the Zelenogradka, the B. Moryanka, the Zapadny canal, the Vostochny canal, the Zlaya, the Instruch, the Pissa
Medium sized	101–500	2,000–50,000	The Pregolya, the Angrapa, the Lava, the Sheshupe
Large	>500	Over 50,000	The Neman

[a]Groups of watercourses were distinguished in compliance with the classification given in [35, 40, 41]

Table 6 Extreme values of watershed slopes [37–39]

River watersheds (subwatersheds)	The rivers Pissa, Prokhladnaya, Instruch, Zlaya, Osa	The Sheshupe river	The Neman delta	The Pregolya river
Slope value	0.023–0.01	0.0008	0.0004	Down to 0.0001 and less

very few rivers have specific runoff from 7 to 9.8 l/s/km^2 in the upper reaches. The water content of small rivers flowing into the Baltic Sea during low water periods is under 1 l/s/km^2.

Most of the watercourses except for the Neman river are mixed-fed. Distribution of different water sources that feed into and create the rivers (as a percentage of annual runoff) is estimated as follows (Table 7).

The mixed-fed rivers receive more snow water in the years with long, snowy and harsh winters, in general not typical for the Kaliningrad Region, and more rainwater after the most frequent warm winters with little snow and thaws. Rather large percentage falls to the share of underground water in any year. The water regime of the rivers is characterized by spring snowmelt flood and summer, autumn and winter rainfall floods. Low water periods are unstable and in mild winters with frequent thaws and in summer during intense widespread rains can be interrupted by long enough floods.

The water temperature is largely determined by climate and weather conditions. During unsteady winters, the floods may occur two to three times due to thaw periods [39]. Since the thaws often occur in winter, the ice regime is unstable. Such a rare feature as two-layer ice is connected with it. The weather conditions in 2010 were not typical for the area. The winter was abnormally cold and snowy. The air temperature in December 2009 was by 0.5–1.5 °C, in January 2010 by 4.0–6.0 °C and in February by 1.0–3.5 °C below normal. In addition, precipitation in December amounted to 185% of the norm, in January it was 30–50% and in February it was three times higher than normal [43]. Some watercourses (the Chistaya, the Motyl, the Spokoynaya) were completely frozen down to the bottom, and two-layer ice was found on some of them. Its appearance was conditioned with the fact that the first weak ice cover appeared before the winter low water period began. It was followed by the thaw during which the river opened in the areas with the higher flow velocity, but fragments of ice plates preserved on the coastal trees, shrub and herbaceous vegetation. After the winter low water period started, the water level in the rivers has fallen pretty dramatically (by 0.5–0.7 m, even by 1 m on the Prokhladnaya river), and freezing up proceeded at the low water level (Fig. 3).

2. All watercourses of the Kaliningrad Region can be conditionally divided into three groups subject to which component of the common catchment area of the Baltic Sea they belong to rivers flowing directly into the Baltic Sea, flowing into the Vistula (Kaliningrad) and into the Curonian lagoons.

Rivers flowing directly into the Baltic Sea are situated on Sambia Peninsula, run northwards and drain hilly moraine landscapes. According to the classification [35], these are "streams" or "very small rivers" (Table 8). Most of them belong to the first fishery category.

Table 7 Distribution of water sources that feed into and create rivers in the Kaliningrad Region [37–39]

Water sources that feed into	Percentage of the annual runoff
Rain	37–48
Snow ("spring runoff")	26–37
Underground water	16–37

Fig. 3 Two-layer ice on the rivers in the Kaliningrad Region, 2010: (**a, c**) the Primorskaya river, (**b**) the Prokhladnaya river

Table 8 Main morphometric characteristics of the estuarine cross sections of the rivers flowing directly into the Baltic Sea, 2010

River	Fishery water category	S_{basin}, km^2	$L_{div.}$, km	L_B, km	$B_{Max.}$, m	B_{mean}, m	L, m	I, m/km
Zelyonaya	–	15.4	16	6	4	3.1	3.2	15.6
Svetlogorka	2	25.4	23	8	4	3.2	15.5	3.10
Chistaya	2	13.5	19	8	3	1.7	9.9	5.95
Motyl	1	11.8	17	5	3	2.4	7.3	6.44
Spokoynaya	1	8.4	18	6	2	1.4	7.0	6.86
Zabava	1	17.0	21	6	4	2.8	12.0	1.42
Aleyka	1	41.8	34	10	6	4.2	12.0	2.92
Medvezhya	1	41.9	33	12	6	3.5	10.0	3.20

S_{basin} basin area, $L_{div.}$ length of the drainage divide, L_B basin length, $B_{Max.}$, B_{mean} basin width maximum, mean, L watercourse length, I slope of the watercourse

Watershed slopes are the largest among river basins of the region. The structure of watersheds is monotonous, their length is 5–8 km (the Aleyka's and the Medvezhya's river basins are 10–12 km long), and width is 2–4 km. Taking into account small catchment areas of these watercourses, some authors [9, 44, 45] consider these watersheds as a drainage basin of the Baltic Sea.

Hydrometric parameters of the rivers during the period of investigations varied little. Water flow was small, especially during low water period (Table 9).

Annual runoff is 0.01 km^3 and less, so they can influence the Baltic Sea only during the snowmelt flood. The Aleyka and the Zabava had the highest water content. In abnormally cold winters, some watercourses may freeze completely down to the bottom (e.g. the Chistaya, the Motyl, the Spokoynaya in 2010).

Rivers flowing into the Curonian lagoon are located within the Polesskaya and the Nizhnenemanskaya lowlands. They carry their waters through landscapes of moraine rolling unevenly drained and periglaciolacustrine flat and undulating plains as well as landscapes of deltaic step-and-platform low plains. Watercourses flowing into the lagoon from the west bring their waters from the Sambian moraine plateau. Most of the territory is drained with the help of the covered drainage system and combined network. Estuarine areas of almost all watercourses are lift drained territories [35] equipped with pump stations. Most of the rivers are small and very small (Table 10).

The only large watercourses flowing into the Curonian lagoon are the Neman river and the Deima river which is a distributary channel of the Pregolya river.

Most of the small watercourses draining the Polesskaya and the Nizhnenemanskaya lowlands are of the second fishery category. The Zelenogradka, the Nemonin and the Zapadny canal belong to the first fishery category. The Deima distributary and the Matrosovka canal have the superior fishery status. Small streams which flow into the lagoon are classified as "small" or "very small" according to their morphometric characteristics. Watersheds are narrow and elongate: basin length is from 5–10 to 20 km or more, and mean basin width is 3–6 km. Slopes are the smallest among the rivers of the region. The river network is overloaded with reclamation and drainage system especially in the estuarine parts. Many areas lie below sea level. Within the Neman delta and the

Table 9 Water flow of some small rivers (m^3/s) at different time periods [1, 5]

River	1993	1994	2009	2010
Summer low water periods				
Chistaya	–	–	–	0.004
Aleyka	–	0.10	0.05	0.06
Zabava	–	0.05	0.04	0.07
Medvezhya	–	–	–	0.0
Autumn flood periods				
Chistaya	0.06	–	–	0.03
Aleyka	0.14	0.18	0.29	0.28
Zabava	0.13	0.04	–	0.27
Medvezhya	0.2	–	–	0.03

Table 10 Morphometric characteristics of watersheds of the small rivers flowing into the Curonian lagoon

River	Fishery water category	S_{basin}, km^2	$L_{div.}$, km	L_B, km	$B_{MAX.}$, m	B_{mean}, m	L, m	I, m/km [44]
Zelenogradka	1	72.0	60	21	9	3.4	21.0	0.82
Trostyanka	–	19.0	17	6	4	3.2	12.0	
Kurovka	2	40.4	38	13	5	3.1	18.0	
Pokosnaya	2	4.1	8	4	1	1.0	3.3	
Lobovka (Kraynyaya)	–	62.0	40	15	5	4.1	13.5	
B. Moryanka	2	76.0	46	11	7	6.9	17.0	
Olkhovka	2	53.0	30	10	6	5.3	12.0	
Guryevka	2	42.5	34	12	6	3.5	23.0	
Zapadny canal and Dunayka river	1	137.0	52	21	9	6.5	15.0	
Vostochny canal and Ovrazhka river	2	98.0	70	30	6	3.3	21.0	
Deima	Superior	463.0	125	28	23	20	37.0	
Nemonin	1	2,375 [44]	175	80	55	43.2	46.0	0.54
Matrosovka	Superior						43.0	
Khlebnaya	2						17.0	
Uzkaya	2						5.5	
Shirokaya	2						15.0	
Zlaya	2						62.0	

S_{basin} basin area, $L_{div.}$ length of the drainage divide, L_B basin length, $B_{MAX.}$, B_{mean} basin width maximum, mean, L watercourse length, I slope of the watercourse

Nizhnenemanskaya lowland, the divides are ill-defined; all watercourses are connected with a complex system of drains and canals. On this basis, they are united in the drainage area tributary to the eastern part of the Curonian lagoon. Small streams on the Polesskaya lowland are combined in the drainage area tributary to the southern part of the Curonian lagoon [9, 44, 46].

Water flow of small rivers is insignificant; however its value is slightly bigger than water flow of the rivers running down into the lagoon from Sambia Peninsula. In spring time, the majority of small streams on the Polesskaya lowland are in the backwater of the Curonian lagoon and have no runoff. The distributary Deima is largely dependent on the wind direction. Westerly winds prevailing in the area create notable surges from the Vistula lagoon, and the whole water flow is discharged into the Curonian lagoon. When the surge from the Curonian lagoon occurs created by northern and partly eastern winds, which happens considerably less often, the entire flow of the river goes into the Vistula (Kaliningrad) lagoon. Only in case of substantial water flow which takes place during snowmelt floods the surges cease to have an effect on the course of the Deima.

Rivers flowing into the Kaliningrad (Vistula) lagoon drain all the main landscape types of the Kaliningrad Region. The Pregolya river system plays the key role here; it flows into the lagoon from the east. The Nelma, the Primorskaya and the Grayevka rivers discharge into the lagoon from north and north-west (watersheds of these rivers are located on Sambia Peninsula); the Prokhladnaya and the Mamonovka rivers empty into the lagoon from southeast. Watersheds of the rivers flowing into the lagoon from east and southeast are transboundary. Table 11 shows morphometric characteristics of the river basins.

The Nelma, the Mamonovka and the Prokhladnaya rivers are of high fishery importance, and according to their morphometric characteristics, they are classified as small rivers. The divides are well defined, with winding divide lines. The watersheds are slightly elongate round shaped, from 10–20 to 44 km long and from 5–8 to 10 km or more wide (the Prokhladnaya river). The slopes have significant value.

Hydrometric parameters fluctuate greatly and are closely associated with both the weather conditions and the aggregate natural features of the area drained [1, 47]. Water flow of the majority of the rivers in 2010 ranged 1–6 m^3/s (extremes: 0.2 m^3/s for the Chernaya river, 6.4 m^3/s for the Prokhladnaya river). Annual runoff was about 0.01–0.04 km^3. The maximum stream velocity is typical for the small rivers draining the elevated landscapes (Sambian moraine plateau, Vishtynets and Warmian uplands); the rivers draining the Polesskaya and the Nizhnenemanskaya lowlands have the minimum water flow and stream velocity.

3. *Hydrochemical conditions* of watercourses develop under the integrated impact of the set of factors (Table 12).

Soil pattern is one of the most important factors that contribute to the formation of the hydrochemical regime of surface waters. Parent materials in the region have high carbonate content [39]. The rocks forming the territory of the region are mostly washed free of highly soluble salts (sulphates, chlorides); therefore they cannot be a reason for an increase in TDS in surface waters.

During the spring, snowmelt floods the rivers that mainly receive soil and surface waters, and *salinity level (TDS) in the rivers* is the lowest within a year.

Table 11 Main morphometric characteristics of watersheds of the small rivers flowing into the Vistula (Kaliningrad) lagoon

River	Fishery water category	S_{basin}, km^2	$L_{div.}$, km	L_B, km	$B_{Max.}$, m	B_{mean}, m	L, m	I, m/km
Nelma	1	186.0	124	24	13	7.8	30.0	1.45
Primorskaya	2	112.5	54	20	9	4.7	25.0	1.37
Grayevka	2	142.5	46	18	11	7.9	30.0	1.73
Mamonovka	1	311	136	26	24	13.0	12.0	1.20
Prokhladnaya	Superior	1,170	172	44	35	26.6	77.0	1.23

S_{basin} basin area, $L_{div.}$ length of the drainage divide, L_B basin length, $B_{Max.}$, B_{mean}, basin width maximum, mean, L watercourse length, I slope of the watercourse

Table 12 Priority factors influencing the formation of hydrochemical composition of water courses in the Kaliningrad Region [29]

Natural factors	Anthropogenic factors
1. Flat terrain	1. Predominance of agricultural areas in land structure
2. Temperate maritime climate with abundant moisture (water surplus)	2. Relatively high average population density, especially in the western part of the region and in the cities, strong prevalence of urban population
3. Dense hydrographic network with a complex drainage system	3. Insufficient level of municipal and industrial wastewater treatment (or complete lack thereof)
4. Prevalence of small and very small rivers	4. Main water consumers in the region are the pulp and paper industry (PPI), housing and community amenities
5. Small slopes of river basins and, as a consequence, backwater and a great influence of high winds causing surge	5. Great number of motor vehicles per population and dense road network in the region
6. Heavy percentage of waterlogged soils, high groundwater level	6. Diffuse runoff from urban and industrial areas
7. Large share of meadows and forests in land structure	7. Transboundary location of major watersheds
8. Peculiarities of soils	

During the summer low water, the streamflow in rivers is supported mostly with groundwater, and its TDS value is more sustainable than the same in subsoil waters. *TDS concentration* in the small rivers increases up to 300–600 mg/dm^3. In autumn water salinity varies widely: from 280 (small rivers draining the Vishtynets upland) to 716 mg/dm^3 (small streams located in the Polesskaya lowland) that is comparable to the TDS concentration during summer low water. The results obtained can be explained with the influence of the weather conditions. There were minor floods in autumn time 2007–2010 (exclusive of 2007) resulting in the chemical composition of water affected by groundwater. TDS in river water reaches the greatest value during winter low water when the frozen rivers receive groundwater with a chemical composition formed in deep aquifers. *TDS* in rivers rises to 550–650 mg/dm^3 reaching the maximum concentration in the small streams on the Polesskaya lowland (more than 900 mg/dm^3). HCO_3^- (38–46 eq. percent) and Ca^{2+} (20–24 eq. percent) ions dominate in river waters, except for the watercourses located within the Polesskaya lowland and the Neman delta where magnesium cations were dominant. The content of Cl^- ions in the streams of the Polesskaya Lowland was 9 eq. percent.

In compliance with the classification [48], the streams mainly have calcium and bicarbonate water, usually of the first water type [$HCO_3^- < (Ca^{2+} + Mg^{2+})$]. There is a significant content of sodium and potassium cations in water as the estuarine areas of the rivers in the region are the zones of active interaction between river and sea water. Small slopes greatly contribute thereto, and, as a result, there is a backwater from the receiving water bodies.

According to the total hardness value (3–6 mg eq./dm^3), water in the most rivers can be classified [49] as "moderately hard"; minimum total water hardness was discovered in the streams belonging to the Neman delta where swamp waters influence greatly. Total hardness undergoes pronounced seasonal fluctuations. The lowest values (2–3 mg eq./dm^3) appeared during snowmelt floods when rivers are mainly meltwater-fed, with the exception of the small watercourses located within the Polesskaya and the Lava-Pregolskaya lowlands (4–5 mg eq./dm^3). Hardness reaches its highest value (5–6 mg eq./dm^3, "moderately hard" water) during stable winter low water (maximum takes place in water of the rivers on the Polesskaya lowland and in the Neman river delta – "hard" water) when rivers are groundwater-fed for the most part.

Comparison of the results obtained showed that TDS and total hardness in small watercourses vary little for the researched period. The year 2009 was exceptional. The rivers which have been investigated in this year revealed a maximum content of mineral substances.

Contents of organic substances were assessed by permanganate index (permanganate oxidation) which was subject to significant and quite natural seasonal fluctuations. The nature of fluctuations was determined, on the one hand, by the water regime and dependent thereon input of allochthonous organic matter from the watershed surface and, on the other hand, by temporal changes in processes of production and decomposition of organic matter. In winter under ice, the value of permanganate index varies from 5–10 (the Aleyka, the Kurovka, the Guryevka) up to 10–15 mgO/dm^3 in the rivers which drain the wetlands (the Nemonin, the Shirokaya, the Deima, the Prokhladnaya). In summer due to floods, the mean permanganate value is higher than in winter, and in most rivers it ranges within 10–16 mgO/dm^3. In the watercourses that suffer severe human impacts (the Grayevka, the Chernaya, the Medvezhya, the Chistaya), permanganate index is heavily affected by the regime of influent of industrial and household wastewater: permanganate value notably increases in summer as a result of insufficient dilution of wastewater with river waters. The amount of organic substances in highly polluted watercourses during low water periods reached catastrophic values – up to 20–30 mgO/dm^3 in winter and up to 65 mgO/dm^3 in summer. The rivers in the Neman delta and on the Polesskaya lowland and the rivers situated in the southwestern part of the Kaliningrad Region (the Prokhladnaya, the Mamonovka river) are fed and influenced by swamp waters that exacerbates the effect of the anthropogenic factor. The acutest impact of the anthropogenic factor appears in the river basins in the western and central parts of the region where the following numerous elements produce an effect on the content of organic matter. These elements include flow of the Kaliningrad bypass channel (the Grayevka, the Chernaya); municipal effluents from the nearby villages disposed directly into such watercourses as the Lobovka river (settlement Khrabrovo), the Malaya Moryanka river (settlement Kashirskoe), the Primorskaya river (city of Primorsk) and the Chistaya river (city of Pionersk); runoff from agricultural land and livestock farms (land use intensity within the Polesskaya and the Lava-Pregolskaya lowlands is over 0.8 [35]); and partly transboundary pollution [34, 50–52] through the Mamonovka and the Prokhladnaya rivers and small streams tributary to the Lava and the Angrapa river basins. During the spring snowmelt flood and autumn floods, the permanganate index of most rivers reaches its maximum (10–18 mgO/dm^3). The exception is the heavily contaminated rivers (the Grayevka, the Lobovka, the Medvezhya, etc.) where the concentrations of organic matter slightly decrease in this period due to diluting with cleaner rain and melt water.

Comparison of the obtained materials with the previous research data showed that the content of organic substances in small rivers reaches maximum values in summer. Thus, there is quite much organic matter dissolved in river water. Permanganate value varies mainly from 5–10 to 15–28 mgO/dm^3, and it is estimated as "increased" exclusive of the rivers on the Vishtynets upland with the "average" organic content.

According to the content of organic matter, the small watercourses of the Kaliningrad Region can be divided into the following groups.

Slightly polluted rivers: the amount of organic substances is minimum during low water periods with some increase during snowmelt and flash floods. The value of permanganate index in such watercourses is classified as "average" (5–10 mgO/dm^3) or "increased" (10–15 mgO/dm^3). These are rivers of the Vishtynets upland (the Pissa river) and the Sambian upland (the Spokoynaya, the Aleyka, the Nelma, the Primorskaya) and the major part of the rivers on the Polesskaya and the Lava-Pregolskaya lowlands, in the Neman river delta.

Polluted rivers: the amount of organic substances is maximum during low water periods with some decrease during snowmelt and flash floods. The value of permanganate index in these waterways is classified as "increased" (10–15 mgO/dm^3) – the Prokhladnaya, the Mamonovka, the Lobovka, the Svetlogorka, the Chistaya and the Zabava.

Heavily contaminated rivers: organic content is catastrophically high. Value of permanganate index is classified as "high" (20–30 mgO/dm^3) or "very high" (above 30 mgO/dm^3) – the Grayevka river, the Chernaya river and the Medvezhya river.

Oxygen conditions largely determine the chemical and biological situation in aquatic ecosystems. Groundwater and swamp and drainage waters make a strong effect on the oxygen content in the small rivers of the Kaliningrad Region, especially within the Polesskaya and Lava-Pregolskaya lowlands, in the Neman river delta. These waters are characterized with reduced oxygen content. In some watercourses (such rivers as the Chistaya, the Medvezhya, the Trostyanka, the Lobovka, the M. Moryanka and others), discharge of municipal wastewater from the nearby settlements causes oxygen deficiency. The *annual cycle of oxygen* in the small rivers has the following features. The oxygen content is low in a winter time, under the ice cover. However, in the absence of ice on some streams, oxygen concentration is high enough due to the increase in oxygen solubility and slow oxidation processes under low temperature. For example, in winter 2010 oxygen content in the rivers Spokoynaya and Zabava was 15.6 mg/dm^3 (104% of saturation) and 12.8 mg/dm^3 (85%), respectively. In spring, at the peak of the algae population, during intensive photosynthesis but under low water temperature (below background summer temperature), there is high oxygen content. In summer, water is less saturated with oxygen than in spring. In conditions of higher water temperature, the solubility of oxygen decreases, and it is very rapidly consumed for the decomposition of dead plants and organic substances in excess received by the rivers from their catchment areas and with domestic wastewater. Oxygen deficiency in some rivers (the Medvezhya, the Grayevka, the Chernaya, the Vostochny canal) grows into a disaster. In autumn, a decrease in water temperature results in an increase in oxygen solubility, and oxidative process intensity drops significantly. In some rivers, there is a supersaturation in this period. In general, the highest within year oxygen content in the small streams of the Kaliningrad Region takes place in autumn. Concluding description of oxygen conditions, it deserves to be noted that the most favourable situation was found in the small rivers on the Vishtynets upland and Sambia Peninsula flowing northwards. The only exception is the Medvezhya river with the relative amount of oxygen in summer often below 10%. Similar conditions were discovered in some rivers of the Polesskaya lowland (the Lobovka, the Pokosnaya, the Guryevka, the Zelenogradka). Oxygen saturation in the small rivers of the Neman delta (the Zayachya, the Uzkaya, the Khlebnaya) does not exceed 30%. The most unfavourable oxygen conditions develop in the rivers of the Polesskaya lowland and of the Neman delta as well as in the estuarine part of the Pregolya river and in the Grayevka river basin. Oxygen saturation of water in the border areas of the transboundary rivers was about 70–80%.

There is a slight shift in pH to the alkaline side. The minimum values (6.8–7.6) are typical for the watercourses in the Neman delta. The greatest pH (7.5–8.1) was found in the rivers on the Sambian and the Vishtynets uplands. In other waterways, pH ranges between 7.5 and 7.8.

Nutrients are dissolved in the river water in a fair amount. The content of nitrogen compounds is one of the important indicators of natural water quality. High nitrogen concentrations reflect recent contamination of the water body, for instance, with household water. Mineral forms of nitrogen appear in large, often in catastrophic, amounts.

Ammonia nitrogen occurs in most streams during autumn and winter in a maximum concentration: in the watercourses that are slightly polluted (rivers of the Sambian upland) and relatively unpolluted (rivers of the Vishtynets upland), it varies from 0.3–0.5 to 0.8–1.3 mgN/dm^3. In spring and summer, while photosynthesis is especially intensive, ammonia nitrogen concentration equals to 1.0 mgN/dm^3 (0.1–0.4 mgN/dm^3 in the rivers of the Vishtynets upland). In summer (June–July), its concentration in most streams was reduced down to 0.1–0.4 mgN/dm^3. There were discovered devastatingly high levels of ammonia nitrogen in some rivers flowing southwards on the Sambian upland (the Chernaya, the Grayevka), in the watercourses of the Polesskaya lowland (the Lobovka, the Zelenogradka) and of the Neman delta: more than 4 mgN/dm^3 in such rivers as the Chernaya and the Grayevka and above 10 mgN/dm^3 in the Lobovka river. In most watercourses concentration of ammonia nitrogen far exceeds the MPC (maximum permitted concentration) for the fishery water bodies [53]. The most favourable conditions appear in the rivers that flow from the Vishtynets upland downhill. The most unfavourable ones are in the rivers draining landscapes of the Sambian upland (such rivers as the Chistaya, the Medvezhya, the Grayevka, the Chernaya) and in the streams of the Polesskaya lowland (especially such rivers as the Zelenogradka, the B. Moryanka, the Lobovka) and in the Neman delta.

Nitrite nitrogen, with rare exception, can be always found in the river water, sometimes in rather large amounts. The minimum content of nitrites (from 0.001–0.005 to 0.015 mgN/dm^3) is typical for the winter period, which is natural. In spring, during mass die-away of aquatic organisms, in the early stages of decomposition of organic matter, and under increased runoff of organic substances and nutrients from the watershed surface, the nitrite content increases to the values ranging from 0.005–0.01 up to 0.02 mgN/dm^3. The highest concentration of nitrite is observed in summer: 0.01–0.02 mgN/dm^3 and more, reaching 0.04–0.1 mgN/dm^3 in highly polluted waterways. In autumn the nitrite levels in the rivers drop to 0.005–0.01 mgN/dm^3, while in such rivers as the Chistaya, the Medvezhya, the Chernaya and the Zapadny canal, it remains rather high – above 0.02 mgN/dm^3. The elevated content of nitrite in the course of the year is indicative of the decline in self-purification capacity of the watercourse. The general background concentration of nitrite in water is influenced by groundwater containing usually much more nitrite, especially in the upper aquifers. The rivers that drain the landscapes of the Vishtynets upland and the Nizhnenemanskaya lowland have the lowest levels of nitrite. Higher concentration of nitrite is a distinctive feature of the river basins situated in the western and central parts of the region. The greatest amount of nitrite is in the rivers of Sambia Peninsula flowing southwards (the Grayevka, the Chernaya), in the Mamonovka river, in the ditches in the area of settlement Ladushkino and in the transboundary rivers of the Lava-Pregolskaya lowland where the extreme high concentrations of nitrite (over 0.05 mgN/dm^3) were defined.

The content of *nitrate nitrogen* has the following characteristic features. The seasonal cycle is indistinct. The lowest concentration is observed in summer, and in most streams it ranges from 0.3–0.5 to 1.0 mgN/dm^3. An exception to this is the Chistaya river that experiences the increased human impact due to municipal wastewater of the city of Pionersk. Annual maximum nitrate concentrations (1.5–2.0 mgN/dm^3 and more) appear in most watercourses in spring exclusive of the rivers on the Vishtynets upland and in the northern part of the Nizhnenemanskaya lowland where the level of nitrate was the lowest for this period (0.2–0.5 mgN/dm^3). In autumn, the nitrate amount differs little from the summer values (0.5–1.0 mgN/dm^3). In winter, while the nitrogen consumption rate is the lowest, decomposition of organic matter and transition of organic nitrogen to mineral forms take place, and nitrate concentration should reach its annual maximum. However, in the rivers of the Kaliningrad Region, the nitrate concentration shows little difference in comparison to the autumn period, slightly rising up to 1.0–1.2 mgN/dm^3. Only in such rivers as the Deima, the Shirokaya and the Vostochny canal, the level of nitrates increases to 2.0–2.5 mgN/dm^3 which may indicate anthropogenic pollution. Disturbances in the seasonal variations suggest a permanent influx of these compounds with household and industrial wastewater of the neighbouring settlements.

Throughout the year, the minimum values are typical for the rivers of the Vishtynets upland, the northern part of the Neman delta, the Sambia Peninsula and the Prokhladnaya river; the greatest amount of nitrate is in the small transboundary rivers of the Lava-Pregolskaya lowland.

The content of *mineral phosphorus* is quite large and varies widely; seasonal fluctuations are strongly pronounced. The lowest amount of phosphate is observed in spring, during a period with an intense photosynthetic activity of phytoplankton and active biochemical oxidation of organic matter. During the summer and autumn floods, there is a significant increase in phosphorus content in the water of the most studied streams up to 0.2–0.6 mgP/dm^3. Interestingly the phosphorus concentration at this time of the year 2010 was substantially higher than during the winter low water. It allows for the conclusion that the dominant factor predetermining input of phosphorus into water bodies was the surface runoff from the catchment area. High intensity of land use (over 0.6–0.8) and large amount of labile phosphorus in soils (from 14 to above 15 mgP per 100 g of soil) contribute to excessive phosphorus entering from the catchment area, especially during periods of high water and floods. A significant amount of phosphorus can enter the rivers with effluents from the populated areas. Phosphate-phosphorus content in the rivers studied exceeds the MPC for fishery water bodies by several times [53]. The most favourable conditions appear in the rivers draining landscapes of the Vishtynets upland and in the Neman delta (except for the main river). The highest concentration of mineral phosphorus is observed in the rivers of the Sambian upland (such rivers as the Grayevka, the Chernaya, the Chistaya, the Medvezhya), in water of the Mamonovka river, within the drainage area in the vicinity of the settlement Ladushkino and in some rivers of the Polesskaya lowland (the Lobovka, the Zelenogradka).

The high content of *iron* is typical for the rivers in the Kaliningrad Region. This is supported primarily by such natural features of the area as a high level of groundwater and the presence of large areas of swamps and wetlands which as a rule contain the high amount of iron in the form of complexes with salts of humic acids. Ratios of ferrous to ferric iron allow estimating the influence of groundwater. Typically, more than 70% of total iron falls to the share of ferric iron, and the proportion of ferrous iron easily oxidized to the ferric iron is only 10–20%. Influence of groundwater is clearly expressed in the watersheds of such rivers as the Neman and the Deima and the small streams of the Polesskaya lowland. Here, the proportion of bivalent (ferrous) iron is 40–45%. The Pregolya river and the rivers of the Sambian upland contain ferrous iron in smaller amounts (20–30%). The minimum content of the ferrous iron was found in the rivers draining elevated rolling and moraine plains (rivers of the Vishtynets and Warmian uplands, of the Lava-Pregolskaya lowland) which indicates the less influence of groundwater. Since the majority of small watercourses in the Kaliningrad Region are chiefly swamp- and groundwater-fed, the seasonal variation is reduced. In most streams, the concentration of total iron is 0.3–0.6 mg/dm^3 in spring and autumn and 0.3–1.0 mg/dm^3 during the low water period.

In general, the highest iron content is typical for the rivers in the western (Sambian upland), south-western (Warmian upland, some watercourses of the

Lava-Pregolskaya lowland) and north-eastern (the Neman delta) parts of the Kaliningrad Region. The lowest level of iron can be usually found in the rivers of the eastern and central districts (Vishtynets upland, the Pregolya river basin).

Seasonal variation of nutrients is barely discernible which is indicative of the intensive anthropogenic load. It appears especially clearly in such rivers as the Medvezhya, the Lobovka and the Grayevka where the seasonal cycle is almost completely disrupted. This proves the limit anthropogenic pressure and loss of self-purification capacity of the watercourse.

The amount of nutrients in the studied streams with the exception of nitrate nitrogen generally does not comply with requirements of the fishery water bodies [53]. Thus, based on the content of organic matter and nutrients, the water quality in such rivers can be assessed as threatened.

The condition of some rivers suggests for their catastrophic pollution. A number of streams (the Lobovka, the Pokosnaya, the Medvezhya, the Chistaya, the Grayevka) basically look more like effluent gutters. Water in these watercourses is grey and has a characteristic smell of sewage. All year round the watercourses were giving off a distinctive smell of hydrogen sulphide. There were whitish settings on the river banks left after foam of synthetic surfactants; the bottom was covered with black mud. There were extremely high levels of nutrients in 2010: concentrations of ammonia nitrogen ($N-NH_4^+$) were 11.063 mgN/dm^3 in the Lobovka river, 2.555 mgN/dm^3 in the Grayevka river and 1,054 mgN/dm^3 in the Pregolya river. Pursuant to the classification [56], the investigated watercourses based on the content of the most parameters were classified as "moderately polluted" and "dirty."

4. Specific features of the watercourses belonging to the major river basins of the Kaliningrad Region can be described as follows.

The Pregolya river basin with its numerous tributaries is the main water system of the Kaliningrad Region. The Pregolya river is a watercourse of the superior fishery category; its tributary, the Pissa river, is also of the superior fishery category; the Putilovka river is of the first category. The waterways within the Lava river basin (such rivers as the Pravda, the Stogovka, the Mazury channel) are of the second fishery category. The Pregolya river basin (catchment area is 13.600 km^2) occupies the bigger part of the region area (15.100 km^2). The river flows in the latitudinal direction westwards from east, in the middle of the Kaliningrad Region, along a wide fluvioglacial valley with rich grasslands. Its low banks are almost free of trees; there are extensive swamps and lakes widely scattered over the floodplain. It receives a great number of tributaries and drainage channels. The Pregolya river is formed by the confluence of the river Angrapa and the river Instruch in the north-western suburbs of the city of Chernyakhovsk. At the southern end of the city of Gvardeysk, there is a large navigable distributary channel (built along the bed of once a small river) separating northwards, the Deima river that discharges into the Curonian lagoon. The Lava river, the biggest tributary, meets the Pregolya in the town of Znamensk. Twenty kilometres downstream of the city of Gvardeysk, the Pregolya river is divided into two branches (arms): the right one (the New Pregolya) and the left one (the Old Pregolya). In Kaliningrad, the Old and the New Pregolya join in two places to form two islands (Oktyabrsky and Kant) in the city centre. The river empties into the Vistula (Kaliningrad) lagoon along the single river bed forming the estuary. The artificially deepened mouth of the Pregolya river meets the Kaliningrad Marine Canal (KMC) connecting the river with the Baltic Sea. According to its morphometric characteristics, the Pregolya river falls into the class of "average" rivers. Its length is 123 km; the depth is 2–3 m, in the lower reaches sometimes to 9–16 m; and the width is 20 m at the city of Chernyakhovsk, 80 m near the city of Kaliningrad. The width of the river valley is 1–1.5 km. The river bottom is sandy, muddy at places. During snowmelt and high rainfall floods, the river overflows for 1–2 km and for up to 5 km within the floodplain depressions upstream of Kaliningrad. Annual runoff is 2.5 km^3, 60% of which goes into the Vistula (Kaliningrad) lagoon and 40% into the Curonian lagoon along the Deima river. Westerly winds prevailing in the area create significant surges from the Vistula lagoon, and the whole flow is discharged into the Curonian lagoon. When the surge from the Curonian lagoon occurs, created by northern and partly eastern winds, which happens reasonably less often, the entire flow of the river discharges into the Kaliningrad lagoon. Only in the event of considerable water flow which happens during the snowmelt floods, surges lose their influence on the distribution of water between the branches [54]. Watercourses in the southern part of the watershed are transboundary and flow from the territory of the Republic of Poland. All of them belong to the group of small rivers. The second largest tributary is the Pissa river that outflows from the Vishtynets (Vištytis) lake. Small river basins are elongate in shape, with the width of about 10–20 km and the length of 20–40 km or more; the divides are well defined, with winding divide lines and significant slope values. The rivers of the Lava subwatershed pursuant to their morphometric parameters are close to the other small rivers flowing into the Vistula lagoon; the Pissa river is comparable to the Prokhladnaya river (Table 13).

Hydrometric parameters of the small rivers belonging to the Pregolya river basin are also mostly comparable to the other small streams flowing into the Vistula lagoon. The rivers associated with uplands have slightly higher flow velocity and discharge due to substantial surface slopes. The Krasnaya and the Chernaya rivers

Table 13 Main morphometric characteristics of the small rivers tributary to the Pregolya river basin

Watercourse	Fishery water category	S_{basin}, km^2	$L_{div.}$, km	L_B, km	$B_{Max.}$, m	B_{mean}, m	L, m	I, m/km [44]
Pravda	2	102	52	25.2	8.6	4	22	1.67
Stogovka	2	281	92	34.4	14	8.8	64	
Putilovka	1	525	130	42	22.4	12.5	58	
Mazury channel	2	–	–	–	–	–	42	
Pissa	Superior	1,374	250	77	27	17.8	98	1.47

S_{basin} basin area, $L_{div.}$ length of the drainage divide, L_B basin length, $B_{Max.}$, B_{mean} basin width maximum, mean, L watercourse length, I slope of the watercourse

flowing from the Vishtynets upland downhill have the maximum discharge. Such rivers as the Pravda, the Stogovka and the Putilovka that originate from the terminal moraine formations of the Baltic moraine ridge and flow across the fluvioglacial plain have slightly less discharge.

The mean flow velocity of the river Pregolya is gradually reduced from 0.5 m/s upstream of Gvardeysk to 0.1 m/s at the Pregolya river mouth; during the snowmelt flood, the flow velocity at Gvardeysk rises up to 1.2 m/s. The mean annual discharge of the river upstream of the division of the stream into two branches is 83 m^3/s. Annual runoff is 2.5 km^3. Specific annual runoff according to the different data ranges from 5.9 to 6.5 l/s/km^2, and the runoff coefficient is about 24–27% [37, 39, 54].

The Pregolya river, along with the other rivers of the Kaliningrad Region, is mixed-fed: snow accounts for approximately 40%, rain for about 35% and groundwater for about 25% of its annual runoff. The water regime of the river is characterized by spring snowmelt flood and high-level periods in summer, autumn and winter. The highest rises in water level occur during late snowmelt floods in case of the rapid snow melting if there were no intense thaws in winter, and significant reserves of snow have been accumulated. The maximum levels throughout the year usually appear in spring. In the years with early snowmelt flood, the level rise is interrupted by a freezing, and the high water period lasts longer. The recession of the snowmelt flood is disrupted by the first rainfall floods. Summer low water, at an average, starts in May and lasts until October, with breaks for the rainfall floods. Relatively low discharge during summer low water is provided by the groundwater and some natural streamflow regulation as a result of storage in lakes in the upper reaches of the river basin. In autumn, there is a smooth increase in discharge and water levels due to continuous extensive rains. Early in winter, when the precipitation falls as snow and the river receives mainly groundwater, a brief low water period sets in. Typically, it has high flash floods during sudden and lengthy thaws. The level of winter floods depends on snow storage, intensities of rainfall and snow melting; the level of such winter floods is close to and sometimes even exceeds the level of spring snowmelt flood. The position of the water level in the lower reaches is highly conditioned by the wind inducing up and down surges. When the water flow in the river is more than 50 m^3/s, and if such water flow occurs together with the westerly gale-force winds, the upsurge creates a high rise in water level which could cause inundation in the area of the city of Kaliningrad. The river flow has no direct effect on the variation of the water level at the estuary as in the lower reaches, the flood waves flatten out over the floodplain [54].

Thermal regime of the river is influenced by heat exchange with the atmosphere. Ice conditions are very unstable due to the diversity of synoptic processes in winter and can be very different from one winter to another. Variability of synoptic processes and return of warm weather caused by these processes impede the rapid freezing of the river. Winter thaws trigger the temporary ice break-ups, ice movements and even complete river clearance of ice. In some mild winters, there is no steady ice cover freezing up on a water surface. In severe winters, the stationary ice remains until the end of March. Variation in timing of complete clearance of ice is about 2 months [54].

The Pregolya river is exposed to intensive pollution. In terms of the number of industrial enterprises located in the Pregolya basin, Kaliningrad has the lead; Chernyakhovsk ranks the second and Gvardeysk the third. Accordingly, their effect on the river is proportional thereto. In 2004, the annual volume of wastewater discharge into the Pregolya river in Kaliningrad exceeded the wastewater discharge of Chernyakhovsk by four times, which, in turn, was by half or even twice bigger than the discharge of Gvardeysk. The river water quality within the boundaries of the cities and between them changed controversial. Downstream from the city of Chernyakhovsk to the city of Gvardeysk, condition of the river improved somewhat due to self-purification, but between the background cross section and the control ones in Gvardeysk, it deteriorated. Similar changes were found in the city of Kaliningrad, but water in the estuary was slightly cleaner than at the control section within the city [13, 20].

The Neman river basin is located within five states (Table 14).

The river Neman has the superior fishery category. It originates from the Republic of Belarus in the south-western spurs of the Minsk upland, 45 km south-west of Minsk; the source of some of its tributaries, in particular, of the Sheshupe river, is situated in the Republic of Poland. The Neman river enters the territory of the Republic of Lithuania at the city of Druskininkai where it gets the name of the Nemunas (Fig. 4).

Downstream to the north, the Nemunas crosses the Baltic moraine ridge flowing along the deep valley among the moraine hills. There are the Kaunas Hydroelectric Power Plant and a reservoir built on the river by the city of Kaunas. The river flow is regulated. Downstream of the 240th km counting from the mouth, there is Velno-Tiltas, a chain of nicks extended for 4 km. In Kaunas, the river again turns sharply to the west. In the vicinity of Smalininkai, the river crosses the state border and furthermore flows in the Kaliningrad Region where it has the name of the Neman. From the city of Sovetsk downstream, the Neman river flows north-westwards, and the river banks are becoming more low and flat. In the lower reaches, the river flows across the undulating swampy lowland which has the name of the Primorskaya on the right bank and of the Nizhnenemanskaya on the left one.

Within the deltaic lowland of the Neman river, the absolute marks drop to 1.44 m below sea level, swamps occupy up to 36% of the area, and the forest coverage is up to 21% [35, 39]. In the river mouth area (downstream of Sovetsk), 48 km upstream

Table 14 Watershed area of the Neman river [51]

Total river basin area, km²	Country	Watershed area, km²	Watershed area, %
98,200 (Sovetsk) 101,000 (mouth) [39–53]	Lithuania	46,795	47.7
	Belarus	45,600	46.4
	Russia	3,132	3.2
	Poland	2,554	2.6
	Latvia	98	0.1

Fig. 4 Watershed of the Neman river (excluding the Republic of Belarus and Republic of Latvia)

from the Neman's mouth, a distributary outflows westwards of the mainstream – the Matrosovka river discharging into the Curonian lagoon. The Nemoninsky canal and the Primorsky canal (7–8 km upstream from the mouth) connect the Matrosovka with the Nemoninka-Nemonin river in two places, and through this, the Matrosovka joins the Deima river (distributary of the Pregolya) via the Polessky canal.

Twelve kilometres upstream from the mouth, near the Lithuanian city of Rusnė, the river Neman (Nemunas) is divided into two branches. One distributary outflowing south-westwards is the river Severnaya (Skirvite). At approximately 3 km upstream from where the Severnaya river meets the lagoon, it, in turn, is divided into the effluents the Severnaya and the Yuzhnaya. Another distributary, the Atmata river, in fact, continues the stream of the main river, and its mouth is considered to be the mouth of the river Neman (Nemunas).

The Neman delta is quite complicated. There are several low-lying islands between the rivers Severnaya and Atmata (the largest ones are Severny, Yuzhny, Bolshoy and Kamyshevy Islands near the left bank of the Yuzhnaya effluent) and many smaller streams, branches, bayous and oxbow lakes. The delta is formed from alluvial deposits which are underlain with peat at the lagoon coast. It is also dissected by a dense network of canals and canalized channels of small delta branches. All of them are locked by sluice gates and do not give runoff into the lagoon. Almost all the rivers in the Neman delta and in the catchment area of the Curonian lagoon are connected with channels and form an integrated drainage system. Most of the rivers are channelized; they receive the water pumped out from the polder lands. The natural state of these rivers has been changed greatly.

The left bank of the delta is protected from inundations during floods with the help of the earth dikes that stretch along the southern shore of the main branch of the Neman, the lagoon coast and on both banks of the Matrosovka for tens of kilometres. The height of the dikes gradually decreases from 5.4 m at the head of the delta down to 2 m at the coast of the lagoon. It took three centuries to erect the dikes and step by step to increase their height. After the catastrophic flood in 1958, the dikes gained height significantly. There is an open drainage network with mechanical pump water lift in the Neman delta [36, 54].

There are many bars and shoals on the stream beds of the rivers Neman and Matrosovka that impede navigation. To maintain the required depth, the river beds were straightened and narrowed by special semi-impoundments which contribute to the increase in flow velocity along the midstream and to decrease in the deposition of river sediments on the bars. The semi-impoundments also provide a more stable position of the channel bed. In natural conditions the channel bed of the Neman river and its delta branches often moved as evidenced by numerous old channels and bedplates of small delta branches, along some of which the tributaries of the Neman river are flowing now. In the vicinity of the city of Sovetsk behind the Queen Louise Bridge, there are another five similar old bridges that confirm meandering of the Neman river bed over hundreds of years.

The river basin of the Neman has a highly branched hydrographical network. The river receives many tributaries throughout its complete length, which, in turn, have tributaries of the second, third and higher orders. The largest tributaries of the main river in the Kaliningrad Region are the right-bank ones (Fig. 4); these rivers flow from the territory of Lithuania (the Minija, the Jūra, the Nevėžis, the Vilnia (with the Šventoji river), the Merkys, etc.); the left-bank tributaries are the Tylzha river (length 44 km) and the Sheshupe river (length 308 km, 114 km of which are within the Kaliningrad Region). There are swamps, lakes, canals, gullies and ravines in the catchment area. The drainage density in the middle and lower reaches of the Neman river is about 1 km per 1 km^2 of the basin; downstream it rises to 1.5 km per 1 km^2 or more due to numerous land reclamation canals. In the middle reaches of the river, the bottom slope has low values, and the swamp and lake percentage is small [55]. The Neman is connected with the Dnieper river through the rivers Shchara (the Dnieper-Neman Canal) and the Pripyat, and via the Pripyat river and the Dnieper-Bug Canal, it is linked to the Bug river. A number of channels and the Matrosovka connect the Neman with the Deima and the Pregolya river, thus forming a system of inland waterways of international importance.

The length of the Neman river is 937 km (107 km within the Kaliningrad Region). The water catchment area is 98.200 km^2. The river valley was formed during the last glaciation. The maximum width of the river during the low water period can reach 180–350 m, while during floods the river overflows to 1–1.5 km wide and up to 5 km at the non-diked right bank of the delta. The depth is 3.4 m in the river pools, 1.5–2.5 m in the shoals. The mean flow velocity is 0.8–1 m/s at the shoals and 0.6–0.8 m/s in the pools. During floods, the flow velocity increases to 1.5 m/s or more. When approaching the mouth of the river, the flow is slowing down progressively [54]. The width of the floodplain in the lower reaches of the Neman river can increase up to 2 km [39].

The river Neman is classified as a plain stream (the mean slope of the river is 0.16‰ in the upper, 0.23‰ in the middle and 0.10‰ in the lower reaches [51]); it is mixed-fed with a predominance of snow water. The share of snow decreases from east to west, with an increase in the contribution of rainwater and groundwater. The role of snow in the river runoff rises in the years with severe, persistent, snowy winters; and in the years with warm or average winters, the rainfall becomes more important. Dryness of the year has a similar influence: during dry years when little rain falls, snow takes a bigger part in the river runoff [55].

The mean annual water discharge upstream of the division into the branches is about 600 m^3/s. Of these, approximately 10% outflows through the distributary Matrosovka, about 55% through the river Severnaya (Skirvite) and about 35% through the Atmata river [54]. Specific runoff (on average for a long-term period) in the territory of the Republic of Lithuania increases downstream from 6.1 to 6.7 l/s/km^2 and within the Kaliningrad Region drops to 6.3 l/s/km^2; the respective increase and decrease in maximum specific runoff is from 8.0 to 8.6 and down to 8.2 and in minimum specific runoff from 4.4 to 4.9 and down to 4.6 l/s/km^2 [39].

Water regime of the rivers in the watershed of the Neman is determined by climatic conditions, features of the underlying surface (topography, forest coverage, swamp and lake percentage and others) and the type of water supply. It is characterized by spring snowmelt flood, summer and winter low water periods and summer, autumn and winter high water periods. The maximum flow at Sovetsk occurs in March. In compliance with the nature of the water supply of the river, its winter runoff (December–February) is minimal and supported by floods. During wet years, when much rain falls, the river at Sovetsk in spring discharges slightly less water than in summer. During the years with normal water amount, spring runoff becomes larger than the summer-autumn one. With decreasing water content of the river, this difference tends to widen, and during dry years the runoff in spring is almost twice as much as in summer and autumn seasons (Fig. 5).

Winter low water is interrupted by high flash floods during frequent and long thaws. The height of the flood waves depends on snow storage and intensity of snow melting. The spring snowmelt flood in the middle and partly in the lower reaches usually starts in March and ends in May. Flood crest falls on the end of March or beginning of April. A feature of the spring snowmelt flood is several waves caused by the nature of the snowmelt. In the years with early snowmelt flood, the level rise is interrupted by freezing period; consequently the snowmelt flood lasts longer. Decline in spring snowmelt flood goes on slowly and usually it is disrupted by the first rainfall floods. Summer low water is intermittent, with breaks for the rainfall

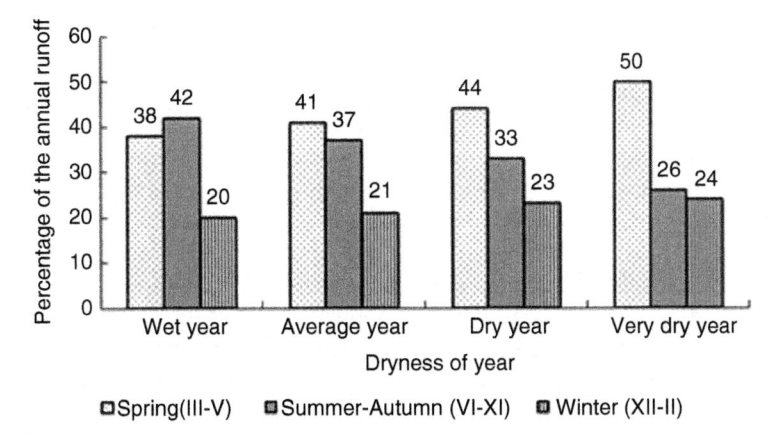

Fig. 5 Estimated distribution of seasonal runoff as a percentage of the annual runoff for the Neman river depending on dryness of the year (according to [39])

floods. Runoff is gradually reduced from June to September but during wet years it may increase in this period. Autumn floods have longer duration compared with the summer ones [54].

Thermal regime of the river is determined by weather and climatic peculiarities and influenced by heat exchange with the atmosphere. Ice conditions are very unstable. The diversity of synoptic processes in winter results in a big difference in ice conditions of each winter. During the milder winters, there is no steady ice cover freezing up on a water surface of the small rivers in the middle and, especially, in the lower reaches of the basin. In severe winters, the stationary ice remains until the end of March. Repeated returns of warm weather impede the rapid freezing of the rivers. Sometimes during a year, freeze-up and clearance of ice happens in certain places more than once. Formation of ice cover in the lower reaches of the river is often accentuated by hanging dams which cause a notable rise in water level and reduced water flow downstream of a hanging dam. Owing to the variability of temperature conditions, ice jams are possible in winter. Uneven destruction of ice cover takes place during thaws, and with returns of frost, the appeared floes and ice pieces freeze up together and predetermine future centres and factors for not only winter but also spring ice jams. The river clearance of ice usually begins in the mid-second half of March. In the event of vigorous spring melting, the ice run is violent and accompanied by heavy and lasting ice jams. Such ice jams are particularly frequent at the cities of Birštonas, Seredžius, Jurbarkas, Smalininkai and Sovetsk. The height of the level rise may exceed 2 m; duration ranges from several hours to 8 days. Ice jams often become catastrophic with a rise in water level by 4 m or more [39]. Location of the ice jams depends on the nature of spring heating. The longer the time gap between the maximum warming and the day when the ice jam appeared, the more upstream it shall be situated [54]. The complete river clearance of ice, as well as ice break-ups, is highly contingent upon the severity of the winter and intensity of the spring processes. Variation in timing of complete river clearance of ice is about 2 months. In some mild winters, the river may completely clear of ice already in the second half of February; very severe winter with long cold spring will delay ice out until the 20th days of April [39].

Water level fluctuations are related to changes in flow, hanging dams and ice jams and surge phenomena. The latter are particularly significant at the mouth of the river during low water periods. The amplitude of the annual average water level in the river near Sovetsk is approximately 150 cm. The maximum annual levels usually occur in spring, but in winter in case of especially high floods during the long thaws, the water level often exceeds the level of spring snowmelt flood. The highest level rises happen during late snowmelt floods in the event of rapid snowmelt if there were no strong thaws in winter and significant reserves of snow have been accumulated. Autumn level rise starts in September or October. The amplitude of the low-level fluctuations in summer and autumn is 0.2–1 m [54].

Catchment area of the river Neman is located in the forest zone having large swamp percentage in many places; it is characterized by the low activity of surface water erosion. Turbidity of water is typically small, on average no more than 50 g/m^3, but on some days may exceed 200 g/m^3. The highest turbidity appears during the spring snowmelt flood when 85–90% of the annual sediment yield outflows [37].

As our observations revealed [8, 14, 18], the fluctuation range of the hydrochemical parameters for the Neman river over the study period (2007–2014) is quite wide. According to the classification [48], TDS concentration is medium (350–430 mg/dm^3) or increased (550–750 mg/dm^3). TDS and water conductivity related thereto reach maximum values during low water periods (January–February) when the proportion of groundwater in water flow of the river increases. These characteristics also have higher values at the station downstream of the city of Sovetsk and at the outfall station (settlement Yasnoe) which may indicate contamination. The minimum values of conductivity and salinity are typical for spring snowmelt flood (March–April) and for autumn-winter rainfall floods (November–December). During all the summer months (June–August), the relevant characteristics stick to the medium level which is connected with heavy summer rainfall floods. Throughout all the seasons, TDS level is increased at the large localities like the cities of Neman and Sovetsk (especially downstream thereof) as well as at the outfall station in the settlement Yasnoe. In compliance with the classification [48], TDS concentration is medium [48], and the river has calcium and bicarbonate water, of the 1st water type [$HCO_3^- < (Ca^{2+} + Mg^{2+})$]; content of major ions along the river changed ambiguously. Pursuant to the classification [49], the river water is mainly moderately hard (3–6 mg eq./dm^3), but downstream of Sovetsk and at the outfall station, it is classified as "hard" (the result of contamination).

Oxygen conditions during all the years under consideration were relatively favourable. The largest amount of oxygen (up to 11–14 mg/dm^3) as a result of photosynthesis and reduced temperature is contained in water in March–May and October–November; there is a supersaturation in these periods, sometimes up to 130–150%. In May, when both the water temperature and oxygen consumption for oxidative processes increase, concentration of oxygen tends to decline, and in June–

September it consistently equals to about 10 mg/dm^3 against the high water temperature (relative amount is approximately 100–80%). There is much oxygen in water also in December as a consequence of lower water temperature (2 °C and below). In January, the concentration of oxygen under ice is 8–7 mg/dm^3; saturation drops down to 60% or less. The value of hydrogen index is slightly shifted to the alkaline side. The average pH is above 8.0; in some periods it was above 9.0 (9.89 max). The single minimum pH value of 6.69 was found at the outfall station downstream of Sovetsk.

The content of organic (easily oxidable) substances was measured by permanganate index. Its mean values in accordance with the classification [48] fall into the "increased" class, vary in different periods from medium (7–8 mgO/dm^3) to increased (11–17 mgO/dm^3) and fluctuate only slightly along the river. The greatest amount of organic substances was found in May, under vegetation growth when oxidation processes are slowed down because of still low water temperature and already rich production of organic matter (permanganate index is 15–17 mgO/dm^3). As in March–April, vegetation development is just beginning; the permanganate index (10–12 mgO/dm^3) has less value than in May. During summer months, despite the high temperature and intense oxidation processes, the oxidation remains significant. Even in the end of winter (January–February), the permanganate index was 11–12 mgO/dm^3. The greatest value of permanganate index over the entire study period (23.12 mgO/dm^3) was found at the near-border cross section (settlement Neman) in March 2014. However according to the values averaged over the relevant period, this station almost does not differ from all the other ones. Rather high levels of permanganate oxidation throughout the year suppose a big amount of organic substances in the Neman water.

The concentration of nutrients in unpolluted water bodies is determined by the natural course of photosynthesis and in some cases may be a consequence of contamination. However, this rule is not always complied with for the Neman river. In March, after the beginning of photosynthesis, the content of nitrate along the river is reduced noteworthily, but in April and May downstream of Sovetsk, it increases again. The latter cannot be explained by anything other than pollution since the concentration of these compounds should decrease under intense photosynthesis which is confirmed also by oxygen conditions. All forms of nitrogen compounds (especially nitrate) are found in the largest quantities during the spring snowmelt flood (March–April) and autumn (October–November) and winter (December) rainfall floods. The increased overland runoff ensures delivery of NH_4^+ и NO_3^- ions from agricultural areas at this time. The concentration of ammonia and nitrate nitrogen in this period is 0.7–1.5 and 2.0–4.5 mgN/dm^3, respectively. The lowest amount of ammonia nitrogen (0.2–0.8 mgN/dm^3) and nitrate (0.2–0.6 mgN/dm^3) is dissolved in water in summer when it is actively consumed by phytoplankton. In winter under ice the NH_4^+ и NO_3^- concentration is 0.5–1.0 and 0.5–1.5 mgN/dm^3, correspondingly. Phosphate-phosphorus is dissolved in water in moderate amount. The highest content of PO_4^{3-} ions (0.03–0.1 mgP/dm^3) is observed in autumn and winter. In spring and summer, the content of phosphate ions is reduced to 0.01–0.07 mgP/dm^3 with a minimum in July (0.01–0.02 mgP/dm^3). The largest PO_4^{3-} concentration during almost all seasons was typical for the outfall station (settlements Yasnoe and Levoberezhnoe). The temporal fluctuations of iron are to some extent consistent with similar variations in the value of TDS and water hardness. This suggests that it is principally supplied by groundwater, and it is a quite intense process as evidenced by the permanent presence of all forms of iron in water including ferrous iron. The greatest content of iron in all forms is specific to the dry period, i.e. during winter and summer low water (0.3–0.5, 0.1–0.2 and 0.2–0.3 mg/dm^3 in June for total, ferric and ferrous iron, respectively). The least amount of iron is dissolved in water during the periods of high water content when the surface (rainfall) flow dominates in water recharge of the watercourse. The lowest concentrations of all forms of iron were found during rainfall floods in summer (July) and autumn (September). In spring, iron content is only slightly less than during low water periods.

In conformity with the comprehensive environmental classification of the surface inland water quality [56], the integrated environmental assessment of the river Neman can be represented as follows (Table 15).

As will readily be observed, the river water is most commonly defined by hydrochemical parameters as "moderately polluted" although proper consideration should be given to that on certain days the situation at some cross sections may be significantly worse.

Chlorophyll a which reflects the peculiarities of photosynthesis was most comprehensively studied by us in 2009–2014 in the area of the Baltic NPP under construction. A complete data was collected in 2014. Our observations showed that the concentration of chlorophyll a in 2014 ranged from 97.58 to 2.32 µg/dm^3. Temporal fluctuations in its concentration were largely determined by the change in water temperature. Along with the warming and consequent increase in the photosynthesis rate, the concentration of chlorophyll a also rose, and the peak in its content occurred at the end of May. Afterwards there was a recession in photosynthesis which lasted until August, after which the decline in chlorophyll amount slowed down, and since the end of August, its value remained almost at the same level, rather low for the Neman river (Fig. 6).

Table 15 Environmental assessment of the Neman river according to its hydrochemical characteristics, average over the period 2007–2014

Water quality class	Characteristic (station)
Quite clean water	O_2 (all stations)
Clean enough water	NO_2^- (all stations), NO_3^- (st. upstream of Sovetsk)
Slightly polluted water	pH (st. settl. Nemanskoe, st. settl. Levoberezhnoe, st. upstream of Sovetsk), $P-PO_4^{3-}$ (all stations)
Moderately polluted water	pH [st. Rusnė, the Skirvite distrib., headwaters, st. settl. Yasnoe, st. downstream of Sovetsk, st. downstream of Neman, st. upstream of Neman (settl. Bolshoe Selo)]; permanganate index, $N-NH_4^+$, $N-NO_3^-$ [st. settl. Yasnoe, st. downstream of Sovetsk, st. downstream of Neman, st. upstream of Neman (settl. Bolshoe Selo), st. settl. Nemanskoe]
Heavily contaminated water	$N-NO_3^-$ (st. settl. Levoberezhnoe)

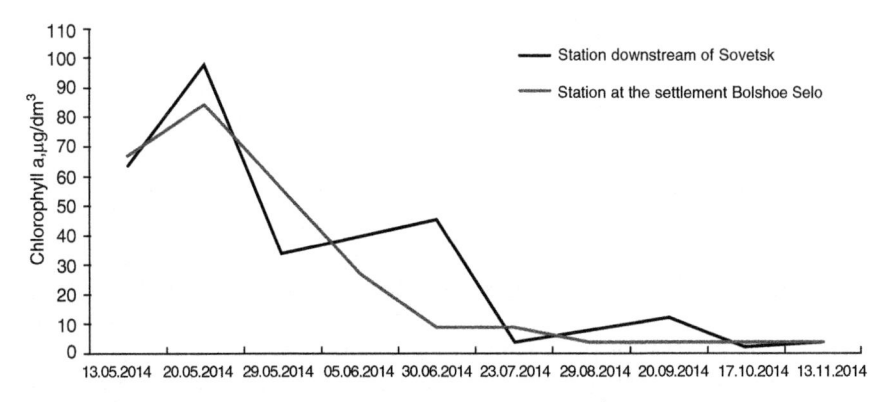

Fig. 6 Concentration of chlorophyll a in the river Neman ($\mu g/dm^3$) in 2014

The station Bolshoe Selo (chosen as a background station for the city of Sovetsk, the area of alleged water discharge from the Baltic NPP) is located downstream from the confluence of the Jūra river (from the territory of the Republic of Lithuania) and the Sheshupe river. The average concentration of chlorophyll a for May–August amounted there to 44.56 and 47.52 $\mu g/dm^3$ at the station downstream of Sovetsk. Smaller chlorophyll content at the station Bolshoe Selo could be attributed to the influence of the Sheshupe river (no data available regarding the Jūra river). The concentration of chlorophyll a in water of the Sheshupe, average for May–August 2009–2014, was low along the entire length of the river (5.80–8.54 $\mu g/dm^3$), and in the outlet area, it was 6.60 $\mu g/dm^3$ [30]. According to the classification of water bodies given in [57], the lower reaches of the river Neman (in the area of the Baltic NPP under construction) were evaluated as eutrophic in May–July 2014, as mesotrophic since August (except for the station downstream of Sovetsk: eutrophic, 20.09.2014).

The river Sheshupe is a left bank and the largest in the territory of the Kaliningrad Region tributary of the Neman river (of the first order); it is a watercourse of the superior fishery category. Its catchment area is situated in three states. The Sheshupe river originates from the lake on the territory of the Republic of Poland. Flowing northwards the river enters the territory of the Republic of Lithuania at the city of Kalvarija. Downstream of the city of Kudirkos Naumiestis, the Sheshupe enters the territory of the Kaliningrad Region where it flows along the state border between Russia and Lithuania for a considerable distance. At approximately 55° north, the river runs away from the border and flows generally westwards.

Along its entire length, the Sheshupe river is very meandering; it receives plenty of small tributaries of both natural and artificial origin (drainage canals), some of which flow through the swampy areas. Most of its tributaries are right bank, flowing from the territory of Lithuania. In the Kaliningrad Region, the river crosses the Sheshupe plain, flowing along the well-developed river bed or in many places along the deep picturesque valley in which banks are often steep and covered with forests.

The length of the river is 308 km; catchment area is 6,120 km². The river bottom slope within the boundaries of the Kaliningrad Region is small and decreases downstream. In the same direction, there is a decline in percentage of lakes (down to 1%) and swamps (down to 7%) in the river basin, but the forest coverage slightly increases (up to 17%).

As well as the main river, the Sheshupe is also classified as a plain stream aided with mixed water, but the distribution of its individual water sources is much more even: the share of rainfall equals to 37–48%, of snow (spring runoff) to 26–37% and of the groundwater to 16–37% [39]. The average annual water flow of the Sheshupe river (according to the long-term data) is 35.4 m³/s [54]. Mean annual specific runoff before entering the territory of the Kaliningrad Region is 5.6, 8.7 and 3.2 l/s/km² for average, maximum and minimum values correspondingly. Downstream, owing to the tributaries and to the decrease in lake and swamp percentage, these numbers increase somewhat.

The Sheshupe river has large water level fluctuations with the amplitude of up to 5–6 m.

Direct thermal vertical stratification of water (not so clear though) is possible in summer if the depth is greater than 1 m; as soon as the autumn cooling begins, the inverse stratification may appear to set in but poorly marked again.

Start of the ice phenomena (frazil and slush ice run) occurs in the first half of December, ice cover is formed after the third week of December, and spring ice run begins in mid-March. Ice jams are possible in spring, during which the water level at the city of Kudirkos Naumiestis may rise by 2 m or more. The duration of ice gorges ranges from few hours up to 8 days [39–53]. The mean annual turbidity of water does not exceed 25 g/m³ [37].

Pursuant to [8, 14] and in conformity with the classification [49], the water hardness varies in different periods within the limits of "moderately hard," "hard" and "fairly hard" water. The alkalinity of water is 5.1–6.9 mmol/dm³ increasing downstream. The pH is slightly shifted to the alkaline side (8.7–8.9). Suspended solids are present in relatively moderate amounts (2.0–7.0 mg/dm³), and their concentration slowly increases downstream [8].

Oxygen conditions are not always favourable, especially in dry 1997.

The value of permanganate index that allows estimating the amount of easily oxidable organic substances ranged between "average" and "increased" at different times in 1997 and 2007.

Under the rainy weather conditions in August 2007, the soil erosion from the catchment area influenced greatly. Nutrients, particularly phosphates, were found in sufficiently large quantities. The concentration of ammonia nitrogen decreased downstream, while the same of nitrate ions increased notably. Fluctuations of the main hydrochemical parameters, one way or another connected with the aquatic life, apparently demonstrated the seasonality during our observations.

To conclude the integrated description of geoecological features of the river basins, it is important to emphasize that they can be divided into four groups with different environmental conditions, such as "favourable," "moderately favourable," "hardly favourable" and "very unfavourable" [23].

Favourable condition. Watercourses with average TDS (300–500 mg/dm^3), moderate total hardness (2–4 but no more than 5 mg eq./dm^3), low and medium organic matter content (permanganate index is 5–10 but no more than 15 mgO/dm^3), favourable oxygen conditions throughout the year (60–90% of saturation, 7–11 mg/dm^3) and fairly low concentrations of nutrients (0.1–0.3 mgN/dm^3 of N-NH$_4^+$, 0–0.01 mgN/dm^3 of N-NO$_2^-$, 0–1 mgN/dm^3 of N-NO$_3^-$, 0–0.1 mgP/dm^3 of P-PO$_4^{3-}$ and 0–0.3 mg/dm^3 of Fe$_{total}$). Seasonal cycle of chemical elements is very obvious. The priority natural factors influencing the water composition are as follows: high level of aquifers providing water recharge of the rivers during low water periods, but these waters have reduced oxygen content and a large amount of ferrous iron. The priority anthropogenic factors are as follows: diffuse runoff from agricultural areas and livestock farms which increases the content of organic substances and phosphorus during snowmelt and rainfall floods. This is the smallest group that includes the rivers draining the Vishtynets upland, such as the Pissa, the Krasnaya and the Mazury channel.

Moderately favourable condition. Watercourses with average TDS (400–500 mg/dm^3), moderate total hardness (4–5 mg eq./dm^3), medium and increased content of organic matter (permanganate index is 10–15 and over 15 mgO/dm^3), favourable oxygen conditions throughout the year (60–90% of saturation, 7–11 mg/dm^3) and concentrations of nutrients (0.3–0.4 mgN/dm^3 of N-NH$_4^+$, 0.01–0.02 mgN/dm^3 of N-NO$_2^-$, 1–2 mgN/dm^3 of N-NO$_3^-$, 0.1–0.2 mgP/dm^3 of P-PO$_4^{3-}$ and 0.3–0.6 mg/dm^3 of Fe$_{total}$). Seasonal cycle of chemical elements is visible. The maximum permissible concentrations were exceeded for N-NH$_4^+$, N-NO$_2^-$ and P-PO$_4^{3-}$, especially during low water periods. The priority natural factors influencing the water composition are as follows: the rivers receive underground and swamp waters depleted in oxygen but overloaded with organic matter and nutrients. The priority anthropogenic factors are as follows: diffuse runoff from agricultural areas and livestock farms providing an increase in content of organic substances and phosphorus during snowmelt and rainfall floods and drainage canals that carry water with high level of organic matter. This group includes mainly the streams located in the eastern and southeastern parts of the Kaliningrad Region (watercourses in the southern part of the deltaic lowland, the river Instruch, the Angrapa, the Golubaya, the Deima distributary, the Prokhladnaya river).

Hardly favourable condition. Watercourses with average (400–500 mg/dm^3) and increased TDS (up to 600 mg/dm^3 and more); moderate total hardness (4–5, up to 6 mg eq./dm^3); increased content of organic matter (permanganate index is 15–20 mgO/dm^3); depleted oxygen amount throughout the year (30–60% of saturation, 5–8 mg/dm^3), especially during summer low water; and high concentrations of nutrients exceeding MPC: 0.4–0.6 mgN/dm^3 of N-NH$_4^+$, 0.02–0.04 mgN/dm^3 of N-NO$_2^-$, 2–4 mgN/dm^3 of N-NO$_3^-$, 0.2–0.3 mgP/dm^3 of P-PO$_4^{3-}$ and 0.6–0.9 mg/dm^3 of Fe$_{total}$. Seasonal variations of nutrients are disturbed; the highest concentrations thereof in watercourses were found in winter and in summer. The priority natural factors affecting the water composition are as follows: the rivers are aided by underground and swamp waters; the receiving water bodies create backwater which slows down the outflow of contaminants from the catchment area. The priority anthropogenic factors are as follows: diffuse runoff from agricultural areas and livestock farms; drainage canals; and high population density on the watersheds causing additional pollution from point sources, i.e. from small and large localities, as well as due to transboundary pollution. This group includes the majority of the small streams draining the western, south-western and southern parts of the region and specifically the small rivers on Sambia Peninsula, the Mamonovka river, the rivers belonging to the Lava river basin (the Pravda, the Putilovka) and the small rivers of the Polesskaya lowland.

Very unfavourable condition. Watercourses with average (400–500 mg/dm^3) and increased TDS (up to 600 mg/dm^3 and more), moderate total hardness (4–5, up to 6 mg eq./dm^3), high content of organic matter (permanganate index is over 20 mgO/dm^3), depleted oxygen amount throughout the year (30–60% of saturation, 5–8 mg/dm^3) down to deficiency (below 30% of saturation, 3–4 mg/dm^3) especially during summer low water, and extremely high concentrations of nutrients exceeding MPC: 0.6–0.9 mgN/dm^3 of N-NH$_4^+$ and more, more than 0.04 mgN/dm^3 of N-NO$_2^-$, over 4 mgN/dm^3 of N-NO$_3^-$, over 0.3 mgP/dm^3 of P-PO$_4^{3-}$ and over 0.9 mg/dm^3 of Fe$_{total}$. Seasonal variations of chemical elements are barely discernible; the self-purification capacity of the watercourses is lost. The priority natural factors affecting the water composition are as follows: the rivers are aided by underground and swamp waters; backwater from the receiving water bodies which slows down the outflow of contaminants from the catchment area. The priority anthropogenic factors are as follows: diffuse runoff from agricultural areas and livestock farms; drainage canals; high population density and heavy human activity on the watersheds causing additional pollution from point sources, i.e. from small and large localities and household waste disposal sites; as well as transboundary pollution. This group with very unfavourable environmental conditions includes the smallest streams suffering heavy anthropogenic impact, such as the rivers on Sambia Peninsula (the Primorskaya, the Grayevka, the Chistaya, the Medvezhya), the rivers on the Polesskaya lowland (the Lobovka, the Zelenogradka, the Malaya Moryanka), the Stogovka river, the watercourses of the Lava river basin (the Pravda, the Putilovka) and the small rivers in the northern part of the deltaic lowland.

5. And, finally, prior to summarizing this article, it is necessary to consider an issue, very important for the Kaliningrad Region, which is related to the presence of a large number of transboundary watercourses in the south, southeast and east of the region (Table 16).

Conditions in these watercourses depend rather upon the economic than on the geographical features of the border areas. One of the immediate problems for the transboundary watercourses is measuring the quality of water coming from the territories of adjoining states and determining the proportion of cross-border flow in a general discharge of contaminants.

It is possible to estimate indirectly the transboundary pollution level based upon water quality at the border and estuarine stations (outfalls). Naturally, as the

Table 16 Transboundary rivers in the Kaliningrad Region

River	Where the river flows from	Where the river flows into	Length, km Total	Length, km Within the Kaliningrad Region	Catchment area, km^2	Annual runoff, km^3
Neman	Lithuania	Curonian lagoon	937	115	101,000	25.10
Sheshupe (Šešupė)		Neman	308	102	6,120	0.52
Angrapa (Węgorapa)	Poland	Pregolya	169	95	3,960	0.72
Krasnaya		Pissa	83	56	412	0.10
Lava (Łyna)		Pregolya	289	56	7,130	1.31
Stogovka		Lava	64	45	281	0.04
Pravda			22	20	102	0.01
Putilovka			58	56	525	0.03
Mamonovka (together with the Vitushka)		Vistula (Kaliningrad) lagoon	51	12	400	0.27
Kornevka		Prokhladnaya	42	29	376	0.02
Rezvaya			33	29	173	0.01

watercourse runs from the border downstream to the mouth, it receives a certain amount of pollutants from the settlements located on its banks; the water quality may deteriorate due to surface runoff from farms, pastures, fields and so forth; along the entire river, there are areas where it receives the bulk of pollution and where the water self-purification takes place [52]. However, the key element is the water quality at the outfall because this is exactly the quality of river water flowing into the receiving water body. Previously, it was shown [23] that in the border areas, water quality of the rivers was significantly worse than at their mouths. For instance, concentrations of nitrite and nitrate nitrogen, phosphate-phosphorus and organic compounds were considerably higher at the border stations than in the mouth sections of the rivers. Studies conducted in 2010 confirmed the above-mentioned regularities. The most adverse conditions were found in the near-border areas of the rivers Sheshupe, Neman, Mamonovka, Stogovka and Putilovka. In accordance with the classification [56], water of the rivers at the border stations shall be evaluated by the content of ammonia nitrogen as "slightly" and "moderately polluted," except for the Omaza river, water which is "heavily contaminated." At the mouth of the rivers, the amount of ammonia nitrogen is substantially greater, and the water falls into the class of "moderately polluted" and "heavily contaminated." However, the nitrite nitrogen content at the border stations was catastrophically high, especially in the rivers of the Lava-Pregolskaya lowland belonging to the watershed of the Pregolya river.

The nutrient load from transboundary rivers calculated according to the formula recommended in [58] showed the following. In general, the load is rising along with the increase in water flow and the catchment area. According to the data of 2010, the load on the receiving water body had the greatest numbers from the rivers with the highest water content, such as the Angrapa, the Krasnaya and the Chernaya. The largest part in the transfer of nitrite and nitrate nitrogen and phosphate-phosphorus belongs to such rivers as the Stogovka, the Putilovka and the Banuwka. Despite the low discharge of these rivers (0.5–1.5 m^3/s), transportation of nutrients to the region is large enough that is evident of their significant contamination [52]. Total outflow of the Sheshupe river from the territory of the Republic of Lithuania contains the highest amount of nitrate nitrogen (1.3 tons/year); the Neman river transports a lot of inorganic phosphorus (3.6 tons/year) and total iron (2.8 tons/year). Thus, the major nutrient load is transported to the Kaliningrad Region by the rivers flowing from the territory of the Republic of Lithuania (owing to the rivers Neman and Sheshupe). Nutrient runoff from the territory of the Republic of Poland is formed by the rivers Angrapa and Lava with tributaries, which flow into the Pregolya river, as well as the Prokhladnaya river and the Mamonovka river flowing into the Vistula lagoon.

Balance between nutrient input from the neighbouring countries and outflow of nutrients from the territory of the Kaliningrad Region was calculated by the ratio of the load values at the border and estuarine stations (outfalls) of the transboundary watercourses. The balance is a difference in the weight of nutrients (in tons/year) that entered the territory of the Kaliningrad Region with the river flow from the territories of the neighbouring states and discharged into the receiving water body through an outfall. The share of nutrient runoff received from the territory of the neighbouring states ranges from 10 to 80% or more of the total nutrient runoff of the transboundary watercourses discharged from the Kaliningrad Region. The Mamonovka river, a larger part of which basin is located in the Polish territory, and the Neman river suffer the heaviest load from the neighbouring countries. The amount of nitrite nitrogen, mineral phosphorus and total iron discharged into the Curonian lagoon by the Neman river differs little from what is incoming from the adjacent territories (Fig. 7).

The balance between nutrient input from the neighbouring states and outflow of nutrients from the Kaliningrad Region reflects two components.

The first is the anthropogenic factor, the intensity of which can be estimated by point and diffuse sources of nutrient runoff. The second is the natural features and the watershed structure at the observation station that indirectly affect the hydrometric characteristics (water discharge and runoff volume). For example, in the area of the border station (settlement Nemanskoe), the Neman river drains the landscapes of elevated moraine plains, and the outfall (settlement Yasnoe) is located in the delta with the delta landscapes. Many parts of the delta are lying below sea level, very swampy, with flat terrain, mostly channelled territory (drainage canals). The small-level difference reduces flow velocity which is besides affected by the backwater of the Curonian lagoon, and the water discharge at the outfall was less than at the border station (438 and 550 m^3/s, respectively). Similar features were found at the cross sections of the rivers flowing from Poland (small rivers belonging to the Lava, the Angrapa, the Mamonovka and the Prokhladnaya river basins).

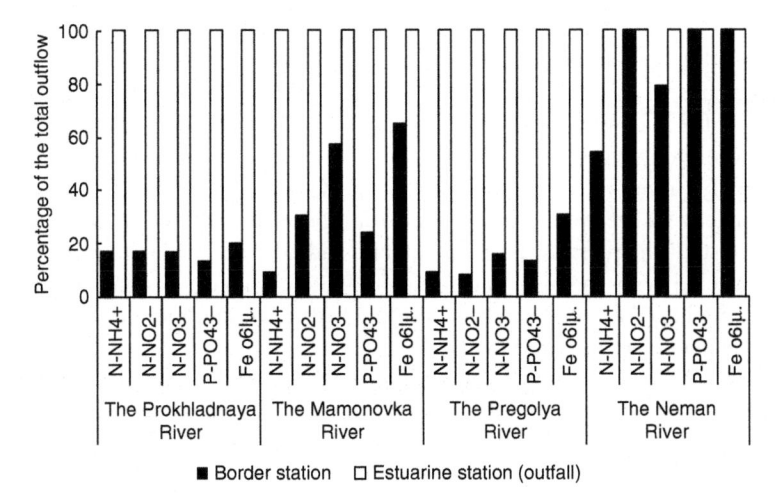

Fig. 7 Contribution of the neighbouring states to the total outflow of nutrients by the transboundary rivers of the Kaliningrad Region, 2010 [23]

Border stations are situated within the landscapes of terminal moraine ridges and plateaus or elevated moraine plains (small rivers belonging to the Lava, the Angrapa river basins, the Mamonovka river and the Prokhladnaya river); and the outfalls are located within glaciolacustrine plains (the Prokhladnaya river) or estuarine sandy lowlands (the Mamonovka river). Taking into account the natural characteristics, it can be concluded that a significant portion of nutrients remains in the Kaliningrad Region as a result of retention mainly due to backwater from the receiving water bodies, large swamp area and flat terrain reducing velocity and amount of nutrient transportation.

Thus, the major nutrient load is transported to the Kaliningrad Region by the rivers flowing from the territory of the Republic of Lithuania (together with the river Neman flow). Nutrient runoff from the territory of the Republic of Poland is formed by the rivers Angrapa and Lava (with their tributaries) that flow into the Pregolya river, as well as by the Prokhladnaya river and the Mamonovka river flowing into the Vistula lagoon.

4 Conclusions

1. The river network is well developed and very dense. The watercourses are classified as plain streams and belong to the Baltic Sea catchment area. Over 95% of watercourses are "small," characterized by a large short-term variability of hydrological and hydrochemical parameters comparable to the seasonal changes. Much of the Kaliningrad Region's territory is occupied by the Pregolya river basin with the Angrapa river and the Instruch river subbasins,

as well as by the basins of its several major tributaries. The Pregolya river is formed by the confluence of the rivers Angrapa and Instruch.
2. As a consequence of small slopes, the estuarine and adjacent parts of the rivers are in the backwater of the receiving water bodies and very dependent on the wind-driven surges. Most small rivers (especially within the Neman delta, the Pregolya river valley, at the coast of the Curonian and Vistula lagoons) are channelized; the water pumped out from the polder lands is dumped therein. Almost all the rivers in the Neman delta and at the Curonian lagoon coast are connected with channels and form the united drainage system. Many rivers flowing into the Baltic Sea have shifting beds near their mouths; displacement movement can reach 200 m and more.
3. Hydrometric parameters fluctuate significantly and are closely associated with both the weather conditions and the aggregate natural peculiarities of the drained area. Ice conditions are very unstable; the formation of two-layer ice is possible.
4. TDS in water is medium (350–430 mg/dm^3) or increased (550–750 mg/dm^3). According to the chemical composition of water, the studied streams belong to bicarbonate class, calcium group, mainly of the first water type [$HCO_3^- < (Ca^{2+} + Mg^{2+})$] with medium and increased TDS value (rivers of the Polesskaya lowland and the Neman delta, the Prokhladnaya river). Pursuant to the total hardness value, the water is "moderately hard."
5. Oxygen conditions are generally favourable, but the acute oxygen deficiency is possible in some small streams in certain periods (depletion down to 30–10% of saturation). In 2010, the most complicated oxygen conditions developed in the rivers of the Polesskaya lowland and of the Neman delta and in the estuarine part of the Pregolya river. In the border areas of transboundary watercourses, the water oxygen saturation was about 70–80%. The pH was slightly shifted to the alkaline side. There is rather much organic matter in the watercourses; the value of permanganate index is "increased." According to the content of organic matter, the small watercourses of the Kaliningrad Region can be divided into the slightly polluted, polluted and heavily contaminated rivers with catastrophically high organic content.
6. The content of nutrients chiefly did not meet the requirements for fishery water bodies. Particularly unfavourable conditions developed in the small rivers flowing into the Curonian lagoon and at the border stations on transboundary rivers where the extremely high values of hydrochemical parameters were found most frequently, especially during low water periods.
7. The Pregolya river is exposed to intensive pollution. The condition of some small rivers is indicative of their catastrophic contamination. A number of streams (the Lobovka, the Pokosnaya, the Medvezhya, the Chistaya, the Grayevka) basically look more like effluent gutters.
8. Oxygen conditions, the content of organic matter and nutrients in most watercourses do not comply with MPC for fishery water bodies, especially during low water periods. The most unfavourable conditions for the listed indicators were found at the border stations of the small rivers in 2010.

9. The rivers were divided into four groups with different environmental conditions: "favourable" (typical only for the small rivers draining the Vishtynets upland), "moderately favourable" (the Prokhladnaya, the Deima, watercourses in the southern part of the delta plain, the Instruch river), "hardly favourable" (majority of the small streams, including the Mamonovka, the rivers flowing into the Baltic Sea, the Zapadny and the Vostochny canals) and "very unfavourable" (such rivers as the Stogovka, the Grayevka, the Primorskaya, the Chistaya, the Medvezhya, the Zelenogradka, waterways in the northern part of the delta plain).
10. The river Neman which enters the Kaliningrad Region already quite polluted and the major watercourses in the delta plain (the Matrosovka canal, the Nemonin river) determine the load on the Curonian lagoon. The main source of nutrient load on the Vistula lagoon is the Pregolya river; the rivers Prokhladnaya and Mamonovka and others also contribute thereto. Small rivers emptying directly into the Baltic Sea build up a load on the sea mostly only during wet periods.
11. Transboundary rivers arrive in the Kaliningrad Region already quite polluted as evidenced by water quality at the border stations. The proportion of nutrient runoff received from the territory of the neighbouring states ranges from 10 to 80% or more of the total nutrient runoff of the transboundary watercourses discharged from the Kaliningrad Region. The amount of nitrite nitrogen, mineral phosphorus and total iron discharged into the Curonian lagoon by the Neman river differs little from what is incoming from the adjacent territories.

References

1. Bernikova TA et al (1997) Necessity to consider short-term changes in hydrological parameters in the study of aquatic ecosystems. In: Bernikov TA, Ryaboy VE (eds) The first congress of Russian ichthyologists: abstracts of the first congress of Russian ichthyologists, September 1997, in Astrakhan. M.: VNIRO, p 105
2. Bernikova TA et al (1996) Illumination as a factor of the hydrological features of aquatic ecosystems. In: Bernikova TA, Ryaboy VE (eds) The study of aquatic biological resources of the Kaliningrad Region: proceedings. Kaliningrad, pp 71–75
3. Bernikova TA et al (1997) Assessment of the ecological condition of some small rivers in the Kaliningrad Region according to the hydrological indicators. In: Bernikova TA, Shibaeva MN, Shkitsky VA (eds) Ecological problems of the Kaliningrad Region: proceedings. Kaliningrad, pp 24–29
4. Bernikova TA et al (2000) Environmental assessment of the Nelma and the Primorskaya rivers. In: Bernikova TA et al (eds) Proceedings of the international scientific and engineering conf. dedicated to the 70th anniversary of KSTU, 17–19 Oct 2000, in Kaliningrad. KSTU, Kaliningrad, Part 1. pp 167–168
5. Bernikova TA et al (1997) Removal of mineral and organic substances by small rivers of the Kaliningrad Region. In: Bernikova TA, Ryaboy VE, Shkitsky VA (eds) Modern aspects of agronomy and environmental management: proceedings. KSTU, Kaliningrad, pp 114–129
6. Bernikova TA et al (2004) Contribution of small rivers to the pollution of the Baltic Sea coastal zone. In: Bernikova TA et al (eds) International coastal conference: proceedings of the XXI international coastal conference. Kaliningrad, pp 208–209

7. Bernikova TA, Shibaeva MN, Tsoupikova NA et al (2004) Contribution of small rivers to the eutrophication of the Curonian Lagoon. In: Bernikova TA, Shibaeva MN, Tsoupikova NA et al (eds) International coastal conference: proceedings of the XXI international coastal conference. Kaliningrad, pp 59–60
8. Bernikova TA et al (2001) Assessment of the Sheshupe River ecological condition. In: Bernikov TA, Shibaeva MN, Shkitsky VA (eds) Hydrobiological society congress of the RAS: proceedings of the VIII congress of the RAS hydrobiological society, 16 Sept 2001, in Kaliningrad, Kaliningrad, pp 31–35
9. Belov NS (2011) Evaluation of geoecological situation in the river basins of the Kaliningrad Region using geoinformation technologies: Author's abstract of PhD thesis in geography: 25.00.36. Belov Nikolai Sergeyevich. Immanuel Kant BFU, Kaliningrad, 24 p
10. Nagornova NN et al (2011) Hydrogeochemical characteristic of the small rivers in the Kaliningrad Region. In: Nagornova NN, Bernikov TA, Tsoupikova NA (eds) Bulletin of the Immanuel Kant BFU, vol 7. Immanuel Kant BFU, Kaliningrad, pp 160–166
11. Bernikova TA et al (eds) (2006) Small rivers of the Kaliningrad Region: problems and possible solutions. In: Bernikov TA, Tsoupikova NA (eds) Comprehensive study of the Atlantic Ocean basin: proceedings. IKSUR, Kaliningrad, pp 50–57
12. Bernikova TA et al (2003) Biogenic load on the Vistula Lagoon by the small rivers. In: Bernikova TA, Shibaeva MN, Shkitsky VA (eds) Ecological and fisheries aspects of the study of coastal seas and inland waters: proceedings. KSTU, Kaliningrad, pp 63–69
13. Matveeva EV (2005) Contamination and self-cleaning processes in the Pregolya River. In: Matveeva EV, Bernikova TA (eds) Innovations in science and education 2006: Proceedings of the IV international scientific conference dedicated to the 10th anniversary of KSTU, 12–20 Oct 2006, in Kaliningrad. KSTU, Kaliningrad, pp 44–46
14. Bernikova TA et al (2007) Hydrological conditions of the Nemunas and the Sheshupe rivers (within the Kaliningrad Region) in March 2007. In: Bernikova TA, Nagornova NN (eds) Innovations in science and education 2007: proceedings of the V international scientific conference KSTU, 23–25 Oct 2007, in Kaliningrad, Kaliningrad, Part 1, pp 58–61
15. Bernikova TA, Nagornova NN, Tsoupikova NA, Gusarova AV (2007) Pollution and water exchange of the Primorskaya Bay with the Kaliningrad Maritime Chanel (KMC). In: Bernikova TA, Nagornova NN, Tsoupikova NA, Gusarova A (eds) Innovations in science and education 2007: proceedings of the V international scientific conference KSTU, 23–25 Oct 2007, in Kaliningrad, Kaliningrad, Part 1, pp 61–64
16. Bernikova TA et al (2013) Hydrological study of watercourses in the Kaliningrad Region. In: Bernikova TA, Nagornova NN, Tsoupikova NA (eds) Water biological resources, aquaculture and ecology of water bodies: proceedings of the first scientific conference, 25–26 Sept 2013, in Kaliningrad, KSTU, Kaliningrad, pp 353–356
17. Bernikova TA et al (2014) Some results of the hydrological study of the watercourses in the Kaliningrad Region. In: Bernikova TA, Nagornova NN, Tsoupikova NA (eds) Bulletin of KSTU. vol 32, pp 74–85
18. Bernikova TA et al (2014) Some results of background hydrological studies of the Neman River in the area under construction of Baltic NPP. In: Bernikova TA, Agarkova TS, Nagornova NN (eds) Environmental safety of nuclear power plants: proceedings of the first scientific-practical conference with international participation, dedicated to the 60th anniversary of the nuclear power industry, 16–17 Oct 2014, in Kaliningrad, Axios, Kaliningrad, pp 5–11
19. Loiko TV (2002) Influence of surges on the content of dissolved oxygen and water oxidation in the Pregolya River. In: Loiko TV (ed) Ecological problems of the Kaliningrad Region and the Baltic Region: proceedings. KSU, Kaliningrad, pp 205–207
20. Matveeva EV (2007) Contamination extension downstream the Pregolya River. In: Matveeva EV, Nagornova NN (eds) Ecological problems of the Kaliningrad Region and the Baltic Region: proceedings. IKSUR, Kaliningrad, pp 71–78

21. Nagornova NN, Malyavkina AN, Bernikova TA, Tsoupikova NA (2008) Monitoring of the organics and nutrients removal by the Prokhladnaya River to the Kaliningrad (Vistula) Lagoon. In: Nagornova NN, Malyavkina AN, Bernikova TA, Tsoupikova NA (eds) Integrated management, development indicators, spatial planning and monitoring of the Coastal South-Eastern Baltic: proceedings of the international conference, 23–28 Mar 2008, in Kaliningrad. Terra Baltika, Kaliningrad, p 111
22. Nagornova NN (2014) Seasonal variability of hydrochemical conditions in the Neman River within the Kaliningrad Region. In: Nagornova NN, Bernikov TA (eds) Ecological safety of nuclear power plants: proceedings of the first scientific-practical conference with international participation dedicated to the 60th anniversary of nuclear energy, 16–17 Oct 2014, in Kaliningrad. Axios, Kaliningrad, pp 67–76
23. Nagornova NN (2012) Geoecological assessment of the small watercourses in the Kaliningrad Region: Author's abstract of PhD thesis in geography: 25.00.36. Nagornova Nadezhda Nikolayevna, Kaliningrad, 21 p
24. Bernikova T, Malyavkina A, Nagornova N (2007) Discharge of organic and biogenic substances from small rivers of the Kaliningrad Region in the Baltic Sea. Proceedings, 6th international conference. Kalmar, part II, pp 569–573
25. Bernikova TA, Shibaeva MN, Shkitsky VA, Tsoupikova NA (2008) The Prokhladnaya River. In: Chubarenko B (ed) Transboundary waters and basins in the South-East Baltic. Terra Baltica, Kaliningrad, pp 80–88
26. Bernikova TA, Shibaeva MN, Shkitsky VA, Tsoupikova NA (2008) The Primorskaya River. In: Chubarenko B (ed) Transboundary waters and basins in the South-East Baltic. Terra Baltica, Kaliningrad, pp 88–96
27. Bernikova TA, Shibaeva MN, Shkitsky VA, Tsoupikova NA (2008) The Mamonovka River. In: Chubarenko B (ed) Transboundary waters and basins in the South-East Baltic. Terra Baltica, Kaliningrad, pp 96–106
28. Nagornova NN (2010) Methodological aspects of identification of the rivers-analogies (the case of the small rivers ecosystems in the Kaliningrad Region). Water resources, ecology and hydrological safety: proceedings of the IV international scientific conference of young scientists and talented students, 6–8 Dec 2010, Moscow. M.: IWP RAS, pp 227–230
29. Tsoupikova NA (2010) Factors of formation of nutrient runoff from the territory of the Kaliningrad Region. In: Tsoupikova NA, Nagornova NN (eds) Innovations in science and education 2010: Proceedings of the KSTU VIII international scientific conference dedicated to the 80th anniversary of the university, 19–21 Oct 2010, Kaliningrad, Part 1, pp 172–174
30. Nagornova NN (2014) Preliminary results of the study of chlorophyll "a" in the Neman-Sheshupe system (within the Kaliningrad Region). In: Nagornova NN, Bernikova TA, Krivopuskova EV (eds) Water biological resources, aquaculture and environment of water bodies: proceedings of the second international scientific and practical conference, 15–17 Oct 2014, Kaliningrad, VPO KSTU, Kaliningrad, pp 177–179
31. Bernikova TA, Malyavkina AN, Nagornova NN, Tsoupikova NA (2008) Hydrology. In: Bernikova TA, Malyavkina AN, Nagornova NN, Tsoupikova NA (eds) Laboratory manual and practical training. M.: Kolos, 304 p
32. State Standard (1990) GOST 17.1.4.02-90 water. Methods of spectrophotometric determination of chlorophyll "a." M.: IPC Standards Publisher, 11 p
33. SP 33-101-2003 Rules and Regulations (to Replace SNiP 2.01.14-83) (2004) Calculation of the main estimated hydrological characteristics. M.: GP TsPP, 70 p
34. Report of the European Economic Commission. Assessment of transboundary waters in the UNECE Region. Preliminary assessment of transboundary water resources in the Baltic Sea Basin. Transboundary Rivers and International Lakes in the Narva, the Salaka, the Gauja/Koiva, the Daugava, the Lielupe, the Venta and the Neman River Basins. ECE/MPWAT/WG2/2007/7 19 Apr 2007
35. Olyonok VV (ed) (2002) Geographical Atlas of the Kaliningrad Region. KSU, Kaliningrad, 276 p

36. Tsybin Yu A (ed) (2004) The scheme of nature protection in the Kaliningrad Region. TENAX Media, Kaliningrad 136 p
37. Vladimir Kupriyanov (ed) (1971) Hydrographic characteristics of the river basins in the European Territory of the USSR. L.: Gidrometeoizdat, 100 p
38. The State Water Cadastre (1988) Reference edition. Long-term data on regime and resources of surface waters. L.: Gidrometizdat, Book 1, vol 4, 88 p
39. Vodogretsky VE (ed) (1969) Surface water resources of the USSR. Lithuanian Soviet Socialist Republic and the Kaliningrad Region of the RSFSR. L.: Gidrometeoizdat, Book 4, vol 3 506 p
40. Kamaletdinova ET (ed) (2004) Report on the status and on environmental protection of the Kaliningrad Region in 2003. NRD under the RF MNR in the Kaliningrad Region, Kaliningrad, 216 p
41. Report on the status and on environmental protection of the Kaliningrad Region in 2002 by the Ministry of Natural Resources of the Kaliningrad Region (2003) OOO "Pechatny Dvor," Kaliningrad, 173 p
42. Bernikova TA et al (2004) Small rivers monitoring problems: the case of the Kaliningrad Region. In: Bernikova TA, Tsoupikova NA, Shibaeva MN (eds) Role of environmental engineering in sustainable operation and development of ecosystems: proceedings of the international scientific-practical conference. M: MSUP, Part 2, pp 31–37
43. Climate Monitor. Weather in Kaliningrad. The air temperature and precipitation, 2005-2014. http://pogoda.ru.net/monitor.php?id=26702. Accessed 29 Apr 2015
44. Domnin DA et al (2007) In: Domnin DA, Chubarenko BV (eds) Atlas of transboundary river basins in the Kaliningrad Region. Terra Baltika, Kaliningrad 38 p
45. Zotov SI (2002) Surface water monitoring system in the Kaliningrad Region and possible ways of its optimization. In: Zotov SI, Velikas YV, Shchagina NV (eds) Environmental problems of the Kaliningrad Region and of the Baltic Region: proceedings. KSU, Kaliningrad, pp 27–37
46. Zotov SI (2006) Assessment of environmental sensitivity of the river basins in the Kaliningrad Region to chemical pollution. In: Zotov SI, Belov NS (eds) Bulletin of the Immanuel Kant Russian State University, vol 1, IKSUR, Kaliningrad, pp 19–22
47. Nagornova NN (2010) In: Nagornova NN, Bernikova TA (eds) Some factors of formation of short-term variability of small aquatic ecosystems (the case of the Kaliningrad Region), vol 1. Bulletin of the People's Friendship University of Russia M. pp 30–38
48. Alekin OA (1970) In: Alekin OA (ed) Basics of hydrochemistry. L.: Gidrometeoizdat, 443 p
49. Baranov IV (1962) In: Baranov IV (ed) Limnological types of lakes in the USSR. L.: Gidrometeoizdat, 276 p
50. Report of the European Economic Commission. Assessment of transboundary waters in the UNECE Region. Preliminary assessment of transboundary rivers flowing into the Baltic Sea and their major transboundary tributaries – the Pregolya, the Vistula and the Oder. ECE/MPWAT/WG2/2007/8 16 Apr 2007
51. Strategic directions to adapt the River Neman Basin for climate change. Edited for the First Working Discussion in the Framework of the Project "Water Resources Management of the Neman River Basin Taking into Account Climate Change Adaptation." 2014. 59 p. http://www2.unece.org/ehlm/platform/display/climateChange/Neman+project+worshop+and+expert+meeting+19-20+March+2013. Accessed 29 Apr 2015
52. Shibayev SV, Nagornova NN, Bernikova TA, Tsoupikova NA (eds) (2011) Geoecological aspects of the nutrients transfer by the transboundary rivers (the case of the Kaliningrad Region), vol 22, Bulletin of KSTU, Kaliningrad, pp 133–141
53. Water quality standards for fishery water bodies, including the maximum permissible concentrations of harmful substances in fishery water bodies: [Approved by the Order of Federal Agency for Fishery dated 18.01.2010 No. 30]
54. Markova LL (1999) Rivers. In: Markova LL (ed) Kaliningrad Region. Nature sketches. Yantarny Skaz, Kaliningrad, pp 69–83
55. Shibayev SV et al (eds) (2008) Fishery inventory of transboundary water bodies of Russia (Kaliningrad Region) and Lithuania. PE Mishutkina, Kaliningrad, 200 p

56. Integrated environmental classification of surface water quality by O.P. Oksiyuk and V.N. Zhukinsky. http://www.ievbran.ru/kiril/Library/Book1/content236/tables.htm. Accessed 29 Apr 2015
57. Bouillon VV (1994) Laws of primary production in the limnological ecosystems. V.V. Bouillon. SPb.: Nauka, 224 p
58. Collected Volume of the Helsinki Commission Recommendations (2008) Reference manual. St Petersburg, 331 p

Formation and Re-Distribution of the River Runoff in the Catchment of the Pregolya River

Dmitry Domnin, Boris Chubarenko, and Rene Capell

Abstract The Pregolya River has a transboundary catchment between Russia, Poland, and Lithuania. The Pregolya River discharge is directed to two receiving reservoirs – the Vistula Lagoon and the Curonian Lagoon. In the article, the calculation of the volume of water flow rate from the Pregolya River catchment was made by numerical model. The annual average flows to the Vistula and Curonian lagoons constitute 1.96 and 1.2 km^3/year, respectively, which in total gives 3.16 km^3/year from the Pregolya River catchment towards the Baltic Sea through both lagoons. The contribution of Polish and Russian national areas to the annual river runoff was determined. The long-term seasonal mean re-distribution of the water flow to both receiving reservoirs (two lagoons) was analyzed. Average monthly values of water balance components for the entire catchment basin of the Pregolya River was drawn up, the annual average of 760 mm of precipitation is balanced by 530 mm of evapotranspiration and 230 mm of flow with surface runoff.

Keywords Numerical simulation, Pregolya River, Runoff, Transboundary catchment, Water balance

Contents

1 Introduction 270
2 Background Information, Tools and Techniques 272
3 Discussion of Results 276
   3.1 Water Discharge in the Outlet Sections of the Pregolya River 276

D. Domnin (✉) and B. Chubarenko
P.P. Shirshov Institute of Oceanology, Russian Academy of Sciences, 36 Nahimovskiy prospekt, Moscow 117997, Russia
e-mail: dimanisha@rambler.ru; chuboris@mail.ru

R. Capell
Swedish Meteorological and Hydrological Institute, Norrköping, Sweden
e-mail: rene.capell@smhi.se

3.2 Surface Runoff as an Indicator of the Hydrological Basin Structure 279
3.3 Cross-Border Discharge Value Assessment 281
4 Main Conclusions 281
References 284

1 Introduction

All the main water systems of the Kaliningrad Oblast (Region) are transboundary. These are the coastal waters of the Baltic Sea, the Vistula Lagoon watershed and the Curonian Lagoon watershed. The main river delivering for the Vistula Lagoon is the Pregolya River which is formed through confluence of the Instruch and Angrapa Rivers[1] (Fig. 1).

The watercourse from Chernyakhovsk to the mouth at the confluence into the Vistula Lagoon, which is actually called the Pregolya River, is 123 km long. The watercourse of the Pregolya and its easternmost tributary, the Angrapa River is 292 km long. Other major tributaries are the Instruch River (101 km), the Lyna-Lava River (289 km), the Pissa River (98 km) (Fig. 1). The area of the Pregolya River is 14.8 thousand km^2 [1].

According to reports, the average annual water discharge of the Pregolya River at the town of Gvardeysk (56 km upstream of the mouth) is 86 m^3/s (or 2.7 km^3/year) [2]. A particular feature is that the mainstream of the Pregolya River divides into two arms (Fig. 1) in Gvardeysk. The main arm, the Downstream Pregolya (or Pregolya proper, 56 km long), carries an average of 60% of water into the Vistula Lagoon, and the second one, the Deyma branch (37 km long, called also as the Deyma Arm or Deyma River) exports an average of 40% of the water volume into the Curonian Lagoon [3]. The average slope of the water surface of the Downstream Pregolya is 0.9 cm/km, and 13 cm/km for the Deyma [4].

All previous studies of the river runoff from the catchment area of the Pregolya River were based on discharge data attributed to the point of division of the main river stream into the arms in Gvardeysk (Fig. 1), where there are points of measurements of the state monitoring network. Location of observation points away from the outlet sections is related to the existence of these two watercourses in the lower reaches, on the one hand, and on the other, to the fact that the Downstream Pregolya and the Deyma are constantly exposed to backwater effect during run-up winds from the receiving bodies of water – the Vistula and Curonian Lagoons [3].

In fact, all the ongoing monitoring points of the state monitoring network available within the catchment area of the Pregolya River are located not at the outlet sections of its tributaries, but at the places suited for measurements (bridges, dams, etc.).

---

[1]Further we shall sometimes use just "Pegolya," "Angrapa," etc. to call rivers.

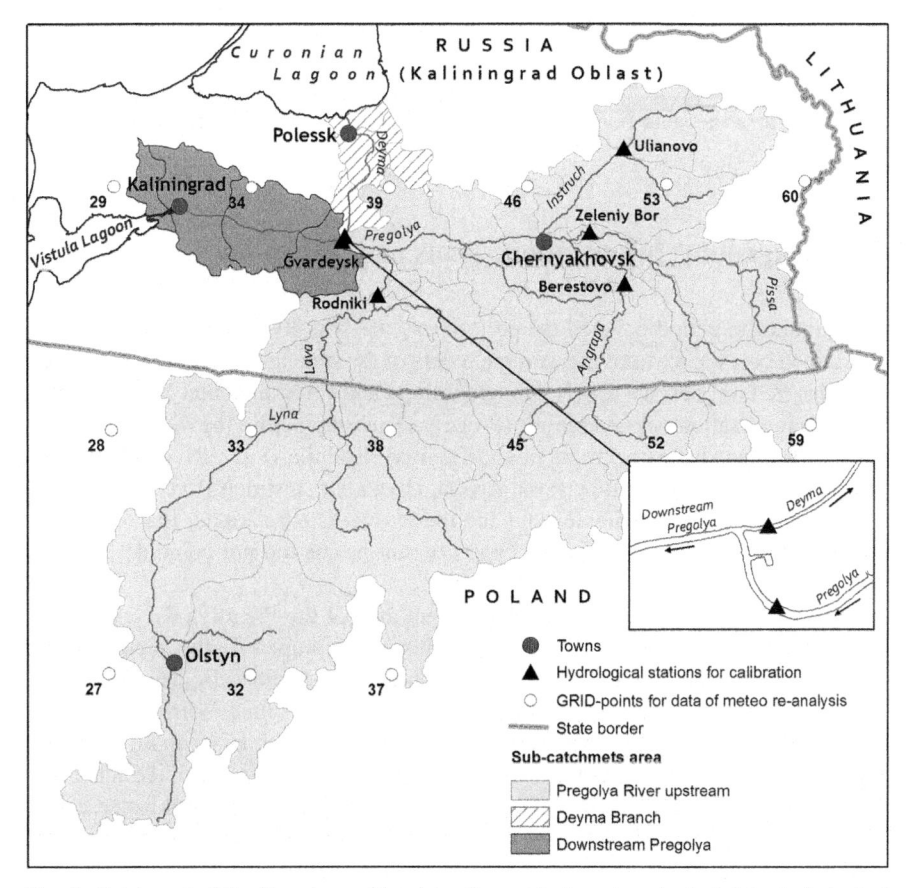

Fig. 1 Catchment of the Pregolya, grid points of re-analysis meteorological data, hydrological monitoring points. The *inset* shows the division of the Pregolya into the arms of the Downstream Pregolya and Deyma branch in Gvardeysk

Therefore, this study aims to obtain estimates of mean annual values of the river water runoff precisely for the outlet sections – at the confluence point of the Pregolya River into the Curonian and Vistula Lagoon, and at the confluence points of its tributaries into the mainstream.

Another important aspect of the study will be an assessment of a cross-border component of the runoff from the catchment area of the Pregolya River [5], the watershed of which is located within three countries: Russia (47% of the area), Poland (52%), and Lithuania (1%). Direct measurements of the runoff in the main watercourses crossing the border are not taken.

Numerical modeling is the most optimal method for assessing of runoff re-distribution of the Pregolya River at the branch point in Gvardeysk and a cross-border component of the flow on a uniform basis. Such an approach for the whole catchment area of the Vistula Lagoon was used previously in solving problems of climate change [6, 7], penetration of brackish water up through the bed of the Pregolya River [8]. In this paper, the emphasis is laid on the estimates of mean annual values of the water discharge running into the coastal lagoons and coming across the state borders from the territory of Poland and Lithuania.

2 Background Information, Tools and Techniques

The simulation period covered the period from 1979 to 2009. To calibrate a model (described below), we used measurement data of the state network of RosHydroMet (Russian Federal Service for Hydrometeorology and Environmental Monitoring) at hydrological stations of Gvardeysk (Pregolya River up before the division into the arms of the Downstream Pregolya and Deyma), Rodniki (Lava River), Berestovo (Angrapa River), Zeleny Bor (Pissa River), Ulyanovo (Instruch River). Due to the heterogeneity of the available data for the above-listed stations, the calibration periods of the model for individual parts of sub-basins did not coincide, but they had a duration of at least 5 years in a row.

As a result of delineation of the catchment area of the Pregolya River, a model installation for the system of 42 sub-basins was composed, the input data of temperature and precipitation for which should be determined by nearby weather stations. However, there are a number of limitations for this. Firstly, data is not always publicly available, and secondly, the data series may contain gaps and, thirdly, the stations are located far from each other. Lack of coverage of the catchment area of the Pregolya River by meteorological measurement points was noted earlier [7]. Therefore, re-analysis data of the European Watch Project (http://www.eu-watch.org/data_availability, ftp European project: ftp://rfdata:forceDATA@ftp.iiasa.ac.at) served as input information about the amount of precipitation (volume per day) and air temperature (daily average).

The data are spatially attached to grid nodes with the step of 30′. The catchment area of the Pregolya River used for simulation comprises 15 nodes (Fig. 1 and Table 1).

At the stage of the input data analysis (for the period of 1979–2009), the procedure of comparing of temperature values was carried out (average per month) and

Table 1 Coordinates of grid points of re-analysis data of Watch project used in simulation of the runoff from the catchment basin of the Pregolya River

ID VLC	Lat	Lon	ID VLC	Lat	Lon
27	53.75	20.25	37	54.75	21.25
26	54.25	20.25	45	54.25	21.75
25	54.75	20.25	44	54.75	21.75
32	53.75	20.75	52	54.25	22.25
31	54.25	20.75	51	54.75	22.25
30	54.75	20.75	59	54.25	22.75
39	53.75	21.25	58	54.75	22.75
38	54.25	21.25			

precipitation (amount per month) obtained according to the re-analysis and direct measurement data. It was found that the average temperature of the measured values was +7.8°C, and +7.5°C according to the re-analysis data. The average monthly amount of precipitation according to the measured data was 68 mm, and 69 mm according to re-analysis data. The correlation coefficient between the measured and recovered data for temperature is equal to 1, and 0.95 for precipitation.

As a primary runoff hydrological model, HYPE model (HYdrological Predictions for the Environment) of the Swedish Hydrometeorological Institute (SMHI) [9, 10] was used. Incoming information for HYPE calculations are data on the land use structure, soil types in the basin, time series of precipitation and temperature; calibration requires time series of the water discharge.

Given the complexity of the catchment system, the whole catchment area was divided for simulation into three constituent parts, with their own created model installations (Fig. 2): the catchment area of the Pregolya River up to the division into the arms of the Downstream Pregolya and Deyma, the catchment area of the Downstream Pregolya, the catchment area of the Deyma River.

The calculated catchment area was 13.1 thousand km^2 for the Pregolya upstream. It was divided into 34 sub-basins: area of the smallest one was 60 km^2, and the biggest one 1.5 km^2. The Downstream Pregolya basin (with an area of 1.1 thousand km^2) was divided into 4 sub-basins (min = 100 km^2, max = 370 km^2), basin of the Deyma branch (area of 0.4 thousand km^2) was also divided into 4 sub-basins (min = 60 km^2, max = 150 km^2).

For calibration of the model installation for the Pregolya upstream, the point of hydrological measurements in Gvardeysk (Fig. 1) was used, which is located before the place of branching of the Pregolya into the arms. For verification, the data of the hydrological measurements at the points located in the largest tributaries of the Pregolya was used: the Lava (Rodniki), Instruch (Ulyanovo), Angrapa (Berestovo), and Pissa (Zeleny Bor) (Fig. 1). It should be noted that all the stations are situated within the Kaliningrad Oblast, as the authors did not have access to long-term hydrological data on the territory of Poland.

The catchment sub-basin of the Angrapa River is formed by two rivers: the Angrapa and Pissa, so when checking the results, they were considered together, and two checkpoints were chosen: at the Angrapa properly (point in Berestovo) up to the confluence of the Pissa River into it and at the Pissa (point in Zeleny Bor). Analysis showed that model solutions closely match the results of measurements at these points, but with the different sets of calibration parameters. As a result, those calibration parameters that provided an optimally high correlation coefficient for the both points (see Table 2) were selected as a uniform optimal set of calibration parameters.

When checking the results for the Instruch River, data of Ulyanovo point at the center of its catchment area was used. A set of calibration parameters used in general for the Pregolya River is poorly suited for the catchment area of the Instruch (correlation coefficient is 0.55). The reason of low correlation is poor coincidence of peaks in the periods of high water during high water and seasonal floods, as well as rather overestimated values of water flow rate during low water. Most likely, the

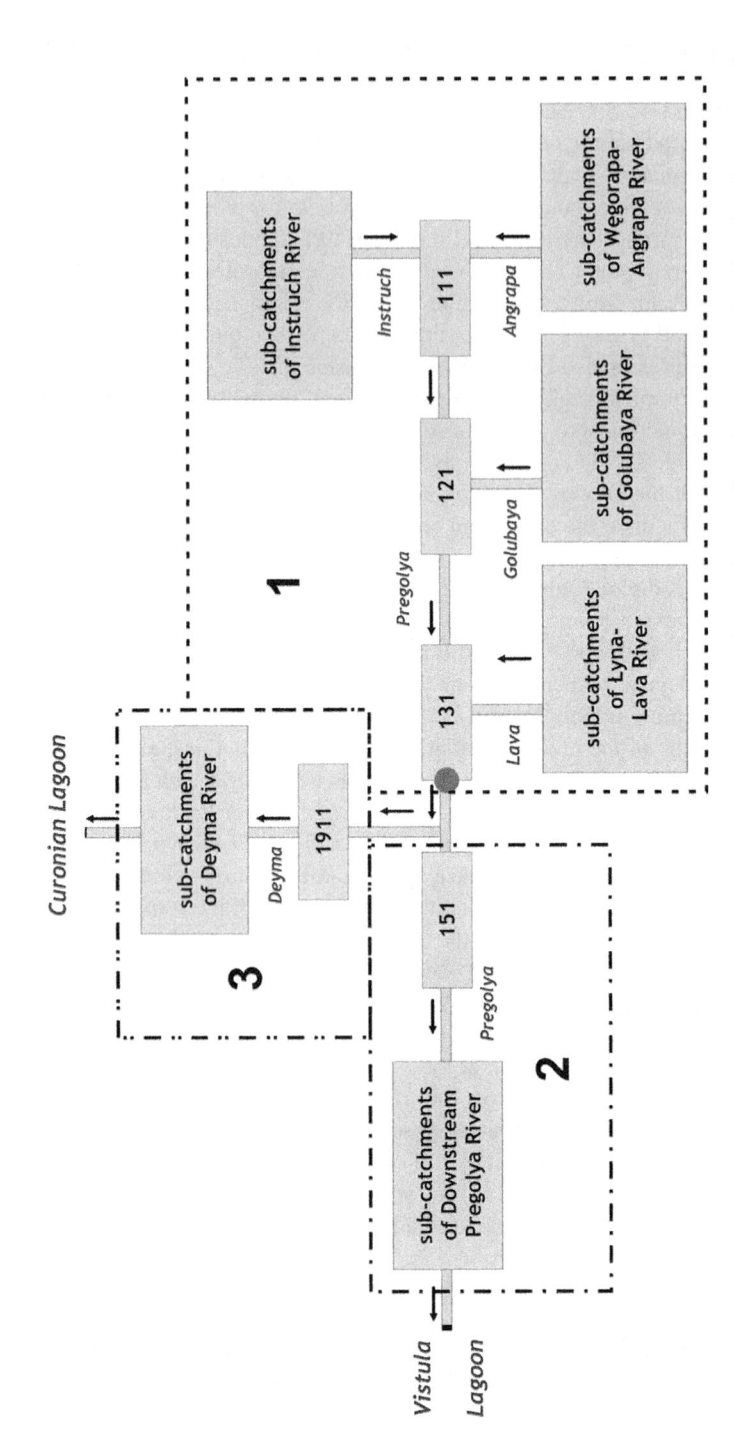

Fig. 2 Scheme of branching of the catchment basin of the Pregolya River on the model installation: *1* – Pregolya upstream, *2* – Pregolya downstream, *3* – Deyma

Table 2 Measured and simulated water discharge at the Angrapa and Pissa Rivers

Sub-basin	Checking period	Average water discharge measured (m³/s)	Average water discharge simulated (m³/s)	Correlation coefficient
Angrapa (Berestovo)	1995–2000	12.6	12.8	0.75
Pissa (Zeleny Bor)	1980–1986	10.6	9.6	0.72
Lava (Rodniki)	1995–2000	44.2	40.4	0.75

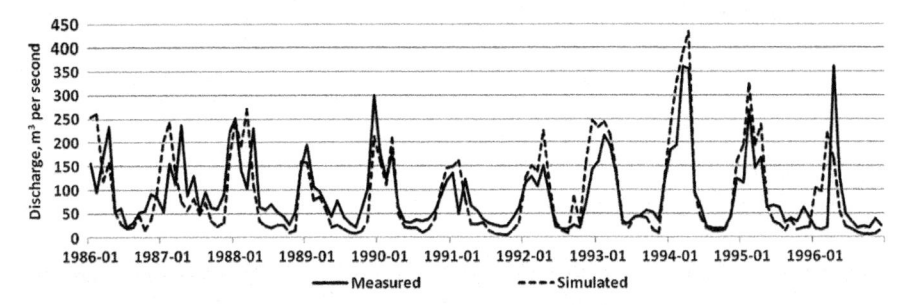

Fig. 3 Variations of the Pregolya River discharge at the Gvardeysk point of measurements during 1986–1996

low ratio of the measured data with the simulation result is associated with the fact that the checkpoint (Ulyanovo) is located in the upper reaches of the river, and its catchment area is small and the effect of local factors is significant.

Catchment area of the Lava River is the largest in the Pregolya River system, the runoff there from largely determines the water discharge of the Pregolya itself. To check the results for the Lava River, data from Rodniki point was used which is located in the Russian part in the lower reaches after all the upstream reservoirs. The calibration parameters used for the Pregolya River were suitable for the sub-basin of the Lava River. The quality of the result expressed as a correlation coefficient equal to 0.75 confirms this. Average values of the water discharge at the checkpoint, obtained in a simulation way are below the value of the measured discharge by 10% (Table 2). However, despite a steady interrelation, the high water peaks, obtained through simulation, are somewhat behind the measured values. Perhaps this is due to the over-regulation of the river, and, accordingly, the water discharge through water storage reservoirs, which the simulation applied cannot take into account.

As a result of calibration and verification, an optimal set of parameters was chosen that made it possible to bring the simulated discharge curve at the flow control section of Gvardeysk to the curve of the measured flow (Fig. 3). The period of 1986–1995 was chosen as a calibration of water discharge for the Pregolya River.

Table 3 Mean, maximum, and minimum discharge for the Pregolya River, obtained by measurements and calibration simulations by the HYPE model (calibration period is 1986–1996, verification period is 2008–2009, outlet section before the point of division into two arms)

Time-series	Correlation coefficient	Mean (m^3/s)	Maximum (m^3/s)	Minimum (m^3/s)
Results of model calibration, 1986–1996				
Measured data	0.79	90	361	15
Simulated data		89	434	5
Results of model verification, 2008–2009				
Measured data	0.92	76	239	18
Simulated data		83	277	10

The correlation coefficient (R) between measured and simulated discharges at the outlet section of the sub-basin 131 (Gvardeysk point) was 0.79. Average measured discharge is 90 m^3/s, the simulated one – 89 m^3/s (Table 3).

3 Discussion of Results

Measured and simulated hydrographs (river runoff), for the most part, are very similar at the calibration section in Gvardeysk (Fig. 3). However, there is some discrepancy, which we attribute to the fact that the main tributary of the Pregolya River – the Lava River is overregulated, and there are six hydroelectric power plants along the entire length there. Therefore, the measured water discharge in the Pregolya River depends essentially on the water discharge from the hydroelectric power stations, whereas the simulated one is determined by direct boundary conditions: air temperature and precipitation.

According to verification (for the period of 2008–2009) of simulation results, the correlation coefficient was higher ($R = 0.92$), and the discharge charts better related to each other. Thus, taking the resulting model installation as satisfactory, the following water discharge from the territory of sub-basins of the main tributaries of the Pregolya River was obtained (Table 4).

3.1 Water Discharge in the Outlet Sections of the Pregolya River

Since flow measurement points at the Pregolya River and its arm Deyma are only located at the point of their division, i.e. 56 and 37 km upstream of their outlet sections at the confluence into the Vistula Lagoon and the Curonian Lagoon, respectively, there is currently no reliable information about the volume of the water runoff falling into the Vistula and the Curonian Lagoons from the catchment basin.

Table 4 Basic hydrological characteristics of the main tributaries of the Pregolya for the period of 1980–2009

River catchment	Area (thousand km^2)	Discharge (m^3/s)	Discharge (km^3/year)	Runoff (mm/year)
Pregolya (Gvardeysk)	13.3	86.4	2.72	202
Lyna-Lava (outlet)	7.0	43.4	1.37	193
Wengorapa-Angrapa with Pissa (outlet)	3.7	22.0	0.69	183
Instruch (outlet)	1.3	9.4	0.30	233
Golubaya (outlet)	0.6	4.5	0.14	248

Table 5 Values of the simulation water discharge, obtained for outlet cross-sections of private basins of the Downstream Pregolya and Deyma for the period of 1980–2009

River catchment	Area (thousand km^2)	Discharge (m^3/s)	Discharge (km^3/year)	Runoff (mm/year)
Downstream Pregolya (outlet)	1.1	10.4	0.33	305
Deyma (outlet)	0.4	3.4	0.11	276

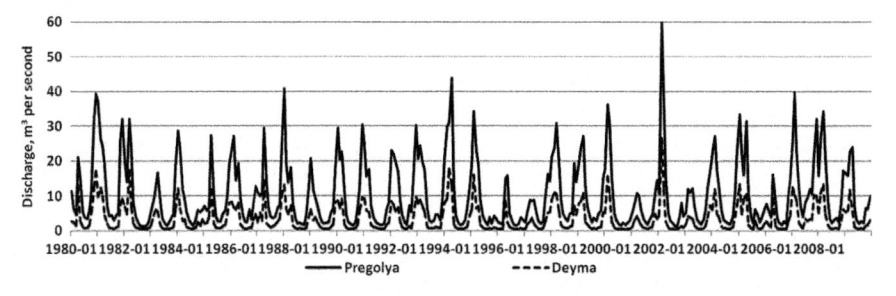

Fig. 4 Diagrams of the water discharge coming only from private catchment areas of the Downstream Pregolya and Deyma for the period of 1980–2009

To solve this problem, model installations for watersheds of the Downstream Pregolya and Deyma were used (Fig. 2), as for individual catchment areas. Calibration and verification of model calculation of the water discharge in the outlet sections (Table 5) of these installations were not carried out due to the lack of field data; the calibration coefficients obtained from the results of calibration for the reaches of the Pregolya upstream were used.

Average discharge of the water flowing properly from the territory of the Downstream Pregolya obtained as a result of numerical simulation is 10.4 m^3/s, average water flow of the Deyma River is 3.4 m^3/s. The Downstream Pregolya is characterized by higher values of water discharge, compared with the Deyma River, during the period of high water and seasonal floods (Fig. 4).

So, as a first approximation, the discharges of the water flowing into the Vistula and Curonian Lagoons were determined by summing the flows separately from these watersheds and the flows coming from the upper reaches of the Pregolya upstream and estimated either in accordance with generally accepted proportion of water runoff along the arms of the Downstream Pregolya and the Deyma [3] or in view of available data on the actual water distribution.

According to direct measurements data of the flow (1901–1956) at two control points in Gvardeysk (before the division of the Pregolya River into arms), on average, 60% of water is carried into the Vistula Lagoon; and its arm Deyma carries away about 40% of the water volume to the Curonian Lagoon [3]. Due to the fact that special model analysis of the water flow distribution between the arms was not conducted, this proportion was adopted for this study. Table 6 shows resulting calculated values at the outlets of the Downstream Pregolya and Deyma.

Considering long-term seasonal mean changes of the water discharge coming into the Vistula Lagoon from the control section of the Pregolya River and the one coming into the Curonian Lagoon from the control section of the Deyma River, it should be noted that the above-described ratio 60/40 typical of the place of division into the arms is retained, only during periods of winter and spring floods. During the summer low water, the amount of water reaching the receiving lagoon water bodies correlates to 50/50 (Fig. 5). Such a fluctuation of the water flow ratios is most likely

Table 6 Water discharge values from the catchment basin of the Pregolya River in the Vistula and Curonian Lagoons at the outlet sections of the Downstream Pregolya and Deyma

River	Discharge before the separation (m^3/s)	Proportions of separation	Own discharge (m^3/s)	Discharge from outlet to lagoon (m^3/s)	Discharge from outlet to lagoon (km^3/year)
Downstream Pregolya	86.4	0.6	10.4	62.2	1.96
Deyma		0.4	3.4	38.0	1.20

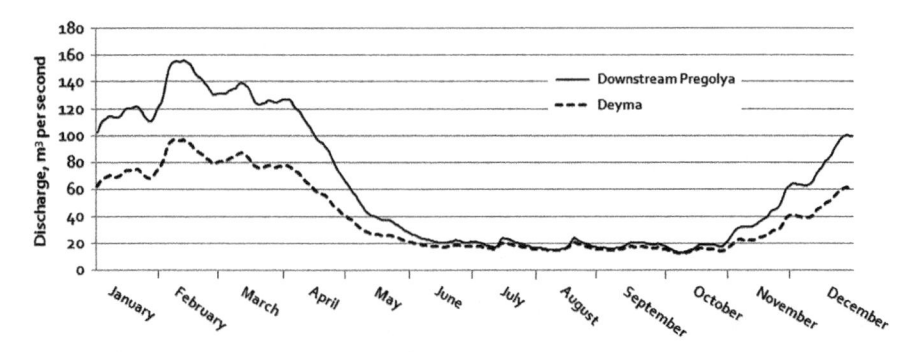

Fig. 5 Long-term seasonal mean change of the water discharge, running through the estuary sections of the Downstream Pregolya River into the Vistula Lagoon and the Deyma River into the Curonian Lagoon according to model calculations for the period of 1980–2009

associated with different sizes of catchment areas: the watershed formed by the Downstream Pregolya is almost three times bigger than the catchment area of the Deyma, which directly affects the degree of evaporation during a warm period of the year.

3.2 Surface Runoff as an Indicator of the Hydrological Basin Structure

Discharge parameter, as described above and expressed in m^3/s, is the main hydrological value characterizing the water amount coming per unit of time. However, this indicator characterizes the water volumes that come from all over the catchment area, and does not provide a spatial differentiation between its separate parts (sub-basins). It is customary to use a specific parameter for such an analysis, expressed in millimeters – surface runoff, characterizing the amount of water flowing from the catchment area within any period of time, and equal to thickness of the conditional water layer, evenly distributed over its area.

Values of surface runoff (Y) in the water balance equation (1) for any catchment area are determined primarily by two opposing factors: the amount of precipitation falling over the territory of the catchment basin (precipitation, X), as well as the amount of water evaporating from the catchment area (evaporation, E):

$$X = Y + E \qquad (1)$$

This equation is satisfied only by averaging over a sufficiently long period of time, because a complete hydrological year includes periods of water accumulation and consumption.

Value range of the average annual surface runoff for the period of 1980–2009 for the catchment basin of the Pregolya River varies from 140 to 340 mm, while the minimum values are characteristic of the upper reaches of the basin, and the maximum ones – for the areas close to outlets (Fig. 6). Annual dynamics of the water balance components for the entire catchment area of the Pregolya River is shown in (Fig. 7). With an average (1986–2009) annual amount of precipitation falling over the catchment area of the Pregolya River, their greatest number accounts for the warm season (more than 80 mm). At the same time, this period is also characterized by the maximum evaporation from the land surface, exceeding the amount of precipitation. This leads to low levels of surface runoff. Thus, out of the annual average of 760 mm of precipitation, 530 mm comes back to the atmosphere by means of evaporation and 230 mm – flow with surface runoff.

Jump of the characteristics of surface runoff over the state border is caused by discontinuous in land use structure.

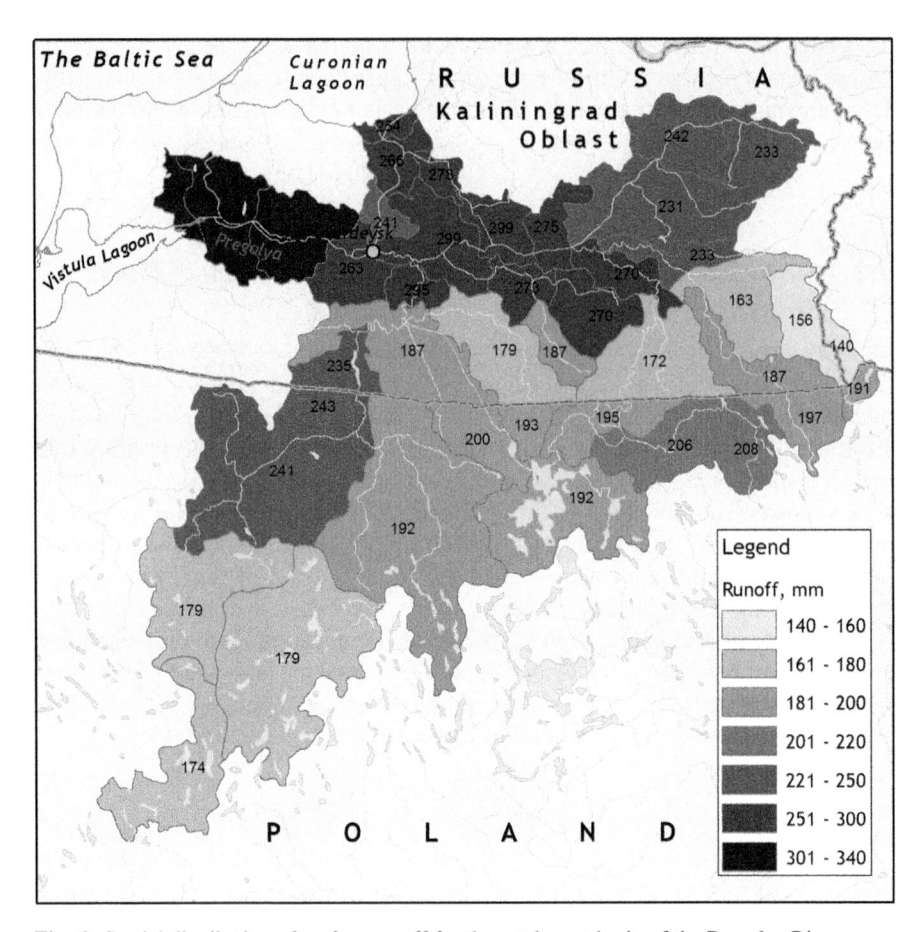

Fig. 6 Spatial distribution of surface runoff for the catchment basin of the Pregolya River

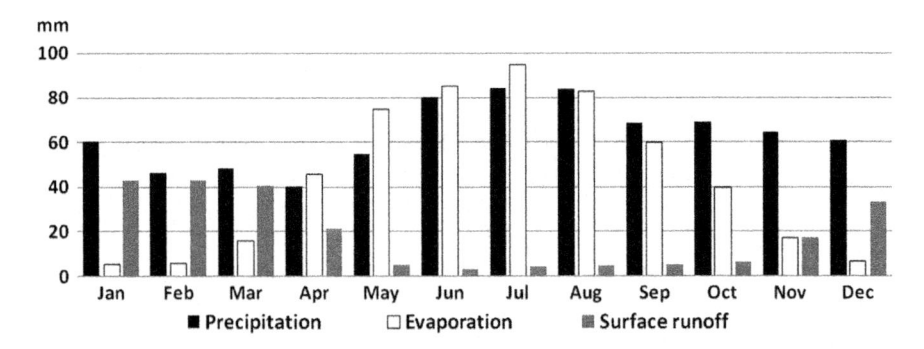

Fig. 7 Average monthly values of water balance components for the entire catchment basin of the Pregolya River, according to calculations data for the period of 1980–2009

3.3 Cross-Border Discharge Value Assessment

Due to the fact that the catchment area of the Pregolya River is located on the territories of different states, the cross-border component of the water mass transfer is of particular interest. Using modeling tool mentioned above the analysis of the water discharge coming from different segments of the hydrographic system was made (Fig. 8). Comparing areas of sub-basins and the water volume coming therefrom, it should be noted that the flow rate is not always increased with increasing area. For example, the upper reaches of the catchment systems have a comparatively lower water content in reference to the lower parts of the catchment area.

In general, 1.46 km^3 of water comes from the territory of neighboring countries through the Kaliningrad Oblast over the watershed of the Pregolya River per year. In addition, next 1.26 km^3 more of surface water is formed on the territory in the Kaliningrad Oblast before the division of the Pregolya River into arms. The flows to the Vistula and Curonian lagoons constitute 1.96 and 1.2 km^3/year, respectively, which in total gives 3.16 km^3/year from the Pregolya River catchment towards the Baltic Sea through both lagoons.

4 Main Conclusions

The watershed of the Pregolya River was divided into 42 interconnected sub-basins in the model installation, the allocation of which was carried out taking into account major tributaries, hydrological stations, as well as the existing state border between the segments of catchment areas in Poland and Russia. This model installation allowed to use data of the permanent hydrological measurements network, assess the discharge rate from the watersheds not included in the constant monitoring, as well as identify a cross-border component of the flow from the side of Poland. Discharge at the outlet sections at the confluence points into the Vistula and Curonian Lagoons was assessed using three model installations that include the catchment area of the Pregolya River before the division into the arms, the catchment area of the Downstream Pregolya, the catchment area of the Deyma River.

Due to the absence of a required set of direct measurement data at the outlet cross-sections, the analysis of average annual values of discharge at the outlet cross-sections of the Pregolya River and its tributaries was carried out based on numerical simulation results through the runoff model HYPE for a period of 30 years (1979–2009).

The correlation coefficient (R) between measured and simulated discharges at the control cross-section of the Pregolya River (measurement point in Gvardeysk) was 0.79. Average measured discharge is 90 m^3/s, the simulated one – 89 m^3/s. Correlation coefficients for separate catchment areas of major tributaries of the

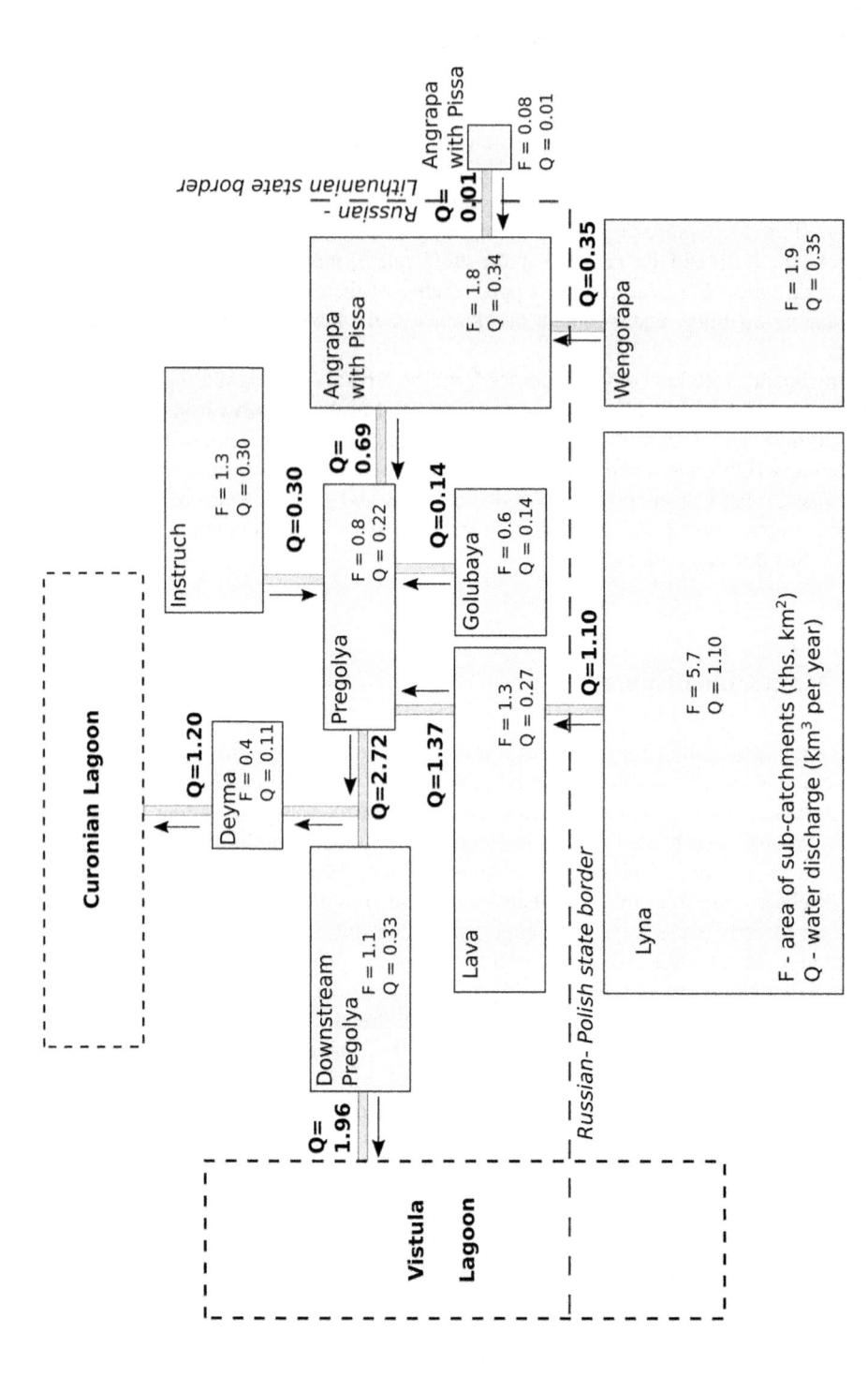

Fig. 8 Hydrographic diagram of the catchment basin of the Pregolya River, taking into account a cross-border component and the runoff from separate sub-basins. Conditional sizes of the sub-basins in the diagram are proportional to their areas

Pregolya are slightly lower. For the basin of the Angrapa River, $R = 0.77$, for the sub-basin of the Pissa River, $R = 0.72$, for the sub-basin of the Lyna-Lava River, $R = 0.75$. A low coefficient is noted for the Instruch River ($R = 0.55$), due to the influence of local factors. However, in general for all the sub-basins, the difference between measurement data of the water discharge and calculation results does not exceed 10%.

In general, 1.46 km^3 of water per year comes to the Kaliningrad Oblast from the territory of watershed of the Pregolya River in the neighboring countries (Poland and Lithuania). In addition, next 1.26 km^3 more of water is formed in the catchment before the division of the Pregolya River into arms. Each individual sub-basin of the Pregolya River major tributaries provides through outlet section at the confluence: the Lyna-Lava River – 1.37 km^3/year, the Wengorapa-Angrapa River – 0.69 km^3/year, the Instruch River – 0.3 km^3/year, the Golubaya River – 0.14 km^3/year. The mean annul flows to the Vistula and Curonian lagoons were estimated as 1.96 and 1.2 km^3/year, respectively, in total it gives the flow of 3.16 km^3/year towards the Baltic Sea through both lagoons from the Pregolya River catchment.

The surface runoff parameter obtained as a result of model calculations shows that the range of its value varies from 140 to 340 mm/year, where the minimum values are typical for the upper reaches of rivers, and the maximum – for downstream ones. The average annual water balance of the catchment basin of the Pregolya River consists of 760 mm of precipitation, 530 mm of evaporation, and 230 mm of surface runoff.

Given that in long-term average 2.72 km^3 of water a year is flown through the control section of the Pregolya River before its dividing into the arms, then, 1.96 km^3 flows directly into the Vistula Lagoon (via the Downstream Pregolya), and 1.20 km^3 into the Curonian Lagoon (via the Deyma branch). Considering the long-term seasonal mean changes in the water discharge coming into the Vistula and Curonian Lagoons, it should be noted that the well-known ratio of 60% and 40% of the Pregolya River flow, respectively, is valid only for the control section of its division into the arms. In the outlet sections this ratio is retained during periods of winter and spring floods; in the low water period in summer the ratio accounts for 50% to 50% (at the expense of evaporation in the watershed of the Downstream Pregolya).

Acknowledgements Collection of calibration data and preparation of the model installation were made under support of State Assignment of the P.P. Shirshov Institute of Oceanology, Russian Academy of Sciences, simulations and analysis were fulfilled under the support of the BONUS Project Soils2Sea "Reducing nutrient loadings from agricultural soils to the Baltic Sea via groundwater and streams" and its complimentary grant RFBR-Bonus 14-05-9173015.

References

1. Domnin D, Chubarenko B, Lewandowski A (2015) Vistula Lagoon catchment: atlas of water use. Exlibris Press, Moscow, p 106. ISBN: 978-5-9900699-4-7
2. Silich MV (1971) Water balance of the Lagoon. In: Lazarenko N, Majevski AL (eds) Hydrometeorological regime of the Vistula Lagoon. Hydrometeoizdat, Saint Petersburg, pp 143–172. (in Russian)
3. Markova LL, Nechay IY (1960) Hydrological essay of the mouth areas of the Neman and Pregolya Rivers. In: Proceedings of the Russian State Oceanographic Institute, vol 49. Russian State Oceanographic Institute, Moscow, pp 118–187 (in Russian)
4. Domnin DA, Pilipchuk VA, Karmanov KV (2013) Formation of the indlow of the brackish waters in the lagoon-estuarine system of the Vistula Lagoon – Pregolya River – Deyma River – Curonian Lagoon as a result of wind-driven waves. Nat Eng Sci 6:206–211. (in Russian)
5. Chubarenko B, Domnin D (2008) International and inner transboundary river basins in the Kaliningrad Oblast, South-Eastern Baltic. Integrated water management: practical experiences and case study. Earth Environ Sci 80:309–321
6. Chubarenko BV, Esyukova EE, Domnin DA, Leytzina LV (2014) Assessment of the impacts of possible climate change on the Vistula Lagoon (Baltic Sea) and its catchment basin. Process Geol Environ 1(1):170–178. (in Russian)
7. Hesse C, Krysanova V, Stefanova A, Bielecka M, Domnin D (2015) Assessment of climate change impacts on water quantity and quality of the multi-river Vistula Lagoon catchment. Hydrol Sci J 60(5):890–911
8. Domnin DA, Sokolov AN (2014) Simulation of the river runoff from the catchment area of the Vistula Lagoon and flowing of the brackish water into the mouth of the Pregolya River. Transactions of the Kaliningrad State Technical University, vol 35. Kaliningrad State Technical University, Kaliningrad, pp 11–20 (in Russian)
9. Donnelly C, Dahné J, Rosberg J, Strömqvist J, Yang W, Arheimer B (2010) High-resolution, large-scale hydrological modelling tools for Europe. In: Proceedings of the 6th world friend conference, Fez, Morocco, Oct 2010. IAHS Publication, p 340
10. Lindström G, Pers CP, Rosberg R, Strömqvist J, Arheimer B (2010) Development and test of the HYPE (hydrological predictions for the environment) model – a water quality model for different spatial scales. Hydrol Res 41(3-4):295–319

Hydrobiological Characteristics of Water Bodies in the Kaliningrad Region

M.N. Shibaeva, E.A. Masyutkina, and S.V. Shibaev

Abstract This chapter summarizes the research on zooplankton and zoobenthos in 81 water bodies in the Kaliningrad region from 2006 to 2014. A complete list of identified zooplankton and mezo-zoobenthos organisms from the current period is provided. The structure and quantitative development of these organisms in different water bodies, as well as their indicative properties, are discussed. Water quality was determined using the saprobiological method of Pantle and Buck with the Sladochek modification.

Keywords Kaliningrad region, Water bodies, Zoobenthos, Zooplankton

Contents

1 Introduction	285
2 Material and Methods	286
3 Results and Discussion	288
3.1 Zooplankton Conditions in Inland Waters	288
3.2 Zoobenthos in the Inland Waters	296
4 Conclusions	313
References	315

1 Introduction

The Kaliningrad region of Russia lies in a zone of excessive moisture, with low-lying and flat terrain. The predominance of clay-loam soil and clay rocks on the surface explains the region's extensive river network and numerous lakes. There are more

---

M.N. Shibaeva (✉), E.A. Masyutkina, and S.V. Shibaev
VPO "Kaliningrad State Technical University", 1, Sovetsky Prospekt, Kaliningrad 236022, Russia
e-mail: msh@klgtu.ru; masyutkinaea@gmail.com

than 4,500 rivers in the region, with a total length exceeding 12,500 km. Most of the rivers are small, with a length of less than 100 km. The largest river in the region is the Neman, with a total length of 937 km and a basin area of 98,200 km². However, only 107 km of this river lies in Kaliningrad territory. The region also has more than 4,000 lakes with a total area of approximately 6.7 km². Most lakes in the region are small, with an area less than 10 ha and a depth less than 10 m. The largest lake in the region is Vishtynetskoe, at more than 1,700 ha in area and 52 m in depth [1–5].

The identification of zooplankton and zoobenthos organisms in the Kaliningrad region can serve as the basis for the ecological mapping of its water bodies. As bioindicators, the properties of zooplankton and zoobenthos provide information not only about the current state of the environment but about also the possibility of future changes. The initial stage of this complex work is to determine the richness of the species, their quantitative development, the establishment of a species structure in a particular body of water, and the occurrence frequency of a species in a body of water. With these data, it will be possible to evaluate the region's water quality.

Species richness is an important factor in the stability of natural ecological systems. The composition of zooplankton and benthic communities is determined by many variables, including the amount of organic matter in the water and soil; the structural features of the reservoir bed; the transparency, depth, and flow at a given point; and hydrochemical and hydrological features. This chapter focuses on the species composition of zooplankton and zoobenthos and provides an exhaustive list of species in water bodies of the Kaliningrad region. It also uses the quantitative development and structures of the species to determine water quality.

2 Material and Methods

This chapter is based on research of complex fisheries conducted by Kaliningrad State Technical University from 2006 to 2014. Over these 9 years, we studied more than 80 rivers and lakes (Fig. 1); in the process, we gathered more than 700 samples of zoobenthos and 500 samples of zooplankton. Sixty percent of these water bodies were never previously investigated, while others were studied on an exploratory basis (one to three times prior). Monitoring was conducted regularly for a small number of water bodies, including the Neman River, Sesupe River, Instruch River, Tylzha River, and Vishtynetskoe Lake. Sampling and processing were carried out using standard methods [6–11].

Samples of zooplankton were taken from lakes using a Juday net or Molchanov bottle. For samples from rivers, we strained 100 L of water through an Apshteyn net. Zoobenthos samples from lakes were taken using a Petersen grab with soil capture area of 1/40 m²; in rivers, we generally used a scraper with a fixed creeping length. The zooplankton and zoobenthos organisms were put in a 4% formalin solution. Species identification was carried out using specialized determinants [12–19], many of which were commercial products that do not allow us to publish the

Hydrobiological Characteristics of Water Bodies in the Kaliningrad Region

Fig. 1 Diagram of the hydrobiological sample selection

biomass absolute values of zooplankton and zoobenthos here. Therefore, the quantitative characteristics are given as relative units of the average values for the water bodies.

The Kaliningrad region is mainly flat, which means that most rivers flow at a low speed. More rapid rivers flow from Warmia and Vishtynetskoe Hills and the Sambia moraine plateau. Slower-flowing rivers are located in the Neman Lowland plains, Polesie moraine plain, and Pregolskaya lake-glacial plain. Depending on their flow velocity, the rivers and streams in the Kaliningrad region were generally divided into three groups for our study, as follows (Table 1) [3]:

- Group 1: Rivers with slow currents (<0.2 m/s) throughout the year: In some periods, these watercourses have no runoff at all. This group is characterized by muddy, silty-sandy soils with plant residues and a shortage of oxygen, especially in summer and winter periods.
- Group 2: Streams with average flow velocity (0.2–0.4 m/s): During spring and other high floods, the flow velocity of some rivers in this group can increased by up to 1 m/s. In summer and winter, flow velocity is reduced to 0.15 m/s. Soils are mainly sandy, partially silty, and in places shingle.
- Group 3: Streams with high flow velocity throughout the year (>0.4 m/s). The soils of these waters are mostly sand and gravel (or clay, in the case of the Sesupe River). These fleeting streams are characterized by a high oxygen content throughout the year.

Because they are the largest water bodies in the Kaliningrad region, with a wide variety of habitats and therefore fauna, Vishtynetskoe Lake and the Neman River are considered separately. A special group also contains small quarry ponds (Orlovskiy and nameless quarries in Sokolniki near Zelenogradsk and in Sokolniki near Pregel), eight lakes (Borodino, Domashnee, Kamishovoe, Krasnoe, Lesistoe, Marinovo, Voronie, Dubovskoe), Zaton pong, and Pravdinskoye reservoir [20].

Table 1 Classification of watercourses by their flow velocity

Flow velocity		
Group 1: <0.2 m/s	Group 2: 0.2–0.4 m/s	Group 3: >0.4 m/s
East Channel	Matrosovka Channel	Wegorapa River
Western Channel	Aleukia River	Zelenaya River
Masurian Canal	Bol. Moryanka River	Krasnaya River
Channel IN-18-8	Bonuvka River	Pissa River
Channel OC-1	Grayevka River	Putilovka River
Guryevka River	Zabava River	Sesupe River
Deima River	Instruch River (up stream of the village Ulyanovo)	
Zayachia River	Lavya River	
Zelenogradka River		
Zlaya River	Muchnaya River	
Instruch River (down stream of the village Ulyanovo)	Nemonin River	
Kalynivka River	Pravda River	
Kornevka River	Pregel River	
Kurovka River	Prokhladnaya River	
Lobovka River	Rezvaya River	
Mamonovka River	Svetlogorka River	
Medvezia River	Stogovka River	
Nelma River	Struga Barytska River	
Nemoninka River		
Primorskaya River		
Spokoynaya River		
Tovarnaya River		
Tropininka River		
Tylzha River		
Ulitka River		
Khlebnaya River		
Chernaya River		
Chistaya River		
Shirokaya River		

3 Results and Discussion

3.1 Zooplankton Conditions in Inland Waters

In the inland waters of the Kaliningrad region, zooplankton are quite diverse. A total of 178 species have been found (Fig. 2): 46 rotifers (Rotifera), 76 from the order Cladocera, 55 from the subclass copepods (Copepoda), and 1 from the

Fig. 2 The diversity of zooplankton in inland waters of the Kaliningrad region

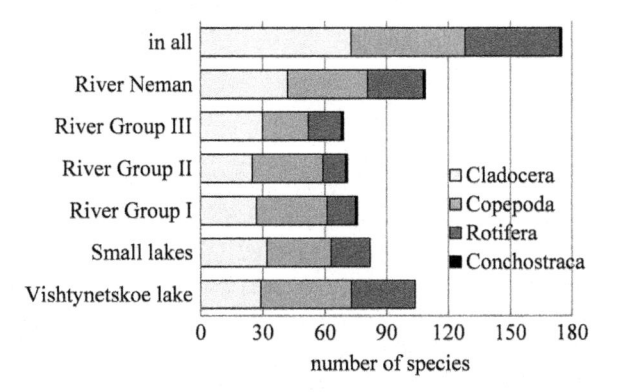

suborder conchostracans (Conchostraca). More than 100 species were found in Vishtynetskoe Lake and Neman River (Fig. 2): Vishtynetskoe Lake had a rich representation of copepods (43 species), whereas Neman River had 44 species of cladocerans. In the small rivers, however, zooplankton diversity was rather low, particularly in fleeting waters.

Only 25 species and forms were common to all groups of waterbodies (Table 2). *Bosmina longirostris* (OF Müller) and *Chydorus sphaericus* (OF Müller) were often found among cladocerans crustaceans, *Mesocyclops leuckarti* (Claus) among copepods, and *Euchlanis dilatata* (Ehrenberg) and *Keratella quadrata* (Müller) among rotifers. Pelagic species and other species typical for slowly flowing waters inhabited Lake Vishtynetskoe [21]: Cladocera *Bythotrephes longimanus* (Leydig), *Leptodora kindtii* (Focke), *Limnosida frontosa* (Sars), *Oxyurella tenuicaudis* (G.O. Sars), *Sida crystallina* (Sididae), rotifers *Conochilus*, and *Polyarthra major* (Carlin). Cladocera crayfish, *Acroperus angustatus* (Sars), were found only in Neman River. Most zooplankton species had an incidence of <25%. Most species were found in just 10% of the water bodies.

In Vishtynetskoe Lake, *Conochilus* rotifers and *Eudiaptomus graciloides* copepods were the most common organisms. Slightly less common were *Thermocyclops oithonoides* and the rotifers *Kellicottia longispina* and *Synchaeta*. The *Keratella quadrata* rotifer was more common in small lakes. Group 1 rivers were more likely to contain *Diacyclops sp*.

In general, the water bodies of the Kaliningrad region are dominated by copepods with regard to abundance and biomass (approximately 54%). We found approximately equal numbers of rotifers and cladocerans (24% and 22%, respectively). However, with regard to biomass, small rotifers were significantly inferior to larger cladocerans. Conchostracans were rare and found in small quantities, accounting for less than 1% of the abundance and biomass of zooplankton (Fig. 3).

The abundance and biomass of zooplankton is naturally higher in lakes than in rivers (Fig. 4), particularly in small lakes. In Vishtynetskoe Lake, these values are approximately 2.5–3 times less than in small lakes. In rivers with slow currents and

Table 2 The species composition of zooplankton found in the studied water bodies of the Kaliningrad Region

No.	Group	Species	Water bodies					
			LV	SL	R1	R2	R3	NR
1	Cladocera	*Acroperus angustatus* (Sars)						+
2		*Acroperus harpae* (Baird)		+		+		
3		*Alona affinis* (Leydig)			+	+	+	+
4		*Alona costata* (G.O. Sars)	+					
5		*Alona guttata* (Sars)					+	+
6		*Alona quadrangularis* (O.F. Müller)	+	+				+
7		*Alona rectangula* (G.O. Sars)	+	+	+	+	+	+
8		*Alona sp.* (Baird)			+		+	+
9		*Alona weltneri* (Keilhack)		+				
10		*Alonella exigua* (Lilljeborg)					+	
11		*Alonella nana* (Baird)	+	+	+			
12		*Alonopsis ambigua* (Lilljeborg)		+				+
13		*Alonopsis elegans* (Kurz)			+			
14		*Bosmina longirostris* (O.F. Müller)	+	+	+	+	+	+
15		*Bosmina sp.* (O.F. Müller)	+		+			
16		*Bythotrephes longimanus* (Leydig)	+					
17		*Cercopagis sp.* (Sars)				+		
18		*Ceriodaphnia affinis* (Lilljeborg)	+	+		+	+	+
19		*Ceriodaphnia laticaudata* (P.E. Müller)		+				+
20		*Ceriodaphnia pulchella* (G.O. Sars)	+	+			+	+
21		*Ceriodaphnia quadrangula* (O.F. Müller)						+
22		*Ceriodaphnia reticulata* (Jurine)	+	+			+	+
23		*Ceriodaphnia rotunda* (G.O. Sars)		+	+	+		+
24		*Ceriodaphnia setosa* (Matile)		+				
25		*Ceriodaphnia sp.* (Dana)	+	+			+	+
26		*Chydorus ovalis* (Kurz)	+	+	+	+	+	+
27		*Chydorus piger* (G.O. Sars)	+					
28		*Chydorus sp.* (Kurz)	+		+	+		
29		*Chydorus sphaericus* (O.F. Müller)	+	+	+	+	+	+
30		*Daphnia cucullata* (G.O. Sars)	+	+		+	+	+
31		*Daphnia longispina* (O.F. Müller)	+	+		+	+	+
32		*Daphnia magna* (Straus)					+	
33		*Daphnia pulex* (Leydig)		+	+	+		+
34		*Daphnia sp.* (O.F. Müller)	+					
35		*Diaphanosoma brachyurum* (Liévin)	+	+	+	+	+	+
36		*Diaphanosoma sp.* (Fischer)			+			
37		*Disparalona rostrata* (Koch)		+	+	+	+	+
38		*Eubosmina coregoni* (Baird)	+	+		+	+	+
39		*Eurycercus glacialis* (Lilljeborg)			+			
40		*Eurycercus lamellatus* (O.F. Müller)		+			+	

(continued)

Table 2 (continued)

No.	Group	Species	LV	SL	R1	R2	R3	NR
41		*Eurytemora affinis* (Poppe)				+		+
42		*Graptoleberis testudinaria* (Fischer)			+		+	+
43		*Ilyocryptus acutifrons* (Sars)						+
44		*Ilyocryptus agilis* (Kurz)			+			+
45		*Ilyocryptus sordidus* (Liévin)					+	+
46		*Ilyocryptus sp.* (G. O. Sars)					+	+
47		*Latona setifera* (O.F. Müller)						+
48		*Leptodora kindtii* (Focke)	+	+				
49		*Leydigia acanthocercoides* (Fischer)				+		+
50		*Leydigia leydigi* (Schoedler)	+	+	+			+
51		*Leydigia sp.* (Kurtz)				+		
52		*Limnosida frontosa* (Sars)	+					
53		*Macrothrix hirsuticornis* (Norman)	+	+	+	+		
54		*Macrothrix laticornis* (Jurine)				+	+	+
55		*Macrothrix sp.* (Baird)			+		+	
56		*Moina macrocopa* (Straus)				+	+	+
57		*Moina sp.* (Baird)	+	+				
58		*Monospilus dispar* (G.O. Sars)						+
59		*Oxyurella tenuicaudis* (G.O. Sars)	+					
60		*Picripleuroxus laevis* (Sars)		+		+		
61		*Pleuroxus aduncus* (Jurine)	+	+	+	+	+	+
62		*Pleuroxus sp.* (Baird)						+
63		*Pleuroxus striatus* (Schoedler)			+			+
64		*Pleuroxus trigonellus* (O.F. Müller)		+			+	
65		*Pleuroxus truncatus* (O. F. Müller)		+				
66		*Polyphemus pediculus* (L.)	+					+
67		*Pseudochydorus globosus* (Baird)		+				
68		*Scapholeberis mucronata* (O.F. Müller)		+	+	+	+	+
69		*Sida crystallina* (Sididae)	+					
70		*Simocephalus exspinosus* (Koch)				+		
71		*Simocephalus serrulatus* (Koch)						+
72		*Simocephalus sp.* (Schoedler)						+
73		*Simocephalus vetulus* (O.F. Müller)		+	+	+	+	
74	Copepoda	*Acanthocyclops americanus* (Marsh)					+	+
75		*Acanthocyclops bicuspidatus* (Jurine)	+					
76		*Acanthocyclops gigas* (Claus)	+		+	+		+
77		*Acanthocyclops languidoides* (Lilljeborg)						+
78		*Acanthocyclops sp.* (Kiefer)	+	+			+	+
79		*Acanthocyclops venustus* (Norman & Scott)				+		+
80		*Acanthocyclops vernalis* (Fischer)			+	+	+	+
81		*Acanthocyclops viridis* (Jurine)	+	+	+	+	+	+

(continued)

Table 2 (continued)

No.	Group	Species	Water bodies					
			LV	SL	R1	R2	R3	NR
82		*Acanthodiaptomus denticornis* (Wierzejski)	+					
83		*Cyclopoida*				+		
84		*Cyclops abyssorum* (G.O. Sars)	+		+	+		+
85		*Cyclops furcifer* (Claus)	+		+	+		
86		*Cyclops insignis* (Claus)	+		+	+		+
87		*Cyclops kolensis* (Lilljeborg)	+	+	+		+	+
88		*Cyclops lacustris* (Sars.)	+		+			
89		*Cyclops scutifer* (Sars)	+	+		+		+
90		*Cyclops sp.* (O.F. Müller)	+	+	+		+	+
91		*Cyclops strenuus* (Fischer)	+	+	+	+	+	+
92		*Cyclops vicinus* (Uljanin)	+	+				
93		*Diacyclops bicuspidatus* (Claus)						+
94		*Diacyclops sp.* (Kieff.)			+			+
95		*Diaptomus castor* (Jurine)	+		+			+
96		*Diaptomus sp.* (Westwood)	+	+		+		
97		*Ectocyclops phaleratus* (Koch)	+	+		+		+
98		*Eucyclops macruroides* (Lilljeborg)	+	+	+	+	+	+
99		*Eucyclops macrurus* (G.O. Sars)	+	+	+	+	+	+
100		*Eucyclops serrulatus* (Fischer)	+	+	+	+	+	+
101		*Eucyclops sp.* (Claus)		+	+			+
102		*Eucyclops speratus* (Lilljeborg)		+	+	+		
103		*Eudiaptomus coeruleus* (Fischer)	+					
104		*Eudiaptomus gracilis* (Sars)			+			+
105		*Eudiaptomus graciloides* (Lilljeborg)	+	+	+	+		+
106		*Eudiaptomus sp.* (Kiefer)	+	+				
107		*Eurytemora affinis* (Poppe)				+		+
108		*Harpacticoida sp.* (G.O. Sars)	+	+	+	+	+	+
109		*Heterocope appendiculata* (Sars)	+					
110		*Macrocyclops albidus* (Jurine)	+	+	+	+	+	+
111		*Macrocyclops fuscus* (Jurine)	+		+	+		
112		*Megacyclops viridis* (Jurine)	+	+				
113		*Mesocyclops leuckarti* (Claus)	+	+	+	+	+	+
114		*Mesocyclops sp.* (G.O. Sars)	+		+	+	+	+
115		*Metacyclops gracilis* (Lilljeborg)	+			+		
116		*Microcyclops sp.* (Claus)	+			+		+
117		*Microcyclops varicans* (Sars)	+					
118		*Paracyclops affinis* (G.O. Sars)	+		+	+	+	+
119		*Paracyclops fimbriatus* (Fischer)	+	+	+	+	+	+
120		*Paracyclops sp.* (Claus)	+	+	+	+		+
121		*Thermocyclops crassus* (Fischer)	+	+			+	+
122		*Thermocyclops oithonoides* (G.O. Sars)	+	+	+	+	+	+

(continued)

Table 2 (continued)

No.	Group	Species	LV	SL	R1	R2	R3	NR
123		*Copepodite calanoida*	+	+	+	+	+	+
124		*Copepodite cyclopoida*	+	+	+	+	+	+
125		*Copepodite harpacticoida*	+	+	+	+	+	
126		*Nauplius calanoida*	+	+				+
127		*Nauplius cyclopoida*	+	+	+	+	+	+
128		*Nauplius harpacticoida*		+	+	+	+	+
129	Rotifera	*Asplanchna herrickii* (De Guerne)	+	+	+			+
130		*Asplanchna priodonta* (Gosse)	+	+	+	+	+	+
131		*Asplanchna sp.* (Gosse)	+		+	+		+
132		*Bipalpus hudsoni* (Imhof)	+					+
133		*Brachionus angularis* (Gosse)	+		+		+	+
134		*Brachionus calyciflorus* (Pallas)	+	+	+	+	+	+
135		*Brachionus calyciflorus spinosus* (Wierz)		+			+	+
136		*Brachionus diversicornis* (Daday)						+
137		*Brachionus diversicornis homoceros* (Wierz)				+		+
138		*Brachionus leydigi* (Cohn)						+
139		*Brachionus quadridentatus* (Hermann)	+	+	+	+	+	+
140		*Brachionus quadridentatus quadridentatus* (Hermann)		+				
141		*Brachionus sp.* (Pallas)	+					
142		*Cephalodella sp.* (Bory de St. Vincent)						+
143		*Colurella sp.* (Ehrenberg)	+					
144		*Conochilus hipporepis* (Schrank)	+					
145		*Conochilus sp.* (Ehrenberg)	+					
146		*Conochilus unicornis* (Rousselet)	+					
147		*Euchlanis dilatata* (Ehrenberg)	+	+	+	+	+	+
148		*Euchlanis incisa* (Carlin)		+				
149		*Euchlanis sp.* (Ehrenberg)	+	+	+	+	+	+
150		*Filinia longiseta* (Ehrenberg)	+	+	+		+	+
151		*Filinia major* (Colditz)	+	+				+
152		*Filinia sp.* (Bory De Saint Vincent)	+	+		+		
153		*Kellicottia longispina* (Kellicott)	+	+			+	+
154		*Keratella cochlearis* (Gosse)	+	+			+	+
155		*Keratella quadrata* (O.F. Müller)	+	+	+	+	+	+
156		*Keratella sp.* (O.F. Müller)	+					
157		*Lecane luna* (O.F. Müller)	+	+				
158		*Lepadella cyrtopus* (Harring)						+
159		*Lepadella sp.* (Bory De Saint Vincent)	+	+				+
160		*Notholca acuminata* (Ehrenberg)		+	+			
161		*Notholca sp.* (Gosse)	+					

(continued)

Table 2 (continued)

No.	Group	Species	Water bodies					
			LV	SL	R1	R2	R3	NR
162		*Platyias quadricornis* (Ehrenberg)	+	+	+			+
163		*Polyarthra dolichoptera* (Jdelson)			+		+	+
164		*Polyarthra major* (Carlin)	+					
165		*Polyarthra sp.* (Carlin)	+	+				+
166		*Polyarthra vulgaris* (Carlin)	+				+	+
167		*Proales daphnicola* (Thompson)				+		
168		*Rotifera* (Scopoli)			+		+	+
169		*Synchaeta pectinata* (Ehrenberg)	+		+		+	+
170		*Synchaeta sp.* (Wierzejski)	+					+
171		*Synchaeta stylata* (Wierzejski)				+		
172		*Trichocerca capucina* (Wierzejski)	+					
173		*Trichocerca pusilla* (Lauterborn)					+	
174		*Trichocerca sp.* (Lamarck)	+					
175	Conchostraca	*Conchostraca sp.* (L.)			+	+	+	+

LV Vishtynetskoe Lake, *SL* small lakes, *R1* Group 1 rivers, *R2* Group 2 rivers, *R3* Group 3 rivers, *NR* Neman River

the Neman River, the abundance and biomass of zooplankton are 35–40 times less than in small lakes, but slightly higher than in the Group 2 and Group 3 rivers. In the Neman River, most zooplankton are found in the zone of backwater from the Curonian Lagoon (in the Yasnoye settlement). The lowest abundance and biomass rates are generally found in Group 2 and Group 3 rivers.

In all groups of water bodies, with only a few exceptions, copepods were the most abundant organisms (Fig. 5). In Group 1 rivers, they accounted for approximately 80% of the total abundance and biomass. In Group 2 rivers, a large proportion of abundance and biomass consisted of copepods. Approximately 40% of the biomass was cladocerans. In Neman River, more than 40% of the organisms were rotifers, with the biomass being dominated by Cladocera. In small lakes and Vishtynetskoe Lake, cladocerans accounted for less than 20%, with a slightly higher abundance and biomass of rotifers. In small lakes, *Keratella quadrata* and *Kellicottia longispina* were prevalent in number and *Kellicottia longispina* on biomass. The numerically dominant genera in Vishtynetskoe Lake were *Conochilus* and *Synchaeta*, with the large species of *Asplanchna priodonta* dominating the biomass. In addition to rotifers, the copepod *Eudiaptomus graciloides* was also dominant in Vishtynetskoe Lake. In small lakes, the dominant species were the cladocerans *Daphnia cucullata* and copepods *Thermocyclops oithonoides*. In rivers with slow current, the copepods of *Diacyclops sp.* dominated in number; in the Group 2 rivers, cladocerans *Bosmina longirostris*; and in Group 3 rivers,

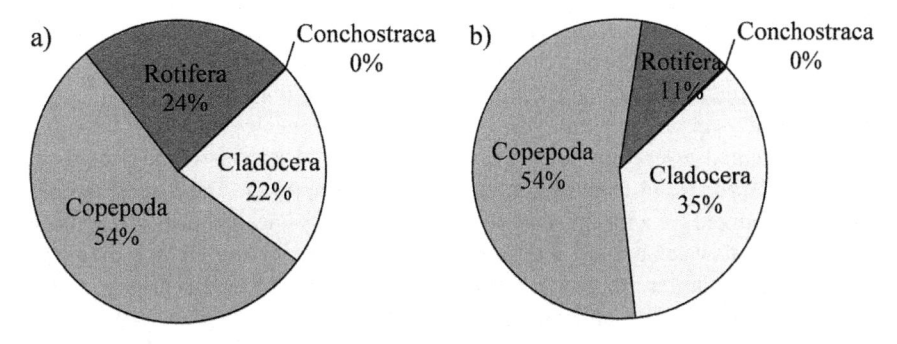

Hydrobiological Characteristics of Water Bodies in the Kaliningrad Region 295

Fig. 3 The ratio of the number (**a**) and biomass (**b**) of individual groups of zooplankton in water bodies of the Kaliningrad region

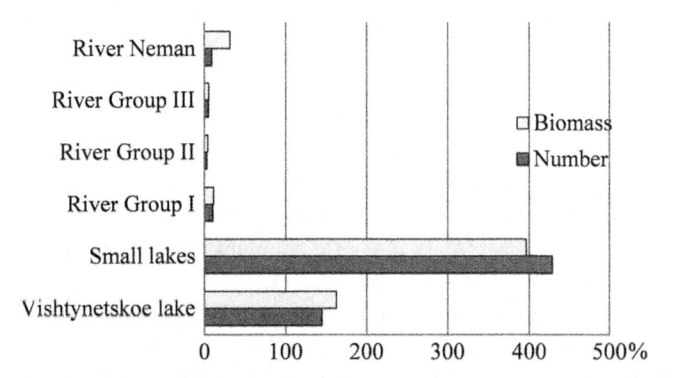

Fig. 4 The ratio of the number and biomass of zooplankton in water bodies of the Kaliningrad region

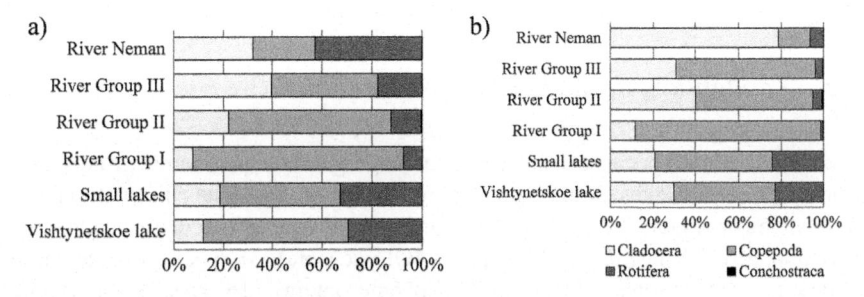

Fig. 5 The structure of zooplankton in water bodies of the Kaliningrad region by number (**a**) and biomass (**b**)

cladocerans *Chydorus sphaericus* and *Bosmina longirostris*. Neman River was dominated by the rotifers *Brachionus calyciflorus* and *Brachionus quadridentatus*. By weight, large but not always numerous species dominated: the copepods *Acanthocyclops viridis* and *Cyclops abyssorum* in Group 1 rivers; the cladocerans

Bosmina longirostris and copepods *Acanthocyclops vernalis* in Group 2 rivers; *Acanthocyclops sp.* and *Mesocyclops leuckarti* in Group 3 rivers; and the cladocerans *Ceriodaphnia affinis* and *Moina macrocopa* in Neman River.

Individual saprobic valence is known for 120 species of zooplankton in 175 investigated reservoirs in the region, including 14 species of oligosaprobes, 59 β-mesosaprobes, 44 α-mesosaprobes, and 3 polysaprobes. The β-mesosaprobic species accounted for approximately 49% of the total species richness of zooplankton in regional water bodies, with α-mesosaprobes accounting for approximately 37%. A slightly larger proportion of α-mesosaprobes were found in Group 1 rivers and small lakes, with somewhat less in Group 2 rivers. Vishtynetskoe Lake contained 10 oligosaprobic species and no polysaprobic groups. In Group 1 and Group 2 rivers, the smallest number of oligosaprobic species was found (Fig. 6).

In all investigated water bodies, α-mesosaprobe species had the greatest numbers. The number of oligosaprobes and polysaprobes was very low (Fig. 7). The only exception was Vishtynetskoe Lake, where oligosaprobic types accounted for 12% of the total zooplankton. Significantly, polysaprobes in Neman River accounted for approximately 7% of the zooplankton population, with approximately 60% of the biomass being α-mesosaprobes. Small lakes and rivers in Group 1 had very similar proportions of zooplankton: approximately 65% of their zooplankton population were α-mesosaprobes and 33% were β-mesosaprobes. Polysaprobes and oligosaprobes accounted for 1–2%.

The Pantle and Buck saprobity index for zooplankton in the water bodies ranged from 1.48 to 1.66. Vishtynetskoe Lake had the lowest value, followed by Group 2 rivers (1.50), small lakes (1.52), Group 1 rivers (1.53), and Group 3 rivers of the third group (1.55). The highest saprobity values were found for Neman River (1.66). Thus, the water quality in Vishtynetskoe Lake and Group 2 rivers was estimated to be a transition from oligosaprobic to β-mesosaprobic, whereas the other groups of water bodies were β-mesosaprobic.

3.2 Zoobenthos in the Inland Waters

We identified 450 species of zoobenthos belonging to nine taxonomic groups in the studied rivers and lakes. Among them were 136 species from the chironomid larvae family Chironomidae, 65 species of mollusks (Mollusca), 50 species of caddis flies (Trichoptera), 39 species of oligochaetes (Oligochaeta), 31 species of dragonflies (Odonata), 24 species of mayflies (Ephemeroptera), 16 species of leeches (Hirudinea), and 13 species of crustaceans (Crustacea). We also classified 76 species in the "other" category, including a few rare water mites (Hydrachnidae), beetles (Coleoptera), bugs (Hemiptera), Diptera larvae (Diptera), stoneflies (Plecoptera), alderfly (Sialidae), and others.

Most zoobenthos species (>200) were found in Neman River, Vishtynetskoe Lake, and fast-flowing rivers (Fig. 8). The smallest number of species were found in small lakes (approximately 130 species). In all groups of water bodies, chironomid

Fig. 6 The ratio of zooplankton species according to their saprobic valence

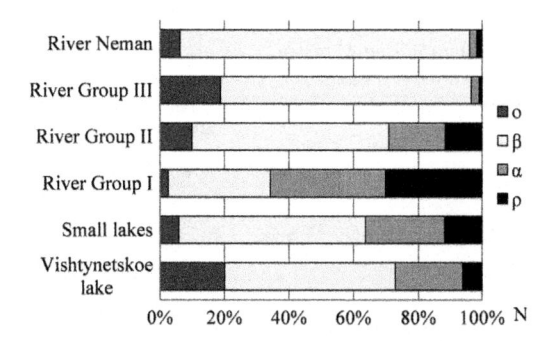

Fig. 7 The percentage of organisms with different saprobic valence values according to zooplankton abundance in the waters of the Kaliningrad region

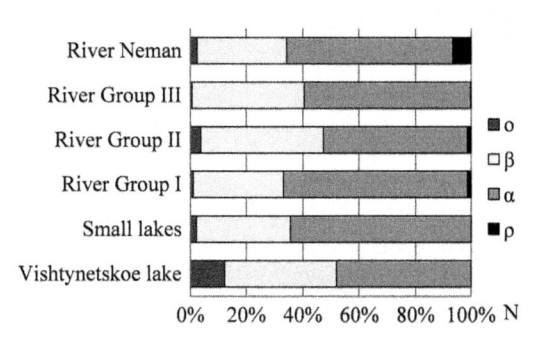

larvae dominated in terms of the number of species [22, 23], accounting for approximately one-third of all zoobenthos. Caddis flies were quite diverse in Vishtynetskoe Lake and fast-flowing rivers. Larvae of mayflies and dragonflies, mollusks, and others were also diverse in this category of rivers. Twenty-six species of oligochaetes were found in Vishtynetskoe Lake and Neman River. In most slowly flowing rivers, there are many species of leeches, oligochaetes, and mollusks.

A total of 58 species of eurybionts were found in all water bodies (Table 2). Among them were chironomids *Cladotanytarsus mancus* (Walk.), *Cryptochironomus defectus* (Kief.), *Microtendipes pedellus* (De Geer), *Polypedilum convictum* (Walk.), *Procladius choreus* (Meig.), Crustaceans *Gammarus lacustris* (G.O. Sars), mayflies *Caenis macrura* (Stephens) and *Ephemera vulgata* (L.), shellfish *Bithynia tentaculata* (L.), *Dreissena polymorpha* (Pall.), and oligochaetes *Potamothrix hammoniensis* (Mich.).

Some species were unique to certain waters (Table 3). For example, the following chironomid larvae were only found in the Neman River: *Kloosia sp.* (Kruseman), *Robackia demeijerei* (Kruseman), *Cricotopus bicinctus* (Meig.), *Orthocladiinae acuticauda* (Pagast), *Paratendipes intermedius* (Tshernovskij), *Virgatanytarsus anduennensis* (Goetghebuer), *Eukiefferiella tschernovskii* (Pankratova), and *Lipiniella arenicola* (Shilova). The chironomids *Chernovskiia ra* (Ulomsky) were found in the Neman River and the estuary of the Sesupe River (which drains into the Neman). Most of these species prefer sandy soils and fast-

Fig. 8 Species diversity of zoobenthos in studied water bodies of the Kaliningrad region

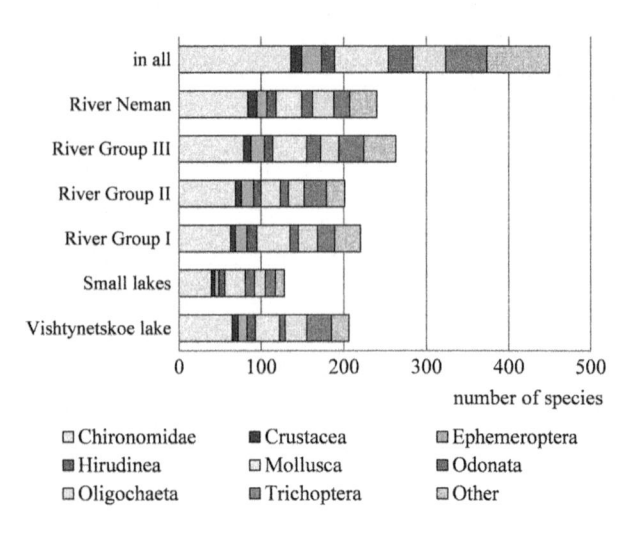

flowing waters. The shellfish *Lithoglyphus naticoides* (C.Pfeiffer) was previously found only in the Neman River, but in recent years it was also found in the estuaries of the Sesupe and Pregel Rivers.

The relict crustacean *Pallasiola quadrispinosa* (Sars) was unique to Vishtynetskoe Lake at depths greater than 15 m. The chironomids *Pseudochironomus prasinatus* (Staeger) have also been repeatedly found only in this lake. The chironomids *Thienemanniella* and *Eukiefferiella, Orthocladius semivirens* (Edw.), *Synorthocladius semivirens* (Kief.), Caddis flies *Apatania auricula* (Forsslund), and water bugs *Aphelocheirus aestivalis* (Fabr.) were found only in fleeting rivers. The leech larvae *Hirudo medicinalis* (L.) and dragonflies *Libellula depressa* (L.) live only in small lakes.

The most common group of invertebrates in the water bodies of the region were mosquito larvae chironomids (Table 4). They accounted for approximately 90–100% of the Lothic and Lenthic systems, among the studied locations. Shellfish were also common (>75%) in all studied reservoirs. The frequency of their occurrence in small lakes was 37%. The average occurrence frequency of crustaceans in all water bodies was approximately 50%. In small lakes, the occurrence of crustaceans decreased to 19%; they preferred Group 2 rivers, where the probability of detection was 93%.

Mayflies are permanent residents of fleeting rivers, where they can be found with a probability of approximately 100%. In other reservoirs, their occurrence decreased to 30–40%. Along the Neman River, their occurrence frequency is less than 20%. Leeches were not common in any water bodies, occurring in only 25–30% of cases. Dragonflies were extremely rare in Vishtynetskoe Lake (6%), but somewhat more common in small lakes, Groups 1 and 2 rivers, and Neman River (20%). In the fast-moving streams, the incidence of dragonfly larvae increased up to 47%.

Table 3 The species composition of zoobenthos found in the studied reservoirs of the Kaliningrad region

			\multicolumn{6}{c}{Group of water bodies}
No.	Group	Species	LV	SL	R1	R2	R3	NR
1	Chironomidae	*Ablabesmyia lentiginosa* (Fries)	+	+	+	+	+	+
2		*Ablabesmyia monilis* (L.)	+	+	+	+	+	+
3		*Ablabesmyia sp.* (Johannsen)		+		+		
4		*Anatopynia plumipes* (Fries)				+		
5		*Brillia longifurca* (Kief.)					+	
6		*Brillia modesta* (Meig.)	+		+	+	+	
7		*Chernovskiia ra* (Ulomsky)					+	+
8		*Chironomus anthracinus* (Zetterstedt)	+					
9		*Chironomus cingulatus* (Meig.)		+	+			+
10		*Chironomus dorsalis* (Meig.)	+	+		+	+	+
11		*Chironomus heterodentatus* (Konstantinov)				+	+	+
12		*Chironomus nigrifrons* (Linevich)						+
13		*Chironomus nigrocaudatus* (Erbaeva)				+		
14		*Chironomus pallidivittatus* (Malloch)		+				
15		*Chironomus plumosus* (L.)	+	+	+	+	+	+
16		*Chironomus sordidatus* (Kief.)			+			
17		*Chironomus sp.* (Meig.)	+	+	+	+	+	+
18		*Chironomus tentans* (Fabr.)	+	+				+
19		*Cladopelma viridula* (L.)	+	+	+	+	+	+
20		*Cladotanytarsus mancus* (Walk.)	+	+	+	+	+	+
21		*Cladotanytarsus sp.* (Walk.)				+		+
22		*Cladotanytarsus vanderwulpi* (Edwards)						+
23		*Clinotanypus nervosus* (Meig.)	+	+	+	+	+	+
24		*Corynoneura scutellata* (Winnertz)		+				
25		*Corynoneura sp.* (Winnertz)				+	+	
26		*Cricotopus algarum* (Kief.)	+	+		+	+	+
27		*Cricotopus bicinctus* (Meig.)						+
28		*Cricotopus fuscus* (Edwards)					+	
29		*Cricotopus latidentatus* (Tshern.)	+					+
30		*Cricotopus silvestris* (Fabr.)	+	+	+	+	+	+
31		*Cricotopus sp.* (Wulp)			+		+	
32		*Cryptochironomus borysthenicus* (Tshernovskij)				+	+	
33		*Cryptochironomus defectus* (Kief.)	+	+	+	+	+	+
34		*Cryptochironomus sp.* (Kieffer)	+	+	+	+	+	+
35		*Cryptotendipes nigronitens* (Edwards)			+	+	+	+
36		*Cryptotendipes sp.* (Lenz)				+	+	
37		*Demeijerea rufipes* (L.)						+
38		*Demicryptochironomus vulneratus* (Zett.)	+		+	+	+	+
39		*Diamesa insignipes* (Kieff.)	+		+			
40		*Diamesa sp.* (Meigen)						+
41		*Dicrotendipes nervosus* (Staeger)	+	+	+	+	+	+
42		*Dicrotendipes tritomus* (Kief.)	+	+	+	+	+	+
43		*Diplocladius cultriger* (Kieff.)				+		
44		*Einfeldia sp.* (Keiffer)						+
45		*Endochironomus albipennis* (Meig.)	+	+	+			+
46		*Endochironomus sp.* (Fabr.)			+			
47		*Endochironomus tendens* (Fabr.)	+	+	+		+	+
48		*Epoicocladius flavens* (Malloch)	+		+	+	+	
49		*Eukiefferiella alpestris* (Goetgh.)					+	
50		*Eukiefferiella bavarica* (Goetgh.)						+
51		*Eukiefferiella claripennis* (Lundbeck)					+	

(continued)

Table 3 (continued)

No.	Group	Species	LV	SL	R1	R2	R3	NR
52		*Eukiefferiella coerulescens* (Kieff.)						+
53		*Eukiefferiella longicalcar* (Kieff.)					+	
54		*Eukiefferiella longipes* (Tshernovskij)					+	
55		*Eukiefferiella sp.* (Thienemann)				+	+	+
56		*Eukiefferiella tschernovskii* (Pankratova)						+
57		*Glyptotendipes barbipes* (Staeg.)	+					
58		*Glyptotendipes glaucus* (Meig.)		+	+		+	+
59		*Glyptotendipes gripekoveni* (Kief.)	+	+	+	+	+	+
60		*Harnischia curtilamellata* (Malloch)			+		+	+
61		*Heterotrissocladius marcidus* (Walker)				+		
62		*Kloosia sp.* (Kruseman)						+
63		*Limnophyes pusillus* (Eaton)						+
64		*Limnophyes sp.* (Eaton)			+	+		
65		*Limnophyes transcaucasicus* (Tshernovskij)				+		
66		*Lipiniella arenicola* (Shilova)						+
67		*Microchironomus tener* (Kieff.)		+		+		+
68		*Microcricotopus bicolor* (Zetterstedt)	+					
69		*Micropsectra praecox* (Meig.)	+		+	+	+	+
70		*Microtendipes pedellus* (De Geer)	+	+	+	+	+	+
71		*Monodiamesa bathyphila* (Kieff.)	+		+	+	+	
72		*Orthocladiinae acuticauda* (Pagast)						+
73		*Orthocladius saxicola* (Kief.)	+			+	+	+
74		*Orthocladius semivirens* (Edw.)				+		
75		*Parachironomus pararostratus* (Lenz)	+	+			+	+
76		*Parachironomus sp.* (Lenz)	+					
77		*Parachironomus vitiosus* (Goetghebuer)				+		+
78		*Paracladopelma camptolabis* (Kief.)	+		+	+	+	+
79		*Paralauterborniella nigrohalteralis* (Malloch)			+			+
80		*Paratanytarsus austriacus* (Kieffer)			+			
81		*Paratanytarsus confusus* (Palmen)					+	+
82		*Paratanytarsus lauterborni* (Kief.)	+	+			+	+
83		*Paratanytarsus sp.* (Thienemann & Bause)	+	+	+	+	+	
84		*Paratendipes albimanus* (Meig.)	+	+	+	+	+	+
85		*Paratendipes intermedius* (Tshernovskij)						+
86		*Pentapedilum exectum* (Kief.)	+	+	+	+	+	+
87		*Polypedilum bicrenatum* (Kieff.)	+	+	+	+	+	+
88		*Polypedilum convictum* (Walk.)	+	+	+	+	+	+
89		*Polypedilum nubeculosum* (Meig.)	+	+	+	+	+	+
90		*Polypedilum pedestre* (Meig.)	+			+		+
91		*Polypedilum scalaenum* (Schrank)	+		+	+	+	+
92		*Polypedilum sp.* (Kief.)				+	+	+
93		*Polypedilum tetracrenatum* (Hirvenoja)	+	+		+	+	+
94		*Potthastia gaedii* (Meig.)				+		
95		*Potthastia longimanus* (Kieff.)	+			+		
96		*Procladius choreus* (Meig.)	+	+	+	+	+	+
97		*Procladius ferrugineus* (Kief.)	+	+	+	+	+	+
98		*Procladius sp.* (Skuse)	+					
99		*Prodiamesa olivacea* (Meig.)	+		+	+	+	
100		*Prodiamesa rufovittata* (Goetgh.)		+	+			
101		*Psectrocladius dilatatus* (Kief.)				+		+
102		*Psectrocladius ishimicus* (Tshernovskij)	+			+		+
103		*Psectrocladius psilopterus* (Kief.)	+	+	+	+		+

(continued)

Table 3 (continued)

No.	Group	Species	LV	SL	R1	R2	R3	NR
104		*Psectrocladius simulans* (Jochannsen)	+	+			+	
105		*Psectrocladius sp.* (Kief.)			+			
106		*Psectrotanypus varius* (Fabricius)	+		+	+		+
107		*Pseudochironomus prasinatus* (Staeger)	+					
108		*Rheotanytarsus exiguus* (Johannsen)					+	+
109		*Rheotanytarsus sp.* (Thienemann & Bause)			+	+	+	+
110		*Robackia demeijerei* (Kruseman)						+
111		*Sergentia coracina* (Zetterstedt)	+					
112		*Sergentia longiventris* (Kief.)	+		+	+	+	
113		*Stenochironomus sp.* (Kief.)			+		+	+
114		*Stictochironomus* "connectens No. 2" (Lipina)	+					+
115		*Stictochironomus crassiforceps* (Kief.)	+		+	+	+	+
116		*Stictochironomus sp.* (Kief.)				+	+	+
117		*Syndiamesa sp.* (Kief.)			+			
118		*Synorthocladius semivirens* (Keiffer)					+	
119		*Tanypus punctipennis* (Meig.)	+				+	
120		*Tanypus vilipennis* (Kief.)	+	+	+	+	+	
121		*Tanytarsus excavatus* (Edwards)	+				+	+
122		*Tanytarsus gregarius* (Kief.)	+	+	+	+	+	+
123		*Tanytarsus lestagei* (Goetghebuer)						+
124		*Tanytarsus lobatifrons* (Kief.)	+	+	+	+	+	+
125		*Tanytarsus medius* (Reiss & Fittkau)			+	+	+	
126		*Tanytarsus mendax* (Kieff.)			+			
127		*Tanytarsus pallidicornis* (Walker)	+		+	+	+	+
128		*Tanytarsus sp.* (Van Der Wulp)	+	+				+
129		*Tanytarsus usmaensis* (Pagast)	+				+	
130		*Tanytarsus verralli* (Goetghebuer)				+	+	+
131		*Thienemanniella flaviforceps* (Kieffer)					+	
132		*Thienemanniella fusca* (Kief.)					+	
133		*Thienemanniella sp.* (Kief.)					+	
134		*Trichocladius inaequalis* (Kief.)			+	+		+
135		*Trissocladius potamophilus* (Tshernovskij)			+	+	+	+
136		*Virgatanytarsus anduennensis* (Goetg.)						+
137	Crustacea	*Argulus foliaceus* (L.)	+					
138		*Asellus aquaticus* (L.)	+	+	+	+	+	+
139		*Astacus astacus* (L.)		+			+	+
140		*Corophium curvispinum* (G. O. Sars)					+	+
141		*Corophium sp.* (Latreille)			+	+	+	+
142		*Gammarus duebeni* (Lilljeborg)						+
143		*Gammarus lacustris* (G. O. Sars)	+	+	+	+	+	+
144		*Gammarus locusta* (L.)	+	+	+	+	+	+
145		*Gammarus pulex* (L.)	+		+	+	+	+
146		*Gammarus sp.* (Fabricius)	+		+	+	+	+
147		*Mysidae sp.* (Haworth)				+		+
148		*Mysis relicta* (Lovén)					+	+
149		*Pallasiola quadrispinosa* (Sars)	+					
150	Ephemeroptera	*Baetis rhodani* (Pict.)			+	+	+	+
151		*Caenis horaria* (L.)	+	+	+	+	+	+
152		*Caenis macrura* (Stephens)	+	+	+	+	+	+
153		*Caenis sp.* (Stephens)			+			
154		*Centroptilum luteolum* (Mull.)	+	+		+	+	+

(continued)

Table 3 (continued)

No.	Group	Species	LV	SL	R1	R2	R3	NR
155		*Cloeon dipterum* (L.)	+	+	+	+	+	+
156		*Cloeon simile* (Eaton)	+	+	+	+	+	
157		*Cloeon sp.* (Leach)			+			
158		*Ephemera lineata* (Eaton)				+	+	+
159		*Ephemera vulgata* (L.)	+		+	+	+	+
160		*Ephemerella ignita* (Poda)	+		+	+	+	+
161		*Eurylophella karelica* (Tiensuu)				+		
162		*Heptagenia flava* (Rostock)					+	
163		*Heptagenia fuscogrisea* (Retzius)	+					+
164		*Heptagenia sp.* (Walsh)	+		+			+
165		*Heptagenia sulphurea* (Mull.)	+				+	+
166		*Leptophlebia marginata* (L.)			+	+		
167		*Leptophlebia sp.* (Westwood)					+	
168		*Leptophlebia vespertina* (L.)				+		
169		*Nigrobaetis niger* (L.)			+	+		
170		*Paraleptophlebia submarginata* (Stheph.)			+	+	+	+
171		*Potamanthus luteus* (L.)	+				+	
172		*Potamanthus sp.* (Pictet)					+	
173		*Procloeon bifidum* (Bengtsson)			+	+	+	
174	Hirudinea	*Batracobdella paludosa* (Carena)			+			
175		*Erpobdella lineata* (O.F.Müller)	+		+		+	+
176		*Erpobdella nigricollis* (Brand.)	+	+	+	+	+	+
177		*Erpobdella octoculata* (L.)	+	+	+	+	+	+
178		*Glossiphonia sp.* (Johnson)	+		+			
179		*Glossiphonia complanata* (L.)	+	+	+	+		+
180		*Glossiphonia concolor* (Apathy)	+		+	+		
181		*Glossiphonia heteroclita* (L.)	+	+	+	+	+	+
182		*Haementeria costata* (Fr. Müller)						+
183		*Helobdella stagnalis* (L.)	+	+	+	+	+	+
184		*Hemiclepsis marginata* (O.F.Müller)	+		+	+	+	+
185		*Hirudo medicinalis* (L.)		+				
186		*Piscicola fasciata* (Koll.)	+		+	+	+	+
187		*Piscicola geometra* (L.)					+	+
188		*Protoclepsis maculosa* (Rathke)		+	+	+	+	+
189		*Protoclepsis sp.* (Livanow)				+		
190	Mollusca	*Acroloxus lacustris* (L.)		+	+		+	
191		*Acroloxus sp.* (Beck)	+					
192		*Amesoda solida* (Normand)			+		+	
193		*Ancylus fluviatilis* (O. F. Müller)					+	
194		*Anisus acronicus* (Ferrusac)					+	
195		*Anisus contortus* (L.)	+		+			
196		*Anisus dispar* (Westerlund)	+				+	
197		*Anisus sp.* (Studer)			+		+	
198		*Anisus vortex* (L.)			+			
199		*Anisus vorticulus* (Troschel)	+					
200		*Anodonta cygnea* (L.)			+			+
201		*Anodonta piscinalis* (Nilsson)				+	+	+
202		*Anodonta sp.* (Lamarck)		+	+			
203		*Anodonta stagnalis* (Gmel)		+		+		+
204		*Anodonta subcircularis* (Clessin)			+	+		+
205		*Bithynia leachii* (Shepp.)	+		+	+	+	+
206		*Bithynia sp.* (Leach)					+	

(continued)

Table 3 (continued)

No.	Group	Species	LV	SL	R1	R2	R3	NR
207		*Bithynia tentaculata* (L.)	+	+	+	+	+	+
208		*Borysthenia naticina* (Menke)	+	+	+	+		+
209		*Choanomphalus rossmaessleri* (Schmidt)						+
210		*Dreissena polymorpha* (Pall.)	+	+	+	+	+	+
211		*Euglesa sp.* (Jenyns)	+	+	+	+	+	+
212		*Hippeutis sp.* (Charpentier)					+	
213		*Lithoglyphus naticoides* (C.Pfeiffer)			+	+		+
214		*Lymnaea auricularia* (L.)	+	+	+		+	+
215		*Lymnaea corvus* (Gmelin)						+
216		*Lymnaea fusca* (C.Pfeiffer)		+		+		
217		*Lymnaea glabra* (L.)		+				
218		*Lymnaea glutinosa* (Mull.)	+		+	+		
219		*Lymnaea intermedia* (Lamarck)		+	+			
220		*Lymnaea lagotis* (Schac.)	+	+			+	
221		*Lymnaea ovata* (Drap.)	+	+	+	+	+	+
222		*Lymnaea patula* (Costa)	+		+		+	
223		*Lymnaea sp.* (Lamarck)	+		+			+
224		*Lymnaea stagnalis* (L.)	+	+		+	+	+
225		*Lymnaea turricula* (Held)	+					
226		*Musculium ryckholti* (Normand)					+	
227		*Neopisidium sp.* (Odhner)			+		+	
228		*Physa fontinalis* (L.)	+		+		+	
229		*Pisidium amnicum* (Mull.)	+	+	+	+	+	+
230		*Pisidium inflatum* (Muhlfeld)				+		
231		*Planorbarius banaticus* (Lang)			+		+	
232		*Planorbarius corneus* (L.)	+		+			
233		*Planorbarius grandis* (Dunker)	+		+			
234		*Planorbis carinatus* (Mull.)			+			
235		*Planorbis planorbis* (L.)	+	+	+		+	+
236		*Pseudanodonta complanata* (Rossmaessler)					+	+
237		*Pseudanodonta kletti* (Rossmaessler)		+		+		+
238		*Segmentina nitida* (O. F. Müller)	+				+	
239		*Sphaerium corneum* (L.)			+		+	+
240		*Sphaerium nitidum* (Clessin)	+	+	+	+	+	+
241		*Sphaerium rivicola* (Lamarck)		+	+		+	+
242		*Sphaerium sp.* (Scopoli)			+		+	+
243		*Theodoxus fluviatilis* (L.)			+	+	+	+
244		*Unio longirostris* (Rossm.)			+			
245		*Unio ovalis* (Montagu)		+	+	+	+	+
246		*Unio pictorum* (L.)		+	+	+	+	+
247		*Unio tumidus* (Philipsson)					+	
248		*Valvata ambigua* (West.)	+	+	+	+	+	+
249		*Valvata depressa* (C.Pfeiffer)	+	+	+	+	+	
250		*Valvata piscinalis* (Mull.)	+	+		+	+	
251		*Valvata planorbulina* (Paladilhe)	+				+	
252		Valvatidae sp.						+
253		*Viviparus contectus* (Millet)		+	+		+	+
254		*Viviparus viviparus* (L.)	+	+	+	+	+	+
255	Odonata	*Aeshna affinis* (van der Linden)					+	+

(continued)

Table 3 (continued)

No.	Group	Species	LV	SL	R1	R2	R3	NR
256		*Aeshna cyanea* (Mull.)		+				
257		*Aeshna grandis* (L.)			+		+	+
258		*Aeshna viridis* (Eversmann)			+		+	+
259		*Calopteryx splendens* (Harris)			+	+	+	+
260		*Calopteryx virgo* (L.)			+	+	+	+
261		*Coenagrion hastulatum* (Charpentier)					+	
262		*Coenagrion ornatum* (Charp.)		+	+		+	
263		*Coenagrion puella* (L.)	+	+			+	+
264		*Coenagrion pulchellum* (Vander Linden)	+		+		+	+
265		*Coenagrion sp.* (Kirby)	+	+	+			
266		*Cordulia aenea* (L.)	+					
267		*Erythromma najas* (Hansemann)		+			+	
268		*Gomphus vulgatissimus* (L.)					+	+
269		*Ischnura elegans* (Vander Linden)		+		+		+
270		*Ischnura pumilio* (Charpentier)		+			+	+
271		*Lestes sponsa* (Hansemann)					+	
272		*Leucorrhinia caudalis* (Charpentier)				+		
273		*Leucorrhinia pectoralis* (Charpentier)		+	+			+
274		*Libellula depressa* (L.)		+				
275		*Libellula quadrimaculata* (L.)	+					
276		*Libellula sp.* (L.)	+					
277		*Onychogomphus forcipatus* (L.)					+	
278		*Ophiogomphus cecilia* (Fourcroy)					+	+
279		*Platycnemis pennipes* (Pall.)		+		+	+	+
280		*Pyrrhosoma nymphula*				+		
281		*Somatochlora metallica* (Vand.)	+		+	+		
282		*Stylurus flavipes* (Charpentier)			+	+	+	+
283		*Sympecma fusca* (Vander Linden)				+		
284		*Sympetrum flaveolum* (L.)					+	
285		*Sympetrum sp.* (Newman)		+				
286	Oligochaeta	*Aulodrilus pigueti* (Kowalewski)	+		+		+	+
287		*Aulodrilus sp.* (Bretscher)				+		
288		*Chaetogaster diastrophus* (Gruithuisen)	+					
289		*Chaetogaster limnaei* (Baer)	+				+	+
290		*Enchytraeidae sp.* (Vejdovský)			+			
291		*Fridericia sp.* (Michaelsen)						+
292		*Isochaetides michaelseni* (Lastock)	+	+	+	+	+	+
293		*Isochaetides newaensis* (Mich.)	+		+		+	+
294		*Isochaetides sp.* (Hrabe)					+	
295		*Limnodrilus claparedeanus* (Ratzel)		+	+	+	+	+
296		*Limnodrilus hoffmeisteri* (Clap.)	+	+	+	+	+	+
297		*Limnodrilus sp.* (Claparede)	+	+	+	+	+	+
298		*Limnodrilus udekemianus* (Clap.)	+		+	+	+	+
299		*Lumbricidae sp.* (пусто)			+			
300		*Lumbriculus sp.* (Grube)				+	+	
301		*Lumbriculus variegatus* (Mull.)	+		+	+		+
302		*Naididae sp.*	+		+			+
303		*Nais barbata* (Mull.)						+

(continued)

Table 3 (continued)

No.	Group	Species	LV	SL	R1	R2	R3	NR
304		*Nais simplex* (Piguet)				+		
305		*Nais sp.* (O.F. Müller)		+	+	+		
306		*Nais variabilis* (Piquet)	+		+		+	+
307		*Ophidonais serpentina* (O. F. Müller)			+			+
308		*Peloscolex ferox* (Eisen)	+	+	+	+	+	+
309		*Potamothrix hammoniensis* (Mich.)	+	+	+	+	+	+
310		*Potamothrix moldaviensis* (Vejdovsky)	+	+		+	+	+
311		*Potamothrix sp.* (Vejdovský & Mrazek)	+	+	+	+	+	
312		*Pristina aequiseta* (Bourne)	+					
313		*Pristina bilobata* (Bretscher)						+
314		*Propappus sp.* (Mich.)	+					
315		*Propappus volki* (Mich.)	+					
316		*Psammoryctides albicola* (Mich.)	+	+				+
317		*Psammoryctides barbatus* (Grube)	+		+	+	+	+
318		*Psammoryctides sp.* (Hrabe)	+			+		+
319		*Rhyacodrilus sp.* (Bretscher)	+		+		+	+
320		*Stylaria lacustris* (L.)	+	+	+	+	+	+
321		*Stylodrilus sp.* (Claparede)	+		+		+	+
322		*Tubifex sp.* (Mull.)			+	+	+	
323		*Tubifex tubifex* (Mull.)	+	+	+	+	+	+
324		*Uncinais uncinata* (Orsted)	+	+			+	+
325	Trichoptera	*Agraylea multipunctata* (Curtis)	+			+		
326		*Agrypnia pagetana* (Curtis)			+	+	+	+
327		*Anabolia soror* (MacLachlan)	+	+	+	+	+	+
328		*Anabolia sp.* (Stephens)				+		
329		*Apatania auricula* (Forsslund)					+	
330		*Arctopsyche ladogensis* (Kolenati)	+					
331		*Athripsodes aterrimus* (Stephens)	+	+			+	
332		*Athripsodes cinereus* (Curt.)	+				+	+
333		*Athripsodes sp.* (Billberg)	+		+			
334		*Brachycentrus subnubilis* (Curtis)			+	+	+	+
335		*Ceraclea annulicornis* (Stephens)				+		
336		*Cheumatopsyche lepida* (Pictet)						+
337		*Cyrnus flavidus* (McLachlan)	+	+		+	+	+
338		*Ecnomus tenellus* (Rambur)	+	+	+	+	+	+
339		*Goera pilosa* (Fabr.)	+					
340		*Halesus interpunctatus* (Zetterstedt)	+		+	+	+	
341		*Halesus radiatus* (Curtis)				+		
342		*Halesus sp.* (Stephens)				+		
343		*Holocentropus picicornis* (Steph.)	+					
344		*Hydropsyche angustipennis* (Curt.)	+		+			
345		*Hydropsyche ornatula* (MacLachlan)			+	+	+	+
346		*Hydropsyche pellucidula* (Curtis)					+	+
347		*Hydroptila tineoides* (Dalman)					+	+
348		*Leptocerus tineiformis* (Curtis)	+					+
349		Limnephilidae				+		
350		*Limnephilus borealis* (Zetterstedt)			+	+	+	
351		*Limnephilus flavicornis* (Fabr.)	+		+	+		
352		*Limnephilus politus* (MacLachlan)	+	+	+	+	+	+

(continued)

Table 3 (continued)

No.	Group	Species	LV	SL	R1	R2	R3	NR
353		*Limnephilus rhombicus* (L.)	+	+	+	+	+	
354		*Limnephilus sp.* (Leach)	+	+		+	+	+
355		*Molanna angustata* (Curtis)	+		+	+	+	+
356		*Mystacides azurea* (L.)	+	+	+		+	
357		*Mystacides longicornis* (L.)	+	+		+	+	
358		*Neureclipsis bimaculata* (L.)	+		+		+	+
359		*Notidobia ciliaris* (L.)	+			+	+	
360		*Oligostomis reticulata* (L.)	+		+	+		
361		*Oxyethira costalis* (Curt.)	+					
362		*Oxyethira sp.* (Eaton)	+					
363		*Phryganea bipunctata* (Retzius)	+	+	+	+	+	+
364		*Plectrocnemia conspersa* (Curt.)	+					
365		*Polycentropus flavomaculatus* (Pict.)	+	+	+	+	+	+
366		*Potamophylax rotundipennis* (Brauer)			+	+	+	
367		*Potamophylax sp.* (Wallengren)					+	
368		*Rhyacophila nubila* (Zett.)			+	+		
369		*Rhyacophila sp.* (Pictet)				+	+	
370		*Semblis phalaenoides* (L.)		+	+	+	+	
371		*Semblis sp.* (Fabricius)						+
372		*Silo pallipes* (Fabr.)	+			+		
373		*Tinoides waeneri* (L.)	+					+
374		*Triaenodes bicolor* (Curtis)					+	
375	Other	*Agabus sp.* (Leach)	+			+		+
376		*Agabus undulatus* (Schrank)			+		+	+
377		*Aphelocheirus aestivalis* (Fabricius)					+	+
378		*Aphelocheirus sp.* (Westwood)					+	+
379		*Atherix sp.* (Meigen)					+	
380		*Atrichops crassipes* (Meigen)					+	
381		*Aulonogyrus concinnus* (Klug)					+	
382		*Brachyptera risi* (Morton)			+			
383		*Ceratopogonidae sp.* (Grassi)	+	+	+	+	+	+
384		*Chaoborus crystallinus* (De Geer)		+				
385		*Chaoborus sp.* (Lichtenstein)	+	+	+	+	+	+
386		*Coleoptera*						+
387		*Collembola* (Lubbock)	+				+	
388		*Colymbetes sp.* (Clairville)	+		+		+	+
389		*Corixa dentipes* (Thomson)						+
390		*Corixa sp.* (Geoffroy)			+		+	+
391		*Corixidae sp.* (Leach)	+	+		+	+	+
392		*Dicranota bimaculata* (Schummel)	+			+	+	
393		*Donacia sp.* (Fabricius)	+		+			
394		*Dytiscidae* (Leach)			+			
395		*Dytiscus sp.* (L.)					+	
396		*Elmidae sp.* (Curtis)					+	+
397		*Elodes minuta* (L.)		+				
398		*Eriocera sp.* (Macquart)				+		+
399		*Gerris lacustris* (L.)					+	
400		*Haliplus sp.* (Latreille)	+					+
401		*Haliplus varius* (Nicolai)	+					
402		*Hebridae sp.* (Amyot and Serville)				+		
403		*Hebrus sp.* (Curtis)					+	
404		*Helophorus sp.* (Fabricius)						+

(continued)

Table 3 (continued)

No.	Group	Species	LV	SL	R1	R2	R3	NR
405		*Hemiptera sp.* (L.)		+	+	+	+	
406		*Hesperocorixa Sahlbergi* (Fieber)			+			
407		*Hydrachna sp.* (Fabricius)	+	+	+	+	+	+
408		*Hydrometra sp.* (Latreille)					+	
409		*Hydroporus sp.* (Clairville)					+	+
410		*Isoperla grammatica* (Poda)					+	
411		*Isotomidae sp.* (Schäffer)				+		
412		*Laccobius sp.* (Erichson)					+	
413		*Laccophilus sp.* (Leach)			+		+	
414		*Lemoniidae sp.* (Hampson)			+			
415		*Leuctra fusca* (L.)			+			
416		*Limnesia sp.* (C.L.Koch)	+		+	+		
417		*Limnochares aquatica* (L.)				+		
418		*Limoniidae sp.* (Rondani)			+	+		+
419		*Microvelia pygmaea* (Dufour)		+				
420		*Naucoris cimicoides* (L.)					+	
421		*Naucoris sp.* (Geoffroy)			+			+
422		*Nematoda sp.* (Rudolphi)			+	+		+
423		*Nemoura cinerea* (Retzius)			+		+	
424		*Nemoura dubitans* (Morton)			+	+		
425		*Nemoura marginata* (Pictet)					+	
426		*Nemoura sp.* (Latreille)			+	+	+	+
427		*Nepa cinerea* (L.)			+		+	
428		*Nepidae* (Latreille)			+			
429		*Noterus clavicornis* (De Geer)						+
430		*Notiphila sp.* (Fallén)			+			
431		*Notonecta glauca* (L.)	+					+
432		*Notonecta sp.* (L.)	+			+	+	+
433		*Paraponyx sp.* (Hbn.)					+	
434		*Peltodytes caesus* (Duftschmid)					+	
435		*Planaria gonocephala* (Duges)	+		+		+	
436		*Planaria sp.* (O.F. Müller)					+	
437		*Platambus sp.* (C.G. Thomson)						+
438		*Plea minutissima* (Leach)	+					+
439		*Psychoda sp.* (Latreille)				+		
440		*Rhagionidae sp.* (Latreille)				+		
441		*Sialis lutaria* (L.)	+	+	+	+	+	+
442		*Sigara falleni* (Fieber)	+			+		+
443		*Sigara sp.* (Fabricius)				+		+
444		*Simulium sp.* (Latreille)	+		+	+	+	+
445		*Stratiomyidae*			+			
446		*Symplecta sp.* (Meigen)		+				+
447		*Tabanus sp.* (L.)	+		+	+	+	+
448		*Tipula sp.* (L.)			+	+		
449		*Tipulidae sp.* (Latreille)	+					+
450		*Xanthoperla apicalis* (Newman)				+	+	

LV Vishtynetskoe Lake, *SL* small lakes, *R1* Group 1 rivers, *R2* Group 2 rivers, *R3* Group 3 rivers, *NR* Neman River

Table 4 The most common types of zoobenthos in the Kaliningrad region's water bodies

Group of water bodies	Names of the species
Vishtynetskoe Lake	*Asellus aquaticus* (L.), *Bithynia tentaculata* (L.), *Dreissena polymorpha* (Pall.)
Small lakes	*Ceratopogonidae sp.* (Grassi), *Potamothrix hammoniensis* (Mich.), *Chironomus plumosus* (L.)
Group 1 rivers	*Microtendipes pedellus* (De Geer), *Asellus aquaticus* (L.), *Bithynia tentaculata* (L.), *Potamothrix hammoniensis* (Mich.)
Group 2 rivers	*Asellus aquaticus* (L.), *Microtendipes pedellus* (De Geer)
Group 3 rivers	*Caenis macrura* (Stephens), *Ephemera vulgata* (L.), *Bithynia tentaculata* (L.)
Neman River	*Lithoglyphus naticoides* (C. Pfeiffer), *Dreissena polymorpha* (Pall.)

With an increase in water flow, the occurrence of oligochaetes decreased. Therefore, in Vishtynetskoe Lake, small lakes, and slowly flowing rivers, their occurrence exceeded 60%. However, in Groups 2 and 3 rivers and Neman River, the rate did not reach 50%. Caddis larvae were rarely found in small lakes or the Neman River (about 20%). In the fast-moving rivers, on the contrary, caddis flies were found in more than 55% of locations; in other water bodies, their rate was approximately 40%. Representatives of other groups could be found in the Vishtynetskoe Lake and Neman River at a probability of 40%; in other bodies of water, this rate was more than 60%.

Larvae of mayflies, dragonflies, caddis flies, and leeches were not found to be dominant groups in any water bodies of the Kaliningrad region. They accounted for less than 5% of the abundance (Fig. 9). Crustaceans were also relatively few at 7%. The numbers of Oligochaeta and other species (mostly from the Diptera family) in some reservoirs reached 30% but averaged 18% and 9%, respectively, across all bodies of water. Chironomid larvae in different bodies of water accounted for 10% to 45% of the population. Shellfish, depending on the type of water body, accounted for 5% to 80% of the zoobenthos, with an average of approximately 31% across all bodies of water.

Despite the fact that most shellfish did not dominate in numbers, their proportion of the biomass was always greater than 90% (Fig. 10). The share of other groups of zoobenthos accounted for approximately 1% of the total weight. Some reservoirs were small and were dominated numerically by chironomid larvae and oligochaetes. However, these were inferior to the biomass dominance of the few large organisms, such as the larvae of dragonflies, caddis flies, and leeches.

The quantitative characteristics of zoobenthos differed significantly between certain groups of water bodies (Fig. 11). The Neman River had the most significant total number of zoobenthos among the studied reservoirs. Approximately 80% of the total number of zoobenthos in the rivers were mollusks, particularly *Lithoglyphus naticoides* (72%) and *Dreissena polymorpha* (7%). There was no such a number of shellfish in any other water body of the region. The number of zoobenthos, except shellfish, in the Neman River was significantly lower compared

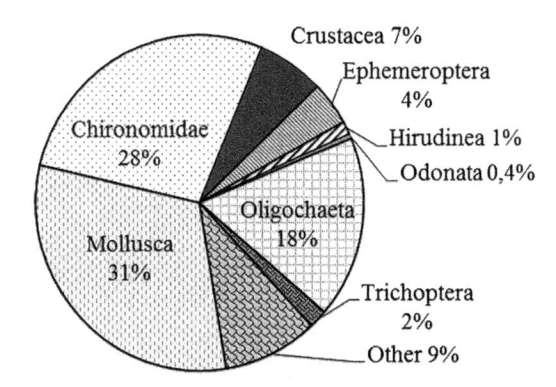

Fig. 9 The ratio of different groups of zoobenthos by number in water bodies of the Kaliningrad region

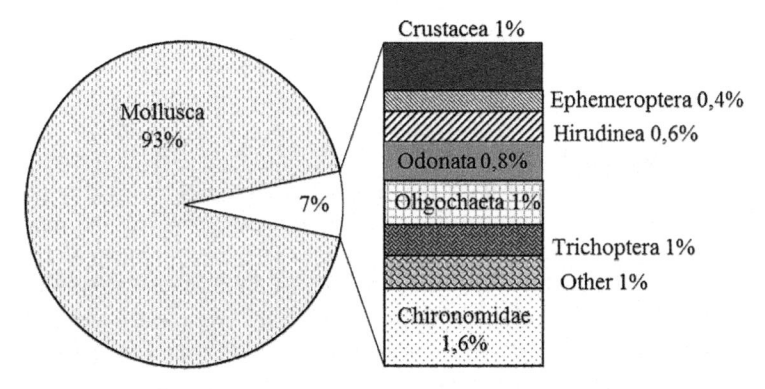

Fig. 10 The ratio of zoobenthos biomass in water bodies of the Kaliningrad region

to other water bodies. Approximately 7% of the Neman River's zoobenthos was made up of chironomids, including *Paratendipes albimanus* and *Polypedilum nubeculosum*. Approximately 3–5% of the river's zoobenthos was crustaceans and oligochaetes (Fig. 12), particularly of the genus Corophium [9].

In Vishtynetskoe Lake, the average total number of zoobenthos was nearly half that of the Neman River (Fig. 11). Here, the zoobenthos consisted of mainly four groups: chironomids, mollusks, crustaceans, and oligochaetes (Fig. 11). The makeup of the organisms in the lake bed was extremely irregular and was determined by a relatively small area of the littoral zone (<30% of the entire area), as well as at a great depth from the coast to the central part [13]. In the littoral zone, the zoobenthos often exceeded 3,000 specimens/m^2; in the central, deepest part of the lake, the abundance was usually not more than 400 specimens/m^2. Shellfish made up approximately 20% of the total number of zoobenthos in Vishtynetskoe Lake. However, in some places where the depth was 10 m, the bottom of the lake was completely covered with zebra mussel. Subdominant species in the littoral and

Fig. 11 The ratio of zoobenthos in water bodies of the Kalinnigrad region

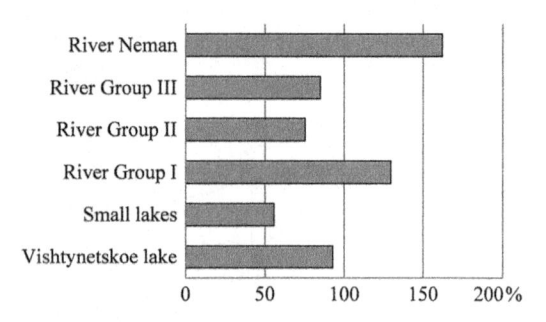

sublittoral zones included the mollusk *Dreissena polymorpha* and crustacean *Asellus aquaticus*. The profundal zone was dominated by the chironomid larvae *Sergentia longiventris* and oligochaetes species *Potamothrix hammoniensis*.

The total number of zoobenthos in small lakes was rather low (Fig. 11), at approximately 50% of the average for all water bodies in the region. The main share consisted of larvae of chironomids, oligochaetes, and organisms from the "other" group, among which midge larvae (*fam. Ceratopogonidae*) were the most common. The predominate species included the oligochaetes *Potamothrix hammoniensis* and *Limnodrilus hoffmeisteri*, chironomid *Polypedilum convictum*, and midge larvae (Fig. 12).

With an increase in water flow, the zoobenthos structure noticeably changed. The number of chironomids and oligochaetes reduced, whereas the number of larvae of mayflies, caddis flies, dragonflies, and other organisms increased (Fig. 12). In slowly flowing rivers, the zoobenthos number was high but did not reach the level of the Neman River. Chironomids and oligochaetes—especially the chironomid species *Polypedilum convictum* and *Chironomus plumosus* and the oligochaete *Potamothrix hammoniensis*—were dominant.

In Group 2 rivers, the total number of benthic species was markedly lower than in Group 1 rivers (Fig. 11). Chironomid larvae and oligochaetes continued to dominate, while the number of shellfish increased (Fig. 12). The species that predominated in slow-flowing rivers were less common. In these rivers, it is difficult to determine the dominant species because they contain zoobenthos that are specific for Groups 1 and 3 rivers.

In the fast-moving rivers, the number of zoobenthos was also high (approximately 85% of the average value for all water bodies in the region). The percentage of shellfish increased to 53% (Fig. 12) in this group of rivers. As a result, the number of zoobenthos in fast-moving rivers without shellfish was significantly less than in Groups 1 and 2 rivers. In addition, there was a small number of chironomids, with a high species diversity in Group 3 rivers. Rheophilic species dominated the zoobenthos here, with numbers that were two or three times higher than in Groups 1 and 2 rivers; these include the larvae of mayflies, caddis flies, and midge of genus *Simulium*.

The largest biomass of zoobenthos was recorded in the Neman River (Fig. 13). This number was significantly lower for small lakes and Vishtynetskoe Lake. In

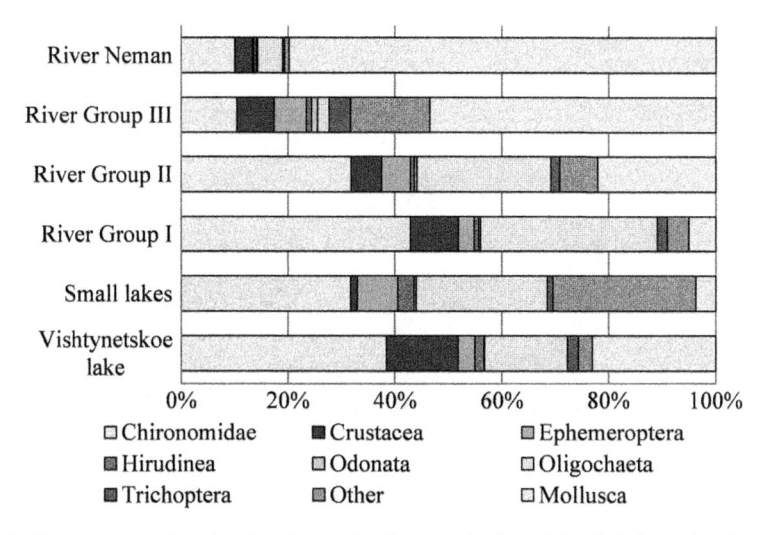

River Neman	
River Group III	
River Group II	
River Group I	
Small lakes	
Vishtynetskoe lake	

0%   20%   40%   60%   80%   100%

☐ Chironomidae ■ Crustacea ☐ Ephemeroptera
■ Hirudinea ☐ Odonata ☐ Oligochaeta
■ Trichoptera ■ Other ☐ Mollusca

Fig. 12 The structure of zoobenthos by number in water bodies of the Kaliningrad region

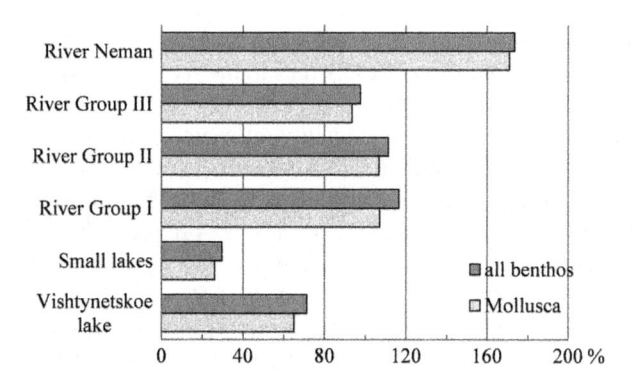

■ all benthos
☐ Mollusca

Fig. 13 The ratio of zoobenthos biomass in water bodies of the Kaliningrad region

general, a river's benthos mass was reduced by 15% with an increasing flow rate. The zoobenthos biomass in small lakes was also rather small. The bulk of the biomass in all water groups was made up by shellfish (more than 90% by weight; approximately 99% in the Neman River). Except for shellfish, the benthos biomass was one of the lowest in the Neman River (Fig. 13). However, the biomass of zoobenthos in the slowly flowing rivers, except for shellfish, was very high (four times higher than in the Neman River). In this group of water bodies, large chironomid larvae and oligochaetes dominated by number and formed the basis of the biomass (Fig. 14).

In other reservoirs, the zoobenthos biomass excepting shellfish was about the same (approximately 1.5–2 times higher than in the Neman River). The biomass of

Fig. 14 The structure of zoobenthos by biomass (excluding shellfish) in water bodies of the Kaliningrad region

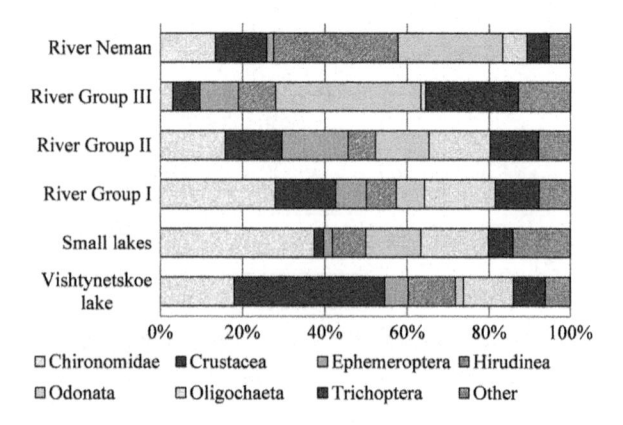

fast-moving rivers was dominated by dragonfly larvae (particularly *Sympetrum flaveolum*) and other organisms; Vishtynetskoe Lake was dominated by crustaceans; and small lakes were dominated by large chironomid larvae (*Chironomus plumosus*). In Group 2 rivers, no group clearly dominated the biomass (without shellfish). Chironomid larvae, mayflies, caddis flies, dragonflies, oligochaetes, and crustaceans made up approximately the same proportion of the biomass. In general, the dominance of chironomid larvae and oligochaetes species in the biomass was replaced by dragonfly larvae and caddis flies in the rivers as flow increased.

Zoobenthos species composition and quantitative development can serve as a reliable means of determining the quality of water, which is widely used in saprobiology [22–24]. The individual saprobic valence is known for 272 of the 450 species of zoobenthos found in the waters of the Kaliningrad region, including 227 species of β-mesosaprobes, 36 species of oligosaprobes, 27 species of α-mesosaprobes, and 5 species of polysaprobes. In all groups of water bodies, β-mezosaprobes accounted for approximately 75% of all species (Fig. 15). Oligosaprobes were found in somewhat larger numbers in Vishtynetskoe Lake and Group 3 rivers (13%). α-Mesosaprobes and polysaprobes were least common in these waters. In small lakes and slow-running rivers, oligosaprobes accounted for only 7–9% of the species; however, these waters had a large number of α-mesosaprobes and all kinds of polysaprobes.

In Vishtynetskoe Lake and Group 3 rivers, the number of oligosaprobes reached the maximum values observed among the studied water bodies—approximately 20% (Fig. 16). α-Mesosaprobes and polysaprobes were also slightly higher in numbers here than in Group 3 rivers. Most zoobenthos in Neman River were β-mesosaprobes (90%). In slowly flowing rivers of Group 1, the smallest number of oligosaprobes were found (3%); however, the number of polysaprobes reached 30% and α-mesosaprobes were at 36%. In small lakes and Group 2 rivers, 60% of the population was α-mesosaprobes (Fig. 16), with 12% polysaprobes. Group 2 rivers had a slightly higher number of oligosaprobes and fewer β-mesosaprobes than in small lakes.

Fig. 15 The ratio of zoobenthos species with different saprobic valence values in water bodies of the Kaliningrad region

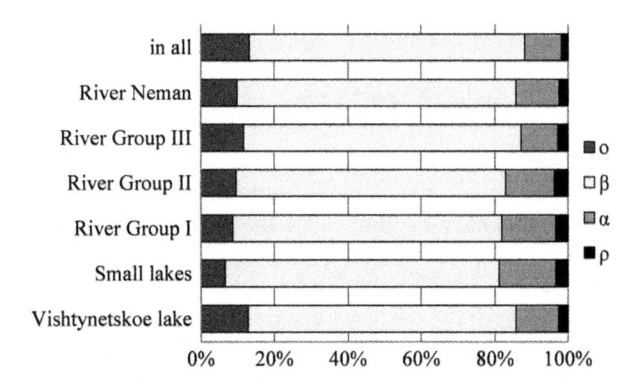

Fig. 16 The ratio of zoobenthos organisms with different saprobic valence values in water bodies of the Kaliningrad region

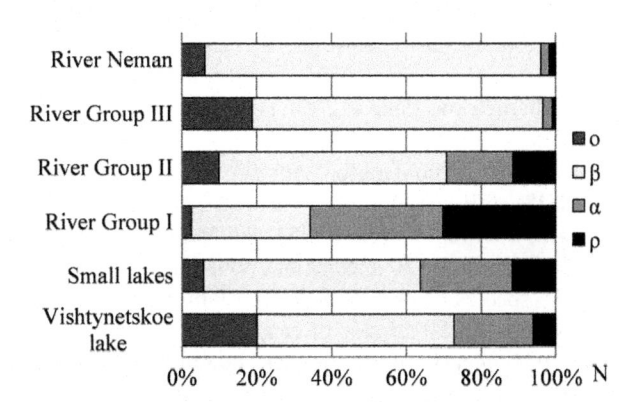

The highest saprobity index value was obtained for slow-flowing rivers in Group 1, at 2.54. Other saprobity index values were 2.47 for small lakes, 2.31 for Group 2 rivers, 2.2 for Neman River and Vishtynetskoe Lake, and 2.0 for Group 3 rivers. The saprobity of Vishtynetskoe Lake markedly increased as the lake's depth increased: from 2.0 in the littoral zone to 2.4 in the deep waters [25].

Thus, we observed ecological laws in action: a reduction in pollution as the water volume of lakes and the water flow of rivers increased. The studied water bodies in Kaliningrad region can be characterized as having moderate organic pollution.

4 Conclusions

1. We identified 178 species of zooplankton organisms, of which 46 species were rotifers (Rotifera), 76 belonged to the order Cladocera (Cladocera), 55 belonged to division of copepods (Copepoda) and 1 was from the suborder conchostracans (Conchostraca). More than 100 species of zooplankton were found in the two

largest water bodies of the region—Neman River and Vishtynetskoe Lake. In other water bodies, the diversity of zooplankton was lower (60–80 species).
2. The abundance and biomass of zooplankton was significantly higher in lakes than in rivers. Small lakes had the highest number of species and biomass. In Vishtynetskoe Lake, these indices were 2.5–3 times lower. With an increase in water flow in rivers, the quantity of zooplankton tended to be zero. The number and biomass in almost all the groups of water bodies were dominated by copepods. The exception to this was Neman River, where copepods were inferior in numbers to rotifers and in weight to Cladocera.
3. In terms of the species diversity of zooplankton, the rivers and lakes of the region were dominated by β-mesosaprobes and α-mesosaprobes (49% and 37% of the total number of species), with 10% oligosaprobes and 2% polysaprobes. α-Mesosaprobes (50–60%) dominated in number, but slightly less than β-mesosaprobes (30–40%). The number of oligosaprobes and polysaprobes was small (1–3%). The greatest number of oligosaprobes were found in Vishtynetskoe Lake (12%). The greatest number of polisaprobes was recorded in the Neman River (7%).
4. The lowest zooplankton saprobity index was recorded for Vishtynetskoe Lake (1.48), followed by Group 2 rivers (1.50), small lakes (1.52), Group 1 rivers (1.53), Group 3 rivers (1.55), and Neman River (1.66).
5. We identified 450 species of zoobenthos in the studied water bodies, including 136 species of chironomid larva family Chironomidae, 65 species of mollusks (Mollusca), 50 species of caddis flies (Trichoptera), 39 species of oligochaetes (Oligochaeta), 31 species of dragonflies (Odonata), 24 species of mayflies (Ephemeroptera), 16 species of leeches (Hirudinea), 13 species of crustaceans (Crustacea), and 76 species that were classified as other, including water mites (Hydrachnidae), beetles (Coleoptera), bugs (Hemiptera), Diptera larvae (Diptera), stoneflies (Plecoptera), and Alderfly (Sialidae). The most diverse zoobenthos were found in the Neman River and fleeting rivers. A smaller variety of zoobenthos was characteristic for small lakes.
6. The quantitative indicators of zoobenthos varied widely among the different water bodies. A large number of zoobenthos was observed in the Neman River, but 80% of the organisms were shellfish. The highest number of organism was noted in slow-flowing rivers that were shorter in length, where chironomid larvae and oligochaetes predominated. In Groups 2 and 3 rivers and Vishtynetskoe Lake, the number of benthic organisms was approximately half that in Neman River. The lowest rates by number were observed in the small lakes.
7. The largest zoobenthos biomass was recorded in the Neman River. This rate was significantly lower in the smaller rivers and Vishtynetskoe Lake. In general, as the flow rate of a river increase, the benthos mass reduced by 15%. The zoobenthos biomass was also low in small lakes. The bulk of the biomass in all groups of water bodies consisted of shellfish.
8. Approximately 75% of the zoobenthos species were β-mesosaprobes, 13% were oligosaprobes, 10% were α-mesosaprobes, and 2% were polysaprobes. By

number, most zoobenthos in water bodies of the area consisted of β-mesosaprobes (30–60%, but up to 90% in the Neman River and Group 3 rivers). The zoobenthos of slow-flowing rivers was made up of nearly equally numbers of β-mesosaprobic, α-mesosaprobic, and polysaprobic species (30–35%). Many α-mesosaprobes also lived in the small lakes, Group 2 rivers, and Vishtynetskoe Lake (20–25%). Polysaprobes made up 12% of the total number of species in the small lakes and Group 1 rivers; however, in other water bodies, their numbers were rather small (1–2%). Oligosaprobe reached their highest numbers in Vishtynetskoe Lake and Group 3 rivers (20%), reaching no more than 5–10% to other water bodies.

9. The highest overall saprobity value for zoobenthos was recorded for Group 1 rivers (2.54), followed by small lakes (2.47), Group 2 rivers (2.31), Neman River and Vishtynetskoe Lake (2.20), and Group 3 rivers (2.00). Thus, most of the studied water bodies in the Kaliningrad region had moderate organic pollution.

Acknowledgements We thank our colleagues from the Department of Ichthyology and Ecology at Kaliningrad State Technical University for their assistance with field research on water bodies in the Kaliningrad region.

References

1. Orlenok V (ed) (2002) Geographic Atlas of the Kaliningrad Region/Ch. KSU Publishing House; CNIT, Kaliningrad, 276 p
2. Abakumov VA (1983) Guidelines for methods of hydrobiological analysis of surface water and sediments. L.: Gidrometeoizdat, 239 p
3. Litvin VM (ed) (1999) Kaliningrad Region: Sketches of nature/comp. D. Ya. Berembeyn; scientific, 2nd edn. Ext. and exp. Yantarnyi Skaz, Kaliningrad, 229 p
4. Tylik KV, Shibaev SV (eds) (2008) Lake Vishtynetskoe. Izd "SP Mishutkina", Kaliningrad, 144 p
5. Regional geography of Russia (2005) Kaliningrad Region: a manual for students enrolled in the geographical specialties. Izd Immanuel. Kant, Kaliningrad, 259 p
6. Shibaev SV, Hlopnikov MM, Sokolov AS et al (eds) (2008) Fishery inventory of transboundary waters of Russia (Kaliningrad Region) and Lithuania. Izd "SP Mishutkina", Kaliningrad, 200 p
7. Abakumova VA (1993) Guidelines for hydrobiological monitoring of freshwater ecosystems. Gidrometeoizdat, St. Petersburg, 240 p
8. Winberg GG (1982) Guidelines for the collection and processing of materials in hydrobiological studies on freshwater. Zoobenthos and its products. L.: Zoological Institute of the USSR Academy of Sciences, 33 p
9. Makrushin AV (1974) Biological analysis of water quality. L.: Publishing House of the Zoological Institute of the USSR Academy of Sciences, 60 p
10. Masyutkina EA, Shibaeva MN (2014) The results of using chironomid (gen. Chironomidae) to determine the water quality of the rivers of the Kaliningrad Region. Ecosystems of small rivers: biodiversity, ecology,. Lecture materials II-nd all-Russian conference, 18–22 November 2014. Institute for Biology of Inland Waters. I.D. Papanina. In two volumes. T. II. Filigrran', Yaroslavl, pp 279–283

11. Matveeva EP (2012) Characteristics of zooplankton community of transboundary river Neman in Kaliningrad Region. In: Matveeva EP, Masyutkina EA, Shibaeva MN (eds) Proceedings of Kaliningrad State Technical University. №24. S.103–110
12. Chekanovskaya OV (1962) Aquatic Oligochaeta fauna of the USSR. O.V. Chekanovskaya. L.: USSR Academy of Sciences, 411 p
13. Tsalolihina SY (ed) (1994–2004) Key to freshwater invertebrates of Russia and adjacent territories. SPb.: Science, T.1–6
14. Lepneva SG (1964) Larvae and pupae of the suborder Kolchatoschupikovyh (Annulipalpia). Fauna of the USSR. Caddisflies. M.-L.: Nauka, vol 1, issue 2
15. Lepneva SG (1966) Larvae and pupae suborder kolchatoschupikovyh. Fauna of the USSR. Caddisflies. M.-L.: Nauka, T. 2, issue 1
16. Pankratova VJ (1983) Larvae and pupae of mosquitoes subfamily Chironominae fauna of the USSR. VJ Pankratov. Diptera, Chironomidae = Tendipedidae. -A.: Nauka, 296 p
17. Pankratova VJ (1970) Larvae and pupae of mosquitoes subfamily Orthocladiinae fauna of the USSR. VJ Pankratov. Diptera, Chironomidae = Tendipedidae. -A.: Nauka, 343 p
18. Pankratova VJ (1977) Larvae and pupae of mosquitoes subfamily Podonominae Tanypodinae and fauna of the USSR. VJ Pankratov. Diptera, Chironomidae = Tendipedidae. L.: Nauka, 154 p
19. Alekseev VR, Tsalolihina SJ (eds) (2010) The determinant of zooplankton and zoobenthos in freshwater of European Russia. SPb. M: Association of Scientific Publications KMK, vol 1, Zooplankton, 495 p
20. Masyutkina EA (2014) Chironomids gen. Chironomidae as an indicator of water quality of small reservoirs of the Kaliningrad Region. In: Masyutkina EA, Shibaeva MN, Matveeva EP (eds) Proceedings of KSTU. №32. C. 54–62
21. Shibaeva MN (2013) Species diversity of zooplankton as an indicator of the ecological status of water bodies of the Kaliningrad Region. In: Shibaeva MN, Masyutkina EA, Matveeva EP, Okhapkina AA (eds) Proceedings of Kaliningrad State Technical University, vol 28, pp 153–163
22. Shibaeva MN (2010) Species diversity of zoobenthos and bioindication of inland waters of the Kaliningrad Region. In: Shibaeva MN, Masyutkina EA, Matveeva EP (eds) Proceedings of KSTU. Kaliningrad, vol 19, p 172
23. Shibaeeva MN, Matveeva EP, Masyutkina EA (eds) (2011) Species diversity of zoobenthos, bioindication and ecological status of Kaliningrad Region lakes. Herald Baltic Federal University, vol 7, Kant, Kaliningrad, pp 91–96
24. Sherbina GH (2010) Taxonomic composition and benthic macroinvertebrates saprobiological importance of various freshwater ecosystems of the North-West of Russia. Ecology and morphology of invertebrates continental reservoirs of water. Collection of scientific works, dedicated to the 100th anniversary of the birth Fedor Dmitrievich Mordukhai-Boltovskiy. Institute of Biology of Inland Waters RAS, Papanina, Science Dagestan Scientific Center, Makhachkala, pp 426–466
25. Masyutkina EA (2014) Evaluation of the ecological state of the lake Vishtynetskoe using different hydrobiological indices, vol 7, Herald Baltic Federal University, Kant, Kaliningrad, pp 66–76

The Structure and Composition of Biological Communities in the Pregolya River (Vistula Lagoon, the Baltic Sea)

E.E. Ezhova, E.K. Lange, M.A. Gerb, O.V. Kocheshkova, J.J. Polunina, and N.S. Molchanova

Abstract The analysis of composition, seasonal and interannual dynamics, and productivity of phytoplankton, zooplankton, and zoobenthos as well as detailed characteristic of aquatic flora of the Pregolya River are given. It is shown that the biota of anthropogenically transformed lower reach recovered partially during the 2000s.

Keywords Macrophytes, Phytoplankton, Pregolya River, Zoobenthos, Zooplankton

Contents

1	Introduction	318
2	Material and Methods	318
3	Results and Discussion	322
	3.1 Case Study 1: Pollution and Water Quality in the Pregolya River	322
	3.2 Case Study 2: Phytoplankton	326
	3.3 Case Study 3: Higher Aquatic Plants and Macroalgae	332
	3.4 Case Study 4: Zooplankton	336
	3.5 Case Study 5: Zoobenthos	341
4	Conclusions	351
	Appendix 1	352
	Appendix 2	357
	Appendix 3	360
	Appendix 4	364
	References	367

E.E. Ezhova (✉), E.K. Lange, M.A. Gerb, O.V. Kocheshkova, J.J. Polunina, and N.S. Molchanova
P.P. Shirshov Institute of Oceanology, Russian Academy of Sciences, Nakhimovsky prospekt, 36, 117997, Moscow, Russia
e-mail: igelinez@gmail.com; evlange@gmail.com; marikegerb@gmail.com; okocheshkova@gmail.com; jul_polunina@mail.ru; nsmolchanova@yandex.ru

1 Introduction

The Pregolya is a lowland river, flowing in the Kaliningrad Region. The river has a length of 123 km and a total catchment area of 15,500 km^2 and belongs to the basin of the Vistula Lagoon of the Baltic Sea. In the upper reach, the width of the Pregolya River is about 20 m, the depth is 2–3 m, and the current speed is 0.5–0.6 m/s, while in the lower reach 80 m, 8–16 m, and 0.1 m/s, respectively. The geographical mouth of the river is connected to the Baltic Sea by the Kaliningrad Sea Canal, which is a continuation of the river in terms of hydrological regime, depths, bottom relief, and other characteristics [1]. The downstream of the Pregolya River and the canal are united by joint hydrodynamic processes and are characterized by wind surges, especially evident in the autumn period. Salinized water from the lagoon can go upstream up to the 16th km and occasionally higher.

Along the Pregolya, draining almost the entire Kaliningrad Region, there are many settlements, including industrially developed Kaliningrad in the river downstream, which adversely affects its ecological state. That is why the documentation of the current ecological state according to hydrobiological features (taxonomic composition, structural and functional characteristics of the biota) is important as a basis for identifying long-term changes under the influence of climatic and anthropogenic factors. There was no regular complex monitoring of the biota state of the Pregolya held until the mid-1990s. In 1995, Atlantic Branch of P.P. Shirshov Institute of Oceanology (AB IO RAS, Kaliningrad) launched a multidisciplinary study of the Pregolya River system, with hydrobiological monitoring as an integral part. The results of this monitoring study are analyzed below.

2 Material and Methods

The paper is based on the 1995–2014 long-term data of the AB IO RAS Laboratory for Marine Ecology. The study area is the part of the Pregolya River from the mouth to the 37th km (Gvardeysk), including the branches: the River Deima from the outlet to the inflow into the Curonian Lagoon, the Novaya Pregolya, and the Staraya Pregolya (Fig. 1).

Samples were taken in 18 transects in the mainstream and riparian parts of the lower reaches of the river up to the 37th km monthly or seasonally. In the regularly surveyed section of the lower reaches (0–17 km from the mouth), 12 transects were made, where, if possible, samples were taken on the left and right banks and on the mainstream. Three transects were located in the branches the Novaya Pregolya and Staraya Pregolya (30–28 and 30o–28o, correspondingly). After the branches confluence, there are five transects (27–24, 22). Two transects are located outside Kaliningrad – 30 and 30o, six ones within the city of Kaliningrad (29, 28, 29o, 28o, 27, 26, 25), and three – downstream of Kaliningrad (24–22). In the period from 1995 to 1998, samples were taken in 12 transects. Once in several years (1999,

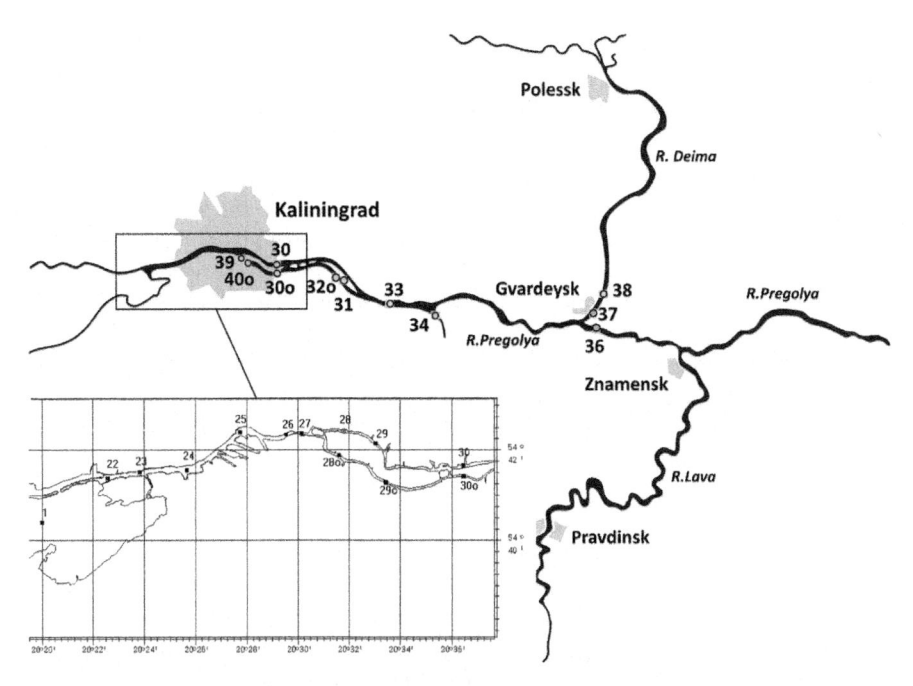

Fig. 1 Study area, the Pregolya River, 1995–2014

2005, and 2011), sampling was carried out on the whole lower flow, with two transects in the middle flow of the Pregolya and also in the River Deima, a tributary of the Pregolya. Six transects were made in 2000 (22, 26, 28, 30, 28o, 30o) and four transects in 2001 (22, 26, 30, 30o) in the Pregolya River (Fig. 1).

Samples were collected in May–July 1995, May–October 1996, and April–December 1997. From 1998 to 2006, samples were taken in the summer period from May to August, as well as in March 2007, in July 2011, and in May and September 2014.

A total of 56 samples of phytoplankton, 278 samples of mesozooplankton, and 1,871 samples of macrozoobenthos were processed. Eighteen botanical sample areas were laid, and twenty-four geobotanical descriptions were performed. Sampling was accompanied by measurement of the main hydrological parameters.

Phytoplankton AB IO RAS data are analyzed, including primary sample processing protocols[1] for 1996–1997, data for 2011 and 2014, and information from published sources.

In 1996–1997, phytoplankton samples with a volume of 1 L were taken with a 1-L bathometer in the surface layer (0–0.5 m) and fixed with 50 mL of a 40% solution of neutral formalin. The condensation was carried out by a sedimentation method to a volume of 100 mL. Identification and counting of algae were performed

---

[1]Protocols of 44 phytoplankton samples, April–October 1996–1997, processed by S. N. Semenova.

with a light microscope MBI-3, magnification 200–900×. In 2011 and 2014, samples were also taken at the surface layer, but samples (0.5 L) were fixed with Lugol's solution with the addition of acetic acid and formalin [2]. Then, they were condensed by a sedimentation method up to 10 mL. Algae were identified and counted in a Nageotte chamber of 0.02 mL under the Ergaval microscope at magnification 256 and 640×. Depending on the type of cell organization, 100 μm of trichome length, a colony, or a cell was taken as a phytoplankton counting unit [3]. The algal volume was calculated by the method of geometric similarity, and the total phytoplankton biomass was calculated from the total volume of countable algal units, assuming that 10^9 μm^3 corresponds to 1 mg of raw biomass [2].

The species and groups of phytoplankton, which constitute more than 10% of the total phytoplankton biomass in the sample, were attributed to the dominant species. The taxonomy of higher taxa was accepted according to algae directory [4]. The nomenclature for algae (genus, species, and intraspecific taxa) corresponds to the modern one accepted in the Baltic Sea International Monitoring Program [5].

To calculate the daily photosynthesis under 1 m^2, the equation $\sum A = A_{opt} \cdot K_s \cdot S$ was used [6], where $\sum A$ is the daily gross primary production, g C m^{-2} day^{-1}; A_{opt}, the rate of photosynthesis per unit volume of water per day at a depth with optimal light conditions, g C m^{-3} day^{-1}; S, water transparency by Secchi disk, m; and K_s, a coefficient equal to 1 [7]. The seasonal and annual primary production of phytoplankton ($\sum\sum A$) is a function of its daily values ($\sum A$) and the vegetation period duration (T) [6]: $\sum\sum A = k \cdot \sum A_{max} \cdot T$, where $\sum A_{max}$ is a maximum value of daily primary production under 1 m^2 over season or vegetation period; k is a coefficient showing the ratio of the area under the curve of the seasonal daily gross primary production to the area of the rectangle with sides $\sum A_{max}$ and T.

Macrophytes During the summer season, water vegetation was surveyed up to Gvardeisk in1999 and 2011, whereas in Kaliningrad in 2011 and 2013. The line transect sampling, techniques of ecological profiles, and sampling plots were applied [8–10]. Only aquatic and semiaquatic plants of the water flora of the river were taken into account. Classification of ecological groups is given according to [10], with changes. When identifying plants and algae, the determinants of higher plants and algae were used [11–13]. Verification of Latin names is carried out according to [14] and http://www.algaebase.org. Indicator properties of plants were analyzed according to [15, 16].

In the analysis of species diversity, in addition to the obtained data, publications of different years were used [17–21]. When allocating associations, a dominant approach was used, with the names of formations and other higher syntaxa given in the tradition of the dominant system [22].

Zooplankton Samples were taken in the river mainstream by the quantitative Juday and Nansen nets ($\emptyset = 14$ and 19 cm) with a mesh size of 100 μm, by vertical hauls from the bottom to the surface. In the riparian biotops, 50 L of river water were filtered by the Apstein net of the same mesh size. Laboratory and statistical processing of samples was performed according to common methods [23, 24].

The zooplankton production was estimated with the data for the transects located in the arms of the Novaya Pregolya and the Staraya Pregolya. The daily zooplankton production was calculated on the basis of the biomass of trophic groups by the physiological method [25, 26]. When translated into calories, the oxy-caloric coefficient (K_2) value of 4.86 cal mL O_2^{-1} was used. For grazing Copepoda, K_2 was 0.2, for Cladocera K_2 was 0.35, for non-predatory Rotifera K_2 was 0.4, and for predatory Copepoda, Cladocera, and *Asplanchna*, K_2 was 0.3. The daily diet was calculated by the formula: $C = P/K_1$. The coefficients of using consumed food for growth (K_1) were 0.22 for peaceful species, 0.16 for predatory and omnivorous copepods, 0.28 for *Asplanchna*, and 0.32 for large Cladoceran predators *Leptodora* and *Polyphemus*. In calculating respiration (R), the temperature correction Q_{10} (the Van't Hoff temperature coefficient) was introduced, showing how much the process speed increases with a temperature increase of 10°C. The values of Q_{10} are taken equal to 2.25 [27].

Macrozoobenthos The collection and processing of benthos samples were performed according to standard methods [28]. Samples were taken with a Petersen grab ($S = 0.025$ m^2), sieved through capron net 0.4 mm mesh size, and fixed with 4% solution of neutralized formalin. For quantitative collection in the riparian locations with rooted vegetation, a frame of 25 × 25 cm joined with a nylon net bag (mesh size 100 μm) was used. Samples were sorted by species under a stereomicroscope MBS-1, weighed by a group method with torsion balance (the accuracy up to 0.5 mg). The number and mass of individuals of every species (groups in the case of oligochaetes and chironomids) in the sample (wet formalin weight, WW) were converted into biomass (g m^{-2}) and abundance (ind. m^{-2}).

Sampled data of 1990–2000 were compared with published ones [29] of June–July 1982 and 1992 in the lower 17 km of the river flow.

The production (P) of the zoobenthos was calculated on the basis of the equations $P = R \cdot (K_2/(1-K_2))$ и $R_{day} = (24 \cdot q_1 \cdot W^b \cdot OK \cdot N)/\gamma$, where P is the production, kJ m^{-2}; R is the exchange rate, kJ m^{-2}; 24 is the number of hours in the day; q_1 and b are coefficients for calculating the exchange rate for each systematic benthic group; W is the individual weight; OK is the oxy-caloric coefficient of 20.3 kJ/mL O; and γ is the temperature correction for the exchange rate [30–32]. Monthly production was calculated by multiplying the daily one by the number of days of the month. The missing values of the monthly production were obtained by averaging the available monthly production data for the season. Summing of monthly and annual P/B coefficients was made with respect to the total one over the period of time of production to the average biomass for the same period of time, converted to the energy equivalent.

3 Results and Discussion

3.1 Case Study 1: Pollution and Water Quality in the Pregolya River

Analysis of prewar archival sources made it possible to conclude that the quality of water in the lower reaches of the Pregolya in the early twentieth century by hydrochemical and hydrobiological data was extremely low [33, 34] (Table 1).

The conclusion of German experts testifies to the highest degree of anthropogenic pollution, including organic substances. At the beginning of the twentieth century, the river was so polluted that there were no conditions for the life of the hydrobionts on the site from the Königsberg paper mill (later "Darita") to *Frisches Haff* (Vistula Lagoon). Throughout the reaches of the river, the species indicating polysaprobic conditions dominated, and at many sites, dead zones were noted.

Table 1 Results of October 3–7, 1911, survey of the Pregolya (by: [33])

Sampling site	SD (cm)	H_2S	COD_{KMnO4} (mg/L)	BOT (mg/L)	Organisms
Pregel[a] at *Pregelkrug* (now Berlin bridge)	110	–	50	908	*Asellus aquaticus, Anodonta* sp., *Valvata* sp., Tubificidae
100 m downstream from the Königsberg paper mill	75	–	>50	450	Small Tubificidae
Near the Catholic church	75	–	200–212		Not detected
Pregel at *Fischmarkt*, near the old university and the cathedral	90	–	No data		Living *Tubifex* sp.
Pregel downstream from *Krämerbrücke*	80	+	200–212		Not detected
Pregel downstream from the chark and gas station	80	–	≪200–212	2	Dead black shells of *Sphaerium corneum*
500 m upstream from the North German paper mill		+	Less higher than previous	0	Not detected
200 m upstream from the paper mill	80	+	Less higher than previous	0	Not detected
400 m downstream from the sewage output of the North German paper mill		+	355	0	Not detected
Pregel at Holstein		+	299		Living *Chironomus* sp., Tubificidae
Sea canal at the buoy 160 (Pregel mouth)	70	–	<299	0	Numerous *Chironomus* sp.

SD transparency by Secchi depth, – absence, + presence of hydrogen sulfide, *COD* chemical oxygen demand (permanganate), *BOD* biological oxygen demand
[a]German name of the Pregolya River

Therefore, it was stated that "Pregel is completely ruined from the fishery point of view" [33, 34].

In the 1920s, for the first time, toxic pollution of the river with arsenic compounds from the effluents of cellulose factories was recorded. According to the *Hartungshen Zeitung zu Königsberg* newspaper, there was a loss among the young cattle from the pastures, irrigated by sewage, with arsenic determined in the tissues [35]. Arsenic was found in the water and sediments of the Vistula Lagoon [34].

In the mid-1920s and 1930s, in addition to toxic pollution of the river and the near-mouth area of the Vistula Lagoon, caused by toxicants from untreated industrial discharge, biogenic pollution with cyanotoxins appeared. In 1924, 1925, 1928, 1931, and 1940, the attacks of "Haff disease" among the population (alimentary toxic paroxysmal myoglobinuria (ATPM) by the modern nosological classification) were registered [36]. In 1924–1925, according to reports of fishermen in the *Frisches Haff*, "fish at a depth of 1.4 m is caught in the network dead," there were outbreaks of disease of cats, livestock, and people consuming fish [35–37]. The cause of ATPM is the effect of the cyanobacteria toxin accumulated in fish tissues; outbreaks in the development of which occur at a certain ratio of nitrogen and phosphorus in eutrophic waters. In 1924–1940, the lower reaches of the Pregolya were characterized by a high degree of eutrophication, caused by the discharge of sewage and industrial wastewater with a high content of labile organics.

During this period, intensive technicalization of agriculture also took place: autumn plowing began to be used, leading to an increase in the removal of phosphorus into water bodies. Also, as a result of the mass use of mineral fertilizers, the productivity of agriculture increased, as well as the supply of livestock wastes to water bodies. The rise in the supply of nutrients to water bodies was also facilitated by the population growth, especially in cities, alongside with the lack of purification and treatment facilities and continued production growth [38].

The nutrient load in the prewar period featured the low impact of agriculture. The agriculture of East Prussia was characterized by the use of agronomic, biological, and mechanical methods with highly developed melioration [39] as opposed to the extensive use of mineral fertilizers. In the 1900s and 1940s, the greatest impact on the river ecosystem was provided by industrial wastewater and sewage water in Königsberg.

In the postwar period from 1946 until the end of the 1980s, the sanitary and ecological state of the river deteriorated. This was due to the restoration of the destroyed German factories and the construction and commissioning of new large enterprises together with the rapid revival of the agrarian sector. From 1948 to 1995, the population of the region increased by 1.6 times, the volume of industrial production by 7 times, and that of agricultural products (milk and meat) by 2.7 and 3 times, respectively. Powerful pulp and paper, fishing, machine building, power and housing, and communal complexes were engaged in production. There was a peak of agricultural load in 1980–1991. Up to 3.2–3.4 million tons of organic fertilizers, 1–2 million tons of mineral fertilizers and lime, and 700 tons of pesticides were annually applied. The condition of the Pregolya in 1984–1986s was assessed as critical. The inadequate ecological situation associated with the development of hydrogen sulfide contamination in the river and high hydrogen sulfide in the air (12–13 times higher than MPC) was observed in the 1990s. The decline in production since the early 1990s contributed to a gradual improvement of the river ecosystem has become possible because of sharp reduction in areas of cultivated land and the amount of mineral fertilizers and a decrease in the number of livestock. In this connection, the introduction of pollutants from land and livestock farms into waterways and groundwater was kept at a relatively low level, which resulted in some improvement in the ecological state of small rivers and groundwater [40].

The watershed of the Pregolya River includes the agricultural land. According to the simulation modeling results, it is shown that in the Pregolya basin, 15% of nitrogen and 1.7% of phosphorus, introduced with mineral fertilizers, are transferred to the rivers with surface runoff while 5% of nitrogen and 0.6% of phosphorus with underground runoff [41]. The flow from arable lands dominates (39% by phosphorus and 71% by nitrogen) with the nutrient load generated from the catchment area upstream Kaliningrad.

In the middle and lower reaches of the Pregolya, self-purification of waters takes place on the section from Chernyakhovsk to Gvardeysk. The Water Pollution Index (WPI) in the Gvardeysk control section line increases to 2.78, which indicates a worsening of the conditions, however, within the MPC limits. As the river passes through the city of Kaliningrad, the state deteriorates sharply, and in the control section line of Kaliningrad, the WPI reaches the value of 7.6 corresponding to "very dirty," grade 6 (Fig. 2).

The section of the Pregolya in the lower reaches is in the Kaliningrad industrial zone and is subject to severe anthropogenic pollution. Two city enterprises, the fuel and cargo complex "State Sea Fishing Port" and enterprise "Tsepruss," are listed as "hot spots" by HELCOM. Wastewater from Kaliningrad, including municipal

Fig. 2 Change in the Pregolya River water quality by the Water Pollution Index (WPI) between Chernyakhovsk and Kaliningrad, 2004 (by: [42])

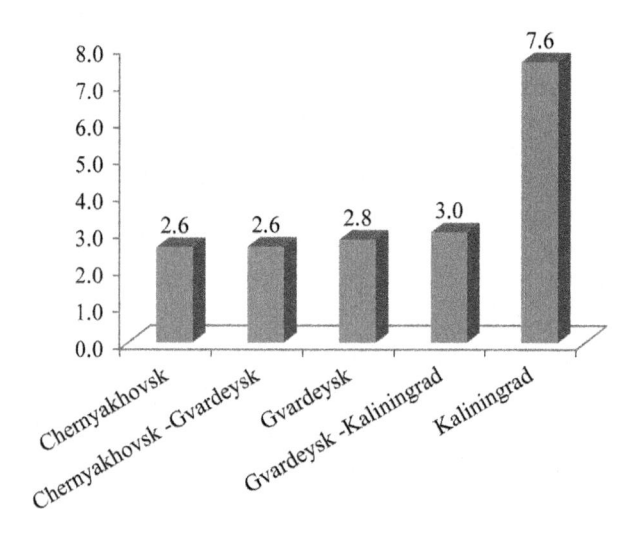

household, industrial effluents, and storm sewage, are discharged into the Pregolya water system not sufficiently purified or without purification.

In 2001–2003, there was a record of concentrations of pollutants and values of water quality indicators (oxygen, hydrogen sulfide, BOD5), classified as extremely high and high pollution [43, 44]. In 2007–2009, there was a tendency to reduce the volume of discharged contaminated sewage into the Pregolya River. In 2008 the river water in Kaliningrad was characterized by 4 and 5 grades of quality as "dirty" and "extremely dirty" [45]. In 2009, the water quality in the Kaliningrad area improved somewhat in the background section line from grade 4A ("dirty") to grade 3B ("very polluted") and in the control section line from class 5 ("extremely dirty") to class 4A ("dirty"). In 2009, in spite of a slight improvement in the quality of water in the Kaliningrad background section line, the repeatability of pollutant concentrations exceeding the MPC (Minimal Permitted Concentration) was as follows: easily oxidable organic substances (by BOD_5), organic substances resistant to oxidation (by COD), ammonium nitrogen, and iron compounds, 100%; chlorides and nitrate nitrogen, 92%; mineralization, 80%; sulfates, 75%; magnesium, 47%; and lignin, 20% [46]. In 2011, due to the suspension of the "Tsepruss" activities and the transfer of its sewage to the city sewage system, the pollution level of the Pregolya decreased somewhat. In particular, from 2009 to 2012, for the first time, there were no recorded cases of high and extremely high pollution of the water body. By the present time, despite the significant level of anthropogenic load, the ecological state of the lower reaches of the Pregolya can be considered the best for the postwar period (Fig. 3).

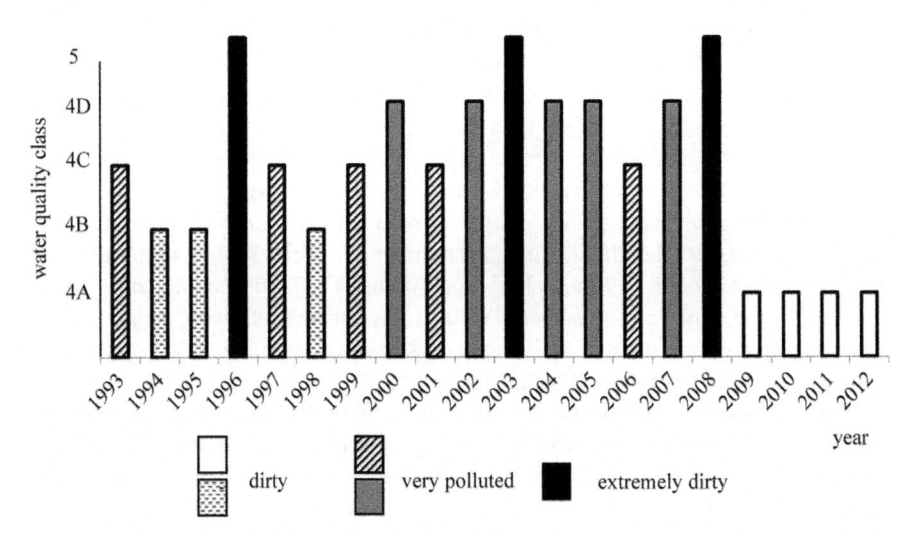

Fig. 3 Dynamics of water quality in the Pregolya River in the city of Kaliningrad, 1993–2012

3.2 Case Study 2: Phytoplankton

The first studies of phytoplankton communities in the Pregolya River were conducted within the framework of an interdisciplinary research of the river water ecosystem held by AB IO RAS in the second half of the 1990s and further study in 2011 and 2014. Observations were made in the lower reaches of the Pregolya within the city of Kaliningrad (from the Berlin bridge to the mouth of the river) [47–50]. In the early 2000s the study of phytoplankton ocenoses of the Pregolya was performed by AtlantNIRO [51, 52]. In these studies, an assessment of the structural and functional state of phytoplankton in the seasonal aspect is made alongside with the estimation of the primary production of phytoplankton, taking into account the species composition, the dominant complex, and the quantitative development of algae.

Taxonomic Composition of Phytoplankton In the growing season of 1996–1997, in the lower reaches of the Pregolya within 16 km from the mouth, phytoplankton was represented by 291 taxa of microalgae from 7 systematic groups (Appendix 1). In the same period of 2002, in the longer river section of 49 km from the source of the River Deima to the mouth of the Pregolya River, 216 species of algae were identified in the plankton. In the autumn of 2005, 136 taxa of phytoplankton were found in the middle and lower reaches of the Pregolya. A total of about 300 species and varieties of microalgae were found during the research of the phytoplankton of the Pregolya River.

In the early 2000s, the ratio of the leading groups, greens (about 40%), diatoms (about 30%), and cyanobacteria (blue-green) (about 20%), remained at the level of the second half of the 1990s. Most of green algae belonged to the order of Chlorococcales, of diatoms to the order of Pennales, and of cyanobacteria to the order of Chroococcales. The golden algae, dinophytes, euglenoids, cryptophytes, and yellow-green algae were less than 10%.

In the second half of the 1990s, extreme values of the *alpha diversity* of phytocenoses in the lower reaches of the Pregolya differed by 5.5 times. On average, 50 or more taxa in the sample were recorded in July–September 1996 and April and September 1997. The least number of species (about 40 on average) was traced in October 1996 and June and August 1997. In general, there was a decrease in the number of taxa from April to June; in the following months, there was a gradual increase with the largest values in September and a decline in October up to early summer values (Fig. 4).

In these years, a section of the river after the confluence of the arms and, especially, in the mouth was characterized by the largest taxonomic diversity of phytoplankton, mainly due to cyanobacterial species from the Vistula Lagoon. In the 2010s, the distribution of the alpha diversity of phytoplankton in the lower reaches of the Pregolya remained similar to that in the second half of the 1990s. In July–August 2011, 8 km from the mouth of the Pregolya, 39 taxa of algae were found in the phytocenosis, while at the mouth the number of taxa increased to 57. In October 2014, 3.5 km from the mouth, there were 38–49 taxa in the phytocenosis of

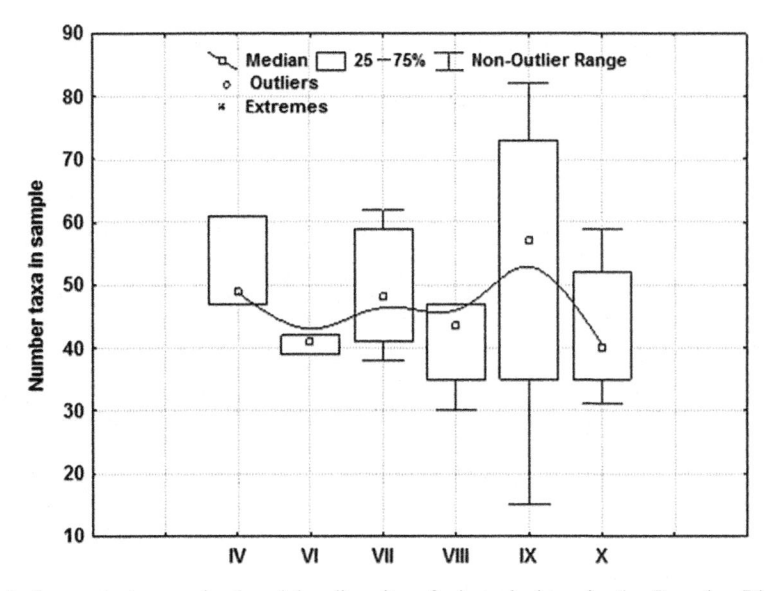

Fig. 4 Seasonal changes in the alpha diversity of phytoplankton in the Pregolya River in 1996–1997 (hereinafter, the curve is constructed by the method of distance-weighted least squares)

the stations, among which the contribution of cyanobacteria was the greatest of more than 40%.

During the research period, green, cyanobacteria and diatoms played the main role in the alpha diversity of the phytocenoses of the Pregolya and its branches, the Novaya Pregolya and the Staraya Pregolya. The diversity of diatoms, many of which belong to benthic species (*Navicula*, *Nitzschia* genera, and other pennate), was naturally higher in the spring (up to 60% of the total), compared with later dates (10–29%). The summer season was characterized by a decrease in the diversity of diatoms by an average of 2.5 times. Also, there was an increase in the number of species of cyanobacteria (an average of three times), which reached the highest values of up to ten taxa in September, against three at the beginning of the summer season. Species diversity of greens remained approximately at the same level (up to 20 taxa/sample) throughout the vegetative period.

Complex of Dominant Species and the Abundance of Phytoplankton In April–October 1996–1997, 25 taxa dominated, mostly small-celled greens, cyanobacteria, diatoms, and golden algae.

In the spring season (April), the most numerous were the centric diatoms: *Stephanodiscus hantzschii* upstream from Kaliningrad (transects 30, 30o) and *Thalassiosira* cf. *Nana* of marine origin in the mouth (transect 22). In addition, the green *Monoraphidium contortum* and, upstream, golden *Chrysococcus rufescens* dominated in the mouth. In summer, the composition of the dominant complex expanded to 22 taxa, among which the most diverse are green algae and cyanobacteria. In the autumn, the dominants changed as follows: the green genus *Chlamydomonas* and most of the diatoms dropped out; the number of cyanobacteria, on the contrary, increased.

In August 2011, small-celled cryptomonads (<10 μm) dominated in the river section from Gvardeysk (transect 36) to the Berlin bridge (transect 30, 30o) and in the Staraya Pregolya (transects 28). Cyanobacteria *Microcystis aeruginosa* and *M. viridis* prevailed in the Novaya Pregolya (transects 30, 29, 28). These algae also dominated after the fusion of the arms (transect 26). In addition, the number of most phytocenoses of the Pregolya within Kaliningrad and especially in the mouth was also determined by small-celled chococci cyanobacteria of the genus *Aphanothece*. At the mouth, their list expanded due to the species of genera *Chroococcus* and *Woronichinia*.

In the second half of the 1990s, the total abundance of algae that varied widely with the extreme values differed 1,400 times. The phytoplankton reached its greatest development in August 1996–1997 and July and September 1997 with a median abundance of more than 1,000 million units m^{-3} and the smallest one of less than 500 million units m^{-3} in September 1996 and October 1996–1997. The course of seasonal dynamics of phytoplankton abundance in the Pregolya corresponded to a single-vertex curve, with a peak in July (Fig. 5).

During all study periods in 1996–1997, a section of the Pregolya from the point of confluence of the arms to the mouth, with about the average of 13,000 million units m^{-3} over the growing season, was characterized by the greatest abundance of phytoplankton due to the vegetation of cyanobacteria and green and diatom algae. The plankton of the branches was mainly characterized by the smaller number of microalgae (about 1,700 million units m^{-3}), except for a leap in the number of phytoplankton in July 1997 and August 1996, which was associated with the mass

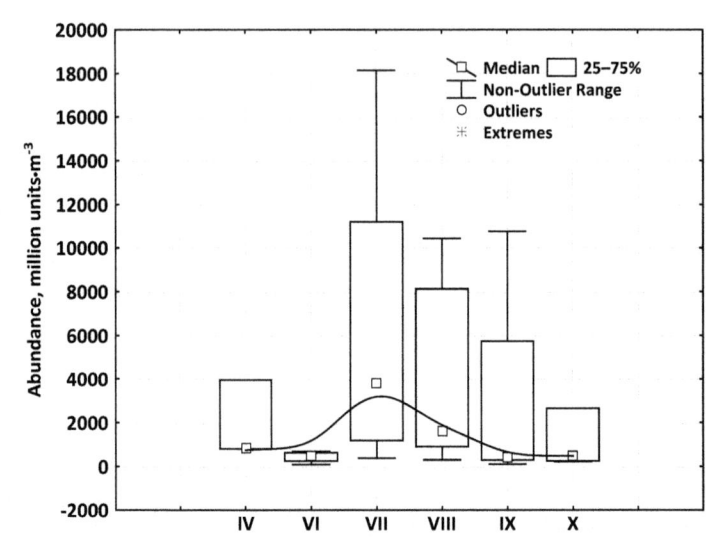

Fig. 5 Seasonal dynamics of phytoplankton abundance in the Pregolya in 1996–1997 (the maximum values of more than 30,000 million units m^{-3} in June and September 1997 not included)

vegetation of the cyanobacterium *Limnothrix planctonica* (transects 29, 29o) and green algae *Polytoma caudatum* (transect 28o), respectively.

As in the late 1990s, in the fall of 2000, in the direction to the mouth of the Pregolya [51], there was a decrease in the number of diatoms, alongside with nine times higher abundance of cyanobacteria (from 30 to 270 million units m^{-3}) and almost three times higher numbers of greens (from 60 to 130 million units m^{-3}). The increase of the phytoplankton numbers from September to October (from 100 to 1,230 million units m^{-3}), registered in 2000, was not observed in the same period in 1996–1997. In October 2014, extremely high values of phytoplankton abundance (more than 150,000 million units m^{-3}) due to small-celled cyanobacteria (<2 μm) introduced into the river during the wind surges from the Vistula Lagoon were recorded at 3.5 km from its mouth. The abundance of phytoplankton in August of the late 1990s and in 2011 was the same, with an average of about 7,000 million units m^{-3}.

Complex of Dominant Species and Biomass of Algae The total biomass of the Pregolya phytoplankton varied over a wide range of values, from 0.10 to 147 g m^{-3} (1,500 times more). In 1996–1997, by the phytoplankton biomass structure, the Pregolya in the lower reaches was divided into the Novaya Pregolya and the Staraya Pregolya, as well as the river section from the point of confluence of the branches to the mouth.

In the spring-summer period, the diatoms prevailed in the plankton of the Novaya Pregolya and the Staraya Pregolya. Among them, *S. hantzschii* dominated with up to 0.25 g m^{-3}, 33%. In the spring, their contribution to the total biomass was higher (92%) than in the summer season (33–44%). Along with the diatoms in this period, green algae and cyanobacteria accounted for 18–24% and 14–19% of the biomass, respectively. The summer vegetation was characterized by a wide range of dominant species. For example, in June and August 1997, vegetating under organic contamination green *Chlamydomonas reinhardii* accounted for 12–30% of the total phytoplankton biomass. In June 1997, there was an outbreak of colonial golden alga *Synura* spp. with 1.56 g m^{-3} (69%) in the Staraya Pregolya. In July 1996, the filamentous cyanobacterium *Limnothrix planctonica* and cryptophyte *Cryptomonas* spp. dominated with 1.34–2.15 g m^{-3}, 44–55% of total biomass and 0.13–0.52 g m^{-3}, 14–18%, respectively. In August 1996, the diatom *Cyclotella kuetzingiana* and *Melosira varians* prevailed in the Novaya Pregolya. In August 1997, the diatom *Aulacoseira granulata* dominated in the branches. In autumn, the summer ratio of systematic groups in the phytoplankton biomass was the same. In September 1996, against a general sixfold decrease in phytoplankton production, small-celled colonial species of cyanobacteria of the genus *Aphanocapsa* (in total) dominated, and *L. planctonica* continued to prevail, as well as benthic large-cell diatom *Cymatopleura solea*. In the autumn season, the level of algal vegetation with the dominance of cyanobacteria decreased to the lowest values. The average vegetation biomass of phytoplankton in the branches was 1.19 ± 0.29 g m^{-3}. Extreme values of 3.89 (July) and 0.22 g m^{-3} (September) were recorded in

1996. A single-vertex curve of the seasonal dynamics of phytoplankton biomass in the Pregolya branches with its maximum in July was registered (Fig. 6).

After the confluence of the branches and up to the mouth (Sects. 27, 25, 22), the Pregolya River, which is subject to the greatest anthropogenic impact and the influence of the inflow of brackish eutrophic waters of the Vistula Lagoon, was characterized by a dozen times greater abundance of phytoplankton (Table 2).

A large variety of values was typical for the intraseasonal biomass as a probable result of different living conditions of microalgae in the Pregolya River. For example, in June 1997, a huge accumulation of phytoplankton was recorded at the mouth of the river (Table 2). At the same period, upstream, the phytoplankton biomass did not rise above 2.26 g m^{-3}. The mouth of the Pregolya River, which is most affected by the inflow of brackish eutrophic waters of the Vistula Lagoon, in 1996–1997, compared to the rest of the Pregolya in its lower reaches was characterized by the predominance of phytoplankton biomass exceeding 10 g m^{-3}, caused by the introduction of colonial cyanobacteria, mainly of the genus *Aphanocapsa*

Fig. 6 Seasonal dynamics of phytoplankton biomass in the Pregolya River in 1996–1997

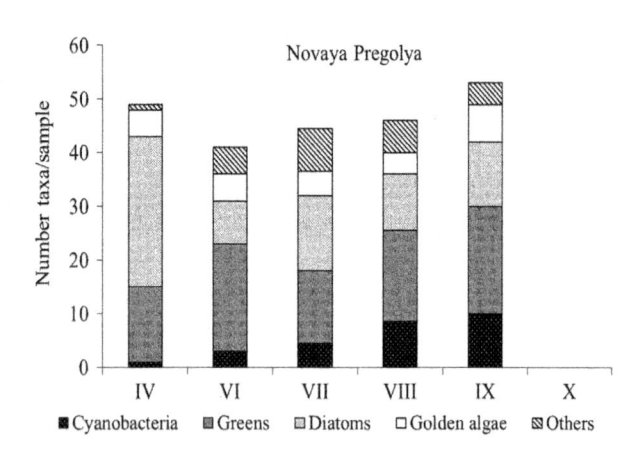

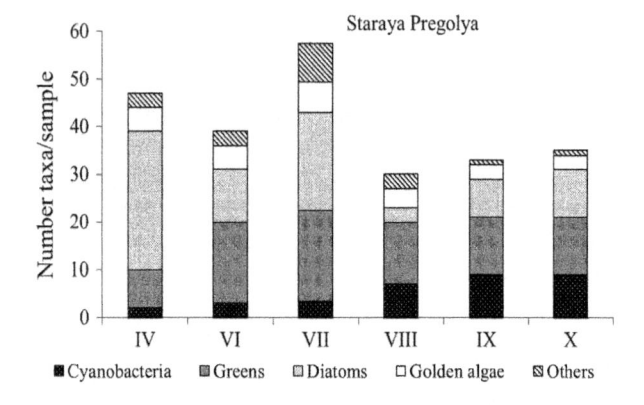

Table 2 Phytoplankton biomass of the Pregolya River section after the branches confluence in different years

Month, year	$X \pm SE$	Median	Min	Max
April 1997	2.59	–	–	–
June 1997	146.74[a]	–	–	–
July 1996	4.93 ± 2.02	4.18	0.92	10.45
August 1996–1997	7.73 ± 1.96	8.65	2.27	11.34
July 2002	9.0	–	–	–
August 2002	~23	–	–	–
September 1996–1997	30.06 ± 13.54	16.38	0.12	10.19
October 1996–1997	7.28 ± 5.10	1.92	0.17	32.18
August 2011	0.70 ± 0.05	0.70	0.65	0.76
October 2014	0.82 ± 0.07	0.82	0.75	0.89

Biomass in g m^{-3}; $X \pm SE$ – mean ± standard error; 2002 – by: [53]
[a]Summer samples only

from the lagoon in the summer-autumn season.[2] At the mouth of the Pregolya River, the phytoplankton biomass reached its highest values, while the course of its seasonal dynamics was described by a two-vertex curve, with a peak in June and September–October (See Fig. 2, paper of E.Lange, this volume).

In August 2011, in the Pregolya after the merger of the arms, diatoms (*Aulacoseira granulata*), different-sized cryptomonads, and cyanobacteria of the genus *Microcystis* (*M. aeruginosa*, *M. viridis*) prevailed in the biomass. The phytoplankton biomass was 0.70 ± 0.05 g m^{-3}. Compared to the late 1990s, the biomass of late-year phytocenoses of the river in 2011 decreased, with the median biomass of algae differed by three to five times (Table 2).

In October 2014, 3.5 km from the mouth of the Pregolya River, cyanobacteria mainly formed the phytoplankton biomass (on average 0.82 ± 0.07 g m^{-3}). Studies were carried out against the inflow of the Vistula Lagoon waters, as indicated by the presence of the brackish green *Pyramimonas* spp. and photosynthetic infusoria *Mesodinium rubrum* in the river plankton. In October 2014, the phytoplankton biomass did not go beyond the range of its values in the corresponding period of 1996–1997 (Table 2).

Assessment of Phytoplankton Production in the Lower Reaches of the Pregolya River In 1996–1997, the value of the primary phytoplankton production at the background stations of the Novaya Pregolya and the Staraya Pregolya for a year under 1 m^2 was similar, with 748 and 875 kcal, and corresponded to the upper limit of mesotrophy (300–1,000 kcal m^{-2} year^{-1}) [6]. The highest seasonal *P/B* coefficient was registered for the spring (80) and autumn period (83) in the Staraya Pregolya and the Novaya Pregolya, respectively. In spring, diatoms predominated

---

[2]In the studies of 1996–1997, chroococcal cyanobacteria were counted by colonies (slime + cells), which leads to an overestimation of the phytoplankton biomass. This mainly affected is the mouth of the river.

in the Pregolya, among which the species *S. hantzschii* comprised up to 33% of the total microalgae biomass in the Staraya Pregolya. This species is distinguished among diatoms by high production of biomass (1.2 g m^{-3} per day) [53]. In September, the phytoplankton of the Novaya Pregolya was characterized by an active vegetation of the small-celled cyanobacterium *Dactylococcopsis planctonica*, which accounted for 44% of the total phytoplankton biomass. In the summer season, both arms of the Pregolya were characterized by a similar *P/B* ratio of 65 and 70 in the Staraya Pregolya and the Novaya Pregolya, respectively, while small, mainly cyanobacteria, greens, goldens, as well as large forms of benthic diatoms prevailed in the plankton. Numerous studies of lake ecosystems have shown that the specific photosynthetic rate is determined primarily by the composition of the algae, as small forms produce more actively than large ones. The speed of turnover of phytoplankton in small lakes is higher, which is associated with greater availability of biogenic elements as a result of wind mixing [54]. The latter statement is also applicable to lowland river systems with shallow depths and low current velocity, including Pregolya.

Thus, in the late 1990s and 2010s, diatoms (all seasons), cyanobacteria, cryptomonades, and, earlier, greens (summer-autumn) formed the structure of quantitative indicators of phytoplankton in the Pregolya River. In the late 1990s, during the whole vegetation season, the Pregolya mouth was the most productive in terms of phytoplankton. The parts of the Pregolya River differed in the level of microalgae vegetation: the biomass almost doubled toward the mouth, and the productivity of the phytocenosis of the New Pregolya was higher than that of the Staraya Pregolya. However, in August 2011, there were no noticeable differences in the amount of microalgae on the studied river section as the entire river water area after the confluence of the branches was under the influence of the Vistula Lagoon waters. In the 2010s, there was a decrease in the quantitative development of phytoplankton compared to that in the late 1990s, which could be a consequence of a decrease in the nutrient load on the Pregolya river system. In the late 1990s, the background value of phytoplankton primary production in the lower reaches of the Pregolya was 812 kcal m^{-2} year, which corresponded to the upper limit of mesotrophic conditions. The relatively high values of the *P/B* ratio, registered in the spring (the Staraya Pregolya) and in the autumn (the Novaya Pregolya), are related to the dominance of diatom *Stephanodiscus hantzschii* and to the cyanobacterium *Dactylococcopsis planctonica*, characterized by a high individual biomass production rate, respectively.

3.3 Case Study 3: Higher Aquatic Plants and Macroalgae

Some information about the historically significant floral findings of the Pregolya River in the prewar period is given in the multivolume edition [55]. In the Soviet period, in 1975–1976, Galanin [18] conducted studies of the vegetation in the Pregolian lowland; his work provides brief information on the coastal-aquatic vegetation of the Pregolya. The current state of the Pregolya flora and vegetation is covered in several publications [19, 21, 56, 57]. In general, they are not sufficiently studied. In particular, there are only fragmentary data on the macroalgae and aquatic moss composition, and the available information on vegetation says little about the structural changes in plant communities, distribution features, and dynamics of overgrowing in the long-term aspect.

Hygrophilous flora of the lower reaches of the Pregolya includes 140 species of vascular plants of 91 genera and 44 families, which is 53.2% of the hygrophilic flora of the Kaliningrad Region. Cryptograms are represented by 2 species, 1 genus, and 1 family; monocotyledons by 48, 25, and 12; and dicotyledonous by 90, 65, and 31, respectively [19]. Aquatic flora, according to our data, is represented by 59 species (8 macroalgae, 1 *Bryophyta*, 1 *Equisetophyta*, 49 *Magnoliophyta*) from 31 families and 43 genera (Appendix 2). The "water core" of the flora, consists of hydrophytes, includes 33 species from 16 families and 20 genera (53.2% of all hydrophytes in the Kaliningrad Region). The families Potamogetonaceae (7), Lemnaceae (4), and Hydrocharitaceae (3) feature the largest number of species. The rest of the families contain one to two taxa. These data do not pretend to be exhaustive; they are preliminary and can be replenished as the water flora of the river is further studied. For example, among aquatic mosses, only *Fontinalis antipyretica* has been found, although the potential number of species may be larger, as well as the number of species of macroalgae and vascular plants.

The ecological spectrum of aquatic flora is represented by three types. The number of species is dominated by hydrophytes (31 or 52.5% of the river's aquatic flora), with 15 species of submerged rooting plants predominating (64.5% of hydrophytes), 5 species of plants floating on the water surface (19%), 5 species of plants floating in the water column (16%), and 6 species of rooting plants with floating leaves. Helophytes are represented by 12 species (21% of aquatic flora); hygrohelophytes comprise 16 species (29%). The data obtained are consistent with the data of [19] (Fig. 7). Such a distribution of plants by ecological groups is typical for aquatic flora of different regions in Russia and Europe, for example, the number of

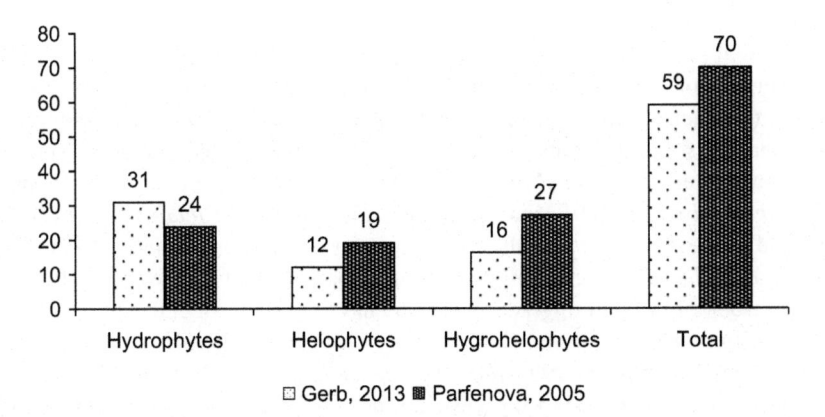

Fig. 7 The species richness of ecological groups of the Pregolya River aquatic flora in 2000s (by: [19] and author's data)

species in ecotypes in the estuary of the Neva River is similar to that of the Pregolya, with 26 species of hydrophytes, 12 helophytes, and 19 hygrohelophytes [58].

The group of Eurasian species prevails in the Pregolya flora with 58 plant species (40%); the Holarctic one is represented by 40 species (29%), the European by 22 species (16%), the European-Siberian by 12 species (9%), and the adventive element by 8 species (6%) [19].

Species diversity and ecological composition of the flora are not the same throughout the lower reaches of the Pregolya River. Species richness of coastal communities is greater in areas less prone to anthropogenic transformations (a transect upstream from Kaliningrad), achieving maximum diversity in the area of 17–27 km from the mouth. The group of hydrophytes is the most variable in the number of species in different river sections, while the water-cut plants (helophytes) and hygrogelophytes are the most stable. *Phragmites australis* predominates throughout the lower reaches, and the codominants *Typha latifolia* and *T. angustifolia* occur more often upstream from Kaliningrad, forming mixed belts. *Potamogeton lucens*, *P. perfoliatus*, and *Ceratophyllum demersum* are dominant among submerged macrophytes; *Lemna minor*, *Spirodela polyrhiza*, and *Hydrocharis morsus-ranae* prevail in plants floating on the water surface; among plants with floating leaves on the surface, *Nuphar lutea* and *Nymphaea candida* dominate; *P. australis*, *Sagittaria sagittifolia*, *T. latifolia*, and *T. angustifolia* prevail in the helophytes. The communities of these plants form almost a continuous belt along the entire lower reaches of the river up to Gvardeysk, replacing each other. In the middle part of the lower reaches of the Pregolya, *Stratiotes aloides* grows, which forms floating curtains. *Sparganium erectum*, *Glyceria maxima*, *Acorus calamus*, *Stratiotes aloides*, and *Batrachium circinatum* are distributed locally. *Batrachium eradicatum* and *Potamogeton nodosus* belong to the rare findings. Closer to the mouth, where the inflow phenomena of brackish waters of the Vistula Lagoon is evident, *P. pectinatus* and *Lemna gibba* are more common. In the industrial zone of the lower reaches of the river, vegetation is less developed or completely absent. River banks are concrete embankments, ports, moorings, and other man-made objects of various enterprises in Kaliningrad.

Perennial species dominate by the life expectancy with 50 species (85% of the total number of flora species). Annuals account for 15% of the total number of species (9 species), all being macroalgae and one vascular plant. In addition to aboriginal coastal water plants, 17 invasive species were noted along the banks all along the river: *Impatiens parviflora*, *Acorus calamus*, *Echinocystis lobata*, *Helianthus tuberosus* [59], and an aquatic plant *Elodea canadensis*. Seventeen rare and protected plants of the Kaliningrad Region were found. *Nymphoides peltata* and *Alisma gramineum*, growing in oxbow lake of the river, are listed in the Red Book of the Kaliningrad Region and have the status of endangered species. The distribution of *A. gramineum* requires clarification in connection with the author's single finding in 1993 [60]. Ten species, of those given by J. Abromeit for the flora of East Prussia [55], were not detected in the Pregolya over the past 100 years, probably due to prolonged anthropogenic impact. At the same time, during the observations held in 1999–2014, a tendency to overgrowing by macrophytes is revealed, which is associated with the improvement of water quality after the shutdown of the pulp and paper industry within the city of Kaliningrad. The development of aquatic vegetation, particularly in the center of Kaliningrad, clearly demonstrates the improvement of the river ecosystem compared to the situation in the 1970s–1990s, when there was no aquatic vegetation at all. In 1999, the development of coastal-aquatic vegetation in Kaliningrad was registered firstly, and it already had a pronounced belt character of overgrowth along the greater part of the low reaches in 2001. In recent years (2011–2013), in the center of Kaliningrad, the downstream dispersal of the water lily and spatterdock belts, both along the banks and in a width, has been visually noted. Compared to 1999, there has been an increase of *P. lucens* in macrophyte communities. This typical mesotrophic and ß-mesosaprobic species of pondweed is now massively growing in the Pregolya river, even in the very center of Kaliningrad.

A set of dominant species of macrophyte-trophic indicators characterizes the state of the aquatic environment of the lower reaches of the Pregolya as a mesotrophic one. At the same time, a high abundance of pleistophytes in the composition of aquatic phytocenoses was noted in Kaliningrad. They are indicators of the eutrophic state (*H. morsus-ranae*, *S. polyrhiza*, and *L. minor*), the local mass development of which is characteristic of the areas in the places of nutrients and organic contamination, most likely from sewage. Among the 59 species identified, only 18 had high incidence and abundance. Basically, these are plastic and tolerant species that inhabit in a wide range of environmental conditions and tend toward mesotrophic and β-mesosaprobic aquatic conditions. According to Sviridenko et al. [16], species as *P. australis* and *P. pectinatus* are eurythropic and eurysaprobic species, while *N. lutea*, *Hydrodictyon reticulatum*, and *Butomus umbellatus* are eurysaprobic. Among the species indicating the trophic state, 73% of macrophytes refer to mesotrophs (47%) and mesoeutrophs (26%). Mesotrophic species show the highest incidence and projective coverage, with the exception of eutrophic *L. minor* and mesoeutrophic *S. polyrhiza*, locally dominant in nutrient-rich areas.

Of the total macrophyte indicators of saprobity, 55% of the species belong to β-mesosaprobionts, 21% to o-β-mesosaprobionts, 11% to eurysaprobes, and 11% to ß-α-mesosaprobes. The composition of dominants of aquatic plants in communities throughout the river lower reaches varies, probably due to the different saprobiological status.

Despite the strong influence of brackish waters of the Baltic Sea on the Pregolya estuary area, the majority of coastal water plants and macrophytes of the river lower reaches are typical freshwater species, their share is 45% of the total list of species, and 21% are conditionally freshwater species. *Lemna gibba*, which prefers brackish water, does not occur more than 16 km above the mouth, while the strongly saline-freshwater *Ulva intestinalis* and euryhaline *Cladophora glomerata* occur in the area from the mouth to the 14th km.

3.4 Case Study 4: Zooplankton

As a result of analysis of long-term data, 109 species and taxonomic groups of zooplankton were identified in the lower reaches of the Pregolya (Appendix 3). Widespread Palaearctic species, which predominantly refer to the filter-feeder species; rotifers *Asplanchna priodonta*, *Brachionus calyciflorus*, *B. angularis*, *Keratella quadrata*; Copepods *Acanthocyclops viridis* and *Eudiaptomus gracilis*; and Cladocera *Bosmina longirostris*, *Daphnia cucullata*, and *Chydorus sphaericus* belong to mass species in the river, most of them being mesosaprobes. The larvae of Polychaeta, Cirripedia, and Bivalvia are registered as mezoplankton.

The occurrence of certain species in different parts of the river is conditioned by their biotopical and hydrological features. Three groups of species are distinguished with respect to salinity: freshwater stenohaline, species that practically do not occur below the confluence of the two Pregolya arms; freshwater euryhaline, found almost throughout the water area; and species of brackish and marine complex. The following types are distinguished with respect to biotopic conditions: a phytophilic complex (found only in the coastal zone of the two Pregolya arms with pronounced coastal vegetation, consists mostly of cladocerans); pelagic species (inhabit the water column of the river, in the medial, most species of rotifers and copepods), and species that inhabit the bottom, which are representatives of the suborder Harpacticoidae, some species of cyclops, and crustaceans. In general, the diversity of zooplankton is higher in the coastal biotope than in the medial river.

In the fall, salty water from the Vistula Lagoon flows into the Pregolya. During strong surges, salty waters reach the 17th km upstream along both river branches and sometimes even higher. In such periods, the copepods *Eurytemora affinis* and *Acartia* spp. appear in the zooplankton of the Staraya Pregolya and the Novaya Pregolya from the Vistula Lagoon.

The spatial distribution of zooplankton in the arms of the Staraya Pregolya and the Novaya Pregolya, after the confluence and in the estuary, is different. In the medial part of the Novaya Pregolya, a higher number of rotifers are registered compared to that in the Staraya Pregolya, while the numbers of inferior crustaceans are higher in the Staraya Pregolya. After the confluence of the arms and in the mouth, there is high zooplankton abundancy, exceeding that in the river's branches (Table 3). This is due to the presence of species *Eurytemora affinis* and species of the genus *Acartia* that live massively in the Lagoon.

Table 3 Average abundance and biomass of zooplankton in the medial part of the Pregolya River (data of 1996–2007)

Index	Novaya Pregolya	Staraya Pregolya	Area after the confluence of the branches	Mouth
N (thousand ind. m^{-3})	32.5 ± 19.9	35.2 ± 25.9	46.6 ± 37.9	79.2 ± 58.2
B (mg m^{-3})	342.8 ± 255.7	442.5 ± 370.7	988.2 ± 946.1	948.1 ± 795.4

The average biomass of the main groups of zooplankton in the different parts of the river in the medial during the growing season is similar to the numbers in these sections of the river (Table 3).

Distribution of zooplankton along the transverse profile of the Pregolya is characterized by a higher density of zooplankton in riparian biotops, compared to mainstream one, which is due to lower flow rates and the presence of diverse habitats for plankton. On average, the number of zooplankton in riparian zone was 51.4 ± 34.8 thousand ind. m^{-3}, which is almost three times higher than in the medial river, with 18.2 ± 15.3 thousand ind. m^{-3}. The number of copepods in riparian zone and river mainstream is similar, with 15.3 and 11.1 thousand ind. m^{-3}, respectively, whereas rotifers and, especially, crustaceans are the most numerous in riparian zone. On average, the zooplankton biomass did not exceed 1 g m^{-3} in river mainstream and 2 g m^{-3} in riparian zone. At the same time, Cladocera accounted for more than 60% of the total biomass in the riparian zone while in the medial for less than 30%. Here, cyclops and calanides (genus *Eudiaptomus*) formed most of the 66%. In the mouth area of the river, the total abundance and biomass of zooplankton are higher, primarily due to copepods from the lagoon. Thus, in the medial, more flowing part of the Pregolya, zooplankton is qualitatively and quantitatively less developed than in the riparian zone, overgrown with macrophytes, which often almost doesn't move. This trend is also known for other rivers [61] and is, with amendments to regional specifics, probably generally true for the distribution of zooplankton of slow-flowing lowland small rivers.

In the seasonal aspect, the following changes in the structure and abundance of the Pregolya zooplankton were noted. In the early spring period, copepods developed massively, primarily the nauplial and copepodite stages of cyclops (up to 60% of the total number of zooplankton) and cyclops *Cyclops vicinus* (22%) and *Mesocyclops leuckarti* (7.6%). The quantitative development of rotifers and crustaceans is extremely insignificant. In April, in the river plankton, along with the juvenile stages of the cyclops, a massive species from the Vistula Lagoon, *Eurytemora affinis*, and its copepodite stages appear, with the proportion of this species reaching 31% of the total abundance and 57% of the total biomass of the zooplankton in the lower reaches of the river. The share of rotifers in April did not exceed 10%. *Ch. Sphaericus* and *B. longirostris* appeared in the group of crustaceans (did not exceed 0.5% of the total number and biomass). In May, *Filinia longiseta* and *Asplanchna priodonta* were present in the group of rotifers. Among the copepods, juvenile stages of cyclops and calanides of *E. affinis* predominated, whereas *M. leuckarti*, *M. crassus*, and *Cyclops furcifer* developed massively in the group of cyclops. The Cladocera *Bosmina longirostris* reached almost 18% of the total abundance and 6% of the biomass. In summer zooplankton, the share of rotifers increased significantly, with *Euchlanis dilatata* (up to 13% of the total zooplankton biomass), *B. calyciflorus*, and *K. quadrata* developing massively. Among the copepods, juvenile stages of cyclops, *Thermocyclops oithonoides*, *Megacyclops viridis*, and *Diacyclops languidoides*, prevailed. In the Cladocera group, *Chydorus sphaericus*, *Bosmina longirostris*, and *Daphnia cucullata* dominated. In the autumn, with the increase in the wind surge events, *Eurytemora affinis* and its juvenile stages dominated in the lower reaches of the river, accounting for 50% of the total population and almost 65% of the total biomass.

The growth in the number and biomass of zooplankton in the Pregolya river occurred in the spring period, reaching its maximum in summer under maximum warming up of water with peaks in abundance in July (130 ± 110 thousand ind. m^{-3}) and biomass in August (958 ± 788 mg m^{-3}) (Fig. 8).

In July, the maximum of the quantitative indicators of zooplankton was determined by the development of rotifers while in August by the abundance of planktonic crustaceans. In September, the value of the indicators decreased, but in October–November, an increase was observed. In autumn, the frequency and intensity of water surges from the Lagoon to the river rise. Crustaceans that live massively in the bay fall into the lower reaches of the river, which leads to an increase in the abundance and biomass of zooplankton. In general, during the year in the lower reaches of the Pregolya, the quantitative indicators of zooplankton varied depending on the water temperature and the surging phenomena.

Average annual indicators of zooplankton abundance in 2002 and 2006 were significantly higher (about 200 thousand ind. m^{-3}), compared to 1996, 1998, 2001, and 2004 (less than 60 thousand ind. m^{-3}). High numbers of zooplankton in 2002 and 2006 are caused by the development of rotifers, while the number of crustacean zooplankton varied slightly during all the years of research. In these 2 years, the highest values of the average summer water temperature over 20°C were recorded. A positive correlation between the zooplankton abundance and water temperature ($R = 0.6$; $p < 0.05$) was found.

The interannual dynamics of the zooplankton biomass differed somewhat from the dynamics of numbers. The maximum biomass values were recorded in 1996 and 2006 with 938 and 773 mg m^{-3}, respectively (Fig. 9). This is due to the high proportion of Cladocera and copepod crustaceans in the total zooplankton biomass in 1996 and in 2006, respectively.

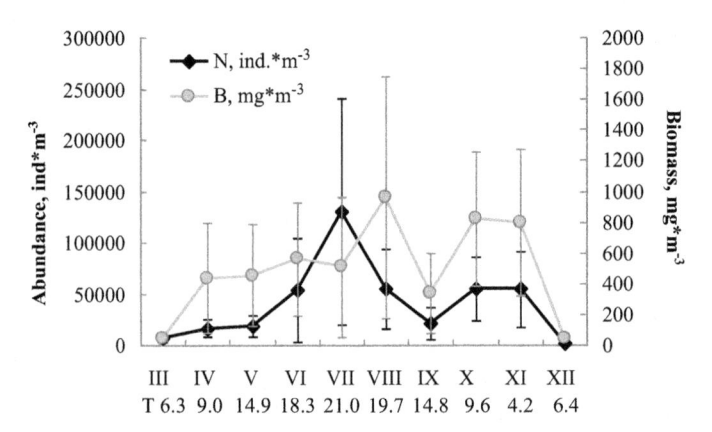

Fig. 8 Seasonal dynamics of zooplankton abundance and biomass and water temperature (T, °C) in the Pregolya River (average long-term data, $X \pm SD$)

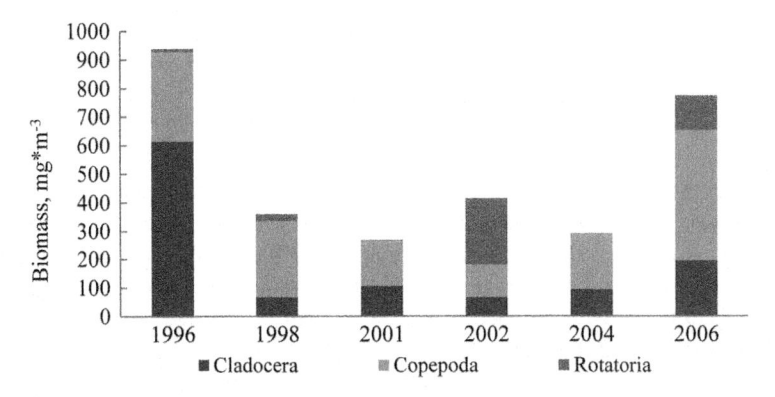

Fig. 9 Biomass of the main groups of zooplankton in the Pregolya River

No significant differences in species composition and quantitative indicators of zooplankton in the 1996–2006 research period were detected. Some variations of interannual abundance and biomass are primarily due to hydrometeorological conditions.

The distribution of *seasonal zooplankton production* in riparian zone of the Novaya Pregolya and the Staraya Pregolya did not differ significantly. In both branches of the river, the maximum production values were observed in July (1.2 and 1.3 kcal m^{-3}) while the minimum values in October (0.05 and 0.02 kcal m^{-3}). The basis of production was formed by grazing zooplankton. In the Novaya Pregolya, microfilter feeders (*Bosmina longirostris*, nauplia) and detritophages (small hidorides) accounted for 58–74% of production in May, June, August, and September. In July, macrofilter feeders from the genus *Daphnia*, *Ceriodaphnia quadrangula*, and the younger copepodites of cyclops accounted for about 55% of the total production. The immature individuals of copepods (67% of *Eudiaptomus gracilis*), together with rotifers (33%), formed the autumn production in the riparian zone of the Novaya Pregolya. In the Staraya Pregolya, the main contribution to the production was attributed to micro- and macrophages from copepods (nauplia and younger copepodites of cyclops), with 66% in May, 55% in August, 51% in September, and 58% in October. The role of detritophages (small hidorides) increased in June (47%) and July (71%). In autumn, in both the Novaya Pregolya and the Staraya Pregolya, zooplankton production was formed by rotifers (42%) and copepodite stages of copepods (58% *Eurytemora affinis*). The total production of grazing zooplankton for the growing season in the Novaya Pregolya exceeded that in the Staraya Pregolya (3.3 and 2.5 kcal m^{-3}, respectively). The production of predatory zooplankton in both arms was low and practically did not differ, with 0.6 and 0.7 kcal m^{-3}. The rate of zooplankton biomass turnover (P/B) was also not different, with 11 and 12 of peaceful and 5 and 6 of predatory.

In May and September, in the Novaya Pregolya River mainstream, rotifers (52% and 29%, respectively) and calanides *Eurytemora affinis* and *Eudiaptomus gracilis* (20% and 47%, respectively) formed the production. In the remaining months, the

Cladocera *Bosmina longirostris*, *Daphnia galeata*, *Daphnia cucullata*, *Ceriodaphnia quadrangula*, and small chydorids accounted for 58–95% of the production. In the river mainstream of the Staraya Pregolya, crustaceans (macro- and microfilter feeders and detritophages) played a significant role in production only in May and June. In July, the production equally consisted of crustaceans and copepods, mainly macrofilter feeders *p. Daphnia*, nauplia, and copepodites of copepods. In the remaining months, 76–97% of the production were formed by nauplia and copepodites of copepods. Rotifers contributed insignificantly to the production of the Staraya Pregolya throughout the growing season. Until October, no significant differences in the total zooplankton production were observed in the river mainstream of the Novaya Pregolya and the Staraya Pregolya, with 1.5 and 1.7 kcal m^{-3}, respectively. In October, with a large amount of *Eurytemora affinis* (up to 98 thousand ind. m^{-3}) brought to the Staraya Pregolya by saline water surges from the Vistula Lagoon, the production increased by 4.6 and 0.4 kcal m^{-3}, respectively, compared to the river mainstream of the Novaya Pregolya.

The ratio of the predatory and grazing zooplankton average biomass for the vegetation season in the Novaya Pregolya is higher than in the Staraya Pregolya, which allows to estimate the riparian zone and river mainstream of the Novaya Pregolya as more eutrophic. Both in riparian zone and river mainstream of both arms, the main part of zooplankton production (78–98%) was formed by grazing zooplankton organisms, mainly micro- and macrofilter feeders and detritophages. The production size was mainly determined by small chydorids (phyto- and detritophagous collectors) and mixed group of juvenile stages of Cyclopoida (euryphagous collectors). The role of rotifers and large forms of crustaceans in production-destruction processes in the Pregolya is insignificant. The amount of energy necessary to maintain the structure of the community (R/B) varies from 21 to 24% in the riparian zone and from 28 to 30% in river mainstream. The highest energy expenditure ensuring the functioning of the community (P/R) is characteristic of the Novaya Pregolya zooplankton. The total production of grazing l zooplankton in the riparian zone of the river during the vegetation season was 5.8 kcal m^{-3} and 8.1 kcal m^{-3} in the medial zone, with the total production of predatory plankton of 1.3 and 0.2 kcal m^{-3}, respectively.

Thus, in 1996–2007, the species diversity of zooplankton in the lower reaches of the Pregolya is 109 species and groups of holoplankton, which is almost three times higher than this figure in the early 1980s [62]. The basis of the zooplankton community of the Pregolya, as in most slowly flowing lowland rivers, is defined by copepods. Also, the structural and quantitative indices of zooplankton of the lower reaches of the Pregolya are significantly influenced by the zooplankton of the Vistula Lagoon, especially in the section of the river below the confluence of its arms. In addition, the zooplankton abundance and biomass in the Pregolya are lower than these indices in the Vistula Lagoon [63] and the Kaliningrad Sea Canal [64] but comparable to the indices of zooplankton in the middle and small rivers of the Northwest Region [61]. The occurrence and quantitative development of individual species and zooplankton as a whole are generally determined by biotopic conditions (the presence of a coastal biotope), different degrees of pollution, and hydrological features (surging phenomena) of certain river water areas. In the coastal zone of the river (riparian zone) within Kaliningrad, and mainly on the arms of the Staraya and the Novaya Pregolya, the zooplankton diversity, abundance and biomass is higher than in the river mainstream part of the river. In the summer season, the average zooplankton abundance and biomass amounted to 97 thousand ind. m^{-3} and 680 mg m^{-3}; the average vegetation values are not less than 20 thousand ind. m^{-3} and 200 mg m^{-3}, respectively. Both in riparian zone and river mainstream of both arms, the main part of zooplankton production (78–98%) was formed by peaceful zooplankton organisms, mainly micro- and macrofilter feeders and detritophages. Production of invertebrate predators in the total zooplankton population, biomass, and production is small, which excludes tensions in the predator-prey chain. In general, the production of zooplankton in the Pregolya is lower than in the Vistula Lagoon [63].

3.5 Case Study 5: Zoobenthos

Conditions for the existence of zoobenthos in the Pregolya are largely determined by the biotopic features of the river at different parts of the water reaches, which are described below. The morphology and hydrology of the lowland area of the middle and lower reaches of the Pregolya are quite specific. The typical features are small slopes along the longitudinal profile, low flow velocity, the dominance of the silty-sandy and sometimes clayey soil, the absence of stony recesses, and the vegetation-anchored banks on most of the river body (with the exception of the last 10 km of the Pregolya and the 2 lower kilometers of the Novaya Pregolya). The middle reaches of the Pregolya can be classified as a classical potamic zone (slow flow with a further increase in the flow size, a good warming of the water, development of macrophyte vegetation along the banks, silting of the bottom) [65, 66].

Transverse zonation is well expressed almost throughout the middle and lower currents, except for the last several, anthropogenically altered, kilometers of the river reaches. Two biotopes, riparian zone and river mainstream zones, are clearly distinguished. Silty-sandy, fairly shallow, riparian zone, 1.5–2.0 m deep, is almost entirely occupied by plant communities. The reed belt, which sometimes comes into the water from the swampy banks and sometimes to the depths of 0.5–0.8 m, is followed by a water lily-pondweed or pondweed-*Ceratophyllum* belt, on average 3–5 m width, to the depths of 2.0 and 2.5 m in some places. In the middle reaches, the water vegetation becomes more diverse, with the appearance of reed mace, water soldier, and other species. Well-developed aquatic vegetation provides microbiotopic heterogeneity and an abundance of plant detritus, the factors that contribute to the development of quantitatively and qualitatively abundant invertebrate fauna.

The mainstream biotope does not have any vegetation, pits, or significant irregularities, with the depths of 4.0–6.0 m, and is separated from riparian zone by a rather sharp depth descent. Bottom sediments differ depending on the river

Table 4 The areas of river biotopes in Staraya Pregolya and Novaya Pregolya within the boundaries of Kaliningrad

River	Square of mainstream biotope (ha)	Square of the right bank riparian biotope (ha)	Square of the left bank riparian biotope (ha)
Staraya Pregolya	21.6	4.35	4.3
Novaya Pregolya	22.6	3.12	5.7

part: silt, silty-sandy, and clayey in some places in the Staraya Pregolya (transect 3o) and in the middle reaches and sandy with a bit of shellfish almost everywhere with an abundance of vegetable detritus in the unpolluted area of the Novaya Pregolya (transect 30).

The boundaries of the riparian zone of the right and left banks were drawn, and their areas were estimated by the planimetric method for both branches of the Pregolya, using a modern topographic map of the interfluve between the Staraya Pregolya and the Novaya Pregolya (Table 4).

Fauna of Bottom Invertebrates of the Pregolya In the perennial collections from the lower and middle reaches of the river, and partly according to the published data, 23 large systematic groups of benthic animals belonging to 7 types have been identified: sponges, coelenterates, flatworms, primary worms, annelids, mollusks, and arthropods. Their representation in different parts of the riverbed is not the same. The population of the upper part of the lower reaches of the Staraya Pregolya and the Novaya Pregolya (11–33 km from the mouth) and the middle reaches of the river (34–60 km from the mouth) is the most rich in species. The total species richness is quite high, with the present list including more than 150 taxa, and the number of species after processing of the collection materials is expected to be substantially higher. The list of species, compiled on the basis of the 1996–2012 materials of AB IO RAS and taking into account the published data [67–72], is given in Appendix 4. It does not claim to be complete, since oligochaetes, chironomids, and other aquatic insects were not always determined in monitoring samplings. The fauna diversity in the middle reaches area is obviously underestimated due to the small number of samplings in this part of the river system.

Most of the hydrobionts that live in the river are freshwater, transpalearctic, or transholarctic; there are also amphiatlantic, European, and cosmopolitan species. The bottom fauna is characteristic for fresh water bodies in the temperate zone of the northern hemisphere. The oligochaetes (34), bivalves (25), gastropods (29), and diptera larvae (17) are the most diverse in the identified fauna. Aboriginal species predominate in the main groups of bottom hydrobionts, both in number of species and in quantitative representation in communities, with an exception of the *Dreissena polymorpha* mollusk, which plays a decisive role in communities in many parts of the riverbed. Among the alien species of benthos, seven species of Malacostraca were noted, mainly representatives of the Ponto-Caspian faunal complex.

Unlike many smaller and more rapidly flowing rivers, in which the larvae of aquatic insects predominate in the bottom population both in the number of species and in the contribution to biomass and abundance, mollusks of fairly diverse species composition are of great importance in the zoobenthos of the Pregolya and the Deima rivers. However, with a relatively large species diversity, a small number of species gain mass development, the composition of which is determined by the specifics of the river part and mainly the mollusks of the genus *Unio*, *Dreissena*, *Viviparus*, *Valvata* and several species of limnaeids, sphaeriids and pisidiids.

Spatial Distribution of Biomass and Abundance of Zoobenthos In the studied section of the Pregolya (middle-lower reaches, including the rivers of Staraya Pregolya and Novaya Pregolya), there is a significant spatial heterogeneity in the zoobenthos distribution. The general regularity of the benthos distribution along the riverbed is a successive change in the type of communities in the direction from the upper sections to the mouth (the unionid type, the Dreissena biocenosis, the biocenosis with dominance of the oligochaetes), characterized with a decrease in the total species richness and in species richness in biocenoses, as well as the total abundance and biomass of the zoobenthos downstream (Figs. 10 and 11). This can be also traced on a river system part, where the Pregolya is divided into branches. The quantitative characteristics of zoobenthos for the Staraya Pregolya and the Novaya Pregolya are similar and, in total, for the riparian-mainstream transect, give similar values, which are lower than for the upstream areas but much higher than in the lower reaches.

There is a noticeable difference between the benthos of the two main biotopes (riparian and mainstream zones), which are well pronounced and are homotopic over a considerable length. This is especially true for the part of the riverbed, which

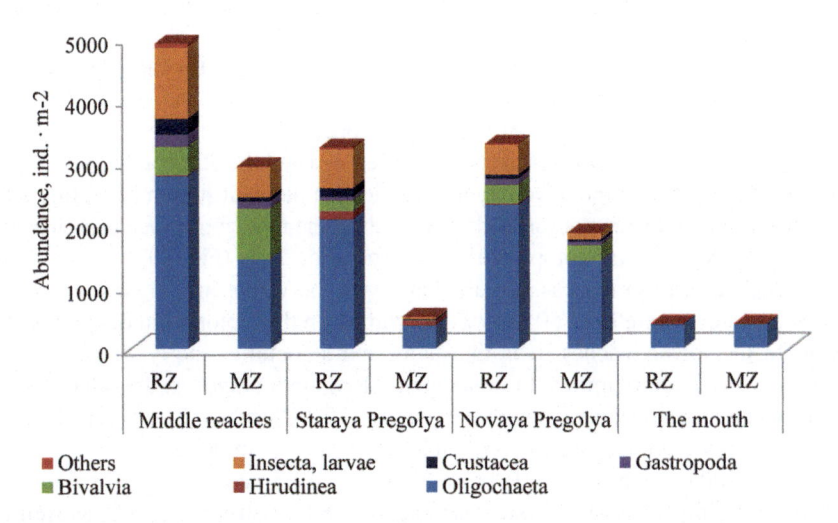

Fig. 10 The structure of the zoobenthos abundance in different parts of the Pregolya River in summer (1997–2000, 2002, 2006) (*RZ* riparian zone, *MZ* mainstream zone)

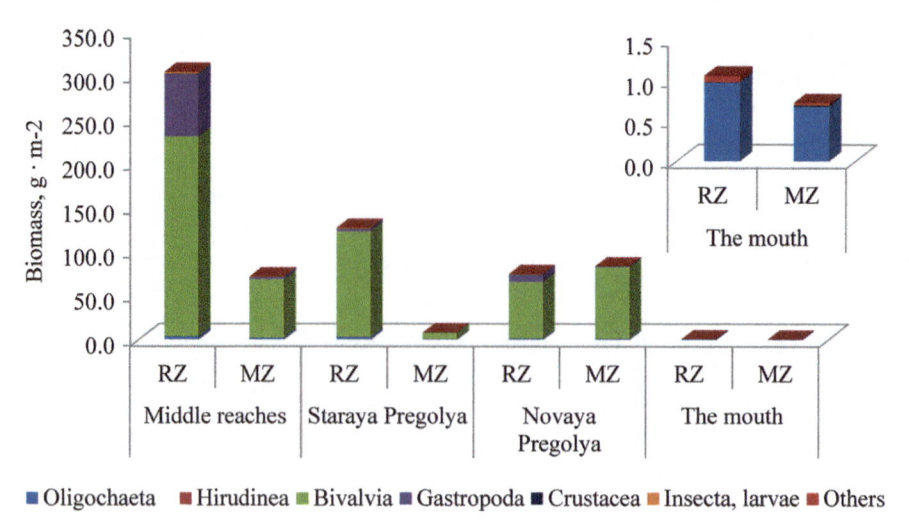

Fig. 11 The structure of the zoobenthos biomass in different parts of the Pregolya River in summer (1997–2000, 2002, 2006, 2007)

has undergone less anthropogenic transformation and pollution and has preserved natural features, namely, the middle reaches, the Staraya Pregolya and the Novaya Pregolya up to the transect 30 inclusive. As a rule, quantitative characteristics for the riparian zone in the Pregolya are always higher than those in the medial, where the nature of the soil and the current are less favorable for many benthic bionts (Figs. 10 and 11). In the lower reaches of the Pregolya after the confluence of the arms, the riverbed is altered anthropogenically (deepened), the riparian zone is not distinct, and the hydrochemical conditions do not contribute to the formation of high biomass.

The abundance (by numbers) of zoobenthos (mean values for each typical area, represented by several stations) varies from 400 to 5,000 ind. m^{-2}, with the maximum in the riparian zone of unpolluted upper areas, and a decrease to 3,300 ind. m^{-2} in the riparian zone of the Novaya Pregolya and the Staraya Pregolya and a fall to 400 ind. m^{-2} in the lower reaches section. In the mainstream zone, throughout the entire length, except for the lower reaches, the total abundance is noticeably less than in the riparian biotope of the corresponding zone (Fig. 10).

Oligochaetes form the basis of abundance on the whole investigated length of the river, contributing from 50 to 97%, depending on the biotope and the part of the river. In the lower reaches, it is the main, and sometimes the only, element of extremely poor communities. In urban areas along both branches, the oligochaetes contribute 65–70%, alongside with bivalves (3–15%), leeches (3–10%), and insect larvae, mainly chironomids (4–10%), starting to play an important role.

In the middle reaches, the role of mollusks in the formation of the population is significantly higher, with bivalves reaching 10–25%, gastropods 2–4%, as well as absolute (Fig. 10) and relative (19–22%) numbers of insect larvae, not only chironomids but also Trichoptera and Ephemeroptera. Along both arms and upstream, *Euglesa supina*, *Neopisidium moitessieranum*, and *D. polymorpha* were noted of the 21 bivalve species. In the coastal area, large specimens of *Unio pictorum* and *U. tumidus* are often registered. Of the 24 species of gastropods, *Lymnaea glutinosa* and *V. Viviparus* and species of the genus *Valvata* were frequently found during the study period.

For the whole data set, the average annual zoobenthos abundance (Table 5) and biomass (Table 6) for the mainstream and riparian biotopes in both branches were calculated and, for comparison, the same indices in the underlying anthropogenized section of the Pregolya after the confluence of the branches.

The total average benthos biomass in the summer season varies from fractions of a gram to hundreds of grams (0.72–304 g m^{-2}, max 2,785 g m^{-2}), differing significantly between the Pregolya in the middle reaches, the two branches belonging to the upper part of the lower reaches, a 16 km long part of the lower reaches of the Pregolya, which is under the influence of Kaliningrad and the estuary area of the river (Table 6).

The unevenness of the quantitative distribution of zoobenthos between these areas is clearly expressed both in abundance and in biomass (Tables 5 and 6). According to summer surveys characterizing bottom communities in their full development, the structure of the zoobenthos biomass for riparian and mainstream zones of the river in the middle and lower reaches was considered.

On the entire investigated river length, the basis of biomass is created by bivalve mollusks, *Dreissena polymorpha* in particular, which contribute 50–97% to the total zoobenthos biomass depending on biotopic conditions. In the upper sections, gastropods (5–25%) also make a significant contribution to the biomass. The composition of bivalves and gastropods is diverse, with 21 and 24 species, respectively, in the middle reaches and 12 and 21 in the Novaya Pregolya and the Staraya Pregolya. However, except three *Unio* and two *Anodonta* species, and also gastropods *Lymnaea glutinosa*, *Viviparus viviparus*, and *Planorbarius corneus* and

Table 5 The average annual abundance of zoobenthos (ind. m^{-2}) in the Pregolya River according to long-term data (1997, 1998, 2002, 2007)

Biotope	Staraya Pregolya	Novaya Pregolya	The lower reaches (from the 16th km from the mouth)	The mouth area
Riparian	3,069 ± 1,013	3,638 ± 7,738	363 ± 176	2,536 ± 814
Mainstream	1,177 ± 378	2,265 ± 515	552 ± 227	3,546 ± 2,127

Table 6 The average annual biomass of zoobenthos (g m^{-2}) in the Pregolya River according to long-term data (1997, 1998, 2002, 2007)

Biotope	Staraya Pregolya	Novaya Pregolya	The lower reaches (from the 16th km from the mouth)	The mouth area
Riparian	242.4 ± 109.9	51.1 ± 16.8	0.84 ± 0.6	1.69 ± 0.6
Mainstream	79.8 ± 73.0	69.9 ± 34.9	1.1 ± 0.3	3.21 ± 1.9

species of the genus *Valvata*, most species do not contribute sufficiently to benthic biomass (Fig. 11).

The indices of quantitative development of *D. polymorpha*, the main dominant species, are high but experience sharp fluctuations by seasons and years. Therefore, in the long-term aspect, the untypical distribution of the total benthos biomass between the riparian and mainstream zones, caused by *Dreissena* in the Novaya Pregolya, is leveled and is also subject to a general pattern, characterized with the biomass in the coastal region higher than that of the midstream zone (Table 6). The total average annual benthos biomass of the riparian and mainstream zones is calculated on the basis of data from long-term regular observations in 1997–2007 and, therefore, demonstrates stable patterns. In connection with the aggregated distribution inherent in the dominant species of *Dreissena*, the biomass data is characterized by a significant spread (Table 6).

Seasonal and Interannual Dynamics of Quantitative Characteristics of Zoobenthos In the upper sections of the lower reaches, where the Pregolya River is divided into two branches, the seasonal dynamics of the number of total benthos and individual groups varied for the Staraya Pregolya and the Novaya Pregolya (Fig. 12a, b). During the study period, the spring peak of abundance due to the reproduction of benthic bionts was well expressed in the Staraya Pregolya. In the Novaya Pregolya, on the contrary, the spring number of zoobenthos was minimal; the maximum number was observed in the autumn (Fig. 12a, b). In Staraya Pregolya, opposite to Novaya Pregolya branch, the population of zebra mussel, contributing significantly to the spring-summer recruitment of zoobenthos because on larvae settling, is developed to a greater extent. In addition, the Novaya Pregolya in the urban area has a less diverse and depleted bottom fauna because of the more than 100 years of the pulp and paper mill (formerly the North German paper mill) presence, whose waste, accumulated in the bottom sediments, worsened the condition of the benthic fauna not only during its existence but also for many years after the production was stopped.

Since the number of the Pregolya benthos is determined mainly by oligochaetes, a "reverse drift" from the lower sections, caused by the brackish waters due to surging phenomena, can contribute to the growth of this group in the cool period of the year (Fig. 12b). The frequency of surges increases from the second half of August and reaches a maximum in the late autumn. In the lower part of the lower reaches (the Pregolya from the confluence of the branches to the mouth), the number decreases from the beginning to the end of the year, with the minimum marked in the autumn (Fig. 12c, d). This is related to two aspects. In the warm period of the year in the contaminated lower reaches of the Pregolya, there is a frequent formation of oxygen-free zones and the release of hydrogen sulfide, the highest in the second half of summer. Fine black pelitic ooze, which is the predominant substrate in this part of the Pregolya bed, is easily agitated and, together with benthic bionts, is taken upstream with heavier brackish waters passing along the bottom during the autumn surges.

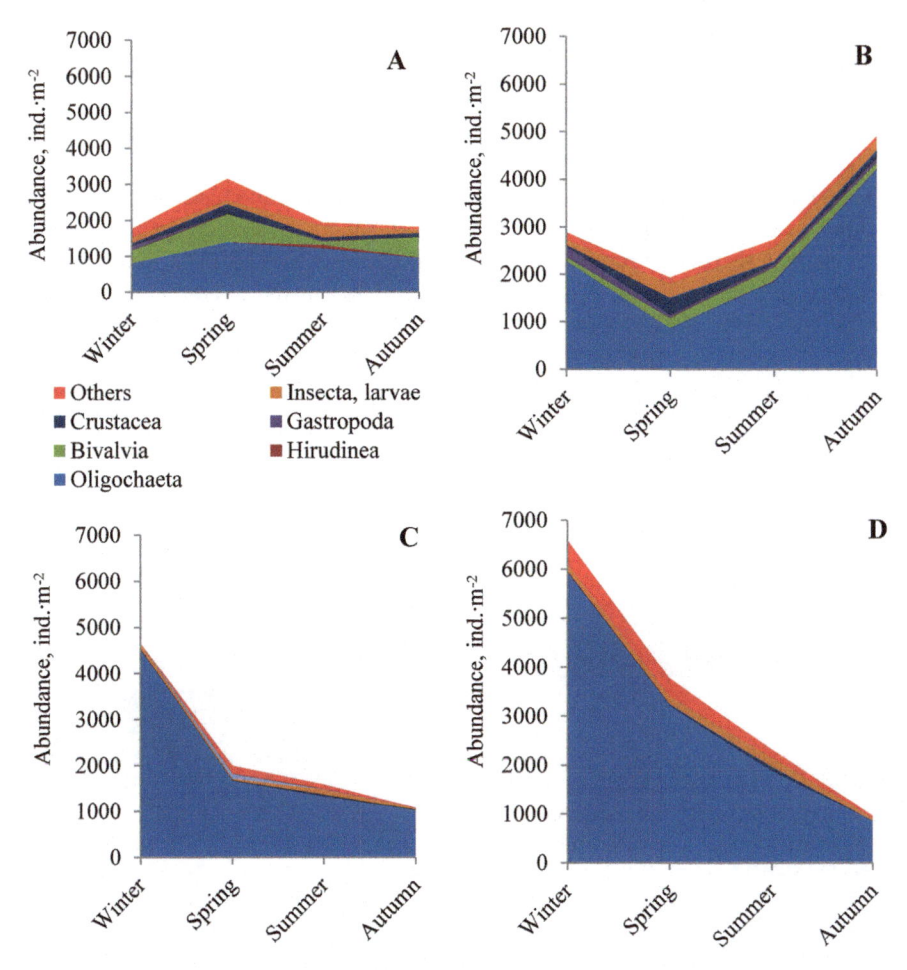

Fig. 12 Average numbers of zoobenthos and its main groups during year seasons (based on 1997, 1998, 2002, 2007 data): (**a**) Staraya Pregolya, (**b**) Novaya Pregolya, (**c**) lower reaches of the Pregolya River, (**d**) the mouth area

The seasonal change in the zoobenthos total biomass also differed markedly in different parts of the river. In the upper regions, on the Novaya Pregolya and the Staraya Pregolya (Fig. 13A, B), it was determined mostly by the features of the life cycle and the entry into reproduction of the main dominant mollusks. In the Pregolya, from the confluence of the arms to the mouth, where until 2007 mollusks were registered only occasionally and accidentally, with oligochaetes being the main contributors to the biomass, the biomass dynamics (decrease from winter to autumn) (Fig. 13C, D), as well as that of the numbers, can be determined by the increase of frequency of oxygen-deficiency conditions by the end of summer and frequent in autumn wind-driven surges, which drift upstream fine silt along with oligochaetes.

The population of is *D. polymorpha* has been dynamically developed in the Pregolya branches, and now it is common to be met also below the confluence of the branches. Samplings in 2011–2014 showed that the zebra mussel, limnaeids, and even swan mussels are now present in the lower 11 km of the river, where at least a fragmented coastal biotope is developed. The reduction in industrial pollution from the late 1990s, obviously, has contributed to the spread of mollusks downstream.

In general, in the late 1990s and early 2000s, there was an increase in quantitative characteristics of macrobenthos [73]. This tendency is especially pronounced if we consider the long-term dynamics of the benthos abundance and biomass in the urban part of the Pregolya (transects 22–27) (Figs. 14 and 15). Since 1998, there

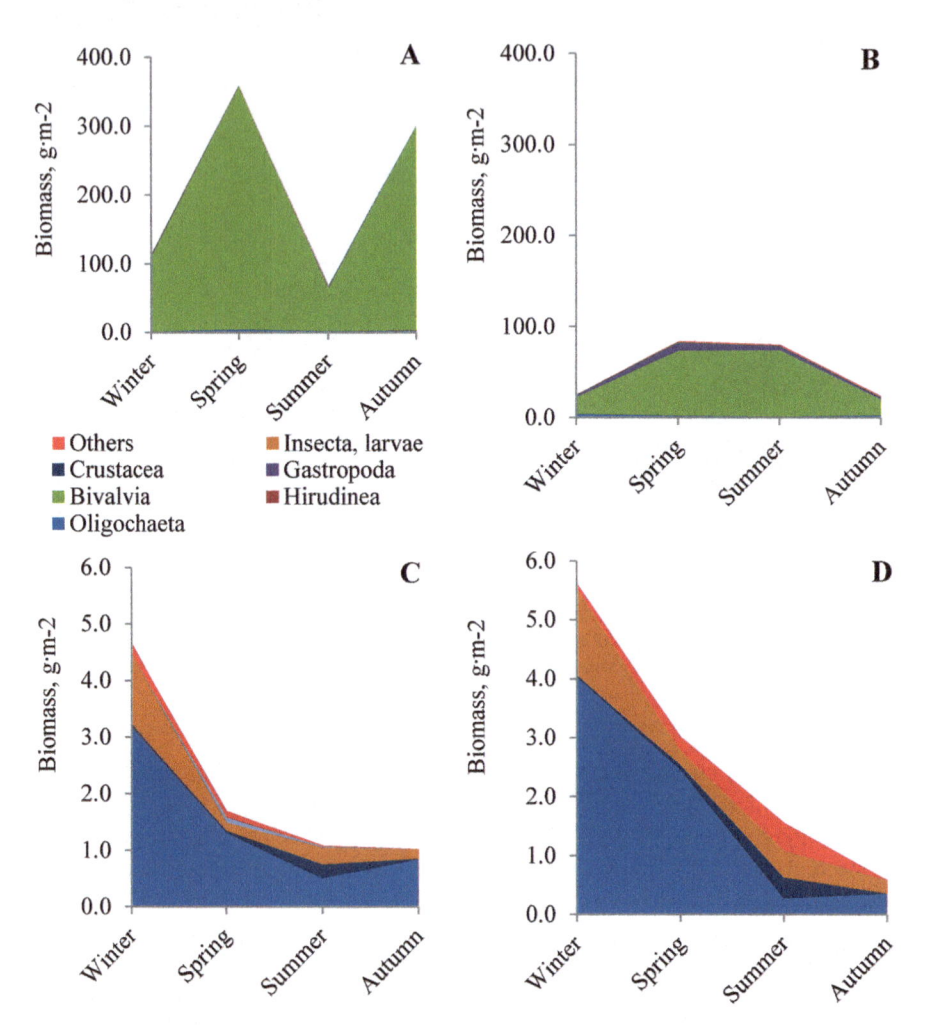

Fig. 13 Average biomass of zoobenthos and its main groups during year seasons (at base 1997, 1998, 2002, 2007 data): (**a**) Staraya Pregolya, (**b**) Novaya Pregolya, (**c**) lower reaches of the Pregolya River, (**d**) the mouth region

was a significant increase in the average biomass and abundance of bottom animals in the most polluted part of the river from 0.34 g m^{-2} and 60 ind. m^{-2} in 1995 to 6.13 g m^{-2} and 3,759 ind. m^{-2} in 2000 (Fig. 14a, b). On the transects located within Kaliningrad above the confluence of the branches, a reliable significant increase in biomass and abundance of benthic animals is also observed in the Novaya Pregolya and the Staraya Pregolya (transects 28, 28o, 30, 30o), although the multiplicity of the increase is not as great as in the lower part of the river, but the absolute values reached by the average abundance and biomass are higher: from 50 to 150 g m^{-2} (Fig. 15a, b).

Evaluation of Zoobenthos Production To calculate the production of zoobenthos, the data on the "Berlin bridge" transect (transects 30 and 30o) in the

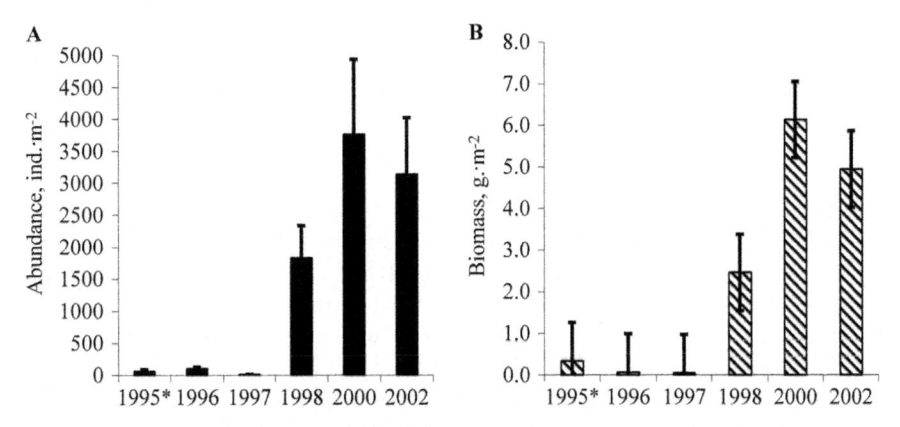

Fig. 14 Average summer zoobenthos abundance (**A**) and biomass (**B**) in the Pregolya River system within Kaliningrad, below the confluence of the Staraya and the Novaya Pregolya, 1995–2002. *Summer samples only

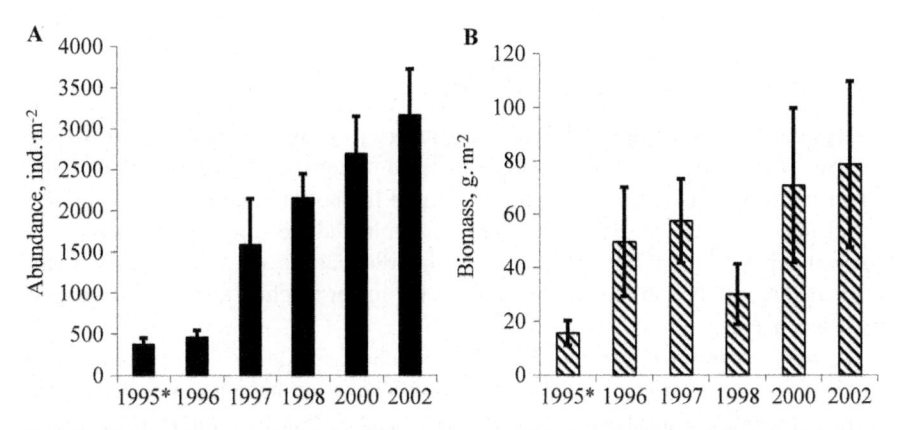

Fig. 15 Average summer abundance (**A**) and biomass (**B**) of zoobenthos in the Pregolya River system within Kaliningrad, averaged for the Staraya and the Novaya Pregolya, 1995–2002. *Summer samples only

Table 7 Average values of production characteristics (P, R, kJ m^{-2}, and P/B-coefficient for the vegetation season) for the period from 1997 to 2006 in the Pregolya River

Transect	Biotope	P	R	P/B
30	Riparian	595.0	1,596.7	4.49
	Mainstream	258.6	662.6	4.21
30o	Riparian	238.7	624.4	3.42
	Mainstream	20.1	55.4	4.76

branches of the Novaya Pregolya and the Staraya Pregolya were used. Calculated data on the macrobenthos production of the Pregolya were obtained over a 10-year period (from 1997 to 2006); the results of processing 107 samples were used. The production of zoobenthos during the vegetation period at the stations studied varied from 20.1 to 595 kJ m^{-2} with a change in the biomass values from 13.1 to 198.8 kJ m^{-2} (Table 7). The production of zoobenthos during the vegetative period in the mainstream zone is higher than in the riparian one (2.3 times higher at the transect 30 and 12 times at the transect 30o).

Obtained data on the zoobenthos production in the Pregolya was compared with the data on 11 small rivers of the Middle Volga Basin [74]. For the latter, the average macrobenthos production during the vegetation season varied from 63.7 to 1,675.3 kJ m^{-2}, with the average annual production varying from 101.2 to 2,050.2 kJ m^{-2}. These values are somewhat higher than those found for the Pregolya. For the available data on the benthos of the Karelian rivers Porja and Kuzreka [75], using Kukharev's formula [75, 76] for the connection between the average biomass and production in the Karelian rivers during the vegetation season, $P_b = (2.11 \pm 0.24) \cdot B$, where P_b и $B_{\text{ср}}$ are given in kJ m^{-2} [76], the values of the average seasonal production during the vegetation season in the small rivers of Karelia were calculated. The productivity values for the rivers Porja and Kuzreka, from 223.3 to 538.9 and 179.6 to 422.4 kJ m^{-2}, respectively, were close to those obtained in the Pregolya River.

Thus, the productivity indicators for the vegetation season are generally quite high and comparable with the average production values known for small rivers and water bodies in the European part of Russia.

Assessment of the Ecological Status of the Pregolya by Macrozoobenthos The indicators of zoobenthos biomass in the studied biotopes and areas of the Pregolya River indicate a high biological productivity of the bottom throughout the middle-lower reaches, with the exception of the contaminated area of Pregolya in the city. In clean and slightly polluted waters of the middle reaches, the basis of the biomass is created by mollusk-filter feeders while in the lower reaches, unproductive course by oligochaetes.

At the present time, riparian biotopes have changed significantly on both banks of the river, with various rooted and floating vegetation having developed significantly, including various species of pondweed, dark green *Ceratophyllum*, black-amoor, common reed, bulrush, yellow and white water lilies, water plantain, and arrowhead. Immersed and floating vegetation (Canada waterweed, parrot's feather, duckweed) was also noted in the mainstream zone of the river [73]. Some of these species were and have been encountered all along the riverbed from transects 30 and 30o to transects 26 and 25, located in the city, where previously there were signs of heavy pollution. This change in the nature of the riparian biotope provides an opportunity for abundant zoobenthos development in the places, where, in 1980–1990, "dead zones" were observed.

Analysis of interannual changes in macrozoobenthos biodiversity, abundance, and biomass in 1990–2007 showed a change in the environmental situation for the better in the 2000s in comparison with the 1990s. The population of the lower areas now comprises of various mollusks, including filter feeders, in particular, species indicators, preferring water moderately contaminated with organic matter: large bivalves *Anodonta cygnea*, *Dreissena polymorpha*, large gastropods *Viviparus viviparus*, *Bithynia tentaculata*, and *Planorbarius corneus*. Earlier, these species were not observed in the urban part of the river, or there were occasional individuals that had been drifted by the current from the upstream river part.

However, the indicators of biodiversity and quantitative development of benthic fauna in the city river part (transects 27–24) remain still low. Most likely, this is due to the accumulated long-term layer of bottom sediments contaminated with toxic substances, as well as to an excessive amount of organic matter, which decomposes with the liberation of hydrogen sulfide [77]. These conditions are extremely unfavorable for the development of the bottom fauna. This situation is predicted to persist for a long time, since in the deeper parts of the riverbed (navigable channel), a peculiar trap for pollution is formed.

The analyzed data on macrozoobenthos indicate a significant improvement in the state of the river ecosystem by the end of the first decade of the twenty-first century in comparison with the situation in the early 1980s and the first half of the 1990s. An increase in the species diversity of the macrozoobenthos has been noted, as well as an increase in the quantitative development of the bottom fauna throughout the lower reaches of the Pregolya River. There is a gradual restoration of biocenosis of bottom invertebrates in the lower reaches of the river, which indicates the improvement of the entire river ecosystem.

4 Conclusions

Data on the ecological state of the Pregolya lower reaches in the past 100 years have indicated a significant improvement from about the mid-2000s. For the period from 1946 to 1982, the ecological situation in the lower reaches can be characterized as severe. The comparison of data for 1982, 1990–1992, and 1995–1998 demonstrates the gradual depletion of river communities of plankton and benthos by 1995, which is explained by chemical pollution and anthropogenic eutrophication. Since 1997 there has been a sharp increase in the quantitative characteristics and qualitative diversity of communities.

The analysis of interannual changes in biodiversity, abundance, and biomass of components of the river biota in 2000–2011 showed an improvement in the ecological state of the investigated section of the river compared to the 1990s. The boundaries of biocenoses with the dominance of mollusk-filter feeders and rooted aquatic vegetation gradually advanced downstream, with appearance of the benthos species that were not observed in the lower reaches in the 1980–1990s.

Currently, despite the still significant level of pollution of the Pregolya waters by municipal sewage, the biota of the river in the lower reaches is in the best condition ever recorded in the scientific press.

Acknowledgments This work was done in the frames of IORAS budget theme 0149-2014-0055 and supported by the Federal Target Program "World Ocean" (theme 01201162712). The authors' team is grateful to Dr. Vadim Paka, Sergei Aleynikov, Viktor Shkurenko, Alexey Kuleshov (all AB IO RAS) and Dr. Galina Tsybaleva (Kaliningrad Technical University), without whom this interdisciplinary biological study of the Pregolya River was never started and never done.

Appendix 1

Taxonomic composition of phytoplankton in the Pregolya River, April–October 1996–1997, 2011

Division Chlorophyta	Division Cyanophyta	Division Bacillariophyta
Actinastrum hantzschii Lagerh.	*Anabaena contorta* Bachm.	*Amphora ovalis* Kutz.
Actinastrum hantzschii v. fluviatile Schroed.	*Anabaena flos-aquae* f. aptekariana Elenk.	*Amphora veneta* Kutz.
Ankistrodesmus acicularis (A. Br.) Korschik.	*Anabaena lemmermannii* P. Richt.	*Asterionella formosa* Hass.
Ankistrodesmus angustus Bern.	*Anabaena spiroides* Kleb.	*Asterionella gracillima* (Hantzsch.) Heib.
Ankistrodesmus longissimus v. acicularis (Chod) Brunnth.	*Anabaena variabilis* Kutz.	*Caloneis bacillum* v. lancettula (Schuiz) Hust.
Ankistrodesmus minutissimus Korschik.	*Anabaenopsis elenkinii* v. Miller	*Chaetoceros muelleri* Lemm.
Ankistrodesmus pseudomirabilis Korschik.	*Aphanizomenon flos-aquae* (L.) Ralfs	*Chaetoceros wighamii* Bright.
Ankistrodesmus pseudomirabilis v. spiralis (T.) L. (=*Monoraphidium contortum* (Thuret) Komàrková-Legnerová)	*Aphanothece clathrata* W. et G. S. West	*Cocconeis placentula* Ehr.
Ankistrodesmus rotundus Korschik.	*Coelosphaerium kuetzingianum* Nag.	*Cocconeis placentula* v. intermedia (Herib. et Perag.) Cl.
Binuclearia lauterbomii (Schm.) Proch.-Lavrenkei	*Coelosphaerium pusillum* van Goor	*Coscinodiscus excentricus* f. fasciculata Hust.
Carteria klebsii (Dang.) France	*Dactylococcopsis elenkinii* Roll	*Coscinodiscus lacustris* Grun.

(continued)

Division Chlorophyta	Division Cyanophyta	Division Bacillariophyta
Chlamydomonas conferta Korsch.	*Dactylococcopsis planctonica* Teiling	*Cyclotella atomus* Hustedt
Chlamydomonas globosa Snow	*Eucapsis alpina* Clem. et Shantz	*Cyclotella comta* (Ehr.) Kutz.
Chlamydomonas gloeocystiformis Dill	*Eucapsis minor* (Skuja) Hollerb,	*Cyclotella Kuetzingiana* Thw.
Chlamydomonas incerta Pasch.	*Gomphosphaeria aponina* Kutz.	*Cyclotella meneginiana* Kutz.
Chlamydomonas monadina v. charkowiensis Korsch.	*Gomphosphaeria lacustris* Chod. (= Snowella lacustris (Chodat) Komárek and Hindák)	*Cyclotella stelligera* Cl. et Grun.
Chlamydomonas noctigama Korsch.	*Gomphosphaeria lacustris* f. compacta (Lemm.) El.	*Cymatopleura solea* (Breb) W. Sm.
Chlamydomonas Reinhardtii Dang.	*Gomphosphaeria pusilla* (van Goor) Komarek	*Cymbella amphicephala* Nag
Chlorella vulgaris Beyer.	*Lyngbya contorta* Lemm.	*Cymbella estonica* Molder
Chlorococcum dissectum Korschik.	*Lyngbya limnetica* Lemm. (=Planktolyngbya limnetica (Lemm.) Kom.-Legn.et Cronb)	*Cymbella lanceolata* (Ehr.) V.H.
Closterium gracile Breb.	*Merismopedia elegans* A.Br.	*Cymbella lata* v. minor Molder
Coelastrum microporum Naeg.	*Merismopedia glauca* (Ehr.) Nag.	*Cymbella tumidula* Grun.
Crucigenia fenestrata Schmidle	*Merismopedia major* (Smith) Geitl.	*Cymbella ventricosa* Kutz.
Crucigenia quadrata Morren	*Merismopedia punctata* Meyen	*Diatoma elongatum* v. tenuis (Ag.) V.H.
Crucigenia rectangularis (A. Br.) Gay	*Merismopedia tenuissima* Lemm.	*Epithemia argus* Kutz.
Crucigenia tetrapedia (Kirchn.) W. et W.	*Microcystis aeruginosa* f. flos-aquae (Wittr.) Elenk.	*Fragilaria bicapitata* A. Mayer
Dictyosphaerium ehrenbergianum Naeg.	*Microcystis* (=Aphanocapsa) *pulverea* (Wood) Forti em. Elenk.	*Fragilaria construens* (Ehr.) Grun.
Dictyosphaerium pulchellum Wood	*Microcystis pulverea* f. conferta (W. G. S. West) El. (= Aphanocapsa conferta (W. et G. S. West) Kom.-Legn. et Cronb.	*Fragilaria crotonensis* Kitt.
Dictyosphaerium simplex Korschik.	*Microcystis pulverea* f. delicatissima (W.et G. S. West) Elenk. (=Aphanocapsa delicatissima W. et G. S. West	*Gomphonema olivaceum* (Lyngb.) Kutz.
Didymocystis planctonica Korschik.	*Microcystis pulverea* f. holsatica (Lemm.) Elenk. (=Aphanocapsa holsatica (Lemm.) Cronb. et Komárek)	*Gomphonema parvulum* (Kutz.) Grun.

(continued)

Division Chlorophyta	Division Cyanophyta	Division Bacillariophyta
Didymocystis tuberculata Korschik.	*Microcystis pulverea* f. incerta (Lemm.) Elenk. (=Aphanocapsa incerta (Lemm.) Cronb. et Komárek	*Gyrosigma acuminatum* v. lacustre Meist.
Eudorina illinoisensis Pasch.	*Oscillatoria agardhii* Gom.	*Melosira arenaria* Moore
Golenkiniopsis solitaria Korschik.	*Oscillatoria curviceps* Ag.	*Melosira granulata* (Ehr.) Ralfs (= *Aulacoseira granulata* (Ehr.) Simons.)
Hormidium flaccidum A. Br.	*Oscillatoria geminata* (Menegh.) Gom.	*Melosira islandica* v. helvetica O. Mull. (= *Aulacoseira islandica* subsp. helvetica (O. F. Müller) Sim.
Kirchneriella intermedia v. major Korschik.	*Oscillatoria limnetica* Lemm.	*Melosira nummuloides* (Dillw.) Ag.
Kirchneriella irregularis (Smith.) Korschik.	*Oscillatoria limosa* Ag.	*Melosira varians* Ag.
Kirchneriella obesa (West.) Schmidle	*Oscillatoria planctonica* Wolosz. (= *Limnothrix planctonica* (Wolosz.) Meffert	*Meridion circulare* Ag.
Lagerheimia ciliata (Lagerh.) Chod.	*Oscillatoria tenuis* Ag.	*Navicula cincta* (Ehr.) Rutz.
Lagerheimia citriformis (Snow) G. M. Smith	*Phormidium tenue* (Menegh.) Lorn.	*Navicula cincta* v. leptocephala (Breb.) Grun.
Lagerheimia genevensis Chod.	*Pseudoholopedia convoluta* (Breb.) Elenk.	*Navicula cryptocephala* Kutz.
Lagerheimia longiseta (Lemm.) Printz	*Rhabdoderma lineare* Hollerb.	*Navicula cryptocephala* v. intermedia Grun.
Lagerheimia marssonii Lemm.	*Rhabdoderma* sp. novae	*Navicula dicephala* (Ehr.) W. Sm.
Lagerheimia octacantha Lemm.	*Romeria gracilis* Koszw.	*Navicula gracilis* Ehr.
Lagerheimia subsalsa Lemm.	*Snowella rosea* (Snow) Elenk.	*Navicula halophila* f. subcapitata Ostr.
Lambertia setosa (Filarszky) Koschik.	*Spirulina gomontiana* (Setch.) Geitl.	*Navicula hungarica* v. capitata Cl.
Lauterbomiella elegantissima Schmidle	*Spirulina laxa* Smith	*Navicula laterostrata* Hust.
Micractinium pusillum Fr.	*Synechococcus aeruginosus* Nag.	*Navicula microcephala* Grun.
Micractinium quadrisetum (Lemm.) G.M.	*Synechococcus elongatus* Nag.	*Navicula natchikae* Boye
Nephrochlamys willeana (Printz) Korschik.	*Synechocystis aquatilis* Sauv.	*Navicula pupula* v. mutata (Krasske) Hust.
Oocystis lacustris Chod.	*Synechocystis crassa* Woronich.	*Navicula pupula* v. rostrata Hust.
Oocystis borgei Snow		*Navicula semen* Ehr.

(continued)

Division Chlorophyta	Division Cyanophyta	Division Bacillariophyta
Oocystis gigas Archer	**Division Euglenophyta**	*Navicula subrhombica* Hust.
Oocystis parva W. et G. S. West	*Euglena acus* Ehrenb.	*Navicula verecunda* Hust.
Oocystis pusilla Hansg.	*Euglena limnophila* v. swirenkoi (Am.) Pop.	*Nitzschia acicularis* W. Sm.
Oocystis solitaria Wittr.	*Euglena matvienkoi* Popova	*Nitzschia angustata* (W. Sm.) Grun.
Oocystis submarina Lagerh.	*Euglena oxyuris* Schmarda	*Nitzschia closterium* (Ehr.) W. Sm.
Oocystis verrucosa Roll	*Euglena pisciformis* Klebs	*Nitzschia denticula* Grun.
Pandorina charkoviensis Korsch.	*Euglena proxima* Dang.	*Nitzschia distans* Greg.
Pandorina morum (Mull.) Bory	*Phacus parvulus* Klebs	*Nitzschia gracilis* Hantzsch
Pediastrum boryanum (Turp.) Menegh.	*Trachelomonas planctonica* Swir	*Nitzschia longissima* (Breb.) Ralfs.
Pediastrum duplex v. comutum Racib.	*Trachelomonas planctonica* v. oblonga Drez.	*Nitzschia palea* (Kutz.) W. Sm.
Pediastrum simplex Meyen	*Trachelomonas superba* v. echinata (Roll) Popova	*Nitzschia paleacea* Grun.
Phacotus coccifer Korsch.	*Trachelomonas volvocina* Ehr.	*Nitzschia parvula* Lewis
Polytoma caudatum Korsch.		*Nitzschia sigmoidea* (Ehr.) W. Sm,
Pteromonas angulosa Lemm.	**Division Chrysophyta**	*Nitzschia sublinearis* Hust.
Pteromonas robusta Korsch.	*Chromulina rosanoffii* (Woronin) Butschli	*Nitzschia tryblionella* v. levidensis (W. Sm.) Grun.
Scenedesmus acuminatus (Lagerh.) Chod.	*Chrysococcus biporus* Skuja	*Pleurosigma elongatum* W. Smith
Scenedesmus acuminatus v. bemardii (Smith) Deduss.	*Chrysococcus omatus* Pasch.	*Rhoicosphaenia curvata* (Kutz.) Grun.
Scenedesmus acuminatus v. biseriatus Reinh.	*Chrysococcus rufescens* Klebs	*Stephanodiscus astrea* (Ehr.) Grun.
Scenedesmus acuminatus v. elongatus G. M. Smith	*Coccolithus huxleyi* Lohm.	*Stephanodiscus astraea* v. minutulus (Kutz.) (= Stephanodiscus minutulus (Kützing) Cleve and Möller)
Scenedesmus apiculatus W. et W. Chod.	*Dinobryon divergens* Imhof	*Stephanodiscus hantzschii* Grun.
Scenedesmus bijugatus (Turp.) Kutz.	*Kephyrion cupuliforme* Conr.	*Surirella angustata* Kutz.
Scenedesmus bijugatus v. altemans (Reinsch) Han.	*Kephyrion doliolum* Conr.	*Surirella ovata* Kutz.
Scenedesmus brasiliensis Bohl.	*Kephyrion sitta* Pasch.	*Synedra acus* Kutz.
Scenedesmus denticulatus Lagerh.	*Kephyrion rubri-claustri* Conr.	*Synedra pulchella* (Ralfs) Kutz.

(continued)

Division Chlorophyta	Division Cyanophyta	Division Bacillariophyta
Scenedesmus granulatus W. et W.	*Mallomonas acaroides* v. moskowiensis (Werm.) Krieg.	*Synedra ulna* (Nitzsch.) Ehr.
Scenedesmus obliquus (Turp.) Kutz.	*Pseudokephyrion circumcisum* Conr.	*Thalassiosira baltica* (Grunow) Ostenfeld
Scenedesmus obliquus v. altemans Christjuk	*Pseudokephyrion depressum* Schmid	*Thalassiosira nana* Lohm.
Scenedesmus opoliensis v. carinatus Lemm.	*Pseudokephyrion latum* (Schill.) Schmid	
Scenedesmus opoliensis v. setosus Deduss.	*Pseudokephyrion pilidum* Schill.	**Division Cryptophyta**
Scenedesmus protuberans Fritsch	*Pseudokephyrion schilleri* Conr.	*Chilomonas cryptomonadoides* Skuja
Scenedesmus protuberans v. aristatus (Chod.) Ded.	*Pseudokephyrion spirale* Schmid	*Chilomonas minor* Czosnowski
Scenedesmus quadricauda (Turp.) Breb.	*Stenokalyx densata* Schmid	*Cryptella cyanophora* Pasch.
Scenedesmus quadricauda v. abundans Kirchn.	*Stenokalyx laticollis* Conr.	*Cryptomonas brevis* Schiller
Scenedesmus quadricauda v. africanus Fritsch	*Stenokalyx monilifera* Schmid	*Cryptomonas compressa* Pasch.
Scenedesmus quadricauda v. armatus (Ch.) Ded.	*Stenokalyx parvula* Schmid	*Cyatomonas truncata* (Fres.) From.
Scenedesmus quadricauda v. dentatus Deduss.	*Synura lapponica* Skuja	*Rhodomonas lens* Pasch.
Scenedesmus quadricauda v. eualtemans Proschk.	*Synura uvella* Ehr.	
Scenedesmus quadricauda v. papillatus Swir.		**Division Dinophyta**
Scenedesmus quadricauda v. setosus Kirchn.	**Division Xantophyta**	*Amphidinium geitleri* H.-P.
Scenedesmus quadricauda v. striatus Deduss.	*Ophiocytium mucronatum* Rab.	*Amphidinium luteum* Skuja
Scenedesmus striatus Deduss.	*Tribonema vulgare* Pasch.	*Glenodinium berolinense* (Lemm.) Lind.
Scenedesmus striatus v. apiculatus Deduss.		*Glenodinium gymnodinium* Penard
Staurastrum tetracerum Ralfs		*Gymnodinium albulum* Lind.
Stichococcus bacillaris Naeg.	**Division Zoomastigophora**	*Gymnodinium blax* Harris
Stigeoclonium tenue Kutzing	*Ebria tripartita* (Schum.) Lemm.	*Gymnodinium colymbeticum* Harris
Tetraedron caudatum (Corda) Hansg.		*Gymnodinium eurytopum* Skuja
Tetraedron incus (Teil.) G. M. Smith		*Gymnodinium fungiforme* Aniss.

(continued)

Division Chlorophyta	Division Cyanophyta	Division Bacillariophyta
Tetraedron minimum (A.Br.) Hansg.		*Gymnodinium saginaturn* Harris
Tetraedron triangulare Korschik.		*Massartia stigmatica* (Lind.) Stein
Tetrastrum glabrum (Roll) Ahlstr. et Tiff.		
Tetrastrum hastiferum (Am.) Korschik.		
Tetrastrum staurogeniaeforme (Schroed.) Lemm.		
Tetraselmis sp.		
Ulothrix variabilis Kutz.		
Actinastrum hantzschii Lagerh.		
Actinastrum hantzschii v. fluviatile Schroed.		
Ankistrodesmus acicularis (A.Br.) Korschik.		
Ankistrodesmus angustus Bern.		
Ankistrodesmus longissimus v. acicularis (Chod) Brunnth.		

Appendix 2

The list of aquatic plant species of the lower reach of the Pregolya River

No	Species	Ecological traits			
		1	2	3	4
Division Chlorophyta					
1.	*Hydrodictyon reticulatum* (Linnaeus) Bory de Saint-Vincent	1.1	e	eus	–
2.	*Cladophora glomerata* (Linnaeus) Kütz	1.2	ме	β	eh
3.	*Oedogonium* sp.	1.4	–	o-β	–
4.	*Ulva intestinalis* Linnaeus	1.2	ме	β-α	ssf
5.	*U. prolifera* O. F. Müller	1.2	–	–	ssf
Division Charophyta					
6.	*Spirogyra varians* (Hassal) Kützing	1.1	о-м	β-α	f
7.	*Spirogyra fluviatilis* Helsa	1.4	мt	β	f
Division Ochrophyta					
8.	*Vaucheria* sp.	1.2	–	β	–

(continued)

| | | Ecological traits ||||
No	Species	1	2	3	4
Division Bryophyta						
9.	*Fontinalis antipyretica* Hedw	1.2	o	o-β	uf
Division Equisetophyta, class Equisetopsida, order Equisetales						
10.	*Equisetum fluviatile* L.	2.1	мt	o-β	f
Division Angiospermae, class Liliopsida						
Fam. Alismataceae Vent.						
11.	*Alisma plantago-aquatica* L.	2.1	мt	β	uf
12.	*Sagittaria sagittifolia* L.	2.1	мt	β	f
Fam. Araceae Juss.						
13.	*Acorus calamus* L.	3	мt	o-β	f
Fam. Butomaceae Rich.						
14.	*Butomus umbellatus* L.	2.1	мe	eus	f
Fam. Cyperaceae Juss.						
15.	*Bolboschoenus maritimus* Palla.	3	мe	β	msf
16.	*Carex acuta* L.	3	o-м	o-β	f
17.	*C. riparia* Curt.	3	мt	β	f
18.	*C. pseudocyperus* L.	3	o-м	β	f
19.	*Eleocharis palustris* (L.) Roem. et. Schult	3	мe	β	slf
20.	*Scirpus lacustris* L.	2.2	мt	β	slf
Fam. Hydrocharitaceae Juss.						
21.	*Elodea canadensis* Michx.	1.2	мe	β	f
22.	*Hydrocharis morsus-ranae* L.	1.4	мt	β	uf
23.	*Stratiotes aloides* L.	1.1	мt	β	uf
Fam. Iridaceae Juss.						
24.	*Iris pseudacorus* L.	3	o	os	f
Fam. Lemnaceae S. F. Gray						
25.	*Lemna minor* L	1.4	e	β-α	uf
26.	*L. trisulca* L.	1.1	e	eus	msf
27.	*L. gibba* L.	1.4	–	β	–
28.	*Spirodela polyrhiza* (L.) Schleid.	1.4	мe	o-β	f
Fam. Gramineae Juss.						
29.	*Glyceria fluitans* (L.) B. Br.	2.2	мt	β	f
30.	*G. maxima* (C. Hartm.) Halmb.	2.2	мt	β	f
31.	*Phragmites australis* (Cav.) Trin. ex Stend.	2.2	eu	eus	ssf
Fam. Potamogetonaceae Dumort.						
32.	*Potamogeton acutifolius* Linc.	1.2	–	–	–
33.	*P. berchtoldii* Fieb.	1.2	мt	β	f
34.	*Potamogeton friesii* Rupr.	1.2	e	β	ssf
35.	*P. pectinatus* L.	1.2	eu	eus	ssf
36.	*P. perfoliatus* L.	1.2	мe	β	ssf
37.	*P. lucens* L.	1.2	мt	β	uf
38.	*P. nodosus*	1.3	–	–	–

(continued)

		Ecological traits			
No	Species	1	2	3	4
Fam. Sparganiaceae Rudolphi					
39.	*Sparganium emersum* Rehm.	2.1	мt	o-β	f
40.	*Sparganium erectum* L.	2.1	мt	β	uf
Fam. Typhaceae Juss.					
41.	*Typha angustifolia* L.	2.2	мe	β	slf
42.	*T. latifolia* L.	2.2	o-м	o-β	uf
Class Dicotyledoneae					
Fam. Apiaceae Lindl.					
43.	*Cicuta virosa* L.	3	мt	o-β	f
44.	*Oenanthe aquatica* (L.) Poir.	3	мe	β	f
45.	*Sium latifolium* L.	3	мt	β	uf
Fam. Cruciferae Juss.					
46.	*Rorippa amphibia* (L.) Besser	3	мt	β	f
Fam. Ceratophyllaceae S. F. Gray					
47.	*Ceratophyllum demersum* L.	1.1	мt	β-α	f
Fam. Haloragaceae R. Br.					
48.	*Myriophyllum spicatum* L.	1.2	мe	β	ss
Fam. Lythraceae J. St. Hil.					
49.	*Lythrum salicaria* L.	3	–	–	–
Fam. Nymphaeaceae Salisb					
50.	*Nuphar lutea* (L.) Smith.	1.3	мt	eus	uf
51.	*Nymphaea candida* J. Presl.	1.3	мt	o-β	uf
Fam. Plantaginaceae Juss.					
52.	*Callitriche palustris* L.	1.2	мt	β	f
Fam. Polygonaceae Juss.					
53.	*Persicaria amphibia* (L.) S. F. Gray	1.3	мe	β	uf
54.	*Rumex aquaticus* L.	3	–	–	–
Fam. Ranunculaceae Juss.					
55.	*Batrachium circinatum* (Sibth.) Spach	1.2	мe	β-α	uf
56.	*B. eradicatum*(Laest) Fries	1.1	мe	β	f
57.	*Caltha palustris* L.	3	мt	β	f
58.	*Thalictrum flavum* L.	3	–	–	–
Fam. Rosaceae Juss.					
59.	*Comarum palustre* L.	3	o-м	o-β	f

Notation: 1 ecological group. Type 1. Hydrophytes: 1.1, floating in the water column; 1.2, submersed rooting; 1.3, submersed rooting with floating leaves; 1.4, floating on the water surface. Type 2. Helophytes: 2.1, low-grass helophytes; 2.2, tall-grass helophytes. Type 3. Hydrohelophytes: 2 trophy indicators: *o* oligotrophic, *o-м* oligomesotrophic, *мt* mesotrophic, *мe* mesoeutrophic, *e* eutrophic, *eu* eurythrophic; 3 indicators of saprobity: *os* oligosaprobic, *o-β* oligo-β-mesosaprobic, *β* β-mesosaprobic, *β-α* β-α-mesosaprobic, *eus* eurysaprobic; 4 indicators of mineralization: *f* typically freshwater, *slf* slightly saline-freshwater, *ss* slightly saline, *msf* medium-saline-freshwater, *ssf* strongly salinous-freshwater, *eh* euryhaline, *uf* ultra-fresh, – no data

Appendix 3

The list of zooplankton species in the low reach of the Pregolya River, 1996–2007

	Species	Staraya Pregolya	Novaya Pregolya	Pregolya within Kaliningrad	The mouth area
Rotifera					
1	*Asplanchna* spp.	+	+	+	
2	*Asplanchna priodonta* Gosse	+	+	+	+
3	*Brachionus angularis* Gosse	+	+	+	+
4	*Brachionus quadridentatus* Hermann	+	+	+	+
5	*Brachionus diversicornis* (Daday)	+		+	
6	*Brachionus diversicornis homoceros* Wierzejski	+	+	+	
7	*Brachionus calyciflorus* Pallas	+	+	+	+
8	*Br. calyciflorus anuraeiformis* Brehm			+	+
9	*Brachionus calyciflorus spinosus* Wierzejski		+		
10	*Brachionus rubens* Ehrenberg		+		
11	*Brachionus urceus* Linne		+		
12	*Cephalodella ventrosa* (Ehrenberg)	+			
13	*Euchlanis dilatata* Ehrenberg	+	+	+	+
14	*Filinia longiseta* Ehrenberg	+	+	+	+
15	*Hexarthra fennica* Levander		+	+	+
16	*Keratella quadrata* O. F. Muller	+	+	+	+
17	*Keratella cochlearis* Gosse	+		+	+
18	*Kellicottia longispina* Kellicott	+	+	+	
19	*Lecane luna* O. F. Muller	+			
20	*Notholca acuminata* Ehrenberg	+			
21	*Platyias quadricornis* Ehrenberg	+	+		
22	*Polyarthra trigla* Ehrenberg	+		+	
23	*Polyarthra vulgaris* Carlin	+	+		
24	*Polyarthra dolichoptera* Idelson		+		
25	*Rotaria rotatoria* Pallas	+	+	+	+
26	*Synchaeta* sp. Ehrenberg	+	+	+	+

(continued)

	Species	Staraya Pregolya	Novaya Pregolya	Pregolya within Kaliningrad	The mouth area
27	*Trichocerca capucina* Wierzejski et Zacharias	+			
28	*Trichocerca pusilla* (Lauterborn)		+		
29	*Trichocerca rattus* (Muller)	+	+		
Total Rotifera		**22**	**21**	**17**	**12**
Copepoda
Cyclopoidae
1	*Megacyclops gigas* (Claus)	+			
2	*Megacyclops viridis* (Jurine)	+	+	+	+
3	*Acanthocyclops vernalis* (Fisher)	+	+	+	+
4	*Diacyclops bicuspidatus* (Claus)	+	+	+	+
5	*D. languidoides* (Lilljeborg)	+	+	+	+
6	*Cyclops furcifer* Claus	+	+	+	+
7	*Cyclops scutifer* Sars	+	+	+	
8	*Cyclops vicinus* Uljanin	+	+	+	
9	*Cyclops strenuus* Fischer				+
10	*Cyclops kolensis* Lilljeborg	+			+
11	*Cyclops lacustris* (Sars)	+			
12	*Eucyclops macrurus* Sars	+	+		
13	*Eucyclops serrulatus* (Fischer)	+	+		
14	*E. liljiborgi* (Sars)	+	+		
15	*E. macruroides* (Lilljeborg)	+	+	+	
16	*Mesocyclops leuckarti* Claus	+	+	+	+
17	*Mesocyclops crassus* Fischer	+	+	+	+
18	*Mesocyclops ogunnus* Onabamiro		+	+	
19	*Macrocyclops albidus* Jurine	+			
20	*Paracyclops fimbriatus* Fischer		+	+	
21	*Thermocyclops oithonoides* (Sars)	+	+	+	+
22	*Thermocyclops crassus* (Fisher)	+	+	+	+
		19	**17**	**14**	**11**
Calanidae
23	*Eudiaptomus gracilis* Sars	+	+	+	+
24	*Eudiaptomus graciloides* Lilljeborg	+	+	+	+
25	*Heterocope appendiculata* Sars	+	+	+	

(continued)

	Species	Staraya Pregolya	Novaya Pregolya	Pregolya within Kaliningrad	The mouth area
26	*Eurytemora affinis* Poppe	+	+	+	+
27	*Acartia longiremis* Lilljeborg	+		+	+
28	*Acartia bifilosa* Giesbrecht				+
29	*Acartia tonsa* Dana			+	+
30	Harpacticoida	+	+	+	+
		6	5	7	7
	Totally Copepoda	25	22	21	18
Cladocera					
1	*Acroperus harpae* (Baird)	+	+		
2	*Alona affinis* (Leydig)	+	+		
3	*Alona costata* Sars	+	+		
4	*Alona intermedia* Sars	+	+	+	
5	*Alona quadrangularis* (O. F. Muller)	+	+	+	
6	*Alona rectangula* Sars	+	+	+	
7	*Alonopsis elongata* (Sars)		+		
8	*Acantholeberis curvirostris* (O. F. Müller)	+		+	
9	*Bosmina coregoni* kessleri Uljanin		+	+	
10	*Bosmina coregoni* Baird	+	+	+	+
11	*Bosmina crassicornis* (P. E. Muller)	+	+	+	+
12	*Bosmina longirostris* (O. F. Muller)	+	+	+	+
14	*Camptocercus lilljeborgii* Schoedler	+	+		
15	*Cercopagis pengoi* (Ostroumov)				+
16	*Ceriodaphnia affinis* Lilljeborg	+	+	+	+
17	*Ceriodaphnia laticaudata* P. E. Muller	+	+	+	+
18	*Ceriodaphnia megops* Sars		+	+	+
19	*Ceriodaphnia reticulata* (Jurine)	+	+	+	
20	*Ceriodaphnia quadrangula* (O. F. Muller)	+	+	+	
21	*Chydorus sphaericus* (O. F. Muller)	+	+	+	+
22	*Chydorus latus* Sars		+		
23	*Chydorus ovalis* Kurz		+		
24	*Daphnia cristata* Sars	+	+		
25	*Daphnia cucullata* Sars	+	+	+	+

(continued)

	Species	Staraya Pregolya	Novaya Pregolya	Pregolya within Kaliningrad	The mouth area
26	*Daphnia longispina* O. F. Muller	+	+	+	+
27	*Diaphanosoma brachyurum* (Lievin)	+	+	+	+
28	*Daphnia pulex* (De Geer)	+			
29	*Disparalona rostrata* (Koch)	+	+		
30	*Eurycercus lamellatus* (O. F. Muller)	+	+		
31	*Evadne nordmanni* Loven				+
32	*Graptoleberis testudinaria* (Fischer)	+	+		
33	*Ilyocryptus sordidus* (Lievin)	+			+
34	*Ilyocryptus agilis* Kurz		+		
35	*Leptodora kindtii* (Focke)	+	+	+	+
36	*Leydigia acanthocercoides* (Fischer)		+		
37	*Leydigia leydigii* (Leydig)		+	+	
38	*Limnosida frontosa* Sars	+			
39	*Macrothrix laticornis* (Jurine)	+	+	+	
40	*Macrothrix hirsuticornis* Norman and Brady		+		
41	*Peracantha truncata* (O. F. Müller)		+		
42	*Ophryoxus gracilis* Sars			+	
43	*Pleopsis polyphemoides* (Leuckart)				+
44	*Pleuroxus aduncus* (Jurine)	+	+		
45	*Polyphemus pediculus*	+	+		
46	*Pseudochydorus globosus* (Baird)		+	+	
47	*Rhynchatolona rostrata* (Koch)	+			
48	*Scapholeberis mucronata* (O. F. Muller)	+	+		
49	*Sida crystallina* (O. F. Muller)	+	+		+
50	*Simocephalus vetulus* (O. F. Muller)	+	+		
Totally Cladocera		**34**	**40**	**22**	**16**
Прочие					
1	*Polychaeta*, larvae	+	+	+	+
2	*Cirripedia*, nauplii		+	+	+
3	*Bivalvia*, larvae	+	+	+	+
		2	3	3	3
Total		**83**	**86**	**63**	**49**

Appendix 4

The list of zoobenthic species, recorded in the middle and low reaches of the Pregolya River, basing on data: AB IO RAS 1996–2014, Shibaeva, Potrebich 1994, 2000; Chepurina and Ezhova (2000), [69], Chepurina [70], Gusev et al. [71]

No п/п	Таксон	1	2	3	No п/п	Таксон	1	2	3
Spongia					**Lumbricidae**				
1	*Spongilla lacustris*	+	+	+	38	*Eiseniella tetraedra* Savigny	+	+	
Hydrozoa					**Hirudinea**				
2	*Hydra* sp.	+	+		39	*Helobdella stagnalis* (L.)		+	
Turbellaria					40	*Erpobdella octoculata* (L.)	+		
3	*Planaria* sp.	+	+		41	*Erpobdella nigricollis*	+		
Nematoda					42	*Glossiphonia complanata* (L.)		+	
4	Nematoda indet.	+	+		43	*Glossiphonia heteroclita* (L.)		+	
Oligochaeta				+	44	*Glossiphonia* sp.		+	
Naididae					45	*Cystobranchus mammillatus* (Malm)		+	
5	*Stylaria lacustris* (L.)	+	+	+	46	*Haementeria costata* (Müller)		+	
6	*Slavina appendiculata* (Udekem)		+		47	*Piscicola geometra* (L.)		+	
7	*Chaetogaster limnaei* Baer		+		**Polychaeta**				
8	*Nais communis* Piguet				48	*Marenzelleria neglecta* Sikorsky et Bick		+	+
9	*Nais pardalis* Piguet	+	+		**Mollusca**				
10	*Dero* sp.	+	+		**Bivalvia**				
11	*Proppappus volki* Michaelsen	+			49	*Anodonta cygnea* (L.)	+	+	
12	*Pristina* sp.	+			50	*Anodonta stagnalis* (Gmelin)	+		
13	*Paranais litoralis* (Müller)	+	+		51	*Anodonta piscinalis* Nilsson	+		
14	*Paranais friči* Hrabĕ	+	+		52	*Dreissena polymorpha* (Pallas)	+	+	
Tubificidae					53	*Unio pictorum* (L.)	+	+	
15	*Spirosperma ferox* Eisen	+	+	+	54	*Unio tumidus* Phil.	+	+	
16	*Aulodrilus limnobius* Bretscher	+			55	*Unio ovalis* (Mon.)	+		
17	*Aulodrilus pigueti* Kowalewski	+	+		56	*Musculium hungaricum* (Haz.)	+	+	

(continued)

No п/п	Таксон	1	2	3	No п/п	Таксон	1	2	3
18	*Limnodrilus hoffmeisteri* Claparede		+	+	57	*Pisidium supinum* A. Schmidt.	+		
19	*Limnodrilus claparedeanus* Ratzel		+	+	58	*Pisidium amnicum* (Müller)	+		
20	*Limnodrilus udekemianus* Claparede		+	+	59	*Pisidium casertanum* (Poli)	+		
21	*Limnodrilus helveticus* Piguet				60	*Pisidium* sp.		+	
22	*Limnodrilus profundicola* (Verril)		+	+	61	*Neopisidium torguatum* (Stelfox)	+		
23	*Isohaetides newaensis* (Michaelsen)		+		62	*Neopisidium tenuilineatum* Stelfox	+		
24	*Ilyodrilus caspicus* Last.		+		63	*Neopisidium moitessierianum* Paladilhe	+		
25	*Ilyodrilus* sp.		+		64	*Neopisidium* sp.		+	
26	*Psammoryctes deserticola* (Grimm.)		+		65	*Sphaerium nitidum* (Clessin)	+		
27	*Psammoryctes barbatus* (Grube)		+	+	66	*Sphaerium rivicola* Lamark	+	+	
28	*Psammoryctes albicola* (Michaelsen)		+	+	67	*Sphaerium corneum* (L.)	+	+	
29	*Ryacodrilus* sp.		+		68	*Sphaerium subsolidum* Clessin	+		
30	*Stilodrillus* sp.		+		69	*Sphaerium* sp.		+	
31	*Tubifex tubifex* (Müller)		+		70	*Amesoda solida* (Normand)		+	
32	*Tubifex* sp.		+	+	71	*Euglesa ponderosa* (Stelfox)	+		
33	*Potamotrix moldaviensis* (Marzek)		+	+	72	*Euglesa supina* Schmidt	+		
34	*Potamotrix hammoniensis* (Mich.)		+	+	73	*Euglesa* sp.	+	+	
35	*Potamotrix heuscheri* (Bretscher)		+	+	**Gastropoda**				
36	*Branchiura sowerbyi* Beddard	+	+		74	*Anisus vortex* (L.)	+	+	
Aeolosomatidae					75	*Acroloxus lacustris* (L.)	+	+	
37	*Aeolosoma quaternarium* Ehrenberg		+		76	*Ancylus fluviatilis* Müller	+	+	
77	*Bithynia leachii* (Sheppard)	+	+		**Branchiura**				
78	*Bithynia inflata* Hansen	+			120	*Argulus foliaceus* L.	+	+	

(continued)

No п/п	Таксон	1	2	3	No п/п	Таксон	1	2	3
79	*Bithynia tentaculata* (Linne)	+	+		**Arachnida**				
80	*Bithynia* sp.	+			**Acarina**				
81	*Choanomphalus rossmaessleri* (Schmidt)	+			121	*Hydrachnellae* sp.	+	+	
82	*Lymnaea stagnalis* (L.)		+		**Aranei**				
83	*Lymnaea peregra* (Müller)	+	+		122	*Argyroneta aquatica* Clerck	+		
84	*Lymnaea glutinosa* (Müller)	+			**Insecta**				
85	*Lymnaea ovata* (Draparnaud)	+			**Diptera**				
86	*Lymnaea auricularia* (L.)	+	+		123	Diptera, indet.	+	+	+
87	*Lymnaea monnardi* (Hartmann)		+		124	Diptera, Brachycera indet.	+	+	
88	*Lymnaea hartmanni* (Studer)		+		125	Diptera, Scathophagidae indet.	+		
89	*Lithoglyphus naticoides* Pfeiffer	+	+		126	Ceratopogonidae indet.	+	+	
90	*Planorbarius corneus* (L.)	+	+		127	*Chironomus balatonicus* Devai.			+
91	*Planorbis planorbis* (L.)	+	+		128	*Chironomus plumosus* L.	+	+	+
92	*Theodoxus fluviatilis* (L.)	+	+		129	*Chironomus tentans* Fabricius		+	
93	*Borysthenia* sp.	+	+		130	*Cryptochironomus* sp.	+		
94	*Viviparus viviparus* (L.)	+	+		131	*Cryptochironomus defectus* Klef.		+	
95	*Valvata antigua* Sowerby		+		132	*Limnochironomus nervosus* (Staeger)		+	
96	*Valvata ambigua* Westerlund		+		133	*Tanypus villipennis* (Kief.)		+	
97	*Valvata piscinalis* (Müller)	+	+		134	*Procladius ferrugineus* Kief.		+	
98	*Valvata depressa* C. Pfeiffer	+	+		135	*Cricotopus sylvestris* (Fabr.)		+	
99	*Valvata pulchella* Studer	+	+		136	*Polypedilum* sp.	+	+	
100	*Valvata profunda* (Clessin)	+			137	*Polypedilum nubeculosum* Meigen		+	
101	*Valvata planorbulina* (Paladilhe)	+			138	*Chaoborus* sp.	+	+	+
102	*Valvata trochoidea* Menke	+			139	Culicidae, indet.			+
Crustacea					**Odonata**			+	
Decapoda					140	Coenagrionidae indet.	+	+	
103	*Astacus astacus* (L.)	+	+		**Heteroptera**				+

(continued)

No п/п	Таксон	1	2	3	No п/п	Таксон	1	2	3
104	*Pacifastacus leniusculus* (Dana)		+		141	*Ilyocoris cimicoides* (L.)	+		
105	*Pontastacus leptodactylus* (Eschscholtz)	+	+		142	*Corixa* sp.	+	+	
106	*Orconectes limosus* (Rafinesque)	+			143	*Micronecta* sp.	+		
Amphipoda					144	*Microvelia* sp.	+		
107	*Gammarus pulex* L.		+		**Trichoptera**			+	+
108	*Gammarus lacustris* Sars		+		145	*Ecnomus* sp.	+		
109	*Gammarus locusta* Gosars		+	+	**Ephemeroptera**				
110	*Obesogammarus crassus* (Sars)	+	+		146	*Caenis horaria* (L.)	+		
111	*Pontogammarus robustoides* (Sars)	+	+		147	*Caenis* sp.	+	+	
112	*Chaetogammarus warpachowskyi* (Sars)	+			148	Leptophlebiidae, indet.	+		
113	*Gammarus* sp.	+	+	+	**Coleoptera**				
114	*Apocorophium lacustre* (Vanhöffen)	+			149	*Haliplus* sp.	+		
115	*Corophium curvispinum* Sars	+	+		150	Polycentropidae, larva	+		
Mysida					**Megaloptera**				
116	*Limnomysis benedeni* Czerniavsky	+	+		151	*Gomphus vulgatissimus* (L.)	+		
117	*Neomysis integer* (Leach)			+	152	*Sialis* sp.	+	+	
Isopoda					**Collembola**			+	
118	*Asellus aquaticus* (L.)	+	+						
Ostracoda									
119	Ostracoda indet.	+	+						

Column **1**, 29–37 km of the Pregolya river and the Deima river (branch of Pregolya); column **2**, 10–16 km of the low reach, Staraya and Novaya Pregolya; column **3**, 0–9 km of the low reach of the Pregolya River

References

1. Chubarenko BV, Shkurenko VI (2001) Physical mechanisms of salt water penetration up the Pregolya River taking into account the influence of the bottom relief. Physical problems of ecology (ecological physics), vol 7. Physics Faculty of Moscow State University, Moscow, pp 80–88 (in Russian)
2. Abakumov VA (ed) (1992) Guidelines by hydrobiological monitoring of freshwater ecosystems. Hydrometeoizdat, St. Petersburg, p 318 (in Russian)
3. HELCOM and Baltic Marine Environmental Protection Commission-Helsinki Commission (1988) Guidelines for the Baltic Monitoring Programme for the third stage; Part D. Biological determinants, vol 27D, p 161

4. Algae handbook (1989) Naukova Dumka, Kiev, p 608 (in Russian)
5. Olenina I, Hajdu S, Edler L, Andersson A, Wasmund N, Busch S, Göbel J, Gromisz S, Huseby S, Huttunen M, Jaanus A, Kokkonen P, Ledaine I, Niemkiewicz E (2006) Biovolumes and size-classes of phytoplankton in the Baltic Sea HELCOM balt. Sea Environ Proc 106:144
6. Boullion VV (1993) Primary production and trophic classification of reservoirs. Methodological issues of studying the primary production of plankton of inland water bodies. Gidrometeoizdat, St. Petersburg, pp 147–157 (in Russian)
7. Methodical recommendations for the collection and processing of materials in hydrobiological research on freshwater reservoirs. Phytoplankton and its products (1984) GosNIORH, Leningrad, p 32 (in Russian)
8. Katanskaya VM (1981) Higher aquatic vegetation of inland waters of the USSR. Methods of study. Nauka, Leningrad, p 278 (in Russian)
9. Raspopov IM (1985) Higher water vegetation of large lakes of the North-West of the USSR. Nauka, Leningrad, p 197 (in Russian)
10. Papchenkov VG (2003) Dominant and determinant classification of aquatic vegetation. Hydrobotany: methodology, methods. Material School of Hydrobotany (n. Borok, 2003). OAO Rybinsk Printing House, Rybinsk, pp 126–131 (in Russian)
11. Tsvelev NN (2000) The determinant of vascular plants of North-West Russia. SPHFA, St. Petersburg, p 781 (in Russian)
12. Pankow H (1971) Algenflora der Ostsee. I. Benthos (Blau-, Grün-, Braun-, und Rotalgen) Von Dr. rer. Nat. habil. Helmut Pankow. Gustav Fischer Verlag, Stuttgart
13. Rothmaler W (2011) Exkursionsflora von Deutschland. B.3. Gefäßpflanzen: atlasband. Spektrum Akademischer Verlag, Berlin, p 753
14. Cherepanov SK (1995) Vascular plants of Russia and neighboring countries (within the former USSR). Peace and Family, St. Petersburg, p 992 (in Russian)
15. Barinova SS, Medvedeva LA, Anisimova OV (2006) Biodiversity of algae-indicators of the environment. Pilies Studio, Tel-Aviv, p 498 (in Russian)
16. Sviridenko BF, Mamontov YS, Sviridenko TV (2011) Use of hydromacrophytes in the complex assessment of the ecological state of water bodies in the West Siberian Plain. Amphora, Omsk, p 231 (in Russian)
17. Matveeva EP (1967) Meadows of the Soviet Baltic states. Nauka, Leningrad, p 335 (in Russian)
18. Galanin AV (1991) The vegetation cover of Pregolskaya Lowland Area in the Kaliningrad Region. Flora and landscape-ecological structure of vegetation. Far Eastern Branch of the Academy of Sciences USSR-272, Vladivostok. http://geobotany.narod.ru/galanin/m8.htm (in Russian)
19. Parfenova YV (2006) Ecological and geographical analysis of vascular plants Pregolya River (Kaliningrad Region, Russia). In: Proceedings of the I (IX) of the international conference of young botanists in St. Petersburg, St. Petersburg, 21–26 May 2006, pp 55–56 (in Russian)
20. Parfenowa YV (2008) Rare species of hydrophilic plants Kaliningrad ponds. Some problems ichthyopathology and hydrobiology: first steps in science, Kaliningrad, pp 105–120 (in Russian)
21. Utkina YV (2012) Hygrophilous vegetation oxbow lakes in the central part of the Pregolya district on the gradient of anthropogenic transformation (Kaliningrad Region of Russia). In: Abstracts of II (X) of the proceedings of international botanical conference of young scientists in St. Petersburg, 11–16 Nov 2012, ETU "LETI", St. Petersburg, pp 102–103 (in Russian)
22. Shennikov AP (1964) Introduction to geobotany. A.A. Zhdanov Leningrad University, Leningrad, p 447 (in Russian)
23. Vinberg GG, Lavrentieva GM (eds) (1984) Methodical recommendations for the collection and processing of materials in hydrobiological research on fresh water bodies. Zooplankton and its products. GosNIORH, Leningrad, p 33 (in Russian)

24. Balushkina EV, Vinberg GG (1979) Dependence between mass and body length in planktonic crustaceans. General basis for the study of aquatic ecosystems. Science, Leningrad, pp 169–172 (in Russian)
25. Sushenya LM (1972) Intensity of respiration of crustaceans. Naukovadumka, Kiev, p 195 (in Russian)
26. Khlebovich TV (1974) The intensity of respiration of different sizes ciliates. Cytology 16 (1):103–115 (in Russian)
27. Vinberg GG (1983) The Van't Hoff temperature coefficient and the Arrhenius equation in biology. J Gen Biol 44(1):31–42 (in Russian)
28. Salazkin AA, Alimov AF, Finogenova NP, Winberg GG (eds) (1983) Methodical recommendations for the collection and processing of material in hydrobiological research on freshwater reservoirs. Zoobenthos and its products. ZIN of the USSR Academy of Sciences, GosNIORH, Leningrad, p 52 (in Russian)
29. Shibaeva MP, Potrebic AV (1994) To the environmental assessment of the estuary area of the river. Pregoals of the zoobenthofauna. Sb. Sci. Works of KSTU. KSTU, Kaliningrad, pp 26–38 (in Russian)
30. Alimov AF (1987) Introduction to production hydrobiology. Hydrometizdat, Leningrad, p 151 (in Russian)
31. Balushkina EV (1987) Functional significance of larvae of chironomids in continental reservoirs. Science LD, Leningrad, p 179 (in Russian)
32. Ivanova MB, Umnov AA (1979) Methods for determining the production of populations of aquatic animals. General basis for the study of aquatic ecosystems. Nauka, Leningrad, pp 119–132 (in Russian)
33. Günther G. The conclusion on the elimination of the runoff of the North German pulp mill in Königsberg and the current state of water in Pregel, with Appendices 1 and 2 of December 27, 1911. The Merseburg archive. Div. 192b. XIXa. specialist. 15. Sewerage, water disposal and treatment in Königsberg, Kaliningrad in 1911–1914 (in Russian)
34. Schiementz P. The conclusion about the pollution of the river. Pregel and Frisch-Huff Bay by the sewage waters of Königsberg and the damage caused to fishing, in connection with this pollution (according to September 1911). Merseburg archive. Section. XIXa. specialist. 15. No. 699. Sewerage, water disposal and treatment in Königsberg (Kaliningrad) in 1911–1914 (in Russian)
35. Evolution of the phosphorus turnover and eutrophication of natural waters (1988) Science, Leningrad, p 205 (in Russian)
36. Sewerage, water disposal and treatment in Königsberg 1878–1929. The Merseburg archive. Div. 192b. XIXa. specialist. 15. No. 3; 2405, 4614, 1929 (in Russian)
37. Jeddeloch R (1939) Haffkrankheit. Ergebnisse der inneren Medizin und Kinderheilkunde, vol 57. Springer, Berlin, pp 138–182
38. Molchanova NS (2014) 100 years of anthropogenic pollution of the Pregol River. Izvestia KSTU 32:170–178 (in Russian)
39. Levina RS, Voloshenko KY (2012) About the possibilities of using the historical experience of soil fertility improvement technologies in East Prussia in the agricultural business of the Kaliningrad Region. Baltic Region 2:109–117 (in Russian)
40. Tsybin YA (ed) (2004) Nature protection scheme of the Kaliningrad Region. TENAX EDIA, Kaliningrad, p 136 (in Russian)
41. Zotov SI (2001) Assessment of the hydroecological situation in the Kaliningrad Region: Sat. Sci. Tr. Environmental problems of the Kaliningrad Region and the Baltic region. KSU, Kaliningrad (in Russian)
42. Matveeva EV, Nagornova NN (2007) Distribution of pollution along the Pregolya River. Sb. Sci. Tr.: environmental problems of the Kaliningrad Region and the Baltic region. Immanuel Kant Baltic Federal University, Kaliningrad, p 143 (in Russian)
43. Report on the state and protection of the environment of the Kaliningrad Region in 2001 (2002) Kaliningrad Printed Yard, Kaliningrad, p 160 (in Russian)

44. Report on the state and protection of the environment of the Kaliningrad Region in 2003 (2004) Amber Tale, Kaliningrad, p 216 (in Russian)
45. Israel YA, Tsyban AV, Chernogayeva GM, Chelyukanov VV, Egorova VI, Zelenova AS (eds) (2009) Review of the state of environmental pollution in Russia in 2008, Moscow
46. On the environmental situation in the Kaliningrad Region in 2010 (2011) Kaliningrad. http://old.gov39.ru. Date of circulation: 21.02.2013 (in Russian)
47. Semenova SN (1998) The current state of phytocen of the water system the river Pregolya – channel – Vistula Lagoon of the Baltic Sea. Research report, AO IO RAS, Kaliningrad, p 32 (in Russian)
48. Semenova SN (2000) The current state of phytoceane of the water system the river Pregolya – channel – Vistula Lagoon of the Baltic Sea. Hydrobiol study in the basin of the Atlantic ocean T.I. Freshwater hydrobiology: Coll. Scientific Papers. AtlantNIRO, Kaliningrad, pp 20–36 (in Russian)
49. Lange EK (2013) Phytoplankton of the Pregolya River. In: Ezhova EE (ed) Biological community of the Pregolya River (basin of the Vistula Lagoon, Baltic Sea). Smartbuks, Kaliningrad, pp 85–109 (in Russian)
50. Lange EK (2014) Estimation of phytoplankton productivity in the lower reaches of the Pregolya River. Proc KSTU 32:153–161 (in Russian)
51. Dmitrieva OA (2001) The composition and abundance of phytoplankton autumn river Pregel Deima and Kaliningrad Region. In: Abstracts of 8th Congress Hydrobiol Society RAS, Kaliningrad, vol 3, pp 36–37 (in Russian)
52. Dmitrieva OA (2005) Species composition and quantitative phytoplankton characteristics of the Pregolya and Deima rivers of Kaliningrad Region. In: The environmental problems of the Kaliningrad Region and the Baltic Region. Kaliningrad, pp 164–169 (in Russian)
53. Yelizarova VA (1993) Experience in the determination of phytoplankton products by the algological method. Methodological issues of studying the primary production of plankton in inland water bodies. Gidrometeoizdat, St. Petersburg, pp 52–58 (in Russian)
54. Trifonova IS (1990) Ecology and succession of the lake phytoplankton. Science, Leningrad, p 184 (in Russian)
55. Abromeit J, Neuhoff W, Steffen H, Jentzsch A, Vogel G (1889–1940) Flora von Ost- und Westpreussen, Bd 1–3. Berlin, 1246 s (in German)
56. Gerb MA (2013) Vegetation and flora of the river Pregolya. In: Zhova EE (ed) Biological communities of the Pregolya River (Vistula Gulf, Baltic Sea). Smartbooks, Kaliningrad, pp 64–84 (in Russian)
57. Gerb MA (2014) Ecological analysis of aquatic flora in lower reaches of the Pregolya River. In: Proceedings of the Kaliningrad State Technical University, vol 32, pp 162–169 (in Russian)
58. Zhakova LV (2008) Macrophytes: higher aquatic plants and macroalgae. In: Alimov AF, Golubkov SM (eds) Ecosystem of estuary of the Neva River: biological diversity and environmental problems. Association of Scientific, Moscow, pp 105–125. Editions of KMC (in Russian)
59. Gubareva IY, Belova OV (2005) Kalashnikov. Invasive species of the flora of the Kaliningrad Region. In: Tez. doc. intern. conf. study of the flora of Eastern Europe: achievements and prospects, KMK, St. Petersburg, 23–28 May 2005, pp 23–24 (in Russian)
60. Sokolov AA (2010) Alisma gramineum. In: Dedkov VP, Grishanov GV (eds) The red book of the Kaliningrad Region. Immanuel Kant University, Kaliningrad, p 126 (in Russian)
61. Krylov AV (2005) Zooplankton of small plain rivers. Nauka, Moscow, p 263 (in Russian)
62. Tsybaleva GA, Potrebich AV (1995) The change in the structure of the composition of the Zooplankton River Pregolya under the influence of pollution. Sb. Sci. Tr: some aspects of the physiology and pathology of hydrobionts. Kaliningrad, pp 69–76 (in Russian)
63. Naumenko EN (2007) The zooplankton of the Vistula Lagoon. KGTU, Kaliningrad, p 210. (in Russian)

64. Polunina JJ, Terekhova TA (2010) Features of zooplankton lothic system "Pregolya River – Kaliningrad Sea Canal (KSC)". In: Proceedings of the Kaliningrad State Technical University, vol 17, pp 25–29 (in Russian)
65. Chertoprud MV, Peskov KV (2003) Geographical parallels of litoreofil communities of small rivers of Eastern Europe and northern Asia. J Gen Biol 64(1):78–87 (in Russian)
66. Chertoprud MV (2011) Variety and classification of rheophilic communities of macrobenthos of the middle belt of European Russia. J Gen Biol 72(1):51–73 (in Russian)
67. Shibaeva MN (2000) Species composition of zoobenthos of small rivers of the Kaliningrad Region. Hydrobiological studies in the basin of the Atlantic Ocean. AtlantNIRO, Kaliningrad, pp 58–73 (in Russian)
68. Chepurina SG, Ezhova EE (2001) Long-term changes in the distribution of mollusks in the lower reaches of the River Pregolya. In: Proceedings of the 8th congress of the Hydrobiological Society of the Russian Academy of Sciences: thesis doc, T.3. AtlantNIRO, Kaliningrad, pp 87–88 (in Russian)
69. Chepurina SG, Ezhova EE (2002) Some structural features of the macrozoobenthos p. Deima (Kaliningrad Region). In: Biology of inland waters: problems of ecology and biodiversity, thesis doc. Proceedings of the 12th international conference of young scientists dedicated to the 50th anniversary of the appointment. Rear Admiral, twice Hero of the Sov. of the Union I.D. Papanin is director of the IVVV. Borok, pp 104–105 (in Russian)
70. Chepurina SG (2004) Characteristics of some benthic communities of the Lava River (Kaliningrad Region). In: Nigmatullina CM (ed) Modern problems of parasitology, zoology and ecology. Materials of the I and II international readings dedicated to the memory and 85th anniversary of the birth of SS. Shulman, KSTU, Kaliningrad, pp 338–347 (in Russian)
71. Gusev AA, Ezhova EE, Gusev DO, Rudinskiy LV (2012) Higher Crayfish (Malacostraca) of water bodies of the Kaliningrad Region. In: Korovchinsky NM et al (eds) Actual problems of studying crustaceans of continental waters (Borok, Borovo, November 5–9, 2012): sat. lectures and reports of the international school-conference. OOO Kostromskaya Printing House, Kostroma, pp 170–173 (in Russian)
72. Ezhova EE, Lyatun MV, Molchanova NS, Lange EK (2011) Estimation of the ecological state of the Pregolya River (Kaliningrad Region) by macrozoobenthos. In: Abstracts of the all Russian conference with international participation ecology of small rivers in the 21st century: biodiversity, global change and ecosystem restoration, Togliatti. Kassandra, Togliatti, 5–8 Sept 2011, p 55 (in Russian)
73. Kamenev AG (1993) Bioproductivity and bioindication of watercourses of the right-bank middle Volga Region: macrobenthos. Mordov Publishing House. University, Saransk, p 228 (in Russian)
74. Khrennikov VV, Shustov YA (1978) On zoobenthos of small rivers of the Kola Peninsula. Operative-inform. Mater. Comp. Issled. Bioresources of Karelia (Questions of ecology, experimental zoology and botany). Petrozavodsk, pp 17–18 (in Russian)
75. Kukharev VI (1992) Functioning of animal communities of macrobenthos of small rivers of Karelia in conditions of contamination. Water resources of Karelia and ecology. Petrozavodsk, pp 98–.110 (in Russian)
76. Kukharev VI (2006) Structural and functional changes in the communities of macrozoobenthos of the watercourses of Karelia under various types of anthropogenic influences. In: The state and problems of production hydrobiology: Sat-na nauch. Work on the mother. Doc. at the intern. conf. water ecology at the dawn of the XXI century, ded. The centenary of the birth. Prof. G.G. Vinberg. Tov-v nauch. ed. KMC, Moscow, pp 242–252 (in Russian)
77. Kharin GS, Kravtsov VA, Stryuk VL, Kharin SG (1995) Integrated research of the Pregolya River. Kaliningrad

Phytoplankton Community of Small Rivers of the Pregolya River Basin

E.K. Lange

Abstract Comparative data analysis for phytoplankton of the Pregolya River and small rivers of the Pregolya River Basin in the late 1990s and in 2000s shows great influence of the Angrapa River water on the structure of phytoplankton of the Pregolya River compared to the Instruch River. Presumably, cyanobacterium *Planktothrix agardhii* is imported to the Pregolya River and further to the Curonian Lagoon via the Deima River from the Masurian Lakes in Poland, where the species has been one of the summer phytoplankton dominants since the 2000s. Information on the structure and quantitative development of phytoplankton of the Pregolya tributaries of the second and third order was obtained for the first time.

Keywords Phytoplankton, Pregolya River, Seasonal dynamics, Small rivers

Contents

1 Introduction 374
2 Material and Methods 374
3 Results and Discussion 375
   3.1 Case Study 1: Pregolya River 375
   3.2 Case Study 2: Tributaries of the Pregolya River 378
4 Conclusions 382
References 383

E.K. Lange (✉)
Shirshov Institute of Oceanology, Russian Academy of Sciences (IORAS), Nakhimovsky prospekt 36, Moscow 117997, Russia
e-mail: evlange@gmail.com

1 Introduction

First data on the Pregolya River phytoplankton were gathered due to regular studies of the river biota launched by P.P. Shirshov Institute of Oceanology, Russian Academy of Sciences, Atlantic Branch (AB IO RAS, Kaliningrad), in the mid-1990s [1, 2]. In the 2000s to 2010s Atlantic Research Institute for Fishery and Oceanography (AtlantNIRO, Kaliningrad) [3, 4] and AB IO RAS [5–7] continued the research on seasonal changes in the species composition, the structure of phytoplankton and its dominant complex, as well as abundance indices in the lower reaches of the Pregolya River. At the same time, there is no information on phytoplankton of the Pregolya tributaries in the literature. In this paper the field data on phytoplankton from watercourses of the Pregolya River system are analyzed with the comparative analysis of their phytoplankton parameters.

2 Material and Methods

The paper analyzes data obtained from primary processing of the Pregolya phytoplankton samples performed by AB IO RAS in 1996–1997[1] and by the author in 2011 and 2014, as well as information available from the published sources. Samplings by AB IO RAS in 2014 in the Pregolya tributaries (the Instruch, Angrapa, Pissa, Krasnaya, Gudishka, Golubaya, Perelesnaya, and Svoboda rivers) served as the material for analysis.

Phytoplankton samples of 1 L were collected by 1-L bathometer in the surface layer and fixed with 40% formalin in 1996–1997. Condensation was carried out by sedimentation to a volume of 100 mL. The identification and counting of algae were carried out under the light microscope MBI-3, magnification 200–900×.

In 2011 and 2014, as in the late 1990s, sampling was conducted in the surface layer. Samples of 0.5 L were fixed by Lugol solution with addition of acetic acid and formalin [8] and concentrated by sedimentation to 10 mL. The organisms were counted in the Nageotte chamber (0.02 mL volume) under the light microscope (Ergaval Carl Zeiss, Jena), magnification range 256–640×. Wet weight biomass of phytoplankton was assessed from cell geometry and using a cell biovolume table [9, 10]. The algae taxonomy corresponded to AlgaeBase (http://www.algaebase.org/).

Algae species and groups that make up more than 10% of the total phytoplankton biomass in the sample were attributed to the dominant.

---

[1]Laboratory processing protocols of 44 phytoplankton samples from the lower reaches of the Pregolya River in April–October 1996–1997 by S. N. Semenova.

3 Results and Discussion

3.1 Case Study 1: Pregolya River

Two hundred and ninety-one species, varieties, and forms of algae from seven taxonomic groups were discovered during the studies of the Pregolya River phytoplankton. Most of the identified species belonged to the greens (105 taxa), predominantly Chlorococcales (78% of total taxa), and then there came the diatoms (78), mainly Pennales (76%) and cyanobacteria (50), among which the most representative was the Chroococcales (54%). The chrysophytes were represented by 24 taxa, and then descending there came 11 dinoflagellates, 11 euglenophytes, 7 cryptophytes, and 2 yellow-greens [2].

Analysis of materials from AB IO RAS showed that river phytoplankton in 1996–1997 featured species richness throughout the vegetation season with an average of 38–55 taxa in the sample. A decrease in alpha diversity from April to June and a further growth with the greatest number of taxa in September following a decrease in October up to the early summer values were registered. In July–October period, characterized by frequent wind surges, the number of cyanobacteria species increased on account of waters rich in small-cell types of these algae from the Vistula Lagoon that entered the river. During almost the entire vegetation period, in the lower reaches of the Pregolya River, from the confluence of its sleeves, the Novaya Pregolya and the Staraya Pregolya, to the mouth, were characterized by the highest species diversity of cyanobacteria, as well as and alpha diversity.

In April, diatoms, amounting to more than 80% of the total biomass, dominated all over the investigated area of the river. In the summer-autumn period, when in most cases their share reduced to 12–60%, the role of small-cell cyanobacteria increased, especially in the area of the confluence of the Novaya Pregolya and the Staraya Pregolya to the mouth, where they accounted for more than 80%. The contribution of the greens to the phytoplankton biomass was significantly lower and, for the most part, did not exceed 35% (Fig. 1).

In total, 18 dominant species were found more than half of which belong to the diatoms. *Stephanodiscus hantzschii* Grun. in Cl. et Grun. prevailed among the diatoms in the Pregolya plankton in the spring and summer period. The summer was characterized by a broad spectrum of dominant species of different systematic affiliation, such as green *Chlamydomonas reinhardtii* P. A. Dang., chrysophyte *Synura lapponica* Skuja, cyanobacteria *Limnothrix planctonica* (Wołłosz.) Meffert, species of genus *Aphanocapsa*, cryptophyte *Cryptomonas* spp., diatoms *Aulacoseira granulata* (Ehr.) Sim., *Cyclotella kuetzingiana* Thw., *Melosira varians* Ag., and *Cymatopleura solea* (Bréb.) W. Sm.

Total phytoplankton biomass varied in a wide range from 0.10 to 147 g m^{-3} and differed 1,500 times. A large variation of values was typical for the intra-seasonal biomass as a probable result of different biotopes in the Pregolya River. The lower reaches of the Pregolya River, which is subject to the greatest anthropogenic impact and influence of the inflow of the Vistula Lagoon eutrophic brackish waters, were

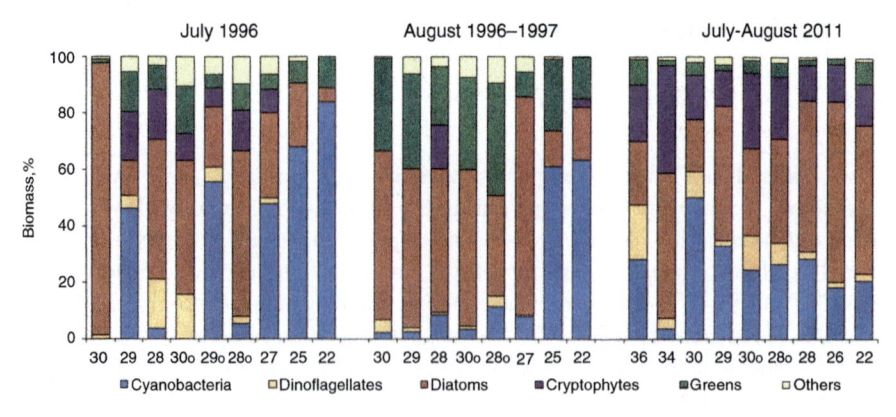

Fig. 1 The proportion of systematic groups in the total biomass of late summer phytoplankton in the Pregolya River in different years (the Staraya Pregolya (stations 30o, 29o, 28o), the Novaya Pregolya (stations 30, 29, 28)), from the confluence of the Staraya Pregolya and the New Pregolya to the Pregolya mouth (stations 27, 25, 22), Ushakovo (station 34), and Gvardeysk (station 36) [6]

characterized by ten times greater quantity of phytoplankton (the average biomass during the vegetation season of 33 ± 23 g m^{-3}) due to the predominance of colonial cyanobacteria of the genus *Aphanocapsa* introduced from the lagoon in the summer-autumn period.[2] The high productivity of phytoplankton (median biomass >1.0 g m^{-3}) was registered in August (both years), July, and September (1997). Lower productivity (<0.5 g m^{-3}) was typical for autumn phytoplankton (1996).

In general, seasonal dynamics of phytoplankton biomass can be described by unimodal curve with a maximum in July (Fig. 2).

In the spring of 2000, as was observed in the corresponding period in the late 1990s, in the lower reaches of the Pregolya River, diatom *S. hantzschii* dominated; at the same time, the complex included other diatoms that were not marked in 1996–1997: *Nitzschia acicularis* (Kütz.) W. Sm., *Synedra ulna* (Nitzsch) Ehr., and *Synedra acus* Kütz. The diatom biomass was 65–80% total biomass. The share of the greens in the total phytoplankton biomass in April–May was low, with no more than 3–12%. In July, as a result of the surge of the Vistula Lagoon waters, there was a three times increase in the average total biomass, mainly due to the group of cyanobacteria. In August, upstream from the town of Gvardeysk and near the Berlin bridge (Kaliningrad), the massive development of green *Chlamydomonas monadina* (Ehr.) Stein with up to 40–20 g m^{-3} was registered, which could be associated with the water heating of up to 24°C and a high nitrate-nitrogen content in water (the average of 366 μg N/L). In September, with the dominance of cryptomonads, the phytoplankton biomass reached 2.3–11 g m^{-3}. In October, the quantity of phytoplankton decreased by four times; at this time diatom *S. hantzschii*

---

[2]One of the reasons for the high biomass is the use of the individual weight of cyanobacteria colonies of the genus *Aphanocapsa* in the calculations, with the colonies being three-dimensional structures of cells embedded in mucilage.

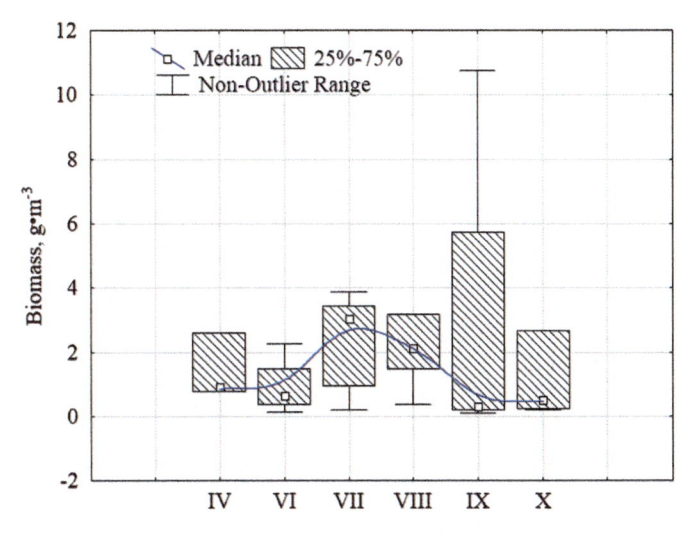

Fig. 2 Seasonal dynamics of phytoplankton biomass in the Pregolya River in 1996–1997 (Hereinafter the curve is constructed by the method of distance-weighted least squares. Maximum values of over 100 g m^{-3} in June and September 1997 are not included) [7]

and *Stephanodiscus minutulus* (Kütz.) Cl. et Möll. (33–70% of total biomass) became dominant along with cryptomonads [4].

According to our data, in May 2014, *M. varians*, *Diatoma vulgaris* Bory, and small-cell cryptomonads formed the basis of biomass, along with *S. hantzschii*. The average total biomass did not exceed 0.47 ± 08 g m^{-3}, which was lower than the corresponding values in the late 1900s (1.42 ± 0.59 g m^{-3}).

Diatoms (*A. granulata*, *M. varians*), cryptomonads of different size and potentially toxic cyanobacteria *Planktothrix agardhii* (Gom.) Anagn. et Kom., and the genus *Microcystis* (mainly *M. aeruginosa* (Kütz.) Kütz. and *M. viridis* (A. Br.) Lemm.) prevailed in late summer plankton in 2011 in the lower reaches of the Pregolya River. The emergence of these cyanobacteria in the dominant complex was new to the lower reaches of the Pregolya River and was not previously observed. On all investigated river section, the biomass of filamentous algae *P. agardhii*, developing in eutrophic waters, was of the same order (biomass 0.10–0.20 g m^{-3}), except for the portion of the river after the merger of its sleeves the Novaya Pregolya and the Staraya Pregolya and near the settlement of Ushakovo, where its biomass reduced to 0.02 g m^{-3} or less. The emergence of this species in the Pregolya River is probably related to its development in the Masurian Lakes in Poland [11], the discharge of which mainly takes place in the river basins of the Vistula and through the Lava and the Angrapa to the Pregolya. The greatest development of the genus *Microcystis* (0.06–0.27 g m^{-3}) was observed in the Novaya Pregolya sleeve and after the merger of the sleeves in the Pregolya areas with the highest content of nitrogen and phosphorus and their ratio [12], the lowest development (less than 0.01 g m^{-3}) – above the city of Kaliningrad and the Staraya Pregolya. In the city, the total biomass of phytoplankton was higher in

Table 1 Late summer phytoplankton biomass in the Pregolya River in different years [7]

Month, year	Number of stations	X ± SE	Median	Min	Max
July 1996	9	3.20 ± 0.99	3.03	0.71	10.45
August 1996	4	4.12 ± 1.85	2.71	1.49	9.57
August 1997	6	3.13 ± 1.67	1.88	0.38	11.34
July 2002	7	9	–	–	–
August 2002	7	~23	–	–	–
July–August 2011	9	0.71 ± 0.13	0.65	0.20	1.52

Note: Biomass in g m^{-3}; Data 2002 [4]

the Novaya Pregolya (1.04 ± 0.29 g m^{-3}), whereas in the Staraya Pregolya and the merger of the sleeves, it was lower and amounted to 0.67 ± 0.04 g m^{-3}. In the river, between Gvardeysk and Ushakovo, the phytoplankton biomass was lower than in the city of Kaliningrad, with 0.32 ± 0.12 g m^{-3}. Compared to the late 1990s, the productivity of late summer phytoplankton in the Pregolya River decreased in 2011, with the median of phytoplankton biomass differed by 3–5 times (Table 1).

The reduction in the vegetation level of phytoplankton in the lower reaches of the Pregolya, registered in 2011, with the biomass of >1 g m^{-3} detected only in 22% of samples against 76% in 1996–1997, could be the result of reducing the nutrient load. In 2011, compared to 2000–2005, the total phosphorus content decreased due to the organic component. In this case, although the total nitrogen content remained the same, organic forms of nitrogen began to prevail over the mineral ones [12].

Thus, the Pregolya phytoplankton reached its greatest development (more than 1 g m^{-3}) in the spring, late summer period, and, rarely, in September, while its lowest (less than 0.5 g m^{-3}) occurred in the autumn season. The phytoplankton succession of the Pregolya River matches a unimodal curve, with a peak in July. The Pregolya River differed in the levels of algae vegetation; toward the mouth their biomass increased about twice; and the productivity of phytoplankton in the Novaya Pregolya was higher than in the Staraya Pregolya. In the second decade of the twenty-first century in late summer, plankton of the lower reaches of the Pregolya potentially toxic cyanobacteria *Planktothrix agardhii* and species of the genus *Microcystis* were found among dominants that were not previously registered. The reduction of phytoplankton quantity during this period compared to the late 1990s could be due to a decrease in the nutrient load in the Pregolya River system.

3.2 Case Study 2: Tributaries of the Pregolya River

The Instruch River (A Tributary of the Pregolya) Seventeen microalgae taxa were registered in spring, of which there were nine diatoms, four cryptophytes, three greens, and one chrysophyte. Cryptomonads (cryptophytes) of different size

(55% of total biomass), diatoms *Nitzschia sigmoidea* (Nitzsch) W. Sm., and *C. solea* (12–13%) prevailed. The phytoplankton vegetation level was relatively low with the biomass of 0.16 g m^{-3}.

In summer phytoplankton consisted of 13 algae taxa with 5 cryptophytes, 3 greens, and 1–2 taxa of diatoms, chrysophytes, and cyanobacteria. In dominant complex were diatom *Cocconeis placentula* Ehr. (75%) and cryptomonads (16% of total biomass). The biomass accounted to 0.19 g m^{-3}.

The Angrapa River (A Tributary of the Pregolya) Forty-nine algae taxa were registered in the spring. The most representative were diatoms (19 taxa). The greens numbered 13 taxa; both cryptophytes and cyanobacteria accounted for 6 taxa and euglenophytes and chrysophytes for 2–3. Taxonomic diversity in the sampling stations is characterized by the value of 32–42 taxa/sample. Diatoms formed 59–72% of the total phytoplankton biomass, while greens numbered up to 16%. Diatoms *S. hantzschii* with up to 14% of the total biomass and *D. vulgaris* (41%) were dominant. The total biomass was 0.63–0.96 g m^{-3} with the average of 0.59 g m^{-3}.

Twenty-three taxa of algae were observed in summer, with 19 diatoms and 1–3 of cryptophytes, greens, and euglenophytes. Taxonomic diversity in the sampling stations ranged from 14 to 19 taxa/sample. Diatoms formed 74–97% of the phytoplankton biomass. Diatoms *M. varians*, *D. vulgaris*, and *C. placentula* prevailed. The total biomass varied from 0.23 to 0.72 g m^{-3} with the average of 0.48 g m^{-3}.

The Pissa River (A Tributary of the Angrapa) In the spring phytoplankton consisted of 46 taxa of algae with 22 diatoms, 8 greens, as well as 3–5 taxa of cryptophytes, euglenophytes, dinoflagellates, and chrysophytes. The alpha diversity ranged from 19 to 39 taxa. Diatoms played the main role, with 65–83% of the total biomass. *Stephanodiscus rotula* (Kütz.) Hendey (up to 34% of the total biomass), *D. vulgaris* (25%), *N. sigmoidea* (22%), and *Tabellaria fenestrata* (Lyngby) Kütz. (16%) dominated among them. Also, chrysophyte *Synura* spp. formed 14% of the phytoplankton biomass which ranged from 0.60 to 0.62 g m^{-3}, with the average of 0.61 g m^{-3}.

Fifty-three algae taxa were registered in summer with 28 diatoms, 11 greens, 9 cyanobacteria, and 1–2 taxa of dinoflagellates, cryptophytes, and euglenophytes. The number of taxa in phytocenosis of the stations ranged from 18 to 39. Diatoms prevailed with 45–91% of the total biomass. *D. vulgaris*, *C. placentula*, and *M. varians* were dominant. At one of the stations, cyanobacteria *Gloeotrichia echinulata* P. G. Richt. also prevailed. As for the phytoplankton productivity, the studied river phytocenoses were comparable, with values of 0.26–0.34 g m^{-3} and the average of 0.30 g m^{-3}.

The Krasnaya River (A Tributary of the Pissa) Spring phytoplankton was represented by 28 taxa, including 16 diatoms, 5 greens, 4 cryptophytes, and 1–2 taxa of chrysophytes, dinoflagellates, and euglenophytes. The total biomass of 0.49 g m^{-3} at 76% was provided by diatoms, among which *N. sigmoidea* (27% of the total biomass), *S. ulna* (20%), as well as chrysophyte *Synura* spp. (12%) dominated.

Fourteen taxa were registered in summer, with seven diatoms and 3–4 taxa of cryptophytes and greens. The total biomass of 0.03 g m^{-3} at 70% was diatoms, among which *Amphora ovalis* (Kütz.) Kütz. and *D. vulgaris* dominated. Also, cryptomonads of different size formed about 20% of the biomass.

The Gudishka River (A Tributary of the Angrapa) Twenty-two taxa were registered there in November. Diatoms, cryptophytes, and euglenophytes with 5–6 taxa showed the highest diversity. There were 1–2 taxa of cyanobacteria, chrysophytes, and dinoflagellates. Chrysophyte *Synura uvella* Ehr. (52%) dominated in biomass. Besides chrysophytes, cryptophytes (*Cryptomonas* complex) determined the biomass by 24%. The total phytoplankton biomass accounted to 0.82 g m^{-3}.

The Golubaya River (A Tributary of the Pregolya) There were 18 algae taxa in November phytoplankton, half of which belonged to diatoms. Euglenophytes, greens, and cryptophytes were only up to 2–3 taxa in the sample. The total biomass of 0.51 g m^{-3} at 26–29% was formed by cryptophytes of *Cryptomonas* complex and diatom *M. varians*. In general, cryptophytes, euglenophytes, and diatoms determined the quantitative development of phytoplankton.

The Perelesnaya River (A Tributary of the Golubaya) Twenty-two algae taxa were registered in the river in November with four euglenophytes, three greens, and also 1–2 taxa of cryptophytes, cyanobacteria, and chrysophytes. Diatoms (ten taxa) prevailed in diversity. The leading role of euglenophyte *Euglena viridis* (O. F. Müll.) Ehr. is evident, with its share of 60% in the total phytoplankton biomass. This species has reached a mass development of 994 thous. cells L^{-1}, 1.54 g m^{-3}, which indicated the contamination of the river water with organic matter. Among diatoms, the species *Nitzschia vermicularis* (Kütz.) Hantzsch prevailed in biomass (17%). In general, phytoplankton abundance was characterized by the biomass value of 2.55 g m^{-3}.

The Svoboda River (A Tributary of the Golubaya) November phytoplankton consisted of 16 algae taxa with 5 euglenophytes, 3–4 diatoms and cryptophytes, as well as 1–2 taxa of cyanobacteria and chrysophytes. Biflagellate forms, referred to the chrysophytes *Ochromonadales gen.* sp. (the length of 10–13 μm, the width of 7–10 μm), formed 87% of the total phytoplankton biomass which was 3.08 g m^{-3}.

Another stretch of the Svoboda River showed mass vegetation up to the "bloom" level of almost single species of euglenophyte *E. viridis*, the abundance of which was about 6,000 thous. cells L^{-1} and the biomass of about 9.20 g m^{-3}. This species, when mass quantity, relates to polisaprobs or alpha-mezosaprobs, being thus an indicator of high organic pollution.

Thus, in late May 2014, the taxonomic diversity of phytoplankton within the Pregolya River system was the highest in the rivers Angrapa and Pissa with up to 39–42 taxa/sample. Phytoplankton of the other studied rivers was lower diversity, with up to 27–29 taxa in the Pregolya and Krasnaya and 17 taxa in the Instruch.

In abundance prevailed cryptomonads (in all rivers), diatoms (except for the Pissa), greens (the Pregolya and Angrapa), cyanobacteria, and a group of flagellate (the Angrapa and Pissa).

Overall, the phytoplankton biomass was determined by diatoms. The plankton species of α-mezosaprob diatom *S. hantzschii* dominated in the Angrapa and the Pregolya, the latter formed by the merger of the Angrapa and the Instruch. Epiphyte β-mezosaprob *D. vulgaris* dominated in the Pregolya and the Angrapa, as well as in the Pissa, the influx of Angrapa, while the benthic-planktonic species β-mezosaprob *N. sigmoidea* prevailed in the Instruch and the Pissa, as well as in the Krasnaya River, the influx of the Pissa. Cryptomonads of different size prevailed in the plankton of the rivers Krasnaya, Instruch, and Pregolya. Chrysophyte *Synura* spp. dominated the phytoplankton of the Pissa River and its tributary the Krasnaya River.

The highest rates of algae vegetation were typical for the Pregolya-Angrapa-Pissa River system, where the maximum abundance of phytoplankton ranged from 1,110 to 1,412 thous. cells L^{-1} and the maximum biomass ranged from 0.62 to 0.96 g m^{-3} (Table 2).

In the rest of the rivers, phytoplankton featured the highest abundance value of 523–568 thous. cells L^{-1} and the highest biomass of 0.16–0.49 g m^{-3} (Table 2). The least productive was phytoplankton in the Instruch, and the most productive was that in the Angrapa. It should be noted that the structure of phytoplankton in the Pissa and the Angrapa is largely determined by the fact that they derive from the lakes Mamry and Vyshtynetskoe, respectively. Phytocenoses of these rivers tend to a higher taxonomic diversity and the dominance in the abundance of cyanobacteria, as well as the highest phytoplankton productivity. The Pregolya River, formed by the confluence of the rivers Instruch and Angrapa, is largely influenced by the Angrapa waters, the flow of which is much higher than that of the second river. This is indicated, in particular, by the dominance of diatoms *S. hantzschii* only in the rivers Angrapa and Pregolya and the presence of potentially toxic cyanobacteria *P. agardhii*, one of the summer dominants in the Curonian Lagoon, apparently, originating from the system of Masurian Lakes, which include Lake Mamry, where this species also contributes to the number of summer dominants.

In the summer the greatest taxonomic diversity showed the phytoplankton of the Pissa (53 taxa), with a relatively large number of species of the greens and cyanobacteria, which was determined by the influence of the lake Vyshtynetskoe from which the Pissa flows, along with a high proportion of diatoms (about 50%). The lake water had the biggest impact on the phytocenosis of the river station, located close to the lake. In addition to a high diversity of phytoplankton with a high

Table 2 Quantitative indicators of the spring-summer phytoplankton in the rivers of the Kaliningrad Oblast in 2014

River	Abundance (thous. cells L^{-1})	Biomass (g m^{-3})
Pregolya	949–1,412/465–1,959	0.37–0.59/0.72–1.30
Angrapa	952–1,252/241–548	0.63–0.96/0.23–0.72
Pissa	885–1,110/225–2,354	0.60–0.62/0.26–0.34
Instruch	568/1,077	0.16/0.19
Krasnaya	523/125	0.49/0.03

degree of limnetic species, cyanobacteria *G. echinulate* dominated in biomass there along with diatoms. In the other rivers, only diatoms markedly varied and were of paramount importance in phytoplankton biomass of all stretch of the river. Among the dominant diatoms, mainly benthic and epiphyte species *D. vulgaris*, *C. placentula*, and *M. varians* were observed. Cryptomonads of different size dominated the phytoplankton in the Pregolya, the Instruch, the Krasnaya, and the Pissa. Most phytoplankton abundance of over 1,000 thous. cells L^{-1} was observed in the rivers Pregolya, Pissa, and Instruch. Phytocenoses of the rivers Pregolya and Angrapa are characterized by relatively high biomass (>0.5 g m^{-3}) (Table 1).

In November phytoplankton of the studied tributaries of the Pregolya River has a relatively high diversity of diatoms (the rivers Golubaya and Perelesnaya) and, to a lesser degree, cryptomonads and euglenophytes (all rivers). In the plankton of the rivers, Golubaya and Gudishka cryptomonads prevailed. Chrysophytes dominated in the phytoplankton of the Gudishka and Svoboda rivers. Mass vegetation of *E. viridis* in the phytoplankton of the rivers Perelesnaya and Svoboda indicated water pollution with organic matter. The rivers Perelesnaya and Svoboda featured the highest productivity of phytoplankton, with the total biomass of 2.50–3.00 g m^{-3}.

4 Conclusions

During the vegetation period, the structure of the Pregolya River system phytoplankton is mainly determined by diatoms. Phytoplankton of the Pissa and Angrapa rivers, flowing from the lakes Vyshtynetskoe and Mamry, respectively, is characterized by a higher taxonomic diversity, the cyanobacteria dominance by abundance, and, in some cases, by biomass and the highest total phytoplankton biomass. The Pregolya River, formed by the confluence of the Instruch and Angrapa rivers, is affected to a greater extent by the latter, the flow of which is considerably higher. In some streams of the second order (the rivers Perelesnaya and Svoboda), the massive development of euglenophytes was registered, indicating water pollution with organic matter.

Acknowledgment The present work was only made possible by the collection of phytoplankton samples by the team of researchers at the Laboratory for Marine Ecology at AO IO RAS: E. Ezhova, M. Lyatun, N. Molchanova, M. Gerb, Y. Polunina, and others. The sampling was conducted during the monitoring of the Pregolya River in different years. Special thanks to S. N. Semyonova for analyzing samples from the 1990s.

References

1. Semenova SN (1998) The current state of the phytocene of the water system is the Pregolya River – the canal – the Kaliningrad Gulf of the Baltic Sea. SRP Report, IO RAS, Kaliningrad, p 32 (in Russian)
2. Semenova SN (2000) The current state of the phytocene of the water system is the Pregolya River – the canal – the Kaliningrad Gulf of the Baltic Sea. Hydrobiol research in the Atlantic Ocean Basin. P 1. Freshwater hydrobiology, AtlantNIRO, Kaliningrad, pp 20–36 (in Russian)
3. Dmitrieva OA (2001) Composition and abundance of autumn phytoplankton of the Deima and Pregolya Rivers of the Kaliningrad Region. In: Abstracts of the 8th congress of the Hydrob Society of the RAS, Kaliningrad, vol 3, pp 36–37 (in Russian)
4. Dmitrieva OA (2005) Species composition and quantitative characteristics of phytoplankton development in the Deima and Pregolya Rivers in the Kaliningrad Region. In: Ecological problems of the Kaliningrad Region and the Baltic Region. Kaliningrad, pp 164–169 (in Russian)
5. Lange EK (2013) Phytoplankton of the Pregolya River. In: Ezhova EE (ed) Biological communities of the Pregolya River (Basin of the Vistula Lagoon, the Baltic Sea). Smartbox, Kaliningrad, pp 85–109. (in Russian)
6. Lange EK (2014) Estimation of phytoplankton productivity in the lower reaches of the Pregolya River. News of KGTU 32:153–161. (in Russian)
7. Lange EK (2014) Characteristics of the state of phytoplankton communities of the Pregoli River (Basin of the Vistula Lagoon, the Baltic Sea). Samara Luke: Probl Reg Global Ecol 23 (2):92–97. (in Russian)
8. Abakumov VA (ed) (1992) Guidance on hydrobiological monitoring of freshwater ecosystems. SPb, Hydrometeoizdat, p 318 (in Russian)
9. HELCOM (1988) Baltic Marine Environmental Protection Commission – Helsinki Commission Guidelines for the Baltic monitoring programme for the third stage, Part D. Biological determinands. 27D, p 161
10. Olenina I, Hajdu S, Edler L, Andersson A, Wasmund N, Busch S, Göbel J, Gromisz S, Huseby S, Huttunen M, Jaanus A, Kokkonen P, Ledaine I, Niemkiewicz E (2006) Biovolumes and size-classes of phytoplankton in the Baltic Sea. HELCOM Balt Sea Environ Proc 106:144
11. Jakubowska N, Zagajewski P, Gołdyn R (2013) Water blooms and cyanobacterial toxins in lakes. Pol J Environ Stud 22(4):1077–1082
12. Tevs OA, Kudryavtsev NG (2013) Nutrient contents in the waters of the Pregolya River. In: Ezhova EE (ed) Biological communities of the Pregolya River (Basin of the Vistula Lagoon, the Baltic Sea). Smartbox, Kaliningrad, pp 51–59. (in Russian)

Flora and Vegetation of the Small Rivers of the Pregolya River System in the Kaliningrad Region

A.A. Volodina and M.A. Gerb

Abstract The study presents the results of floristic surveys of small rivers belonging to the Pregolya River system in the Kaliningrad Region, carried out in 2014–2015. The vegetation of the examined rivers is represented by plant communities, being widespread in the region. *P. australis, S. fragilis, S. alba, A. glutinosa, S. sagittifolia, S. sylvaticus, P. arundinacea* appear to be dominating species in the riparian area. *C. acuta, R. amphibia, S. erectum, B. umbellatus* have high frequency of occurrence. Water vegetation is mainly represented by the communities with the dominance of *P. pectinatus, P. nodosus, S. emersus, N. lutea, M. spicatum*. For the first time for the region, the data on the growth of the rare sensitive algae species (*H. rivularis, A. chalybaea*) and the findings of *C. elegans, D. glomerata, V. frigida, V. bursata, V. canaulicularis*, whose status is unclear due to insufficient study of macroalgae in the region, are published. In the Krasnaya River, there was found *B. trichophyllum*, listed in the Red Book of the Kaliningrad Region as a requiring attention species. There was found a new regional habitat of *E. telmateia* – an endangered species listed in the Red Book of the Kaliningrad Region. Published scientific data for these rivers have not been available so far.

Keywords Macrophyte flora and vegetation, Pregolya River systems

---

A.A. Volodina (✉)
P.P. Shirshov Institute of Oceanology, Russian Academy of Sciences, 36, Nakhimovsky Prospekt, Moscow 117997, Russia

Immanuel Kant Baltic Federal University, 14, A. Nevskogo Street, Kaliningrad 236041, Russia
e-mail: volodina.alexandra@gmail.com

M.A. Gerb
P.P. Shirshov Institute of Oceanology, Russian Academy of Sciences, 36, Nakhimovsky Prospekt, Moscow 117997, Russia
e-mail: marikegerb@gmail.com

Contents

1	Introduction	386
2	Material and Methods	386
3	Results and Discussion	388
	3.1 Case Study 1: The Composition of Macrophyte Flora and Vegetation	388
	3.2 Case Study 2: The Botanical Description	392
4	Conclusions	401
Appendix		402
References		410

1 Introduction

Flora and vegetation of the rivers of the Kaliningrad Region, their current checklist and composition of plant communities are not adequately investigated. Historical data about the findings of certain species of water plants in the late nineteenth to early twentieth centuries can be found in the works on the flora of East and West Prussia [1], but they are fragmentary and cannot present a comprehensive view on the water vegetation of the rivers in the region. In the postwar period, there were very few publications, devoted to rivers: on the Pregolya River [2–5], on the Deima River [6], on the Neman River, on the Sheshupe River [2]. Data on flora and vegetation of other small rivers of the region for the past 80 years were not covered in the scientific press. Topicality of these studies is evident in view of the need to document the diversity of biota during the period of climatic fluctuations. These studies are also relevant because they have an important regional significance in connection with the fishery value of these rivers and their influence on the hydrological regime in the Kaliningrad Region.

2 Material and Methods

The current study presents the results of the floristic survey of the rivers, carried out in May, July, October 2014 and May 2015 based on the area accounting method at 14 stations in accordance with the procedures adopted in Russia [7, 8]. In total, 35 vegetation descriptions were made, and the photograph collection of about 1,300 images of individual species of macrophytes and plant communities was composed. Vegetation was classified using the method of dominants, the names of associations and formations are given in the traditions of the dominant system [9, 10]. Herbarium specimens of quality samples of macrophytes are stored in the hydrobiological collection of the marine ecology laboratory of the AB of the IORAS. Names of taxa are specified by the List of Vascular Plants of Russia [11] and the World Register of Marine Species [12]. The results are not exhaustive and are preliminary in connection with the initial stage of research.

Brief Description of Water Courses

All watercourses belong to the catchment area of the Pregolya River of the Baltic Sea basin. The Deima River, being a typical lowland river, is navigable along the whole length, and has an average depth of 3–5 m. The other rivers are characterized by alternation of shallow (0.2–1.5 m) and relatively deep (3–7 m) areas. Unlike the other rivers, the Krasnaya River, the Pissa River, and the Angrapa River are foothill rivers characterized by a big difference of heights [13]. In some areas with a rocky bottom, the fast flow rate is expressed in numerous rifts. All these affect the checklist of water vegetation species of these rivers.

The Pissa River: Station 21

The stream outlet of the Pissa River, flowing from Vishtynets Lake, is located at an altitude of 172 m above sea level. Shallow-water areas alternate with rifts, the flow is fast and reaches 0.7 m/s, the depth is 0.2–0.5 m, the oxygen content for the time of the survey ranged from 98% in spring to 70% in autumn. The bottom is sand-and-shingle.

The Pissa River: Station 22

In the downstream, it has the width of 25 m, and the depth of 0.5–1.5 m. From May to October, the oxygen content ranged from 94 to 75%. The flow rate is 0.3–0.9 m/s. The soil is clay-silt containing gravel. The river is surrounded by high banks, covered with deciduous forest.

The Angrapa River: Station 10

On shallow areas, the sandy bottom changes to the rocky one with boulders in the rifts, overgrown with algae and water mosses. Higher water vegetation is developed at the depths of 0.4–1.5 m. The flow rate in the sandy areas is 0.2–0.7 m/s. The oxygen content is 90–91%.

The Angrapa River: Station 11

Rocky rifts alternate with sandy areas, where Potamogetonacea communities are spread. The flow rate is 1–1.4 m/s. The oxygen content in the water is the biggest of all stations and reaches 98–105%.

The Instruch River: Station 9

The maximum overgrowing of the riverbed on sediment with detritus was marked here. The low flow rate of 0.2 m/s and the presence of decaying plant residues cause the low oxygen content (52–64%).

The Krasnaya River: Stations 12 and 12a

The examined stations are located near the bridges. The river flows in the deciduous forest, and shade prevents riverside aquatic vegetation development. In the areas with silty-sandy bottom, there is no water vegetation at the examined sites. The oxygen content varies from 79 to 85%. Communities of water mosses and algae are found on rocky shallow rifts.

Table 1 Characteristics of environment conditions in the investigated rivers, 2014

River	Coordinates	Bottom	Transparency (m)	Flow velocity (m/s)	O_2 content (%)
Instruch, st. 9	N 54° 39′ 16″ E 21° 47′ 51.1″	Argilo-arenaceous with detritus	0.6–1.4	0.2	52–64
Pissa, st. 22	N 54° 39′ 17.9″ E 21° 58′ 9.2″	Argilo-silty, breakstone, sand, pebble	Transparent to the bottom	0.3–2.8	75–94
Pissa, st. 21	N 54° 27′ 9.9″ E 22° 42′ 2.5″	Sand and pebble	Transparent to the bottom	2.8	70–98
Angrapa, st. 11	N 54° 36′ 15.9″ E 21° 57′ 56.8″	Sand and pebble	Transparent to the bottom	1–1.4	98–105
Angrapa, st. 10	N 54° 22′ 30.8″ E 21° 59′ 51.2″	Arenaceous (sandy)	Transparent to the bottom	0.2–0.7	90–91
Krasnaya, st. 12	N 54° 56′ 55.3″ E 22° 19′ 0″	Silty-sandy	Transparent to the bottom	0.3 0.4	79–85
Deima, st. 3	N 54° 50′ 59.2″ E 21° 11′ 3.5″	Mud, detritus	Transparent to the bottom	Surface flow not found	64

The Deima River: Station 3

The low flow rate and the low oxygen content (64%) are noted in the water. Near the bank under the silt bottom with detritus, there is a stone riprap at the depths of up to 1 m. A considerable area of the river surface is covered with duckweed communities, indicating the high content of biogens in the water.

The Lava River

The bed is narrow and winding, the river is shallow, there are pits with depths of 5–7 m. Banks are steep.

Table 1 summarizes some of the hydrophysical characteristics for the geobotanical description stations of the examined rivers (Table 1).

3 Results and Discussion

3.1 Case Study 1: The Composition of Macrophyte Flora and Vegetation

In total, 178 plant species (16 – macroalgae; 6 – Bryophyta; 4 – Equisetophyta; 152 (species) Magnoliophyta: 37 – Liliopsida, 115 – Magnoliopsida) were identified in the rivers and in the riparian forest and meadow biotopes (Appendix). *Asteraceae, Ranunculaceae, Poaceae* families are represented by the largest number of species, that is typical of the flora of the forest zone of northern Europe. The water flora of the rivers includes 62 macrophyte species, the main proportion of which is numbered by angiosperms (74%, 46 species) with the largest number of species in

Lemnaceae and *Potamogetonaceae* families. The remaining taxonomic groups are represented by a small number of species (16 – macroalgae from four divisions: Rhodophyta – 3, Charophyta – 3, Chlorophyta – 6, Ochrophyta – 3); two species of Bryophyta; 1 – Equisetophyta. Hydrophytes with the prevailing number of submerged hydrophytes constitute 55% (34 species) in the checklist of species. In the course of investigations, 28 species of helophytes, forming the coastal aquatic vegetation, were found. On the banks of the rivers flowing through the alder forests, 40 species of flowering plants and three species of mosses (*Conocephalum conicum* (L.) Dumort, *Pellia epiphylla* (L.) Corda, *Plagiomnium undulatum* (Hedw) T. Kop.) were found. *Equisetum sylvaticum* L., *E. telmateia* Ehrh. are met in the deciduous forests on the banks of the examined rivers. On the banks of the rivers: Pissa, Krasnaya, Angrappa flowing along the meadow communities in addition to helophytes, there are, represented mainly by plants of Poaceae, Asteraceae families and by herbs (Heteroherbosa). In the identified checklist of flora, two macroalgaes species are rare in the Kaliningrad Region (*Hildenbrandia rivularis*, *Audouinella chalybaea*), and *Batrachium trichophyllum* is listed in the Red Book of the Kaliningrad Region as being in need of attention and control of the population [14]. The adventive fraction of the flora is represented by four invasive species (*Elodea canadensis*, *Acorus calamus*, *Solidago serotinoides*, *Echinocystis lobata*). On the banks of the Angrapa River, the Instruch River, the Pissa River, the Lava River, *E. lobata* is often found and is rapidly spread, and in some areas of the Lava River and the Angrapa River the species abundantly grows on *Phragmites australis*. In general, the basis of the flora of the examined rivers is composed by the species of natural hygrophilous and hydrophytic communities with a relatively small number of ruderal and stranger species, as well as species from neighboring phytocenoses.

In the water communities, 12 species are considered to be dominants, and seven species are considered to be dominants in the coastal aquatic communities. The other species belong to assectators. The number of species in the community ranges from 1 to 7. The maximum number of species is observed at Station 10 in the Angrapa River in the community with *Potamogeton nodosus*, due to biotopical and environmental conditions being favorable for the development of water vegetation. The minimum number of species was recorded in the Pissa River Station 21, where the flow rate was 2.8 m/s, and the depth is 0.10–0.20 m. In the course of investigations, the six common species, found at all stations, were revealed – these are widespread in the region species: *Potamogeton pectinatus*, *Butomus umbellatus*, *Scirpus lacustris*, *Phragmites australis*, *Sparganium emersum*, *Filipendula ulmaria*.

River Vegetation
Areas of alder forests are found along the banks of all the rivers. Willow species (*Salix alba*, *S. fragilis*, *S. viminalis*) and others grow both in the open areas of the rivers and in the forested areas. The two latter species form thickets. At a considerable distance, the Angrapa River, the Krasnaya River, the Pissa River flow through the deciduous forests, and the typical nemoral species (*Hepatica nobilis*, *Pulmonaria obscura*, *Ficaria verna*, *Anemonoides nemorosa*, *Anemonoides ranunculoides*, *Actaea spicata*, *Asarum europeum*, *Mercurialis perennis*, *Ranunculus lanuginosus*)

grow along the banks of these rivers. In the Angrapa River, a new habitat in the Kaliningrad Region of *Equisetum telmateia*, the plant of the one rarity category (endangered species), listed in the Red Data Book of the Kaliningrad Region, was detected [15]. Along the banks of the Krasnaya River, the areas of old-growth coniferous forests still remain. Rivera in alluvial meadows with abundant grasses and some rare plant species (*Polemonium caeruleum*) are located in the floodplain of the river. Under the high banks of the river in the groundwater exit locations, spring swamps with rare and protected plants in the region are often formed [16].

River Vegetation Classification
Riparian vegetation is developed in the areas of the low floodplain of the riverbed having the constant excessive moisture. It mostly consists of dense vegetation (*Phragmites australis, Salix alba, S. fragilis,* and *Carex acuta*) and others. Groundwaters are constantly kept on the surface. This type of habitat is almost isolated floristically from the rest ones. *P. australis, C. acuta, Rorippa amphibia, Galium palustre, Equisetum fluviatile, Ranunculus repens, Caltha palustris, R. palustris, Filipendula ulmaria, Lythrum salicaria, Epilobium palustris, Myosotis palustris* species are common.

Class 1. Semi-aquatic Vegetation
Formation group of hygrophytes and hygrohelophytes. Formation of *Carex acuta*.
  Ass.: Carex acuta purum.
Class 2. Helophytic Vegetation
Formation group of helophytes. Formation of reeds – *Phragmites australis*. Ass.
  Phragmites australis purum, *Phragmites australis* – heteroherbosa, *Phragmites australis* + *Typha latifolia*, *Phragmites australis-Calystegia sepium* (Lava, Angrapa), *Phragmites australis* – Echinocystis lobata (Angrapa, Pissa);
Formation of *Phalaroides arundinacea*. Ass. *Phalaroides arundinaceae* subpurum.
Class 3. Aquatic vegetation
Water vegetation in most rivers is represented by attached, submerged, and floating plants. From algae attached to boulders, the species of *Cladophora* genus are most often found. The species of *Oedogonium, Spirogyra genera* are found in helophyte and hydrophyte communities. In total, 13 plant formations were detected, and 39 plant associations were described (Table 2).

The cortical algae, sensitive to organic pollution *Hildenbrandia rivularis*, and the water moss *Fontinalis antipyretica* were found in the shallow rocky areas with fast-flowing of the Pissa River, the Angrapa River, and the Krasnaya River. The habitats of indicator species of water purity such as *Audouniella chalybea, Batrachospermum gelatinosum* in the Angrapa River and *Draparnaldia glomerata, Chara inconnexa* in the stream outlet of the Pissa River, typical of ponds with running and cool water, were identified in the region for the first time.

Plant associations of *Nuphareta lutei* and *Sagittarieta sagittarifoliumi, Phragmiteta australis* formations appear to be the most common ones (Table 2). Pondweeds communities (*Potamogetoneta nodosi, Potamogetoneta lucensi*) are

Table 2 Aquatic plant communities in the rivers, 2014

Formations	General communities	Rivers and stations
1	2	3
Cladophoreta rivularii	*Cladophora rivularis + Fontinalis antipyretica*	Krasnaya
	Cladophora rivularis purum	Krasnaya, Pissa (st. 21, Yagodnoe)
Fontinalieta antipyreticae	*Fontinalis antipyretica* purum	Angrapa, Krasnaya, Pissa (Yagodnoe)
Acoruseta calamusi	*Acorus calamus – Spirodela polyrhiza*	Instruch
	Acorus calamus – Nuphar lutea	
Butometa umbellati	*Butomus umbellatus – Potamogeton nodosus*	Angrapa (st. 10)
	B. umbellatus + Sagittaria sagitifolia – Potamogeton nodosus – Potamogeton pectinatus	Angrapa (st. 10)
Lemnetaminori	*Lemna minor + Spirodela polyrhiza – Lemna trisulca*	Gudishka, Svoboda
Numpharetalutei	*Nupharlutea* purum	Instruch, Deima, Lava, Pissa, Krasnaya
	Nuphar lutea + Nymphaea candida	Deima
	Nuphar lutea – Potamogeton lucens	Deima, Lava, Angrapa
	Nuphar lutea – Potamogeton nodosus	Angrapa
	Nuphar lutea – Potamogeton pectinatus – Spirodela polyrhiza	Instruch, Deima, Pissa
	Nuphar lutea – Sagittaria sagittifolia – Potamogeton lucens	Deima
	Nuphar lutea + Myriophyllum spicatum	Deima, Instruch
Potamogetoneta nodosi	*Potamogeton nodosus + Potamogeton pectinatus*	Angrapa (st. 10), Pissa (st. 22)
	Potamogeton nodosus subpurum	Pissa (st. 22), Angrapa (st. 10, 11), Lava
	Potamogeton pectinatus purum	Pissa (st. 22), Angrapa (st. 10, 11)
Potamogetoneta lucensi	*Potamogeton lucens – Myriophyllum spicatum*	Deima, Lava
Rorippeta amhibiae	*Rorippa amphibia* purum	Pissa (st. 22), Instruch
	Rorippa amphibia – Spirodela polyrhiza	Instruch
	Rorippa amphibia – Nuphar lutea	Pissa (st. 22)
Scirpeta lacustrii	*Scirpus lacustris – Nuphar lutea*	Instruch, Lava, Pissa
	Scirpus lacustris subpurum	Angrapa (st. 10), Instruch, Lava, Pissa (st. 22)
Sagittarieta sagittarifoliumi	*Sagittaria sagittifolia* purum	Angrapa (st. 10, 11), Instruch, Lava, Pissa (st. 22), Deima
	Sagittaria sagittifolia + Butomus umbellatus	Instruch, Lava
	Sagittaria sagittifolia – Hydrocharis morsus-ranae – Spirodela polyrhiza	Angrapa (st. 10), Instruch, Deima, Lava

(continued)

Table 2 (continued)

Formations	General communities	Rivers and stations
	Sagittaria sagittifolia – Potamogeton nodosus – Mougeotia	Angrapa (st. 10, 11)
	Sagittaria sagittifolia – Potamogeton nodosus Sparganium emersum – Mougeotia sp.	Angrapa (st. 10)
Sparganieta erecti	*Sparganium erectum – Spirodela polyrhiza*	Pissa, Gudishka
	Sparganium erectum – Sagittaria sagitifolia – Potamogeton nodosus – Spirodela polyrhiza	Angrapa (st. 10)
Phragmiteta australis	*Phragmites australis* purum	Everywhere
	Phragmites australis – Nuphar lutea + Nymphaea candida	Deima
	Phragmites australis – Nuphar lutea	Deima, Angrapa (st. 11), Instruch
	Phragmites australis – Sagittaria sagittifolia – Nuphar lutea + Nymphaea candida	Deima
	Phragmites australis – Nuphar lutea – Spirodela polyrhiza + Hydrocharis morsus-ranae	Instruch, Deima, Lava
	Phragmites australis – Sagittaria sagittifolia – Nupharlutea + Nymphaea candida	Deima
	Phragmites australis – Scirpuslacustris	Lava, Angrapa (st. 10)
	Phragmites australis – Scirpus sylvaticus	Angrapa (st. 10)

often found. Unattached plants of *Spirodela polyrhiza*, *Lemna minor* are common in areas with slow current and along the banks among helophytes.

Reedbeds are at least fragmentary developed on the low banks of the rivers. The belt of riparian vegetation is also formed by coastal aquatic *Carex acuta*, *Phalaroides arundinacea*, *Scirpus lacustris*, *Rorippa amphibia*, *Acorus calamus*, *Filipendula ulmaria*, etc., typical of North-West of Russia and Europe.

3.2 Case Study 2: The Botanical Description

The Pissa River: Station 21 The fragments of plants from Lake Vishtynetskoe are found in the stream outlet of the river. *Draparnaldia glomerata* (Fig. 1) and *Ulothrix zonata* are abundant on boulders and rocks in spring, and *Cladophora rivularis* is abundant in summer and autumn. *Vaucheria bursata* (Figs. 2 and 3) is found in the moist soil of the bank. Higher water vegetation is missing (Fig. 4). *Cladophora rivularis* and *Fontinalis antipyretica*, *Calliergon* sp., *Hildenbrandia rivularis* grow in the area near the waterfall (Yagodnoye Village) in the water on boulders 2 km from the stream outlet. Terrestrial forms of *Vaucheria caulicinalis* (Figs. 5 and 6) and *Conocephalum conicum* grow in the moist soil of the steep bank.

Fig. 1 *Draparnaldia glomerata*, actual size, Pissa River, st. 21

Fig. 2 Pissa River, st. 21

The Pissa River: Station 22 Near the banks, the belt of macrophytes are fragmentary represented by *Phragmites australis*, *Typha latifolia*, *Sparganium erectum* (Fig. 7). There is *Scirpus lacustris* in shallow areas. Hydrophytes are represented by communities with dominance of *Potamogeton pectinatus* (Fig. 8), *Sparganium emersum*. Terrestrial forms of *Vaucheria frigida* (Roth) C. Agard. (Figs. 9 and

Fig. 3 *Vaucheria bursata* (O. F. Muller) C. Agard, Pissa River, st. 21

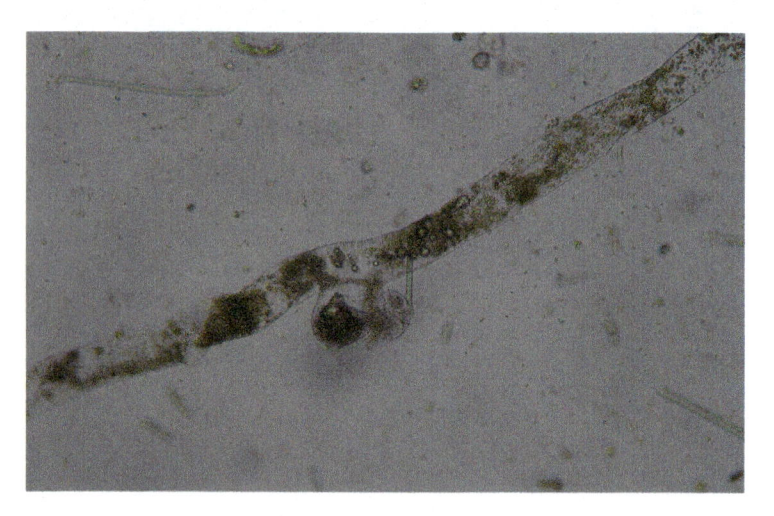

Fig. 4 *V. bursata*, magnification × 600

10) and *Pellia epiphylla* (Fig. 11) and *Conocephalum conicum* grow on the wet banks.

The Angrapa River The shallow areas in the middle stream (St. 10) and the area in the lower stream after the influx of the Pissa River (St. 11) were examined. At St. 10, where the depth reaches 1.5 m, vegetation is represented by hydrophytes with domination of *Potamogeton pectinatus*, *P. nodosus* (Fig. 12), and *Sparganium*

Fig. 5 *Vaucheria canalicularis* (Linnaeus) T. A. Christensen, Pissa River, v. Jagodnoe

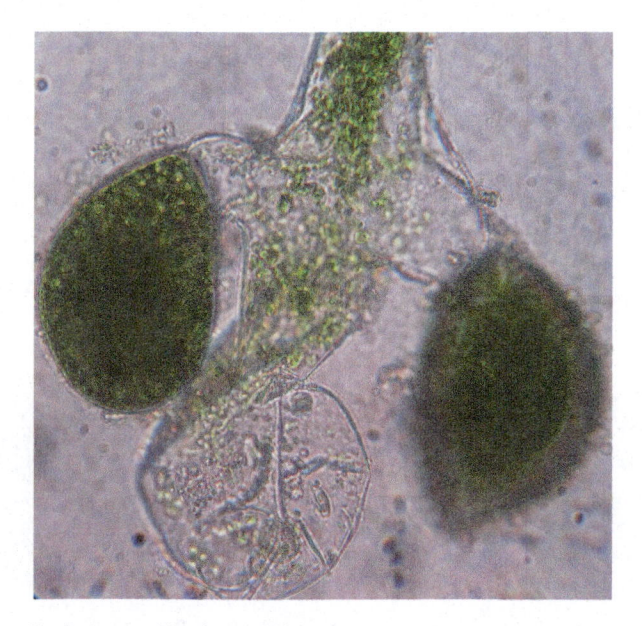

Fig. 6 *V. canalicularis* magnification × 800

emersum (Fig. 13). There is *Cladophora rivularis* on boulders and rocks. *Scirpus lacustris, S. sylvaticus, Sagittaria sagittifolia, Butomus umbellatus, Sparganium erectum* are found in the ripal. The two species of oligosaprobic *Hildenbrandia rivularis,* and *Audouinella chalybaea* sensitive to organic pollution, were identified

Fig. 7 The lower flow of the Pissa River, st. 22

Fig. 8 Community of *Potamogeton pectinatus* L. on the Pissa River, st. 22

in the Kaliningrad Region for the first time. Both species are indicators of low content of biogenic elements in the water. *B. gelatinosum* being a typical mesosaprobe was found in the benthos samples.

Flora and Vegetation of the Small Rivers of the Pregolya River System in the... 397

Fig. 9 *Vaucheria frigida* (Roth) C. Agard, Pissa River, st. 22. 06.05.2015

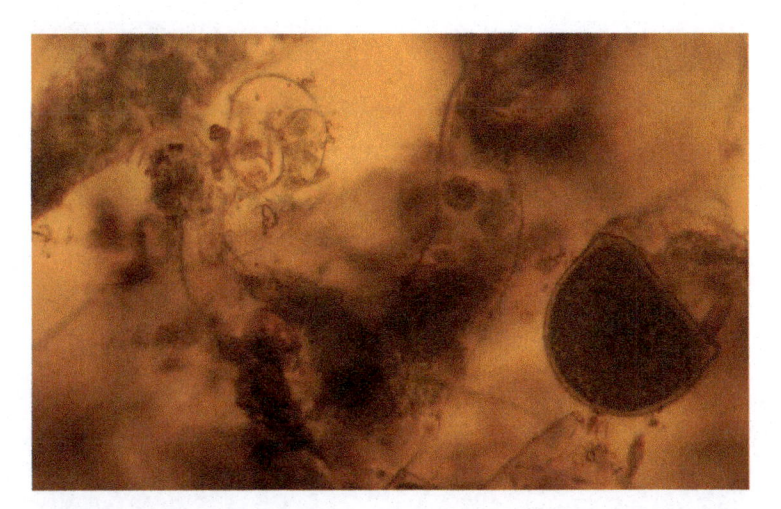

Fig. 10 *V. frigida* magnification × 800

At Station 11, the coastal communities are represented mainly by reedbeds with *Echinocystis lobata*. Associations of *Potamogeton pectinatus* purum, *Potamogeton nodosus* subpurum, *Nuphar luteum – Potamogeton pectinatus*, *Rorippa amphibia* are found in the medial of the river. From algae, *Oedogonium* sp. attached to the underwater parts of plants and *Mougeotia* sp., floating on the water surface among the higher plants, were found. *B. gelatinosum* was found in the benthos.

Fig. 11 *Pellia epyphilla* on the Pissa River bank, st. 22

Fig. 12 *Potamogeton nodosus*, on the Pissa River bank, st. 22

The Instruch River The site with the slow flow rate and the rapidly overgrown riverbed was examined (Fig. 14). *Salix alba, S. fragilis, Phalaroides arundinacea* grow along the banks. *Nuphar lutea + Myriophyllum spicatum, Nuphar lutea – Potamogeton pectinatus – Spirodela polyrhiza* are the dominating communities.

Fig. 13 *Sparganium emersum* L., Angrapa River, st. 10

Fig. 14 Overgrown riverbed of Instruch River

Hydrophytes are represented by 11 species of vascular plants and 6 species of algae (*Cladophora glomerata*, *Rhizoclonium riparium*, *Oedogonium* sp., *Chaetophora elegans*, *Spirogyra* sp., *Mougeotia* sp.) Lemnaceae (*Lemna trisulca*, *Spirodela polyrhiza*, *Lemna minor*) are abundant on the water surface which indicates a high level of water trophicity. *Acorus calamus – Spirodela polyrhiza, A. calamus – Nuphar luteum, Rorippa amphibian – Spirodela polyrhiza* communities are common along the bank. The medial of the river is overgrown with gelo-hydrophytic communities *Scirpus lacustris – Nuphar lutea*.

Fig. 15 The Krasnaya River, st. 12

Fig. 16 *Cladophora rivularis*, Krasnaya River

The Krasnaya River Due to prevailing conditions of shading (Fig. 15), few hydrophytes are found in the river, though heliophilous *Sagittaria sagittifolia, Butomus umbellatus* grow in small open areas. *Cladophora rivularis + Fontinalis antipyretica, Cladophora rivularis* purum (Fig. 16), *F. antipyretica* purum communities dominate in the rifts. *Hildenbrandia rivularis, Calliergon cordifolium, F. antipyretica* were found on boulders in the shade. Hydrophytes are represented by five species of vascular plants. *Lemna minor, L. trisulca, Nuphar lutea* grow in areas with the slow flow rate, *Sparganium emersum, Batrachium trichophyllum* are found in shallow waters. Flooded banks are covered with *Conocephalum conicum, Pellia epiphylla,* and *Vaucheria frigida.*

The Lava River The Lava River is the least examined river of our study. In total, nine species of vascular plants and two species of macroalgae were found. Water communities are represented by *Nuphar lutea – Potamogeton lucens, Potamogeton lucens – Myriophyllum spicatum, Scirpus lacustris – Nuphar lutea, Sagittaria sagittifolia + Butomus umbellatus, Phragmites australis – Nuphar lutea – Spirodela polyrhiza + Hydrocharis morsus-ranae* associations. *P. australis, S. lacustris* dominate in the coastal water communities.

The Deima River The downstream of the river near the village Sholokhovo was examined. The banks are represented by flooded water meadows, overgrown with *P. australis*. The ripal and the medial of the river were overgrown with aquatic vegetation. *Phragmites australis – Carex acuta* dominate. Water communities are composed of 13 species of vascular plants. Associations with prevalence of *Nuphar lutea, Spirodela polyrhiza, Nuphar lutea – Ceratophyllum demersum – Spirodela polyrhiza* are spread. In the vicinity of Sholokhovo village, Parfyonova [6] describes the community *Sparganium erectum + Butomus umbellatus – Spirodela polyrhiza – Ceratophyllum demersum + Batrachium circinatum* in which there were 16 species and *Phalaroides arundinaceae + Urtica dioica – Carex acuta* with the participation of 40 species of vascular plants [6].

4 Conclusions

The vegetation of the examined rivers is represented by plant communities, being widespread in the region. *P. australis, S. fragilis, S. alba, A. glutinosa, S. sagittifolia, S. sylvaticus, P. arundinacea* appear to be dominating species in the riparian area. *C. acuta, R. amphibia, S. erectum, B. umbellatus* have high frequency of occurrence. Water vegetation is mainly represented by the communities with the dominance of *P. pectinatus, P. nodosus, S. emersus, N. lutea, M. spicatum*. For the first time for the region, the data on the growth of the rare sensitive algae species (*H. rivularis, A. chalybaea*) and the findings of *C. elegans, D. glomerata, V. frigida, V. bursata, V. canaulicularis*, whose status is unclear due to insufficient study of macroalgae in the region, are published. In the Krasnaya River, there was found *B. trichophyllum*, listed in the Red Book of the Kaliningrad

Region as a requiring attention species. There was found a new regional habitat of *E. telmateia* – an endangered species listed in the Red Book of the Kaliningrad Region.

Acknowledgements We would like to express our gratitude to Ezhova E. E. (Laboratory of Marine Ecology of the AB of the IORAS) for her valuable comments on the manuscript of this paper.

Appendix

Checklist of flora of the small rivers of Pregolya River systems of Kaliningrad Region (Angrapa, Instruch, Krasnaya, Pissa, Deima, Lava)

No		Angrapa	Instruch	Krasnaya	Pissa	Golubaya	Deima	Svoboda	Perelesnaya	Gudishka	Lava
1	2	3	4	5	6	7	8	9	10	11	12
Divisio Rhodophyta											
1	*Hildenbrandia rivularis* (Liebmann) J. Agardh	+		+	+						
2	*Audouinella chalybaea* (Roth) Bory de Saint-Vincent	+									
3	*Batrachospermum gelatinosum* (L.) DC	+									
Divisio Chlorophyta											
4	*Ulothrix zonata* (F. Weber & Mohr) Kützing				+						
5	*Cladophora rivularis* (Linnaeus) C. Hoek		+		+			+	+		+
6	*Rhizoclonium riparium* (Dillwyn) Kützing		+	+							
7	*Oedogonium* sp.	+	+					+	+		+
8	*Chaetophora elegans* (Roth) C. Agardh		+								
9	*Draparnaldia glomerata* (Vaucher) C. Agardh				+						
Divisio Charophyta											
10	*Chara inconnexa* Allen				+						
11	*Spirogyra* sp.		+		+			+			
12	*Mougeotia* sp.	+	+								
Divisio Ochrophyta											
13	*Vaucheria terrestris* (Vaucher) de Candolle	+		+	+						

(continued)

No		Angrapa	Instruch	Krasnaya	Pissa	Golubaya	Deima	Svoboda	Perelesnaya	Gudishka	Lava
14	*Vaucheria bursata* (O. F. Müller) C. Agardh. (V. sessilis (Vaucher) de Candolle)				+	+					
15	*Vaucheria canalicularis* (L.) T. A. Christinsen		+		+						
16.	*V. frigida* (Roth) C. Agardh				+						
Divisio Bryophyta											
Class marchantiopsida											
1	*Conocephalum conicum* (L.) Dumort			+	+						
2	*Marchantia polymorpha* L.			+							
3	*Pellia epiphylla* (L.) Corda			+	+						
Class Bryopsida											
4	*Fontinalis antipyretica* Hedw.	+		+	+						
5	*Calliergon cordifolium* (Hedw.) Kindb.			+	+			+			
6	*Plagiomnium undulatum* (Hedw) T. Kop.			+	+						
Divisio Equisetophyt, Class Equisetopsida, Ordo Equisetales											
Fam. Equisetaceae rich. Ex DC											
1	*Equisetum arvense* L.				+						
2	*E. fluviatile* L.				+					+	
3	*E. sylvaticum* L.			+							
4	*E. telmateia* Ehrh.	+									
Divisio Angiospermae (= Magnoliophyta) *Class Monocotyledoneae (= Liliopsida)*											
Fam. Alismataceae vent.											
1	*Alisma plantago-aquatica* L.		+		+				+		
2	*Sagittaria sagittifolia* L.	+	+		+		+				+
Fam. Alliaceae											
3	*Allium ursinum* L.			+							
Fam. Araceae Juss.											
4	*Acorus calamus* L.	+	+								
Fam. Butomaceae rich											
5	*Butomus umbellatus* L.	+	+				+				+
Fam. Cyperaceae Juss.											
6	*Carex acuta* L.	+	+			+	+	+	+	+	
7	*C. nigra* (L.) Reichard						+			+	
8	*Scirpus lacustris* L.	+	+		+						+
9	*S. sylvaticus* L.	+						+			

(continued)

No		Angrapa	Instruch	Krasnaya	Pissa	Golubaya	Deima	Svoboda	Perelesnaya	Gudishka	Lava
Fam. Hydrocharitaceae Juss.											
10	*Elodea canadensis* Michx.		+		+		+				
11	*Hydrocharis morsus-ranae* L.	+	+				+				+
12	*Stratiotes aloides* L.		+								
Fam. Iridaceae Juss.											
13	*Iris pseudacorus* L.	+	+		+					+	+
Fam. Lemnaceae S. F. Gray											
14	*Lemna minor* L.	+	+	+		+	+	+	+	+	+
15	*L. trisulca* L.	+	+	+				+		+	
16	*L. gibba* L.					+					
17	*Spirodela polyrhiza* (L.) Schleid.	+	+		+	+	+	+	+	+	+
Fam. Poaceae Barnhart (Gramineae Juss.)											
18	*Arrhenatherum elatius* (L.) J. Presl. & C. Presl.	+	+								+
19	*Bromus inermis* (Leiss.) Holub										+
20	*Calamagrostis epigejos* (L.) Roth	+								+	+
21	*Dactylis glomerata* L.		+		+						+
22	*Deschampsia cespitosa* (L.) Beauv.						+				
23	*G. maxima* (C. Hartm.) Halmb.	+					+			+	+
24	*Holcus lanatus* L.										+
25	*Phalaroides arundinacea* (L.) Rauschert	+			+	+	+	+		+	+
26	*Phleum pratense* L.	+									+
27	*Phragmites australis* (Cav.) Trin. Ex Stend.	+	+	+	+	+	+	+	+	+	+
28	*P. Pratensis* L.	+									+
Fam. Potamogetonaceae Dumort											
29	*Potamogeton nodosus* Poir	+			+						+
30	*P. pectinatus* L.	+	+		+		+				+
31	*P. perfoliatus* L.		+				+				
32	*P. lucens* L.	+			+		+				+
Fam. Sparganiaceae Rudolphi											
33	*Sparganium emersum* Reh	+	+	+	+		+				+
34	*S. erectum* L.	+	+		+		+			+	+
Fam. Trilliaceae											
35	*Paris quadrifolia* L.			+							
Fam. Typhaceae Juss.											
36	*Typha latifolia* L.	+	+		+					+	+
37	*T. angustifolia* L.										+

(continued)

No		Angrapa	Instruch	Krasnaya	Pissa	Golubaya	Deima	Svoboda	Perelesnaya	Gudishka	Lava
V. II. Class Dicotyledoneae (Magnoliopsida)										
Сем. Aceraeae Juss.										
1	Acer platanoides L.	+			+						
2	A. negundo L.	+									
Fam. Adoxaceae										
3	Adoxa moschatellina L.			+							
Fam. Apiaceae Lindl.										
4	Aegopodium podagraria L.	+	+	+	+						
5	Angelica archangelica L.							+			
6	Anthriscus sylvestris (L.) Hoffm.	+	+	+							+
7	Chaerophyllum aromaticum L.	+			+						
8	Sium latifolium L.		+				+				+
9	Pimpinella saxifraga L.			+							
Fam. Araliaceae										
10	Hedera helix L.			+							
Fam. Aristolochiaceae										
11	Asarum europeum L.			+	+						
Fam. Asteraceae Dumort.										
12	Achillea millefolium L.	+	+								+
13	Arctium tomentosum Mill.				+						
14	Artemisia absinthium L.				+						
15	A. vulgaris L.	+	+		+						+
16	Bidens tripartita L.	+								+	
17	Cichorium intybus L.										+
18	Cirsium arvense (L.) Scop.	+	+			+	+	+		+	+
19	Cirsium oleraceum (L.) Scop.			+	+			+			
20	Hieraciums p.										+
21	Petasites spurius (Retz.) Rchb.				+						
22	Senecio paludosus L.				+						
23	Solidago serotinoides A. Et D. Löve.		+		+				+		+
24	Sonchus palustris L.	+							+		+
25	Tanacetum vulgare L.		+		+						
26	Taraxacum officinale Wigg. s.l.	+	+	+	+						+
Fam. Balsaminaceae A. Rich.										
27	Impatiens noli-tangere L.			+	+						

(continued)

No		Angrapa	Instruch	Krasnaya	Pissa	Golubaya	Deima	Svoboda	Perelesnaya	Gudishka	Lava
Fam. Betulaceae S. F. Gray.											
28	*Alnus glutinosa* (L.) Gaertn.	+	+	+	+		+		+	+	
29	*B. pubescens* Ehrh.				+						
30	*Carpinus betulus* L.	+		+	+				+		
31	*Corylus avellana* L.	+		+	+						
Fam. Boraginaceae Juss.											
33	*Myosotis palustris* (L.) L.	+	+	+	+		+	+	+		
33	*Pulmonaria obscura dumort.*			+	+	+					
34	*Symphytum officinale* L.	+		+	+		+	+	+		
Fam. Brassicaceae Burnett (Cruciferae Juss.)											
35	*Alliaria petiolata* (Bieb.) Cavara et Crande	+		+	+						
36	*Erysimum cheiranthoides* L.	+									
37	*Rorippa Amphibia* (L.) Besser	+	+		+	+	+				+
38	*Rorippa palustris* (L.) Besser		+		+						
39	*Cardamine amara* L.	+		+							
40	*Cardamine pratensis* L.			+			+				+
Fam. Campanulaceae											
41	*Campanula latifolia* L.			+							
42	*Phyteuma spicatum* L.			+	+						
Fam. Cannabaceae Endl.											
43	*Humulus lupulus* L.	+		+	+						
Fam. Caryophyllaceae Juss.											
44	*Moehringia trinervia* (L.) Clairv.				+						
45	*Silene dioica* (L.) Clairv.	+									
46	*S. holostea* L.	+		+	+				+		
47	*S. nemorum* L.	+		+	+						
Fam. Celastraceae											
48	*Euonymus europeus* L.	+		+							
Fam. Ceratophyllaceae S. F. Gray											
49	*Ceratophyllum demersum* L.		+				+				
Fam. Chenopodiaceae vent.											
50	*Chenopodium album* L.				+						
Fam. Convolvulaceae Juss.											
51	*Calystegia sepium* (L.) R. Br.	+						+		+	
Fam. Cucurbitaceae Juss.											
52	*Echinocystis lobata* (Michx.) Torr. Et gray	+	+		+	+			+		+

(continued)

No		Angrapa	Instruch	Krasnaya	Pissa	Golubaya	Deima	Svoboda	Perelesnaya	Gudishka	Lava
Fam. Cuscutaceae											
53	Cuscuta europea L.			+							
Fam. Euphorbiaceae Juss.											
54	Mercurialis perennis L.		+	+							
Fam. Fabaceae Lindl.											
55	Astragalus glycyphyllos L.		+	+							
56	Vicia sepium L.		+								
Fam. Fagaceae A. Br.											
57	Quercus robur L.	+			+						
Fam. Fumariaceae											
58	Corydalis solida L.			+							
Fam. Geraniaceae Juss.											
59	Geranium pratense L.		+	+							
Fam. Grossulariaceae DC.											
60	Ribes nigrum L.			+							
Fam. Haloragaceae R. Br.											
61	Myriophyllum spicatum L.	+	+		+		+				+
Fam. Hypericaceae Juss.											
62	Hypericum perforatum L.										+
Fam. Labiatae Juss. (= Lamiaceae Lindl.)											
1	2	3	4	5	6	7	8	9	10	11	12
63	Galeobdolon luteum huds.	+		+	+	+			+		
64	Glechoma hederacea L.	+		+	+						
65	Lamium album L.	+			+						+
66	L. purpureum L.	+			+	+					
67	Lycopus europaeus L.								+		
68	Mentha aquatica	+			+						
69	Scutellaria galericulata L.			+							
70	Stachys palustrisl.			+		+				+	
Fam. Lythraceae J. St. Hil.											
71	Lythrum salicaria L.	+	+				+		+	+	+
Fam. Nymphaeaceae Salisb.											
72	Nuphar lutea (L.) Smith.	+	+		+		+				+
73	Nymphaea candida J. Presl.	+	+				+				
Fam. Oleaceae Lindl.											
74	Fraxinus excelsior L.	+		+	+	+			+		+
Fam. Onagraceae Juss.											
75	Epilobium hirsutum L					+		+		+	+
	E. palustre L.		+		+						

(continued)

No		Angrapa	Instruch	Krasnaya	Pissa	Golubaya	Deima	Svoboda	Perelesnaya	Gudishka	Lava
Fam. Orobanchaceae											
76	*Lathraea squamata* L.			+							
Fam. Oxalidaceae											
77	*Oxalis acetosella* L.		+	+							
Fam. Papaveraceae Juss.											
78	*Chelidonium majus* L.		+	+							
Fam. Plantaginaceae Juss.											
79	*Plantago major* L.	+									
Fam. Polygonaceae Juss.											
80	*Persicaria amphibia* (L.) S. F. Gray										+
81	*P. mitis* (Scranc) Opiz et Assenov								+		+
82	*P. scabra* (Moench) Mold.										+
83	*Rumex obtusifolius* L.	+		+							+
Fam. Primulaceae vent.											
84	*Lysimachia nummularia* L.		+	+	+			+	+		
85	*L. vulgaris* L.			+					+		
Fam. Ranunculaceae Juss.											
86	*Actaea spicata* L.			+							
87	*Anemonoides nemorosa* (L.) Holub			+							
88	*Anemonoides ranunculoides* (L.) Holub		+								
89	*Batrachium trichophyllum* (Chaix) Bosch.		+								
90	*Caltha palustris* L.			+			+				
91	*Ficaria verna* huds.	+		+	+			+	+		
92	*Hepatica nobilis*			+	+						
93	*Ranunculus auricomus* agg. L.			+							
94	*Ranunculus lanuginosus* L.	+		+	+						
95	*R. repens* L.	+			+		+				
96	*R. sceleratus* L.	+									
97	*Thalictrum flavum* L.		+	+			+				
Fam. Rosaceae Juss.											
1	2	3	4	5	6	7	8	9	10	11	12
98	*Alchemilla* sp.						+				
99	*Crataegus monogyna* Jacq.			+							
100	*Filipendula ulmaria* (L.) Maxim.	+	+	+	+	+	+	+	+	+	+
101	*Fragaria vesca* L.			+	+						

(continued)

No		Angrapa	Instruch	Krasnaya	Pissa	Golubaya	Deima	Svoboda	Perelesnaya	Gudishka	Lava
102	*Geum rivulare* L.	+		+			+				
103	*Malus domestica* borckh.			+							
104	*Padus racemosa* (Lam.) Gilib.	+		+	+	+			+		
105	*Potentilla anserina* L.		+								
106	*P. reptans* L.		+								
107	*Rubus caesius* L.	+	+		+			+			+
108	*R. idaeus* L.	+									
109	*Sorbus aucuparia* L.				+						
Fam. Rubiaceae Juss.											
110	*Galium aparine* L.			+	+				+		
111	*G. palustre* L.				+		+	+	+	+	
Fam. Salicaceae Mirb.											
112	*Populus tremula* L.								+		
113	*Salix alba* L.	+	+	+		+		+	+		+
114	*S. caprea* L.					+		+			
115	*S. fragilis* L.	+	+		+	+				+	+
116	*S. purpurea* L.		+								
117	*S. viminalis* L.		+			+				+	+
Fam. Sambucaceae Batschex Borkh.											
118	*Sambucus nigra* L.			+	+			+	+		+
119	*S. racemosa* L.			+	+						
Fam. Saxifragaceae Juss.											
120	*Chrysosplenium alternifolium* L.			+					+		
Fam. Scrophulariaceae Juss.											
121	*Melampyrum nemorosum* L.				+						
122	*Scrophularia nodosa* L.				+			+	+		
123	*V. beccabunga* L.	+	+	+	+						
124	*V. chamaedrys* L.				+						
Fam. Solanaceae Juss.											
125	*Solanum dulcamara* L.					+					
Fam. Tiliaceae Juss.											
126	*Tilia cordata* Mill.	+		+	+				+		
Fam. Thymelaeaceae											
127	*Daphne mezereum* L.			+							
Fam. Ulmaceae											
128	*Ulmus laevis* pall.	+		+	+						
Fam. Urticaceae Juss.											
129	*Urtica dioica* L.	+	+	+	+	+	+	+	+		+
Fam. Valerianaceae Batsch											
130	*Valeriana officinalis* L.	+	+	+							
Fam. Violaceae Batsch.											
131	*Viola odorata* L.				+						

References

1. Abromeit J, Neuhoff W, Steffen H, Jentzsch A, Vogel G (1889–1940) Flora von Ost- und Westpreussen, Bd 1–3. Berlin, 1246 s (in German)
2. Galanin AV (1991) The vegetation cover of Pregolskaya lowland area in the Kaliningrad Region/flora and landscape-ecological structure of vegetation. Far Eastern Branch of the Academy of Sciences, Vladivostok. USSR-272 (in Russian). http://geobotany.narod.ru/galanin/m8.htm
3. Gerb MA (2014) Ecological analysis of aquatic flora in lower reaches of the Pregolya River, vol 32. News KGTU, pp 162–169 (in Russian)
4. Parfenova YV (2006) Ecological and geographical analysis of vascular plants Pregolya River (Kaliningrad Region, Russia). In: Proceedings of the I(IX) of the international conference of Young Botanists in St. Petersburg, St. Petersburg, 21–26 May 2006, pp 55–56 (in Russian)
5. Utkina YV (2012) Hygrophilous vegetation oxbow lakes in the central part of the Pregolya district on the gradient of anthropogenic transformation (Kaliningrad Region of Russia). In: Abstracts of II(X) of the international botanical conference of young scientists in St. Petersburg, 11–16 Nov 2012, Proceedings II(X) international botanical conference of young scientists in St. Petersburg, ETU "LETI", Saint-Petersburg, 11–16 Nov 2012, pp 102–103 (in Russian)
6. Parfenova YV (2008) Hygrophilous vegetation of Deima River (Kaliningrad Region, Russia). Bulletin of the Russian State University, Immanuel Kant Series of Natural Sciences, vol 1. Immanuel Kant Russian State University, Kaliningrad, pp 77–83 (in Russian)
7. Katanskaya VM (1981) Higher aquatic vegetation inland waters of the USSR. Methods of study. Nauka, Leningrad, p 278 (in Russian)
8. Papchenkov VG (2001) The vegetation cover of reservoirs and waterways of the middle Volga. CMP MUBiNT, Yaroslavl, p 214 (in Russian)
9. Alekhin VV (1986) Theoretical problems and phytocenology of steppes. Moscow State University, Moscow, p 213 (in Russian)
10. Shennikov AP (1964) Introduction to geobotany. A.A. Zhdanov Leningrad University, Leningrad, p 447 (in Russian)
11. Cherepanov SK (1995) Vascular plants of Russia and adjacent states (the former USSR)–SPb. Mir & Semya-95, St. Petersburg, p 992 (in Russian)
12. WoRMS (World Register of Marine Species). www.marinespecies.org
13. Tsybin YA (ed) (2004) Scheme of nature protection of the Kaliningrad Region. TENAX Media, Kaliningrad, pp 40–42 (in Russian)
14. Dedkov VP, Grishanov GV (eds) (2010) The red book of the Kaliningrad Region. In: I. Kant Russian State University, Kaliningrad, pp 307–311, 318–324 (in Russian)
15. Gubareva IY (2010) In: Dedkov VP, Grishanov GV (eds) Equisetum Telmateia [Red book Kaliningrad Region]. I. Kant Russian State University, Kaliningrad, p 203 (in Russian)
16. Sokolov AA (2014) Natural complex "Romintskaya Forest". Nature of the Kaliningrad Region. Istock, Kaliningrad, pp 54–56 (in Russian)

The Protection Conditions of the Groundwater Against Pollution in the Kaliningrad Region

Galina Mikhnevich

Abstract The chapter lists the results of the assessment of groundwater protection against pollution studying the case of the upper intermoraine aquifer. The work is based on extensive factual material. Among the main criteria of protection from pollution, the study considers power, lithological composition, and permeability of rocks composing the upper waterproof and the correlation of the level of pressure and groundwater. An attempt to quantify assessment of groundwaters protection is taken. The research defined the regularities of the spatial differentiation of the protection conditions of the Moscow-Valdai aquifer in the Kaliningrad Region and isolated areas characterized by varying degrees of protection of the upper intermoraine aquifer. The article will define the areas of potential water-use conflicts in the oil-extraction areas and in the areas of construction materials excavations.

Keywords Groundwater, Pollution, Protection, Vulnerability

Contents

1 The Concept of "Protection/Vulnerability" of the Groundwater from Pollution: Basic Approaches to Protection Assessment and Its Mapping 412
2 The History of the Groundwater Protection Studies in the Russian Federation and the Kaliningrad Region 415
3 Methods of Assessing the Protection of Groundwater of the Upper Subsurface Pressure (the Top Intermoraine) Aquifer 417
4 Characteristics of the Upper Intermoraine (the Moscow-Valdai) Aquifer 421
5 Characteristics and Regularities of Spatial Distribution of the Factors That Determine the Protection of the Upper Intermoraine Aquifer 426

G. Mikhnevich (✉)
Institute of Environmental Management, Urban Development and Spatial Planning, Immanuel Kant Baltic Federal University, 14, A. Nevsky Street, Kaliningrad, Russia
e-mail: mi78galina@mail.ru; GMikhnevich@kantiana.ru

6	Territorial Differentiation of the Kaliningrad Region on the Degree of the Upper Intermoraine Aquifer Protection	431
7	Possible Areas of Water-Use Conflicts	434
References		435

1 The Concept of "Protection/Vulnerability" of the Groundwater from Pollution: Basic Approaches to Protection Assessment and Its Mapping

Geological environment can provide a certain degree of protection from natural and anthropogenic pollution. This assumption gives the background for the concept of groundwater vulnerability first formulated and tested in practice in France in the late 1960s to early 1970s [1, 2]. Vulnerability is understood as "the natural properties of the groundwater system, which depend on the ability or the sensitivity of the system to deal with natural and man-made impacts." [3] Committee on technology at the US National Research Committee has identified groundwater vulnerability as *tendency* or *probability* to achieve certain pollutant concentration in the groundwaters after entering the zone above the upper aquifer. The Committee then identified two main types of vulnerabilities: *specific vulnerability* (meaning any pollutant or several pollutants) and *inherent vulnerability* which does not depend on the properties and behavior of specific pollutants [4].

Previously, the USSR and currently Russia often interpreted the term "protection" as the opposite one to the term "vulnerability": the less protection, the greater vulnerability, and the reverse. Considering this, we can say that *protection* of groundwaters is the property of a natural system that allows you to save the composition and quality of groundwater corresponding requirements of their practical use over the forecast period [5]. The protection (or vulnerability) of groundwaters depends on many factors, which can be roughly divided into three groups: natural, anthropogenic, and physical and chemical [5, 6]. Natural factors include groundwater depth; the availability of semipermeable layers, strength, and water permeability of the rocks overlying aquifers; sorption properties of rocks; hydrodynamic conditions that determine the direction and speed of filtration; and water exchange rate of different horizons of groundwaters. Anthropogenic factors include the availability and conditions of storage of pollutants on the surface, the distribution of wastewater area including irrigation, and various types of ground surface disturbances (various wells, mines, quarries, wells) that determine the possibility for the penetration of pollutants into aquifers. Among physical and chemical factors, there are sorption and migration properties of the pollutants and the nature of the interaction between pollutants, rocks, and groundwaters.

The difference in the natural conditions, the quality and quantity of information, and the possibility of monitoring the groundwater status became the basis for designing a large number of protection assessment methodologies. As a result, all the methodologies can be divided into two major groups:

1. Methodologies that provide a qualitative assessment of the territory and mapping of groundwater vulnerability or protective properties excluding the characteristics and properties of specific pollutants and aiming to determine the degree of factor influence on the aquifer vulnerability, which allow us to compare the territories in their groundwater protection levels
2. Methodologies that provide a quantitative assessment and mapping of the protective properties of the natural system, based on the calculation of a particular pollutant penetration time in the aquifer taking into account the natural properties of water-bearing aquiclude and pollutant migration properties [5, 7]

Usually it is the degree of protection of nonpressure water or high-pressure aquifer waters which is assessed [6–18]. Much of the research deals with the analysis of groundwater protection. Conditions of groundwater occurrence are such that precise quantification of protection and large-scale mapping are simply enough because the water in the unsaturated zone moves in one direction – from the surface to the level of groundwater. Interstratal waters protection gets assessed much more seldom. Due to the complexity of mathematical models describing the process of water or moisture movement from the surface through the overlying aquifer sediments, complete quantification of protection is difficult and only possible for point objects [10].

In the USSR, the first groundwater protection assessments appeared in the late 1970s to early 1980s of the last century. V. Goldberg is the author of one of the most well-known and acknowledged methodologies of groundwater protection assessment in Russia which is point rating and considers the impact of various environmental factors on the groundwater vulnerability [19]. Another point rating methodology is the one of calculating the protection of groundwater and artesian water proposed in the guidelines for the compilation of ecological and geological maps issued by VSEGINGEO [20]. V. Goldberg's methodology has a lot of modifications, depending on local hydrogeological conditions and data sets.

A more precise quantification of groundwater protection from pollution can be implemented through calculating the time of contaminants penetration of into the aquifer. The mathematical tools and research results are described in a methodological guidelines for groundwater protection [8], in the works of Goldberg [6, 19, 21], Belousova [7, 9, 10], Proskurina [22], Rogachevskaya [23], Abakumov [24], Kolomeev [25], Minyaeva and Karimova [26], Titushkina [27], Stenemo [28], Ambarref et al. [29], and others.

Research of pressure water protection is carried out less frequently, and most of it is about qualitative assessment. An example of evaluation of upper confined aquifer protection is a methodology proposed in the late 1970s of the last century by a VSEGINGEO team of scientists under the guidance of Goldberg [6, 8] and based on the analysis of two indicators – the power of the upper aquifer and the ratio of the levels of the studied pressure aquifer and overlying unconfined aquifer with the calculation of the pollutant penetration time. The methodology suggested by A. Schwartz demonstrates a similar approach to protection factors [30].

Given all their differences, almost all methodologies make it possible to represent the result of protection assessments in cartographic form. According to the A. Belousova, the medium scale (1:200,000, 1:100,000) seems to be the most suitable as it better meets the demands of qualitative and quantitative protection assessment [7, 10]. The role of each factor and the choice depends on the basic geological and anthropogenic environment of the study area, the objectives of the study, and the scale of maps. For example, natural factors are different for pressure and phreatic water. In evaluating protection *of phreatic water*, primarily important factors are the power and lithology of vadose zone, groundwater recharge, groundwater exchange time, and water conductivity of phreatic aquifers. At the same time, the degree of influence of each factor will vary depending on the geological and hydrogeological conditions [5–8, 10]. The data about the structure, permeability, and filtration properties of the aeration zone are important for the study of phreatic water protection, but they are often in short supply. In these cases, a regional assessment is characterized by the lithological composition of the vadose zone, and several categories of rock can be defined: well-permeable, permeable, semipermeable, and almost impenetrable rocks. An important factor to determine the phreatic groundwater protection is the volume of infiltration, which can be quantified through the results of the regional assessment and groundwater flow mapping.

For *pressure water* the main protection factors are the correlation of the estimated levels of the aquifer and overlying unconfined aquifer, which determines the possibility of contaminated water migration from above; capacity and lithology of the upper impermeable layer defining the possible volume of contaminated water overflow from above; and water conductivity, the amount of infiltration and estimated time of water exchange in a certain aquifer. Obviously the first two factors are fundamental, as they combine to characterize the potential danger of contaminated water penetration from above to the discharge horizon [5, 8, 19, 20]. In addition to the abovementioned factors, the study of groundwater protection sees importance of sorption capacity of both water-bearing and the overlying impermeable rocks.

Given this, we can draw the following conclusions:

1. Most high-quality score-based methodologies are based on the expert approach, when some factor is given a greater role (and score), another is valued less on the background of expert experience without specific quantitative evaluation. The results obtained by those methodologies can be used as the initial stage in quantity assessment of groundwater protection in the regional studies [5].
2. The second stage is evaluating the time required for penetration of a contaminant into the aquifer. This value, as well as the migration rate of the contaminant can be reliably determined only with more detailed studies, if there is some evidence of the sorption properties of the vadose zone and aquifer and migration parameters available.
3. Regional assessment and groundwater protection mapping are held separately for the main aquifers, whose waters are used now or will be used for water supply in the future. The reason for this is that the degree of influence of

environmental factors is different for pressure and groundwater. In the disturbed conditions, as compared to the natural ones, the influence of the main factors on the groundwater protection can vary greatly.

The most important practical result of the assessment and mapping is the ability to compare different areas in terms of groundwater contamination protection and decide which area is better protected, where there is a great danger to contamination of freshwater supply wells, which primarily requires protection measures. Timely evaluation of groundwater protection allows taking the necessary measures to prevent the water contamination. Protection area maps can be used by water management organizations for planning measures to improve the environmental situation.

2 The History of the Groundwater Protection Studies in the Russian Federation and the Kaliningrad Region

The VSEGINGEO studies, headed by V. Goldberg, became the first stage in the assessment of the natural groundwater protection. In 1980, the groundwater protection map for the European part of the USSR (scale 1:1,500,000) came out. This map identified the groundwater of the Kaliningrad Region as vulnerable. Antipollution protection of prospect water supply pressure aquifer f in the Kaliningrad Region is also assessed on the ecological and geological map of the Northwest Federal District (scale 1:9,000,000) as weak [31].

Special medium-scale research of groundwater protection in the Kaliningrad Region was conducted by the scientists of the Faculty of Geography of the Kaliningrad State University in the late 1980s to early 1990s of the last century [32–35]. The research resulted in the maps of the natural groundwater protection (ground and pressure waters), made up for some of the administrative districts of the region (Chernyakhovsk, Slavsk and Krasnoznamensk), as well as for the city of Kaliningrad.

V. Goldberg's scoring method of groundwater protection assessment interpreted by G. N. Yeltsina, who took into account local hydrogeological conditions and the factual data, became the background for evaluating groundwater protection. Following Goldberg [6, 21], it was proposed to identify the lithologic variations in the aeration zone as "a," "b," and "c":

"a" – poorly permeable rock: sandy loam, sandy clay loam, filtration coefficient $k = 0.1$–0.01 m/day.

"b" – the alternation of low-permeable and impermeable rocks, medium loam, $k = 0.01$–0.001 m/day.

"c" – impermeable rocks: glacial clays and heavy loams, filtration coefficient $k < 0.001$ m/day.

Protection evaluation scoring was conducted the following way; five graduations of groundwater depth were defined: (N), ≤10 m, 10–20 m, 20–30 m, 30–40 m, and >40 m; each graduation attributed its own score from 1 to 5. Cross sections of low-permeable sediment capacity graduations (m_0) are given in the Table 1. The sediments are classified as "a" and "c" by the prevailing lithological varieties and the total capacity of low-permeable sediments. The total score is determined by the points for the capacity of the vadose zone and the capacity of available low-permeable rocks. For example, if the groundwater depth is 14 m (2 points) and a layer of sandy loam is 3 m (2 points) and a layer of clay is 7 m (8 points), then the total score is 12 points. Total scores distinguish between six categories of groundwater protection: I ≤ 5, II = 5–10, III = 10–15, IV = 15–20, V = 20–25, and VI > 25. The least protected are characterized by conditions corresponding to category I, the best, to category VI.

Tsunker's formula was accepted as the basis for scoring [6, 30]. The time to reach the level of groundwater by the wastewater percolating from surfaces in a section of vadose zone and the constant reservoir water level is calculated on the following formula:

$$t = \left(\frac{nH}{k}\right) \cdot \left[\frac{m}{H} - \ln\left(1 + \frac{m}{H}\right)\right], \quad (1)$$

where H is the height of the column of wastewater in the reservoir; k and m, respectively, are the filtration coefficient and capacity of the vadose zone; n is the lack of saturation of the vadose zone rocks [6, 30].

The scoring baseline for the evaluation is calculated on Tsunker's formula: (1) the time of filtration t_1 through the vadose zone of well permeable rocks ($k \sim 2$ m/day) with a capacity of 10 m. The filtration time t_2 through the vadose zone of 20 m capacity and the same rocks are about twice as long ($t_2 \sim 2t_1$). Thirty-meter vadose zone takes three times longer time ($t_3 \sim 3t_1$). Matching points between gradation capacity of low-permeable rocks and there lithology are defined as follows. According to Tsunker's formula, the following equivalence of a filtration time t through the layers of different capacity m (m) and permeability k (m/day) is accepted:

$$t_{m=10}, k > 1 \sim t_{m=2}, k = 10^{-2} \sim t_{m=1}, k = 10^{-3} \sim t_{m=0.5}, k < 10^{-3} \quad (2)$$

Given a certain equivalence of contaminant filtering time, it can be assumed that filtration through a bed of rock with a capacity of 10 m with $2k = $ m/day is

Table 1 Capacities of low-permeable sediments (m_0) and the corresponding points [6]

		Capacities of low-permeable sediments (m_0) (m)									
		<2	2–4	4–6	6–8	8–10	10–12	12–14	14–16	16–18	18–20	>20
Lithological groups	a	1	2	3	4	5	6	7	8	9	10	12
	b	1	3	4	6	7	9	10	12	13	15	18
	c	2	4	6	8	10	12	14	18	18	20	25

approximately equal to the one t through the rock layer with a capacity of 1 m and $k = 10^{-3}$ m/day or through a bed with a capacity of 0.5 m when $k = 2 \times 10^{-4}$ m/day.

This assumption made it possible to use value *of aquiclude equivalent capacity while assessing the protection of pressure water*. The natural pressure water protection was supposed to be characterized with the upper aquiclude equivalent capacity index (M_c'), estimated in meters and characterized for its strength. Equivalent capacity of the upper aquiclude is calculated using the formula:

$$M_c' = M_c + 0.5 M_b + 0.1 M_a \qquad (3)$$

where M_c' is the equivalent capacity of the upper aquiclude, M_c capacity of "c" rock group, M_b capacity of "b" rock group, and M_a capacity of "a" rock group. The higher the upper aquiclude capacity, the greater the degree of protection of waters in the first subsurface pressure aquifer [32, 33, 36].

Thus, the studies outlined the main methodological approaches to the assessment of the groundwaters in the Kaliningrad Region, the processes leading to their change and demonstrated the vulnerability of the regional hydrogeological system. However, the economic development of the region leads to increased consumption of groundwaters and emergence of new sources of pollution, and it requires advanced studies into the groundwater pollution protection. Therefore, in 2003–2011 there were a number of studies to assess the natural protection of the upper subsurface pressure aquifer groundwater, which ensures water supply for the most significant share of the region's population.

3 Methods of Assessing the Protection of Groundwater of the Upper Subsurface Pressure (the Top Intermoraine) Aquifer

Some aspects are considered while choosing a methodology for assessing the groundwater pollution protection:

- The work covers the entire territory of the Kaliningrad Region, which involves a considerable amount of factual data (data about 1,500 wells) and medium-scale mapping of the groundwater protection conditions.
- The work represents the first attempt at determining the protective properties of the natural system for pressure groundwater; therefore, it is considered more logical to assess the inherent protection, without taking into account specific types of pollution.
- We study the natural protection of the upper subsurface pressure intermoraine aquifer, called the Moscow-Valdai (IIms-IIIvd). Groundwater protection is not addressed, as the issue was studied in detail in the 1980s of the previous century.

- Within the research we also assess protection from pollution that is formed on the earth's surface or in groundwater and moves to the Moscow-Valdai aquifer vertically from top to bottom. Cases of a lateral pulling substandard water or overflow of the lower aquifers are not studied.

It is also important that to quantify the protection of groundwater is not enough data on the physical and mechanical properties of rocks. Since impermeable deposits date back to glacial engineering and geological formation, their physical and mechanical properties vary greatly, not only in size but also in the vertical direction. The permeability of the clay rocks forming the aquiclude depends on the mineral composition and the temperature and filter solution salinity [37]. Therefore a full quantitative assessment involves the study of large-scale application of mathematical models, experimental filtration research, and computer data processing. However, the main protection criterion is the time for a pollutant to move to the groundwater level, so it is necessary to try to give it some approximate definition.

Time of pollutant filtering from the surface into the aquifer is calculated of time of filtration through poorly permeable sediments forming the upper aquitard. If the section of overburden sediment includes groundwater, the calculation of pollutant filtration time becomes more complicated. It is necessary to determine the pollutant filtration time to the groundwater level and the filtration time from the groundwater horizon into the pressure water horizon through the separating aquitard when the groundwater level is higher than the level of pressure water.

The filtration time to the level of groundwater from surface reservoirs of wastewater with a constant level is calculated on Tsunker's formula (Formula 1). Constant flow wastewater discharge on the land surface is calculated on the formulae defined in the works of Goldberg [8], Goldberg and Gazda [6], Shtengelov [38], and the ones used to calculate the sanitary protection zones [39]. When defining zones of sanitary protection for groundwater, it is necessary to calculate the time t_n needed for pollution to go through the vadose zone with the capacity m_n with a coefficient of filtration K_n and active (effective) porosity n_n:

1. with minor infiltration values ($W < K_n$):

$$t_n \approx \frac{m_n \cdot n_n}{\sqrt[3]{W^2 \cdot K_n}}; \qquad (4)$$

2. at infiltration with full saturation of pores (intense infiltration, leaks, etc. $W > K_n$):

$$t_n \approx \frac{m_n \cdot n_n}{K_n}. \qquad (5)$$

At minor quantities of natural infiltration, the time of movement through the vadose zone is much longer than of intensive filtration.

Time of pollution migration from overlying aquifer through overlying impermeable layers is calculated on the following formula:

$$t_n = \frac{m_n^2 \cdot n_n}{K_n \cdot \Delta H}, \qquad (6)$$

where ΔH is the difference of pressures between the groundwater and pressurized aquifers (groundwater level is higher than the pressure one). This formula is used in the design of sanitary protection zones for interstratal aquifers [38, 39] and in determining pressure water protection by Goldberg [8]. The filtration time through the impermeable aquifers is usually very long and greatly exceeds the filtration time up to the groundwater level, so this value can be neglected.

Due to the given reasons, traditional ways of calculating the time required for pollutants to reach the aquifer cannot be used in the research; it applies semiempirical approach to protection assessment. A certain methodology proposed by G. N. Yeltsina in the late 1980s was suggested to be used as the basic one. It determined groundwater protection, considering the capacity and lithological composition of the upper confining layer or the vadose zone as well as the above-described one [32, 34, 35]. The natural protection of the pressure aquifer is characterized with an indicator of the reduced power of the upper aquiclude (M_c'), which is calculated in meters and refers to the reliability of the aquiclude (formula 3). The reduced power describes the aggregate capability of the upper confining layer, its lithology, and filtration properties [32, 34, 35]. At the same time, it was decided to transform that approach, introducing such components as the calculation of a relative time of a pollutant penetration into the aquifer and the ratio of pressure and groundwater levels into the evaluation of protection.

The Calculation of a Relative Time of a Pollutant Penetration into the Aquifer The value of the reduced capacity of the upper aquifer can be used to determine *relative time* of a pollutant penetration from the surface into the groundwater:

$$t \sim \frac{M_c'}{k_c}. \qquad (7)$$

The value of $M_c' = 10$ m with $k_c = 0.001$ m/day and less provides for a mean pollutant penetration time equal to 10,000 days or 27 years, i.e., standard lifetime of an intake service. The choice of 30-year pollutant penetration time as the criterion of groundwater protection depends on both the definition of protection ("... preserving the quality for the forecast period, i.e. for a life time of the water intake facility") and the possibility of full or partial decontamination of the majority of known pollutants. When compiling protection maps, it is possible to use tenfold gradation M_c': 10, 20, 30 m, etc., which approximately corresponds to isochrones 30, 60, 90, etc. years. Moreover, within the ranges outlined with a 30-year isochrone, it is necessary to select shorter relative intervals of a pollutant penetrating into the aquifer [40, 41].

Such an approach is justified by the following considerations. Protection is determined by the presence of low-permeable sediments, mainly clay loams and clays with filtration coefficient of less than 10^{-3} m/day. The reduced power reflects the power of rocks overlying the aquifer through the power of the low-permeability rocks. Terms of occurrence of the Moscow-Valdai aquifer are such that it is overlaid by the sediments with reduced capacity of 10 m or more. The difference in levels may reach 20–30 m, but often the levels of pressure and groundwater differs by 5 m at such capacity of aquiclude. The porosity of clay rocks reaches 0.5, active porosity can be taken as 0.05 [42]. Then, the time of pollutant filtering calculated by the formula (6) will be 10,000 days (with $k = 10^{-4}$ m/day). This is indirectly confirmed by the Schwartz [30]. According to his calculations, the 10-m thick layer of permeable rock (loams) determines the infiltration period of approximately 4,000 days. According to the equivalence given by V. Goldberg and S. Gazda, $t_{m=1}, k = 10^{-3} \sim t_{m=0.5}, k < 10^{-3}$ (see formula 2), this means that the filtering time through 10-m layer of clays will be twice as long as filtering time through the loam (8,000 days), which roughly corresponds to our own calculations [6].

Calculation of relative time of pollutant penetration into the aquifer is similar to the overflow coefficient calculation. The essential difference is that the filter coefficient does not correlate with the capacity of aquiclude, but with the value of the integral parameter of reduced power of the upper confining layer.

Joint account of lithological and hydrodynamic factors enabled to identify the following factors *of protection category of the Moscow-Valdai aquifer*:

- Unprotected – low-capacity aquiclude ($M_c' < 10$ m), poorly defined in areas; in some areas there is no aquiclude (hydrologic gaps) pressure water levels that are below groundwater level ($H_2 < H_1$). Relevant pollutant penetration time (t) is less than 30 years.
- Partly protected – pressure waters are overlaid with the continuous aquiclude of reduced capacity, the value $M_c' = 10$–20 m, and the level of pressure water is lower or the same depth as the groundwater ($H_2 \leq H_1$). Relevant pollutant penetration time (t) is 30–60 years.
- Protected – pressure waters are confined with high capacity ($M_c' > 20$ m) and solid aquiclude; levels of pressure water (H_2) are above the groundwaters (H_1). $t > 60$ years [41].

In addition the protection map displays the water intake wells and factors causing groundwater pollution. The proposed method is simple, does not require additional costs to conduct research and uses the available hydrogeological data. Results of the study of the natural groundwater protection can make the background for large-scale quantitative protection assessments. Before getting down to the classification of protection conditions of the upper intermoraine aquifer, we will consider its main characteristics.

4 Characteristics of the Upper Intermoraine (the Moscow-Valdai) Aquifer

Fresh underground waters of the Kaliningrad Region are the main source of the regional water supply: 53% of people in the cities and 99% in rural are supplied with water from underground sources. Water is provided from the Upper Cretaceous, Paleogene, and Quaternary aquifers: the Moscow-Valdai (formerly known mid-Russian-Valdai) and Oka-Dnieper (previously Lithuanian-Central Russian) intermoraine aquifers. The quaternary aquifers with 63% of water consumption are of the aquifers with the highest operational importance [43, 44]. The highest anthropogenic burden falls on the upper intermoraine (the Moscow-Valdai, mid-Pleistocene) aquifer which is widespread in the southern part of the region. The aquifer distribution area is more than 9,300 km^2 (70% of the regional territory). The aquifer distribution and the capacity of water-bearing rocks are represented on the map (scale 1:200,000), compiled by the author on the drilling data, and demonstrated in a simplified form in Fig. 1.

Sand and gravel-pebble material of different particle size are water-bearing sediments. Power aquifer varies widely, from a few meters to 102.1 m (Svetlogorsk), typically accounting for 10–15 m (Fig. 1). The distribution of aquifer capacity is seen a number of regularities. The Moscow-Valdai sediments are not found in northern part of the region (around Lower Neman Plain) and in the north-eastern part of the region (the area in the Neman Valley). Part of the water-bearing sediments of the aquifer are preserved in the deep incision in the pre-quaternary layer. The Moscow-Valdai aquifer is not widespread and promising within the Polessk plain: there are no water-bearing sediments near the southwestern coast of the Curonian Lagoon and

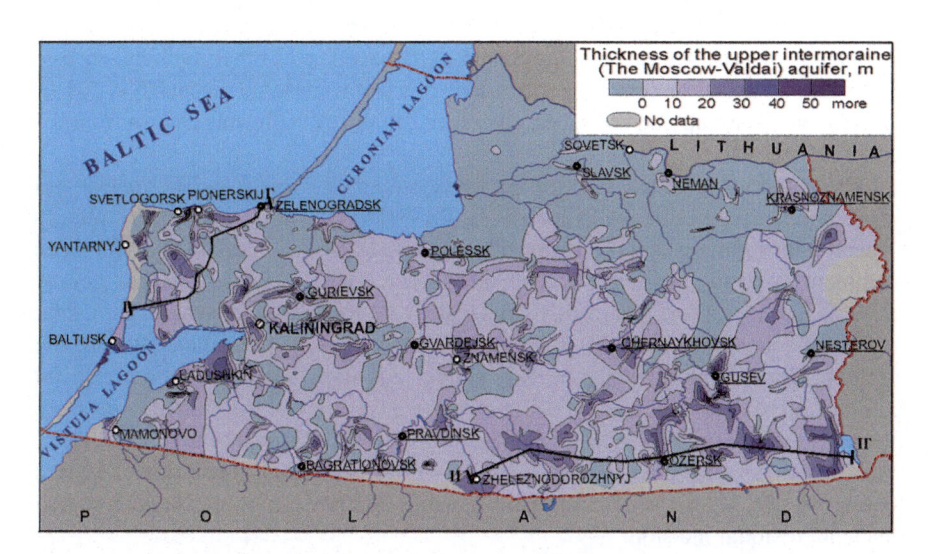

Fig. 1 Map of distribution of the upper intermoraine (the Moscow-Valdai) aquifer

further south to the village. Polyany, they are sporadically spread in the vicinity of Polessk. The mosaic spread of water-bearing sediments is typical for two areas similar in the geological structure: Kaliningrad Peninsula and the southwestern part of the Kaliningrad Region. A particular area where the interglacial marine structures have remained, is the area of the Vistula Spit (maximum capacity is 62.5 m) (Fig. 1) [45]. The rest of the Kaliningrad Region is characterized by a common though not universal distribution of the Moscow-Valdai aquifer sediments (Fig. 1). Widespread occurrence of sediments of this age results from the fact that the interglacial aquifer often includes water-glacial deposits lying below or above the actual interglacial lake, sea, or river. In terms of hydrogeology, these sediments form a single aquifer. In addition, sediments of different ages can hardly be differentiated on lithologic data, and the complex of sediments is often an evolutionary one (runoff valley – river valley – proglacial pond). The capacity of the Moscow-Valdai sediments does not increase in any area. Capacity increases locally from a few meters up to 60 m (Gusev), while values up to 10 m dominate. The absence of the Moscow-Valdai deposits should be considered in the complex as a result of erosion activity of water flows in the interstage or postglacial times (the Pregolya valley), as a consequence of exaration activities of the Late Pleistocene (Valdai) glacier or as a feature of the local sedimentation in interglacial stage [46].

The depth of occurrence of the aquifer varies from a few meters up to 70 m, but in much of the area, it lies at a depth of 20–30 m (Figs. 2, 3, and 4). The absolute level of the aquifer roof depends on the elevation of the modern relief and ranges from −8.9 to 124.2 m (Figs. 2 and 3). The most common level is of 0–30 m. The aquifer is low pressure, the pressure magnitude ranges from a few meters up to 50 m or more, depending on the depth of the aquifer. Piezometric levels of water are +1... −28 m from the ground surface. Depending on the modern terrain hypsometry, the absolute marks of piezometric levels vary from 1 to 140 m. Changing water abundance of the aquifer is determined by particle sizes of water-bearing deposits, as well as the conditions of bedding, layer volume, and its connection to other aquifers, surface water bodies, and watercourses. Performance of water wells ranges from 0.5 to 10 L/s. Specific yields usually do not exceed 2–7 L/s. The filter coefficients change from 0.2 to 53 m/day [47].

The aquifer is limited from top by the sediments of *late Neo-Pleistocene* and *Holocene*. Impermeable rocks overlying the Moscow-Valdai aquifer are mainly glacial in origin, except in areas where the aquifer is overlaid by Holocene sediments (marine, alluvial, deltaic, etc.). The rocks of glacial origin belong to Valdai sub-aquifer. Regionally, there are various combinations of the Valdai sediment deposition. Sometimes the aquifer is overlaid by a complete chain of the Valdai sediments, consisting of two glacial aquifers, separated by interglacial aquifer; more often interglacial sediments are absent. Also, the most frequent are the sediments of only one glacial aquifer: artesian Ostashkov. The total capacity of these deposits varies from 0 to 88.4 m (Fig. 4). The maximum capacity of the rocks overlying the aquifer is characteristic of areas of course-moraine sediments and ancient erosional incision.

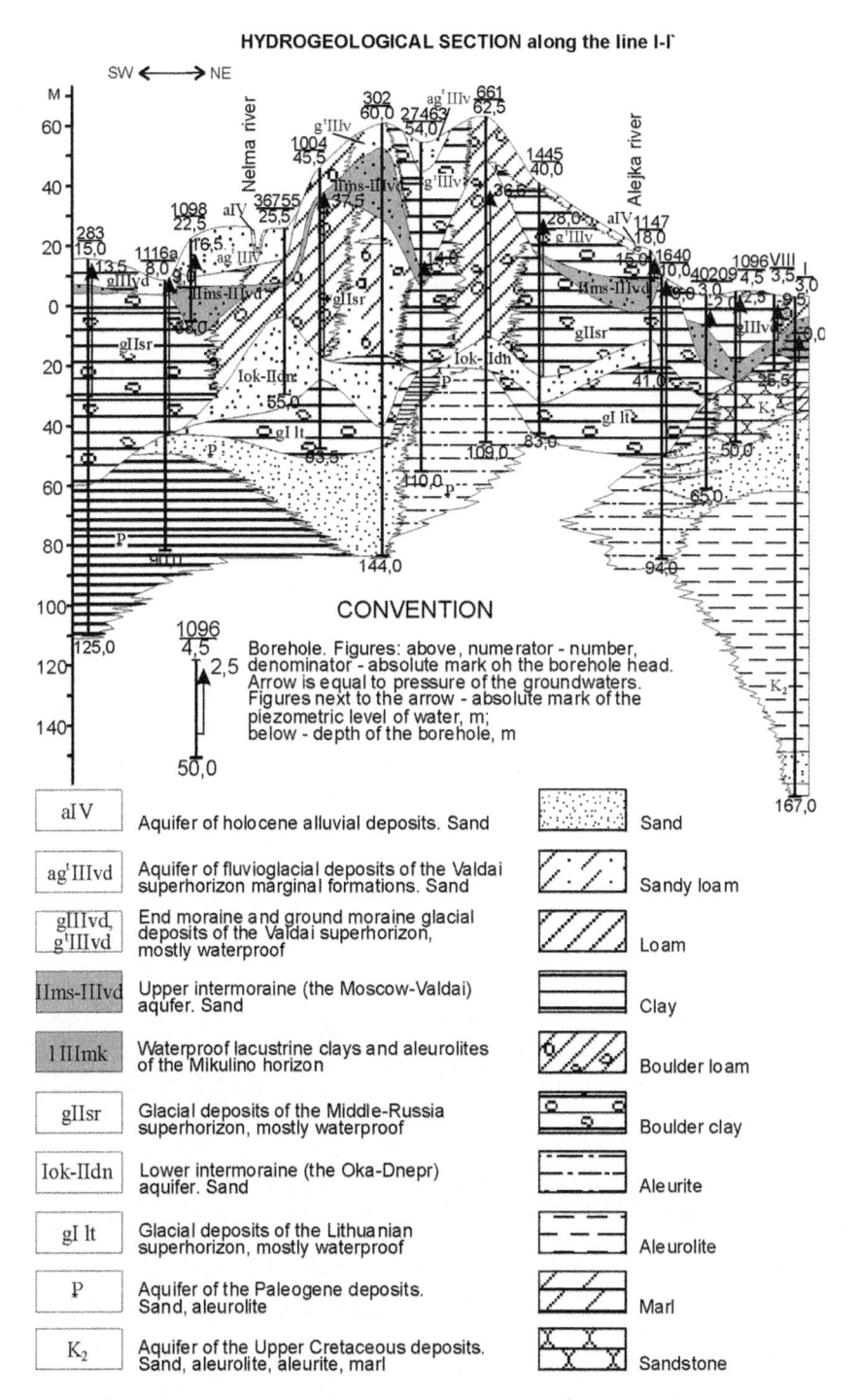

Fig. 2 Hydrogeological section of Kaliningrad Peninsula on I–I′ line. Position of the section line is shown in Fig. 1

HYDROGEOLOGICAL SECTION along the line II–II`

Conventions

alV	Aquifer of holocene alluvial deposits. Sand
lgIIIvd	Aquifer of Valdai superhorisonts glaciolacustrine deposits. Sand. Glaciolacustrine clay of Valdai superhorizon
ag^{t}IIIvd	Aquifer of fluvioglacial deposits of the Valdai superhorizon marginal formations. Sand
g^{t}IIIvd	End moraine glacial deposits of the Valdai superhorizon, mostly waterproof
gIIIvd	Ground moraine glacial deposits of the Valdai superhorizon, mostly waterproof
IIms-IIIvd	Upper intermoraine (the Moscow-Valdai) aqufer. Sand
l IIImk	Lacustrine clays and aleurites of the Mikulino horizon
gIIsr	Glacial deposits of the Middle-Russia superhorizon, mostly waterproof
Iok-IIdn	Lower intermoraine (the Oka-Dnepr) aquifer. Sand
l II lch	Lacustrine clays and aleurites of the Likhvino horizon
gI lt	Glacial deposits of the Lithuanian superhorizon, mostly waterproof
N_1	Miocene waterproof rocks. Clay
P_1	Aquifer of the Paleocene deposits. Sand, aleurite
K_2	Aquifer of the Upper Cretaceous deposits. Sand, aleurolite, aleurite, marl

Sand
Loam
Clay
Aleurite
Aleurolite
Marl

The boundary between fresh and brackish waters

Piezometric level of upper intermoraine aqufers groundwaters

Borehole. Figures: above, numerator - number, denominator - absolute mark of the borehole head, m. Fill color of the rectangle indicates the chemical composition of groundwater in the tried-depth range. Arrow is equal to pressure of the groundwater. Figures next to arrow - absolute mark of piezometric level of water, m; form left - salinity of water, gr/l; from right - discharge, l/sec - depression, m; below - depth, m

Chemical composition of groundwater

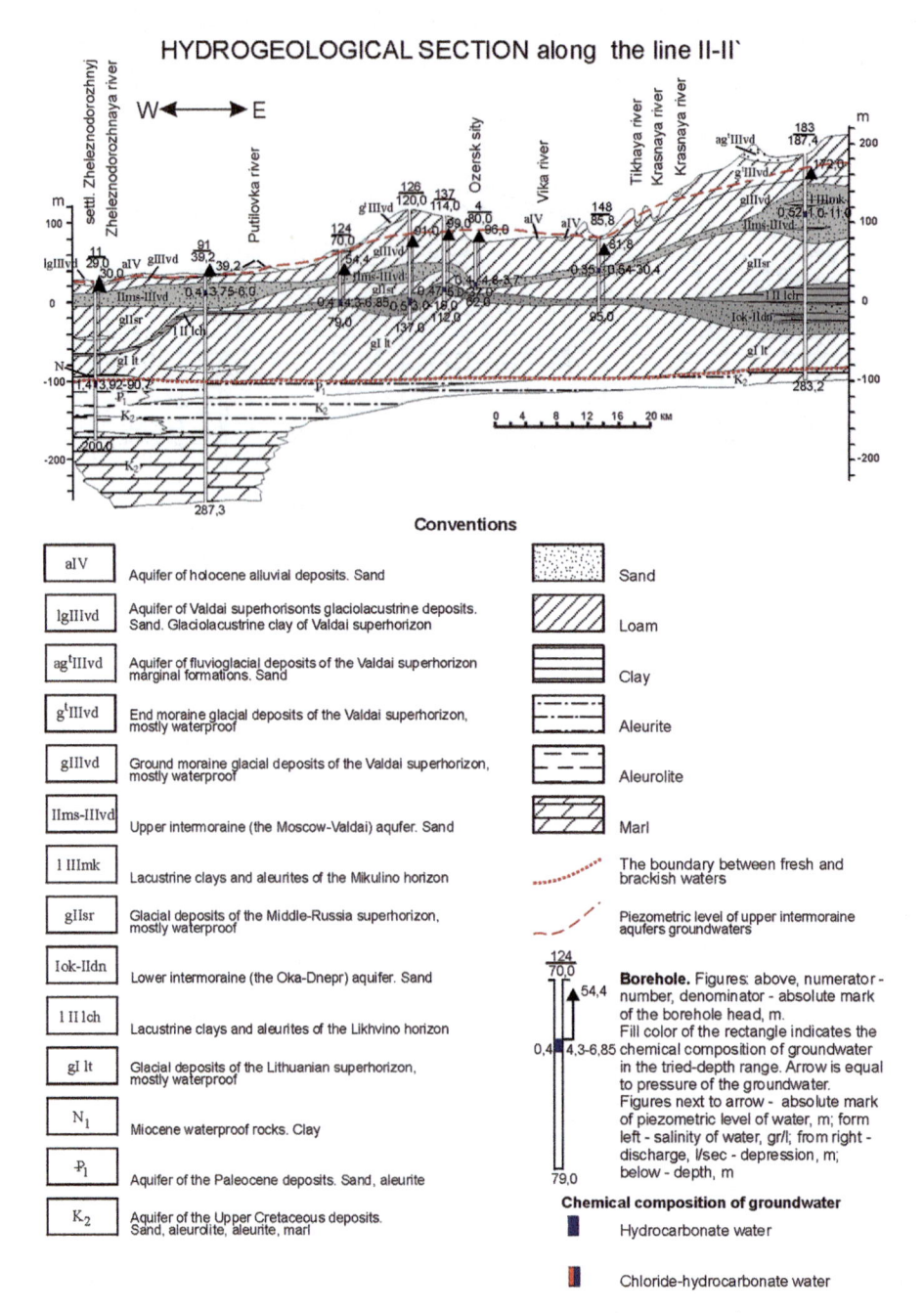

■ Hydrocarbonate water

■ Chloride-hydrocarbonate water

Fig. 3 Hydrogeological section of Kaliningrad Peninsula along the line II–II'. Position of the section line is shown in Fig. 1

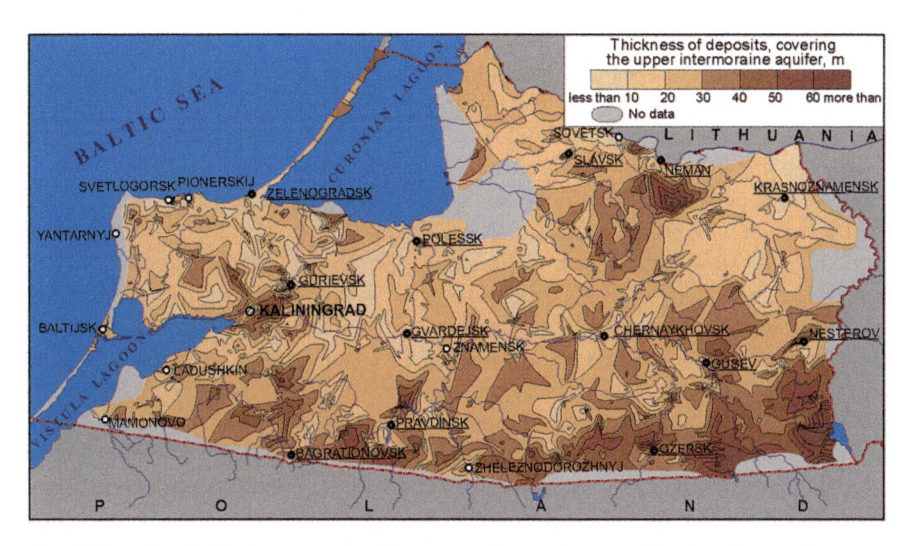

Fig. 4 Capacity distribution map of sediments, overlaying the upper intermoraine aquifer

The lithological composition of sediments moraine is quite diverse: they are represented by a variety of boulder clay, loam, sandy loam, containing inclusions in the form of layers and lenses of sand, and outwashed clays [48]. In the area of course-moraine sediments, the section demonstrates increase in the proportion of sand and sand and gravel rocks. The edge formations of the Baltic highland observe thick masses (up to 50 m) of fluvioglacial sand and gravel sediments. Fluvioglacial sediments of Ostashkov (Curonian) aquifer extensively develop in the region and compose mainly plain areas. These sediments are mainly represented by glaciolacustrine bright clays with unclear layer capacity of 3–4 m, as well as small- and fine-grained sands (2–3 m thick). Glaciolacustrine sands lie beneath clays and overlay them.

Holocene sediments are represented by alluvial, lacustrine, swamp, aeolian, marine, and deltaic facies [48]. Marine sediments formed in different stages of evolution of the Baltic Sea are common along the sea coasts and flooded areas. They are represented by clays, sands, gyttja, peat, and sapropel. Marine sediments of a modern coast compose beach areas up to 100 m wide with a capacity of 5 m. The lake sediments are developed in depressions of moraine, glacial, and lacustrine-deltaic relief. Sediments are represented by silt loam, humus sandy loam, silt, sapropel, calcareous clay, and silty sands. Alluvial sediments compose the flood-plain and river terraces, made up of fine-grained sands with layers of clay and inequigranular sands. Maximum capacity of alluvium (up to 20 m) is observed in the valleys of the rivers Neman, Pregel, Deima, and Instruch. Deltaic sediments in the Kaliningrad Region are developed in the Neman estuary and represented by clays, silt loam, and humus sands. Bog sediments, peat, silt, sapropel, and fine-grained silty sand, have a capacity of 2–3 m (sometimes up to 13 m) and are spread along the eastern and southern coasts of the Curonian Lagoon and in depressions of plains and river valleys. Aeolian deposits are developed on the Curonian and Vistula Spits, as well as in the Neman Delta and the rivers Neman and Sheshupe [48].

In most of the aquifer area, the lower aquiclude forms moraine sediments of the middle Pleistocene ages as boulder clay, loam, and rarely sandy loam. In the northern part of the region clays, silicified siltstone and marl of the Paleogene and Upper Cretaceous age make up the aquiclude.

The Moscow-Valdai aquifer is closely linked to groundwaters, which is confirmed by the similarity of their chemical composition and the close proximity of levels. Groundwaters in the Kaliningrad Region date to various genetic complexes of late glacial and postglacial sediments: (1) marsh, (2) aeolian, (3) alluvial, (4) glaciolacustrine, (5) marine, (6) marginal glacial formations, and (7) ground moraine [47]. The extensive development of the earth's surface clay sediments is not conducive to the infiltration of precipitation and the creation of high-capacity layers, so the actual groundwaters in most parts of the region are missing or have sporadic distribution. It is solely the groundwaters of alluvial deposits which are widely used for the drinking water supply of large settlements.

5 Characteristics and Regularities of Spatial Distribution of the Factors That Determine the Protection of the Upper Intermoraine Aquifer

Value Distribution of the Upper Aquiclude Capacity The analysis of the author's map of the reduced capacity values of the upper aquiclude of the Moscow-Valdai aquifer (M_c') showed that its value is usually 10 m and higher (Fig. 5). In most of the

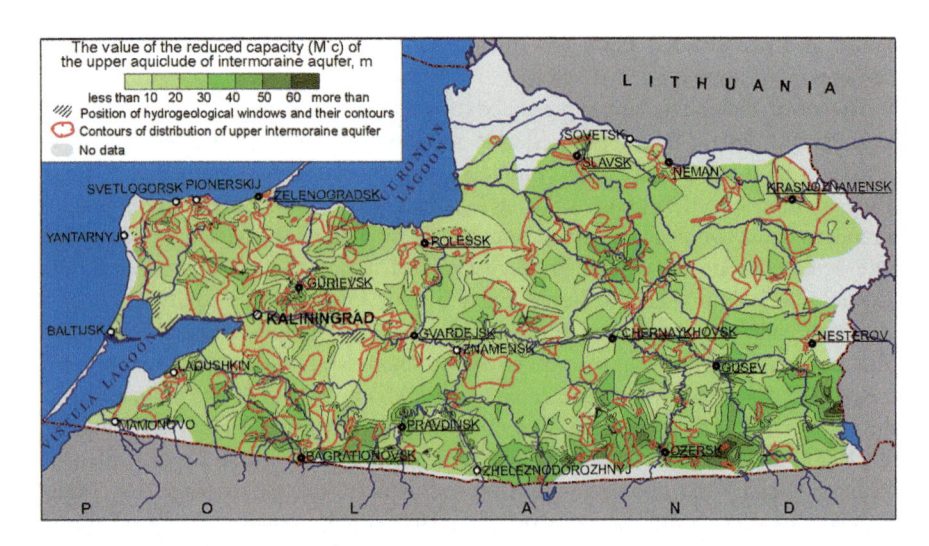

Fig. 5 Map of the reduced capacity of the upper aquiclude

Kaliningrad Region, M's value is 10–20 m, demonstrating a slight tendency to increase while advancing to the south of the region, following an increase in capacity of glacial sediments, forming the upper aquitard. The maximum $M_c{}'$ values are observed in the development areas of the periphery-moraine sediments and can reach 50–70 m in Vishtynets highland, 30–50 m in Warmia highland, and 30–35 m in Sambia Peninsula (Fig. 5). This is explained by the presence of multimeter low-permeability strata of glacial origin sediments.

The minimum values of the reduced power ($M_c{}' < 10$ m) are typical for plain areas: flat alluvial, alluvial-marine, marine, lacustrine, and fluvioglacial wavy plains. This is the stratum of alluvial, fluvioglacial, and lacustrine-glacial sediments, composed of sandy clay, sandy loam, and light loam with inclusion of boulders, pebbles, gravel, and sand. Hydrogeological windows are typical. Together with the low capacity of the upper aquiclude, this creates the conditions for light filtering into the aquitard and leads to its pollution [41, 49, 50]. There are some regularities in the distribution of areas characterized by $M_c{}'$ value <10 m and covering approximately 1/5 of the territory of the Kaliningrad Region (Fig. 5).

The first group of areas with low values $M_c{}'$ is located in periphery-moraine area (central part of Kaliningrad Peninsula, near Bagrationovsk, Ozersk, Gusev, to the west of Lake Vištytis) (Fig. 5). They exist due to fluvioglacial or permeable periphery moraine sediments which shape hydrogeological windows and allow unhindered filtration of pollutants from the surface to the Moscow-Valdai aquifer. Areas with weak protection do not occupy large areas, except Vishtynets highland, where the areas of the south of Gusev (city intakes are located there) can be considered the most exposed ones (Fig. 5).

The second group of areas is marked around the valleys of the rivers Pregel, Neman, Lava, etc. (Fig. 5). Reducing the power of low-permeable glacial sediments or their complete destruction due to erosion activities of rivers, the accumulation of permeable sediments in the valleys (sandy loam, sand, and gravel) promotes active interaction of river and groundwaters. During seasonal floods, polluted river water is likely to penetrate into the aquifer. A special attention should be given to the area between Kaliningrad and Gvardeysk (Fig. 5). Water intakes Ozerki and Velikolukskoye which provide Kaliningrad and Gvardeysk with drinking water are located here.

The third group of areas is located around the shores of the seas and bays and often includes estuaries of some rivers – the Deima, Mamonovka, Nelma, and Primorskaya (Fig. 5). The minimal $M_c{}'$ values are defined by the active processing of coastal areas due to repeated transgressions (especially Littorina). The Moscow-Valdai aquifer is overlaid with low-capacity layer of low-permeable rocks (glacier or glaciolacustrine origin) and/or directly with marine and lake-marine and sand formations [41, 49, 50].

Reduces power $M_c{}'$ values $= 10$–20 m are typical for flat glacial lake plains, flat and slightly convex lowland bog plains, and partly for hilly relief of ground moraine. Impermeable aquifer rocks are represented by sandy clays, medium loam, and sandy loam with inclusion of gravel, pebbles, and boulders and boulder loams of various origins (ice, lake-glacial, and fluvioglacial). The capacity of impermeable rocks is low, which causes water pollution, even in the absence of hydrogeological windows. The inclusion of hydrogeological windows contributes to more pollution (e.g., in Sovkhoznoye, Ladushkin).

Reduced power M_c' values = 20–30 m are typical for distributions of hilly relief of ground moraine and flat glaciolacustrine plains. The upper aquiclude is composed of complex clay, loam, and sandy loam glacial and glaciolacustrine origin. The lithological windows are defined by sand lenses. Hydrogeological windows are to some extent offset with a thick layer of low-filtering sediments, but the water may be polluted (e.g., in Vyshevoye) [41, 49, 50].

The maximum values of the upper aquiclude-reduced capacity (M_c' = 30 m and more) are typical for hilly areas of ground and peripheral moraine. The formation is composed of glacial boulder clay loams, sometimes turning into boulder clay with gravel and pebbles. Groundwater pollution is likely due to hydrogeological windows.

According to the data on the distribution of the upper aquiclude-reduced capacity values, a map of isochrones of pollutant penetration from the surface to the Moscow-Valdai aquifer was designed (Fig. 6). It should be noted that this map does not display the exact time intervals, but their approximate values allow us to estimate the rate of pollutant infiltration. The disintegration time of pollutants varies widely. Such pollutants as solutions of some mineral salts (chlorides, sulfates, nitrates, etc.) or long-lived radioactive isotopes are very persistent and very slowly disintegrate. Other pollutants are also quite resistant, but with a limited lifetime. A large group of pesticides is characterized with a large interval of disintegration time – from several months to 5–10 years [10]. The least resistant is bacterial contamination – from 30–50 to 200–300 days [6, 8, 10]. Thus, we can

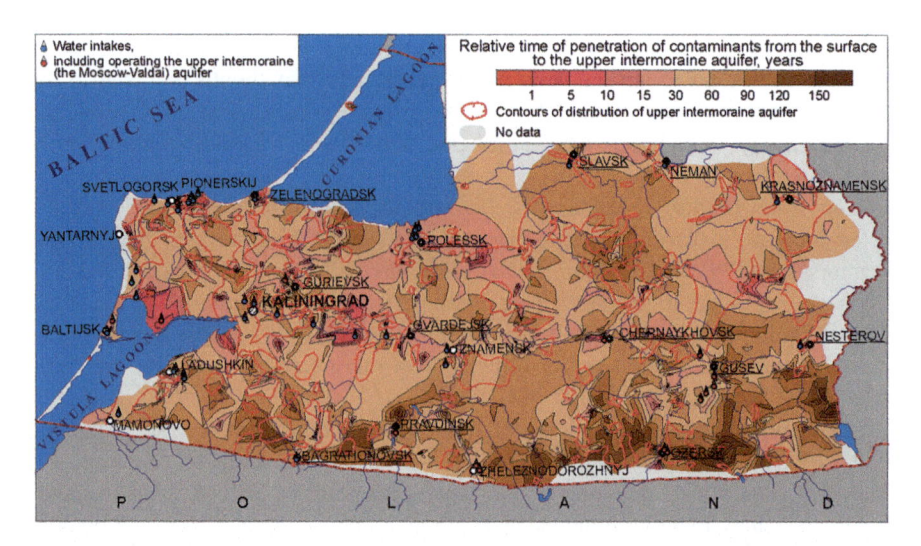

Fig. 6 Map of isochrones of pollutant penetration into the upper intermoraine aquifer

speak of different degrees of groundwater protection in relation to bacterial pollution, a separate group of the least resistant pesticides, etc.

Some certain gradations in the map allowed approximately delineate areas prone to various types of pollution, such as microbial pollution ($t < 1$ year), to pollution with pesticides ($t = 5$–10 years). At the same time, most of the region has a low probability of pollution within the estimated lifetime of the intakes ($t > 30$ years). The calculated time of penetration of pollutants ranges from less than 1 year and up to 220 years (Fig. 6) [41]. The longer the penetration of pollutants, the greater the likelihood that they will be subjected to various physical, chemical, and biological transformations and as a result either defused or destroyed.

The areas contoured with a 90-year isochrone are located in the southeast region (Vishtynets Highland), in the area of the hilly relief of the ground and periphery moraines (Fig. 6). Similar smaller areas can be found in the Warmia and Sambia Highlands, the Instruch ridge. The penetration of pollutants into groundwater within 60–90 years is typical for the ground and periphery moraine distribution areas and flat areas of glaciolacustrine plains (Fig. 6). The time of pollutant penetration over large areas of Sambia Highland, Sheshupe, and Pregel Plains is from 30 to 60 years (Fig. 6). The pollutant penetration time of under 30 years is typical for areas of flat alluvial-marine and marine plains (Zelenogradsk, Svetly, Polessk, Pribrezhny), river valleys (district of Ozersk, Gvardeysk, Chernyakhovsk, Pravdinsk, Ozerki, Znamensk), and local areas of water-glacial and peripheral-moraine origin (northwestern Sambia highland, Gurievsk, Lublino, Vesnovo, Vishtynets highland near Chistye Prudy) (Fig. 6) [41].

Direct filtration of polluted water into the aquifer is likely in the areas of "hydrogeological windows." Pollutant penetration time can range from several days to several months. The area for which pollution penetration time is under 1 year makes up to only 0.05% of the total area of the studied aquifer (4.4 km^2). The territory of the most exposed areas is around 60 km^2 (0.64% of the area of the aquifer) with a certain pollutant penetration time up to 5 years: Svetly, Svetlogorsk, Chernyakhovsk, Primorsk, Pribrezhny, Nivenskoe, Pereslavskoye, Ozerki (Fig. 6) [41]. However, the rapid pace of migration from the surface allows to replenish groundwater reserves and set water intakes in those areas intakes that will supply populated settlements with drinking water, so the selected areas should be closely monitored.

Analysis of the Combination of the Level of Groundwater and Interstratal Water An important indicator of the pressure water protection is the balance between the pressure and the upper groundwaters. If the pressure water levels are significantly higher than the levels of groundwater, then in case of the continuous aquiclude of high capacity which ensures the stability of the level difference, the studied area can be considered highly protected from all types of pollutants. Indeed, regardless of the nature of the pollutants, their penetration of groundwater into the pressure water aquifer under the above-described conditions is not possible. If the water pressure levels are close proximity to the groundwater levels or lower in volume, then the vertical movement of pollutants from the groundwater aquifer into the pressure water one is highly possible. The speed and, hence, the vertical overflow time will depend on the magnitude of the downward vertical gradient, capacity, and permeability of water pressure [5, 8, 30].

Levels of free-flow groundwater in the Kaliningrad Region are located at a depth within the first meters of the earth's surface. Deeper occurrence of groundwater level is only possible for aeolian sediments water and waters of ground moraine sediments. The Moscow-Valdai intermoraine aquifer is characterized by weak pressure, up to its complete lack at the aquifer entry points. Piezometric levels of the Moscow-Valdai aquifer in their smooth shape repeat the landscape, having the highest abs. Marks on the hills (up to 140–160 m). The depth where the piezometric levels can be found ranges on average from -30 to the first meters at the earth's surface (Figs. 2 and 3). In rare cases, spontaneous spouting of wells is possible (Figs. 2 and 3).

A low pressure of the upper intermoraine aquifer and the shallow location of groundwater levels make the combination of levels encourage the overflow of nonpressure water into low-pressure one. The analysis of groundwater level is represented in the diagram in Fig. 7. As it is shown in Fig. 7, vertical movement of groundwater in intermoraine aquifers almost universally develops. In those areas where the diagram shows the predominancy of the level of the pressure water of the Moscow-Valdai aquifer above the nonpressure groundwater, the value of dominance usually ranges from 0.5 (Primorskaya Plain) up to 7 m (Sambia moraine plateau) [41].

Under the conditions of active operation of the Moscow-Valdai aquifer such a combination shows the potential for polluting the pressure aquifer through the

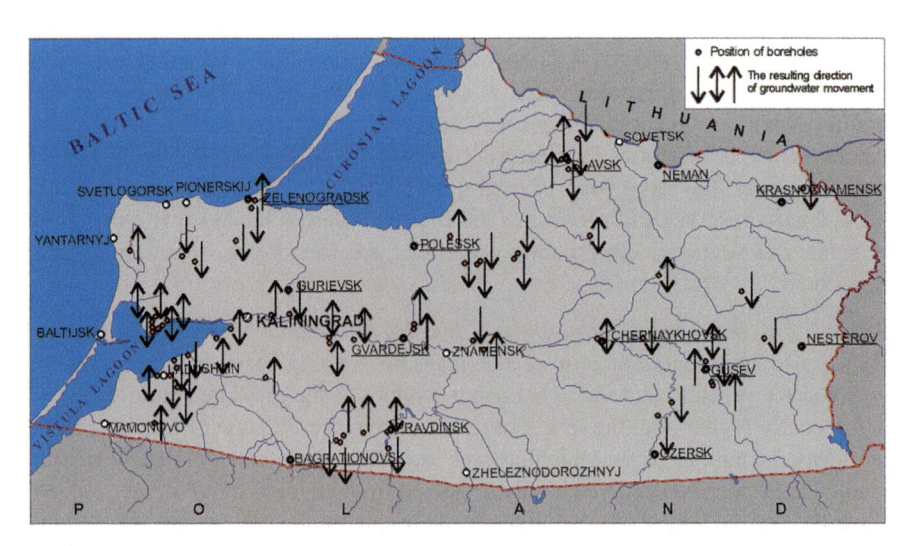

Fig. 7 Correlation of groundwater and interstratal water levels in the Moscow-Valdai aquifer. An *upward pointing arrow* indicates the upward movement of water and the overflow of interstratal waters into ground ones; a *downward pointing arrow* corresponds to the downward movement of water and the overflow of groundwater into interstratal ones; *double-headed arrow* indicates the level of a single aquifer of groundwater and interstratal waters

groundwater overflow. The intake of the Moscow-Valdai aquifer water will cause lowering the piezometric level and as a result will trigger vertical movement of groundwaters. Depression craters where the level lowers from a few up to 20 or more meters are formed everywhere in the areas of intensive and prolonged use of groundwater [44, 51]. In some cases, seasonal flooding or high water can adversely affect the quality of groundwater and interstratal water: their quality deteriorates, especially in microbiological and organoleptic characteristics [52]. The sharp increase in the supply of groundwater due to a large volume of rainfall and rapid snowmelt, which occurs both directly and indirectly, through a connection with the river waters, results in a significant increase in the groundwater level. At close proximity between the groundwater and interstratal water levels, such a combination might not benefit the interstratal waters. The values of the average annual level of ground and pressure waters of the upper intermoraine aquifer change simultaneously, and graphics repeat each other, albeit in a smooth way. In some cases, alluvial groundwater and the Moscow-Valdai aquifer water form a single aquifer in the valleys of the rivers Pregel, Pissa, Instruch, Lava, and others. The lack of water pressure, free-flow nature of groundwater, presence of multiple sources of pollution, and active exploitation of groundwater have resulted in river pollution in Ozerkovskaya, Znamenskoye, Chernyakhovskoye fields, etc.

Thus, the territory of the Kaliningrad Region is characterized by the following features of combinations of interstratal pressure and nonpressure groundwater levels. In the river valleys and on the slopes of the hills (Fig. 7), the low-lying aquifers may in some cases have a higher piezometric surface – there is an overflow and groundwater recharge due to lower-lying interstratal waters. In the watershed areas, pressure in aquifers decreases with depth, and there is a water overflow from groundwater into interstratal aquifer. This idealized regional scheme could be interfered with due to local geological, climatic, and anthropogenic influences. In general, the hydrodynamic conditions, causing the active movement of polluted groundwater vertically down into the interstratal waters, do not contribute to the protection of the Moscow-Valdai aquifer groundwater.

6 Territorial Differentiation of the Kaliningrad Region on the Degree of the Upper Intermoraine Aquifer Protection

The joint analysis of various environmental protection factors in relation to the Moscow-Valdai aquifer in the Kaliningrad Region identified three categories of protection of groundwater (Fig. 8) [41].

1. The category *of unprotected* groundwater of the Moscow-Valdai horizon is characterized by calculated reduced power values for the upper aquifer of less than 10 m, by numerous lithological windows, and water pressure levels below ground level (Fig. 8). Relevant pollutant penetration time (t) is less than 30 years. As early as in the late 1980s, the relationship between the reduced

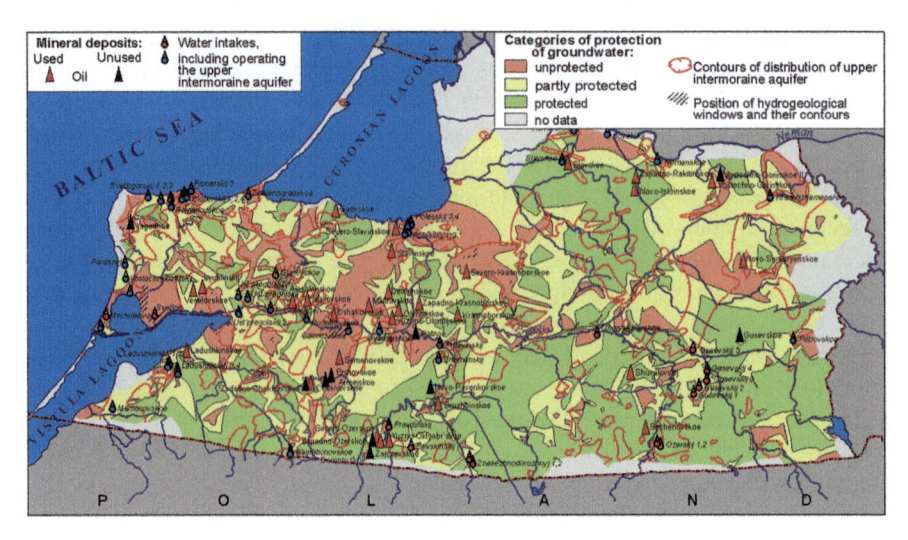

Fig. 8 Map of intermoraine upper aquifer groundwater protection

power value of an aquifer (M's) and chemical (nitrogen) pollution was established [33, 35]. This category of protection is characterized by the following kinds of pollution. When water pressure capacity of less than 10 m in the case of monolithic aquiclude the waters can be polluted with nitrites. The aquiclude provides for a mild though faster infiltration than in the aquiclude with a capacity of more than 10 m. The time happens to be sufficient for the implementation of some primary nitrification: NH_4^+ is oxidized to NO_2^-. It is *nitrite type* of pollution which occurs. A significant ammonium nitrite contamination is likely to happen due to the presence of windows. It is the most rapid penetration of pollutants into the aquifer, and it results in *ammonium-nitrite* pollution.

Within this category of protection in the areas limited with isochrone of 1 year microbiological pollution is likely to occur (standard lifeterm of coliform bacteria is from 30 days to 1 year). Resistant and very persistent pesticide pollution is likely in the areas limited with isochrones of 1, 5, and 10 years. Pollution with non-sorbing substances (macro components, petroleum products) is possible on the entire territory of unprotected waters. The groundwater of an area of 1,700 km² (18% of the area of the aquifer) is considered unprotected (Fig. 8). Unprotected groundwaters are mostly found on the territory of Svetly, Baltic, Svetlogorsk, Gvardeysk, Polessk, Zelenogradsk, and Guryevsk municipalities [41].

2. Categories *of partly protected* groundwater are characterized with $M_c' = 10-20$ m values, pressure water is overlaid with continuous aquiclude, and the pressure water levels are located lower than the groundwaters or at about the same level. Relevant pollutant penetration time (t) is 30–60 years. Defining the category of partly protected waters at a considerable capacity of the upper aquiclude results from the fact that the realistic pollutant penetration is shorter than the conventional

time due to unfavorable hydrodynamic conditions. The territories which are characterized as partly protected, occupy 3,500 km^2 (38% of the area of the aquifer) (Fig. 8). The largest areas of partly protected groundwater can be found in Gvardeysk, Polessk, Bagrationovsk, Gusev, Chernyakhovsk, Krasnoznamensk, Zelenogradsk, and Guryevsk municipalities [41].

The protection is reduced in the case of lithological windows; in agricultural areas, water is often largely polluted with ammonia. Pollutant filtering is accelerated. The soil-cover complex absorbs lower amount of fertilizers, and the time of filtration through the vadose zone is not enough to complete nitrification. Ammonium ion NH_4^+ easily penetrates through windows, forming mainly *ammonium type* contamination. The probability of pollution increases with the deterioration of the quality of groundwater. Pollution with oil and macro components is highly possible.

3. *The category of protected* groundwater is characterized by the maximum value of the upper aquiclude reduced capacity ($M_c' > 20$ m), lack of hydrogeological windows, and dominance of the pressure water level over the groundwater one. Relevant pollutant penetration time (t) is 30–60 years. Of the territory (4,100 km^2) of the Moscow-Valdai aquifer distribution, 44% is characterized as protected from contamination (Fig. 8). These are predominantly the southern territories of the region: Pravdinsk, Ozersk, Bagrationovsk, Nesterov, Gusev, and Chernyakhovsk municipalities [41].

Having analyzed the chemical pollution of this category, it can be concluded that the aquiclude provides for slow infiltration of pollutants. Soil-cover complex absorbs the maximum amount of fertilizer and nitrification process is more complete: NH_4^+ manages to get oxidized to NO_3^-. Nitrates as the most readily soluble form are taken away with groundwater, and the water of the Moscow-Valdai aquifer remains clean. The disintegration period for the group of very persistent pesticides is 10–15 years, so this category of protection ensures the quality of water [41].

Moreover, if the reduced power of upper aquiclude M_c' is set to 10 m or more, it is possible to confidently speak about the antibacterial protection of the territory [41]. Time of *E. coli* penetration to the aquifer level will equal to around 10,000 days (30 years), which far exceeds the lifetime of the bacteria (300 days) and facilitates the amortization period of intake service. Thus, in most cases, the natural properties of the geological environment prevent possible pollution, the water should stay clean.

Therefore, the structure of the sediments overlying the aquifer controls qualitative pollution change. These findings are confirmed by the concentration of areas with identified contamination within the most exposed areas. An example can be found in the areas of chemical contamination of interstratal water in Lazovskoye, Svetly, Znamensk, etc. Similar results were obtained on nitrogen (ammonium) contamination. Half of the monitoring points, where the contamination has been observed for a long time, are located in vulnerable areas, 42% of the points are located on the partly protected area, and 8% are located in well-protected areas.

This scheme of protection study is applicable for microbial pollution as well. In order to test the proposed methodology of protection assessment and mapping the archive data for 2007–2011 provided by Rospotrebnadzor (URL: http://39.rospotrebnadzor.ru/s/39), about 50 cases of violating sanitary in the waters of the Moscow-Valdai aquifer were chosen [53]. Boreholes and wells are located in the territory of Guryevsk, Zelenogradsk, Bagrationovsk, Pravdinsk, Ozersk, Gusev, Nesterov, Chernyakhovsk, Neman, Svetlogorsk, Baltiysk, Svetly, and Kaliningrad municipalities. We analyzed the location of the wells which recorded some pollution and compared it with a map of the natural protection of the Moscow-Valdai aquifer. Half the cases of pollution coincide with the location of wells and boreholes in the areas which were identified as vulnerable. Approximately 1/3 of negative samples were selected in the areas marked on the map as partly protected. About 15% of nonstandard samples were taken from the areas characterized as well-protected. The best results on the correlation of contamination and the defined protection categories were obtained in Zelenogradsk, Guryevsk, and Bagrationovsk municipalities. Thus, the mapping of protection areas can be used to predict the occurrence of microbiological pollution in the waters of the Moscow-Valdai aquifer.

In some cases, there is a good correlation between the categories of protection and water quality of other aquifers. Since the Moscow-Valdai aquifer is widespread, but not universal, and the protection area map covers most of the area, extrapolating the protective characteristics of the upper impermeable aquifer on areas in which the Moscow-Valdai aquifer is missing, it can give good results in correlation with the microbiological indicators of water in other aquifers, as well as the top subsurface one. Full matching of negative samples Upper Cretaceous aquifer and the identified areas of unprotected groundwaters in the central and north-eastern part of the region (the northern part of the Guryevsk municipality, north-eastern part of the Zelenogradsk municipality, Neman and the southern part of Slavsk municipality) can serve as the most graphic example. However, the use of the Moscow-Valdai protection maps is not acceptable for some other aquifers. Similar conclusions can be drawn regarding the groundwater. Despite the fact that most of the wells intake substandard groundwater, located in unprotected areas, it is impossible to apply the Moscow-Valdai protection map for them. These matches can be easily explained: if the top of the strata overlying the upper intermoraine aquifer contains some permeable sediments, where the groundwater concentrates, it negatively affects its protective properties. When analyzing the correlation of groundwater and pressure water levels, these areas are defined as unprotected.

7 Possible Areas of Water-Use Conflicts

Potential hazards are located in the vicinity or directly on unprotected areas, on mineral deposit developments, municipal solid waste landfills, stocks of mineral, and organic fertilizers. In these cases, the land use conflicts are highly likely, the conflicts of territorial priorities of nature protection and economic development of the area within a certain land use regime, expressed in actual or potential degradation of natural complexes or their major components, reducing biodiversity, productivity, and value of landscapes, as well as the effective implementation of the territory's economic functions. The hydraulic connection between the rivers and the waters of the Moscow-Valdai aquifer is manifested particularly actively in the areas with weak protection, and it threatens with the penetration of pollutants from the rivers Pregel, Lava, Instruch, Sheshupe, Neman, and Deima. Particular attention should be focused on the intakes which receive the water of the Moscow-Valdai aquifer and located in the areas characterized as unprotected or partly protected: the intake of Gvardeysk, Chernyakhovsk, Pravdinsk, Sevskoye, and Gusev (Fig. 8).

The analysis of the possibility of water conflicts in the areas of the mineral deposit developments. The greatest potential hazard to the natural systems and groundwater in particular are created by the oil fields. One-third of the region's oil fields (Malinovskoye, Semenovskoye, Isakovskoye, Zapadno-Ushakovskoye, Slavinskoye, Severno-Slavinskoye, Severno-Krasnoborskoye) are located in the areas where the groundwaters of the studied aquifer are not protected, and it increases the risk of contamination (Fig. 8).

Developments of construction materials deposits are less dangerous compared to oil extraction, which is due to the relative geochemical inertness of these minerals. However, removing the overburden rock and subsequently operating the useful clay formation reduce the capacity of impermeable clays rocks overlying the aquifer. Sand and gravel deposit developments lead to a removing an overburden rock and exposing the productive layer, which as a rule encloses groundwater or interstratal waters. Thus, production of building materials reduces the degree of protection of the aquifer. The groundwater of intermoraine upper aquifer is not protected from pollution in the areas of sand and gravel deposit developments in Komsomolskoye, Kashtanovskoe, Ushakovskoe, Rovnoye, Kuibyshevskoye, and Pushkarevo; the estuary of Pregel, Kamenka, Fevralskoye, Sholokhovskoye, and Dmitrievka; construction sands in Lesnoye II; and clays in Moskvino and Sovkhoznoye (Fig. 8).

Natural resource conflicts are most likely to occur at the exploitation of groundwater intakes which are located in unprotected the Moscow-Valdai aquifer (Chernyakhovsk, Guryevsk, Svetly deposit developments, the 1st section of Zheleznodorozhniy and the 3rd section of Svetlogorsk development deposits, Mechnikovo section) (figure) (Fig. 8). Some intakes (e.g., Svetlogorsk, Svetly) have indicated groundwater pollution with oil products and nitrogen compounds.

References

1. Margat J (1968) Vulnerabilite des nappes d'eau souterraine a la pollution: Bases de la cartographie. Doc. BRGM, 68 SCL 198 HYD, Orleans, 123 pp
2. Albinet M, Margat J (1970) Cartographie de la vulnerabilite de a la pollution des nappes d'eau souterraine. Bull BRGM, 2eme serie, section 3, Orleans 4:13–22

3. Vrba J, Zaporozec A, International Association Hydrogeologists (1994) Guidebook on mapping groundwater vulnerability, vol 16. Heise, Hanover, 131 pp
4. Witkowski AJ, Vrba J, Kowalczyk A (2007) Groundwater vulnerability assessment and mapping, vol 11. Taylor and Francis, London, 260 pp. AH-selected papers
5. Zektser IS (2001) Podzemniye vody kak komponent okryzhaushei sredy (Ground water as a component of the environment). Nauchny Mir, Moscow, 328 pp
6. Goldberg VM, Gazda S (1984) Gidrogeologicheskiye osnovy okhrany podzemykh vod ot zagryazneniya (Hydrogeological basis for the protection of groundwater against pollution). Nedra, Moscow, 262 pp
7. Belousova AP (2001) Kachestvo podzemnykh vod. Sovremennye podkhody k otsenke (The quality of groundwater. Current approaches to assessment). Nauka, Moscow, 342 pp
8. Metodicheskoye rukovodstvo po okhrane podzemnykh vod ot zagriazneniya (Methodological guidebook on the protection of groundwater against pollution) (1979) SEV: Postoyannaia komissiya po geologii (CMEA: Standing Committee on Geology), Moscow, 63 pp
9. Belousova AP (2005) Resursy podzemnykh vod i ikh zashishennost' ot zagryazneniya v basseine reki Dnepr i otdelnykh ego oblastei: Rossiyskaya territoria (Groundwater resources and their protection against pollution in the Dnieper basin and its separate areas: the Russian territory). Lenand, Moscow, 168 pp
10. Belousova AP, Gavich IK, Lisenkov AB, Popov EV (2006) Ekologicheskaya gidrogeologiya: uchebnik dlya vuzov (Environmental hydrogeology: the textbook for high schools). ECC "Akademkniga", Moscow, 397 p
11. Elokhina SN (1982) Uchet zashishennosti i zagryazneniya podzemnykh vos pri dolgosrochnom planirovanii (Account of the groundwater protection and pollution at long-term planning). Prot Nat Waters Urals 13:95–98
12. Zektser IS, Karimova OA, Buzhuoli ZH, et al. (2004) Regionalnaya otsenka uyazvimosti presnykh podzemnykh vod (Regional vulnerability assessment of fresh groundwater: methodological aspects and practical application). Water Resour 31(6):645–650
13. Melnichuk NL (1997) Prirodnaya zashishennost' i okhrana podzenmykh vod v kriolitozone severnogo Pribaikaliya v usloviakh tekhnogeneza (Nature protection and conservation of groundwater in permafrost of the northern Baikal region in technogenesis). Geogr Nat Resour 3:105–112
14. Wei CH (2005) Regionalnaya otsenka resursov podzemnykh vod i ikh zashishennosti ot zagriazneniya na ravnine Ping-Tong, Taiwan (Regional assessment of groundwater resources and their protection against pollution in the Plain of Ping-Tong, Taiwan). PhD in Geology thesis (25.00.36), Moscow, 120 pp
15. Anastasiadis P (2003) Vulnerability of groundwater to agricultural activities pollution. In: Proceedings of the 8th international conference on environmental science and technology, Lemnos Island, Greece, 8–10 Sept 2003, vol B, pp 24–30. http://www.docstoc.com/docs/3947072/International-Conference-on-Environmental-Science-and-Technology-September-Lemnos-island. Accessed 21 June 2009
16. Civita MV, De Maio M (1998) Mapping groundwater vulnerability in areas impacted by flash flood disasters. In: 13th ESRI European user conference, Firenze, Italy, 7–9 Oct 1998, 12 pp
17. Huddleston JH (1996) How soil properties affect groundwater vulnerability to pesticide contamination. Oregon State University Extension Service, Corvallis, p 4. http://www.agcomm.ads.orst.edu. Accessed 19 June 2009
18. Verschmutzungsempfindlichkeit des Grundwassers (Ausgangs Maßstab 1: 50000). Digitaler Umweltatlas Berlin (Ausgabe 1993). http://www.stadtentwicklung.berlin.de/umwelt/umweltatlas/d205_01.htm. Accessed 12 Oct 2009
19. Goldberg VM (1976) Girogeologicheskiye prognozy kachestva podzemnykh vod na vodozaborakh (Hydrogeological forecasts of groundwater quality in water intakes). Nedra, Moscow, 152 pp
20. Ostrovsky VN, Ostrovsky LA (eds) (1996) Metodicheskiye rekomendatsii po sostavleniyu ekologo-geologicheskikh kart masshtaba 1:200000 - 1:100000 (Methodological guidelines

for the preparation of ecological and geological maps of scale 1: 200,000 - 1: 100,000). VSEGINGEO, Moscow, 61 pp
21. Goldberg VM (1987) Vzaimosviyaz' zagryazneniya podzemnykh vod i prirodnoi sredy (The relationship of groundwater contamination and the environment). Gidrometeoizdat, Leningrad, 228 pp
22. Proskurina IV (2009) Regionalnaya otsenka opasnosti zagriazneniya podzemnykh vod kak komponenta okruzhaushei sredy (na primere Brianskoy oblasti) (Regional assessment of the danger of groundwater pollution as a component of the environment (case study of the Bryansk region)). Abstract. PhD in Geography thesis (25.00.36), Institute of Water Problems, RAS, Moscow, 20 pp
23. Rogachevskaya LM (2002) Regionalnaya otsenka uyazvimosti gruntovykh vod vostochnoi chasti Dneprovskogo basseina k radionuklidnomy zagriazneniyu (Regional groundwater vulnerability assessment in the eastern Dnieper artesian basin to radionuclide contamination). Abstract. PhD in Geology thesis, Institute of Water Problems, RAS, Moscow, 23 pp
24. Abakumov AS (2007) Migratsia organicheskikh soedineniy cherez slabopronitsaemye tolshi na territorii poligona tvergykh bytovykh otkhodov v s. Parabel' Parabelskogo raiona Tomskoi oblasti (The migration of organic compounds through poorly permeable strata in the territory of the landfill in Parabel village of Parabel district of the Tomsk region). In: Problems of geology and mineral resources development. TPU, Tomsk, pp 172–174
25. Kolomeev MP (2002) Poluempiricheskaya model dlya otsenki bystrogo zagryazneniya gruntovykh vod radionuklidami s poverkhnosti zemli (Semi-empirical model for estimating the rapid contamination of groundwater with radionuclides from the surface of the earth). In: Questions of assessment and prediction of environmental pollution, caused by emergencies, Environmental pollution series, vol 23(165). Gidrometeoizdat, St. Petersburg, pp 83–96
26. Minyaeva JV, Karimova OA (2009) Otsenka zashishennosti podzemnykh vod dvumya nezavisimymi podkhodami i analiz ikh primenyaemosti (Assessment of groundwater protection by two independent approaches and analysis of their applicability). Hydrology and Karst: collection of academic papers, vol 18. PSU, Perm, pp 148–153
27. Titushkina ZY (1991) Modelirovaniye dimaniki intensivnykh zagriazneniy podzemnykh vod (Modeling the dynamics of intensive pollution of groundwater). Models in management of natural resources. KSU, Kaliningrad, pp 13–19
28. Stenemo F (2007) Vulnerability assessments of pesticide leaching to groundwater. Doctoral thesis, Swedish University of Agricultural Sciences, Department of Soil Sciences, Uppsala, Sweden, 39 pp. http://diss-epsilon.slu.se:8080/archive/00001408/01/Thesis_Stenemo.pdf. Accessed 21 June 2009
29. Ambarref M, Aassine S, Bernoussi AS, Haddouchi BY (2007) Cartographie de la vulnerabilite a la pollution des eaux souterraines: application a la plaine du Gharb (Maroc). Rev Sci Eau 20(2):185–199. http://www.erudit.org/revue/rseau/2007/v20/n2/015812ar.pdf. Accessed 21 June 2009
30. Schwartz AA (1996) Ekologicheskaya gidrogeologiya (Environmental hydrogeology). Publishing house of the St. Petersburg State University, St. Petersburg, 34 pp
31. Ekologicheskaya i geologicheskaya karta: Severo-Zapadny federalny okrug (masshtab 1:9000000) (Ecological and geological map: North-western Federal District (scale 1: 9,000,000)). http://sevzapnedra.nw.ru. Accessed 20 June 2009
32. Barinova GM, Yeltsina GN (1985) Vliyanie antropogennykh faktorov na sostav prirodnykh vod i prognozirovaniye ikh kachestva (na primere Kaliningradskoi oblasti) (Human impact on the composition of natural waters and their quality forecasting (the case of the Kaliningrad Region)). In: Report for XXVII Union hydrochemical meeting, Novocherkassk, pp 24–27
33. Barinova GM, Yeltsina GN, Zotov SI, Sergin SY (1985) Ob otsenke i prognozirovanii sostoyaniya podzemnykh vod v svyazi s khozyaistvennoi deyatel'nostiu (On the estimation and forecasting of the state of groundwater in connection to economic activity). Problems of environmental monitoring and ecosystem modelling, vol 8. Gidrometeoizdat, Leningrad, pp 240–245

34. Yeltsina GN (1994) Podzemniye vody. Otsenka i prognozirovaniye ikh kachestva v svyazi s khozyaistvennoi deyatel'nostiu (The groundwater. Evaluation and prediction of their quality in relation to economic activities). In: The main fields of scientific and technical support of development in the Kaliningrad Region. Abstracts. Regional academic research conference. KSTU, Kaliningrad, p 29
35. Orlenok W (ua) (1994) Natur, Wirtschaft und Ökologie der Stadt Kaliningrad. Institute für Landeskunde, Leipzig, p 1–20
36. Yeltsina GN, Mikhnevich GS (2005) Prirodnaya zashishennost' podzemnykh vod Kaliningradskoi oblasti (The natural protection of groundwater in the Kaliningrad Region). In: Orlyonok VV (ed) Ecological problems of the Kaliningrad Region and the Baltic sea region: collection of research papers. KSU, Kaliningrad, pp 67–73
37. Goldberg VM, Skvortsov NP (1986) Pronitsaemost' i filtratsia v glinakh (The permeability and filtering in clays). Nedra, Moscow, 160 pp
38. Shtengelov RS. Poiski i razvedka podzemnykh vod: konspekt lektsiy (Research and exploration of groundwater: lecture notes) (electronic resource). In: Everything about geology. Unofficial server of Geographical Faculty of Moscow State University. http://geo.web.ru/db/msg.html. Accessed 1 Dec 2009
39. Oradovskaya AE, Lapshin NN (1987) Sanitarnaya okhrana vodozaborov podzemnykh vod (Sanitary protection of groundwater intakes). Nedra, Moscow, 167 pp
40. Mikhnevich GS (2010) Otsenka zashishennosti podzemnykh vod Kaliningradskoi oblasti ot zagriazneniya (Assessment of groundwater protection from pollution in the Kaliningrad Region). Bulletin of I. Kant BFU, natural sciences series, vol 1. I. Kant RGU Press, Kaliningrad, pp 93–101
41. Mikhnevich GS (2011) Geologicheskaya otsenka prirodnoi zashishennosti podzemnykh vod ot zagriazneniya (na primere sistemy verkhnego mezhmorennogo vodonosnogo gorizonta Kaliningradskoi oblasti) (Geoecological estimation of natural protection of groundwater against pollution (the case study of the system of the upper intermoraine aquifer in the Kaliningrad Region)). Abstract. PhD in Geography thesis, I. Kant BFU, Kaliningrad, 23 pp
42. Manstein AK (2002) Maloglubinnaya fizika: uchebnoye posobiye (Shallow geophysics: a textbook). Institute of Geophysics SB RAS, Novosibirsk, 135 pp
43. Karpov IV (2005) Sovremennoe sostoianiye i sovremenniye problemy izucheniya i ispol'zovaniya podzemnykh vod Kaliningradskoi oblasti (Current state and current problems of research and use of underground waters of the Kaliningrad Region). Proceedings of the international workshop on sustainable management and protection of water resources, priorities to ensure drinking water supply in the Kaliningrad Region. KSU, Kaliningrad, pp 14–23
44. Polyakova LS (ed) (2010) Informatsionny bulleten' o sostoyanii nedr na territorii Kaliningradskoi oblasti za 2009 god (Gosudarstvenny monitoring sostoyaniya nedr) (News-letter on the state of the subsoil in the Kaliningrad Region in 2009 (state monitoring of subsurface condition)), vol 14. FSUE "Sevzapgeologiya", Gusev, 129 pp
45. Znamenskaya OM, Cheremisinova EA (1970) Noviye danniye o morskikh otlozhaniyakh raiona Nizhney Visly (New data on the marine sediments of the Lower Vistula Region). Vestnik LGU 18:92–101
46. Mikhnevich GS (2003) Vliyaniye dochetvertichnogo reliefa Kaliningradskoi oblasti na raspredeleniye moshnostey chetvertichnykh otlozheniy i sovremenniy relief (Effect of pre-Quaternary relief of the Kaliningrad Region on the distribution of Quaternary sediments capacities and modern relief) (electronic resource). In: Scientific notes of RGO (Kaliningrad branch), vol 2. KSU, RGO, Kaliningrad, pp 3B-1–3B-12. 1 electron optical disk (CD-ROM)
47. Gidrogeologiya SSSR. Kaliningradskaya oblast' RSFSR (Hydrogeology in the USSR. The Kaliningrad Region of the RSFSR), vol XLV. Nedra, Moscow, p 158
48. Zagorodnykh VA, Dovbnya AV, Zhamoida VA (2001) Stratigraphia Kaliningradskogo regiona (Stratigraphy of the Kaliningrad Region) (GS Kharin, ed). Kaliningrad, 226 pp

49. Mikhnevich GS, Gritsenko VA (2008) Prognoz izmeneniy kachestva podzemnykh vod Kaliningradskoi oblasti (Forecasting changes in the groundwater quality in the Kaliningrad Region). Natural and Technical Sciences, vol 4(36). Sputnik, Moscow, pp 246–250
50. Mikhnevich GS, Gritsenko VA (2009) Poluempiricheskaya otsenka prirodnoi zashishennosti podzemnykh vod Kaliningradskoi oblasti (Semi-empirical evaluation of the natural protection of groundwater in the Kaliningrad Region). In: Trukhina VI, Pirogova YA, Pokazeeva KV (eds) Physical problems of ecology (environmental physics), vol 16. MAKS Press, Moscow, pp 213–222
51. Polyakova LS, Kucher EN, Lebedeva NG (eds) (2009) Informatsionny bulleten' o sostoyanii nedr na territorii Kaliningradskoi oblasti za 2008 god (Gosudarstvenny monitoring sostoyaniya nedr) (Newsletter on the state of the subsoil in the Kaliningrad Region in 2008 (state monitoring of subsurface condition)), vol 13. NW GGP "Sevzapgeologiya", Gusev, 87 pp
52. Gosudarstvenniy doklad "O sanitarno-epidemiologicheskoy obstanovke v Kaliningradskoi oblasti v 2009 godu. Electronniy resurs (State report "on the sanitary-epidemiological situation in the Kaliningrad Region in 2009" (electronic resource)) (2010) Office of the Federal Service for Supervision of Consumer Rights Protection and Human Welfare in Kaliningrad Region, Kaliningrad, 458 pp. http://39.rospotrebnadzor.ru/s/39. Accessed 5 July 2010
53. Upravleniye Federalnoy sluzhby po nadzoru v sfere zashity prav potrebiteley i blagopoluchiya cheloveka po Kaliningradskoi oblasti (Federal Service for Supervision of Consumer Rights Protection and Human Welfare in the Kaliningrad Region) (2006) [official website]. http://39.rospotrebnadzor.ru. Accessed 28 Feb 2015

Current Status of the Lake Vistytis in Kaliningrad Region

S.V. Shibaev, A.V. Sokolov, K.V. Tylik, T.A. Bernikova, M.N. Shibaeva,
E.A. Masyutkina, N.N. Nagornova, A.V. Aldushin, and S.K. Zaostrovtseva

Abstract The paper analyzes the basic parameters of the ecosystem of the unique Kaliningrad Region oligotrophic Lake Vistytis. It shows its sufficiently stable conditions for most of the considered parameters. However, there are manifestations of local expressions of meso- and even eutrophy associated with poorly controlled development of recreational areas. The paper confirms high fishing status of the water body and offers recommendations on its rational use.

Keywords Hydrology, Ichthyofauna, Kaliningrad Region, Zoobenthos, Zooplankton

Contents

1	Introduction	442
2	General Characteristics of the Lake	442
3	Hydrology and Hydrochemistry	443
4	Zooplankton and Zoobenthos	446
5	Ichthyofauna	450
6	Fish Community Structure	450
7	Evaluation of Fish Concentration Density Using Hydroacoustics	452
8	Fisheries	454
9	Fishery Characteristics of the Main Fish Species of Lake Vistytis Vendace	459
10	Parasite Fauna of European Vendace	462
11	Whitefish	464
12	Roach	467
13	Perch	468

S.V. Shibaev (✉), A.V. Sokolov, K.V. Tylik, T.A. Bernikova, M.N. Shibaeva,
E.A. Masyutkina, N.N. Nagornova, A.V. Aldushin, and S.K. Zaostrovtseva
Department of Ichthyology and Ecology, Faculty of Bioresources and Nature Usage,
Kaliningrad State Technical University, 1, Sovietsky prospect, 236000 Kaliningrad, Russia
e-mail: shibaev@klgtu.ru; shibaev.s@gmail.com; sokolov@klgtu.ru; tylik@klgtu.ru;
bernikovy@gmail.com; msh@klgtu.ru; masyutkinaea@gmail.com; nagornova@klgtu.ru;
aldushin.andrey@gmail.com; zaostrov@klgtu.ru

14	Pike	471
15	Burbot	473
16	Tench	473
17	Bream	475
18	Eel	475
19	Conclusions	478
References		479

1 Introduction

Lake Vistytis is a unique water body for the Kaliningrad Region. It is the only lake of the Masurian Lake District (Poland) and the lake districts of Lithuania, situated on the territory of Russia. It is the largest in size and the deepest of all the lakes of glacial origin in the Kaliningrad Region. According to the agreement between the Russian Federation and the Republic of Lithuania on the Lithuanian-Russian state border on October 24, 1997, the lake belongs to cross-border waters. In accordance with the contract, 80.7% of the area belongs to the Kaliningrad Region.

In terms of productivity of all trophic levels, the lake is traditionally qualified as the oligotrophic type reservoir, the most vulnerable to the increase of anthropogenic load. However, in recent decades, its recreational value has significantly increased; more and more leisure facilities are built in the coastal part of the lake. Huge man-caused impact can seriously affect its trophic status.

In addition, the lake has a high fishery value due to the presence of the fish fauna of valuable whitefish species – vendace and whitefish, as well as common freshwater fish complex (roach, perch, pike, etc.). Fishing activities here grew rapidly in the 1970–1980s and after a re-start began to develop in recent years, with the prospects for further development.

The purpose of the present work is assessment of the environmental state of the Lake Vistytis on the basis of a long-term integrated monitoring determining the prospects of sustainable fisheries.

2 General Characteristics of the Lake

Lake Vistytis is located in the southeastern part of the region at an altitude of 172 m above sea level. Its area is 1.8 thousand hectares. The volume of water mass is 368 million cubic meters, maximum depth 54 m, average 20.0 m, maximum length of the lake 9.1 km, maximum width 4.4 km, average width 2 km, and coastline 25 km. The bottom contour is complex (Fig. 1).

There are 10–12 small streams flowing into the lake (only five of them flow all year round) and two small rivers are flowing from the south. Flow from the lake is from the river Pissa that through river Angrapa connects the river with the Pregel and then with the Vistula Lagoon [1–4].

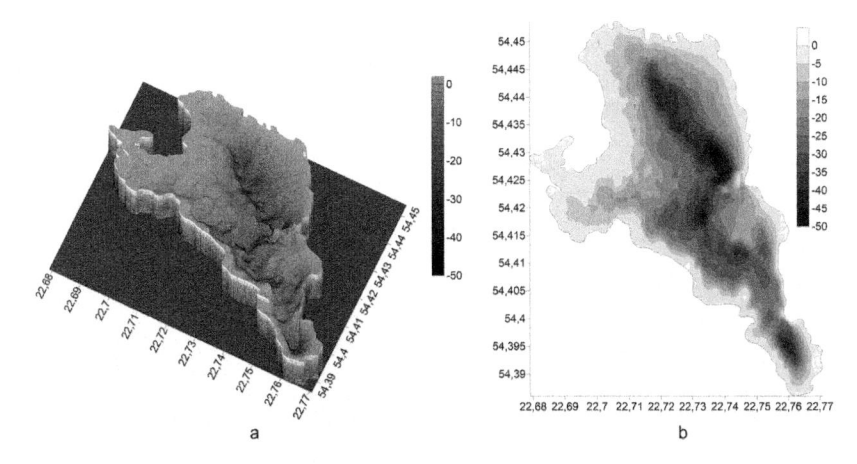

Fig. 1 Contour of the bottom of Lake Vistytis: (**a**) three-dimensional image, (**b**) bathygraphic outline

3 Hydrology and Hydrochemistry

A comprehensive study of the hydrological and hydrochemical parameters of Lake Vistytis has been conducted with varying frequency and specification of the individual parameters since 1976 [1–3, 5–7]. Daily stations, longitudinal and cross sections, were performed; the coastal areas were examined in details. Location of stations is shown in Fig. 2.

All observations were based on standard procedures. Water samples were taken at standard levels (at least on the surface, above and below the transition layer, at the bottom) with the help of a deep-tipping sampler. The water temperature is measured by deep-sea reversing thermometer, in the last decade, using the instrument WTW-3751 and conductivity and pH value using unit WTW Multiline P3. Chlorophyll "a" was determined by a spectrophotometer UNICO 2100. Primary biological products in the oxygen modification were studied in 1975 and 1982 using light-and-dark-bottle method. In addition, starting from 2009, chlorophyll "a" has been defined.

In accordance with the classification [8], the reservoir refers to the group of deep lakes. Thermal processes in the lake develop according to the classical scheme, characteristic of deep freshwater. In winter, there is an inverse stratification. At the end of spring, a straight stratification starts to form, and there appears a transition layer. Together with the appearance of the transition layer, three specific layers are formed: epilimnion, metalimnion, and hypolimnion. Regardless of the warm epilimnion and the vertical gradients in the thermocline, the lower boundary of the latter (with very few exceptions) does not fall below 13–15 m.

In accordance with the classification [9], mineralization of water is changing from "low" (less than 200 mg/dm^3) to "average" (less than 300 mg/dm^3). The water is classified as hydrocarbonate, belonging to calcium group and the second type of

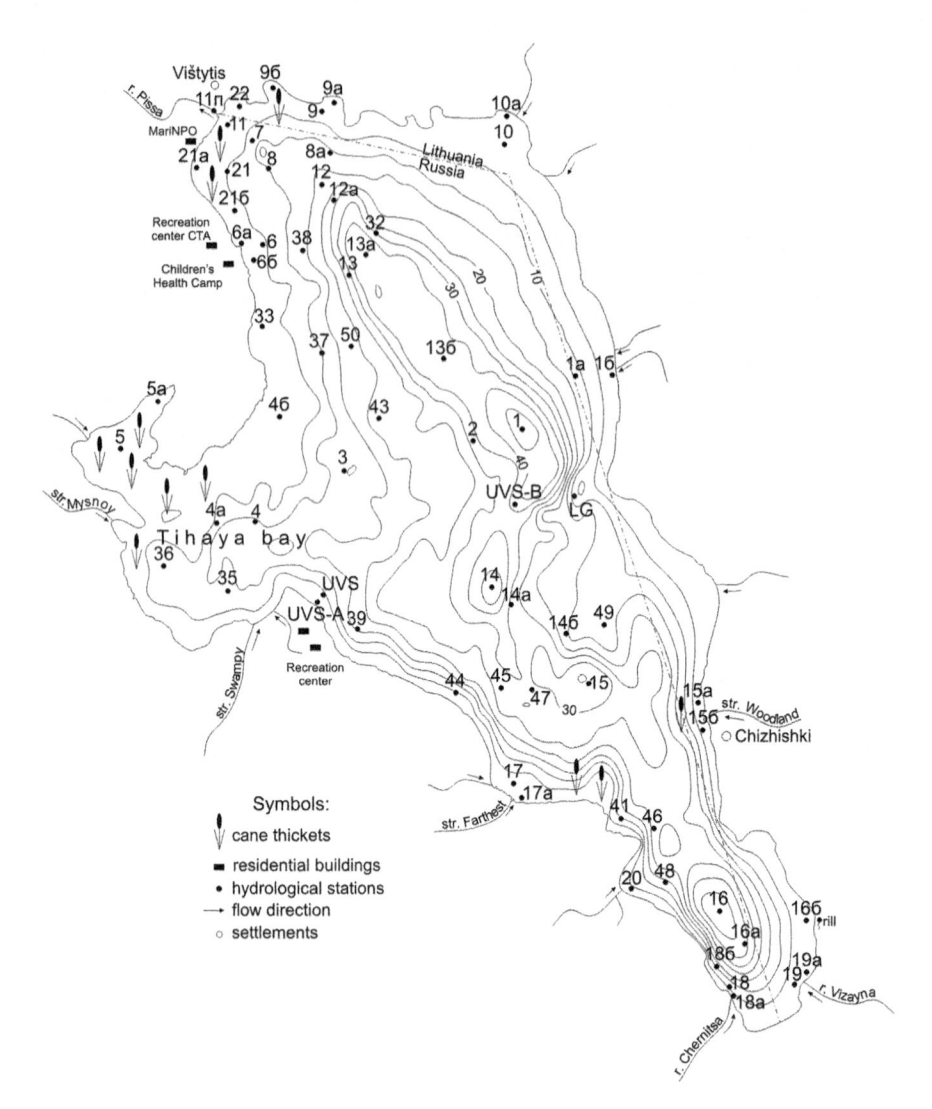

Fig. 2 Location of hydrological stations on Lake Vistytis (dash line, the state border of the Russian Federation)

waters (the content of sodium ions is less than the sum of calcium and magnesium ions). According to the classification [8], water in the lake is soft (total hardness value is 2.1–2.6 mmol/dm^3).

Distribution and seasonal dynamics of hydrochemical characteristics of the lake are largely related to seasonal restructuring of water stratification.

Gas conditions can be generally assessed as prosperous and fitting oligotrophic status of the lake. Seasonal variations in the epilimnion are smoothed; oxygen saturation in the epilimnion during the year is close to 100%. The amount of oxygen in the hypolimnion in summer is drastically reduced, in which below 25 m a rapid drop in oxygen concentration with depth sometimes up to 40% of saturation and below usually occurs in bottom layers.

Nutrients are subject to consumption and regeneration. Their distribution and seasonal dynamics mainly corresponded to seasonal hydrological and hydrobiological processes and, at some stations, also seemed to reflect the degree of water pollution.

Permanganate oxidation allows judging the quantity of organic substances in the water. Its value in the deep stations according to classification [9] falls into the class of "small" or "medium" (in the southern part, it is usually higher). Off the coast, it is "average".

The nature of long-term variability of organic matter does not have a stable trend: in the value of permanganate oxidation at different stations, including the stations of different depths, one can identify periods of raising and lowering oxidation.

Figure 3 clearly shows the periods of increase of organic matter. If we exclude abnormally high peaks in 1990 and 2008 (the data is mainly on Duck Bay and the mouth of the river Chernitsa, respectively), it can be argued that 2010–2012 differ in particularly serious accumulation of organic matter.

Observations made in 2011–2014 allowed to clarify and refine conclusions obtained before. Fundamental changes in hydrochemical conditions in the lake did not happen. Most of the specified indicators that we analyzed meet MPC for all categories of fisheries, but during some extremely hot periods, the concentration of oxygen does not meet the necessary requirements.

However, since the late 1980s and early 1990s of the last century, eutrophication began to appear in some parts of the lake. Contamination by organic matter and nutrients, which began in the coastal zone in the extreme southeast (station 19), gradually spread to the deep part of the lake (surface on station 16a).

Currently, for the entire coastal area, including Tikhaya Bay, first and foremost, for the northern and northeastern shallow water, for eastern part of the lake along

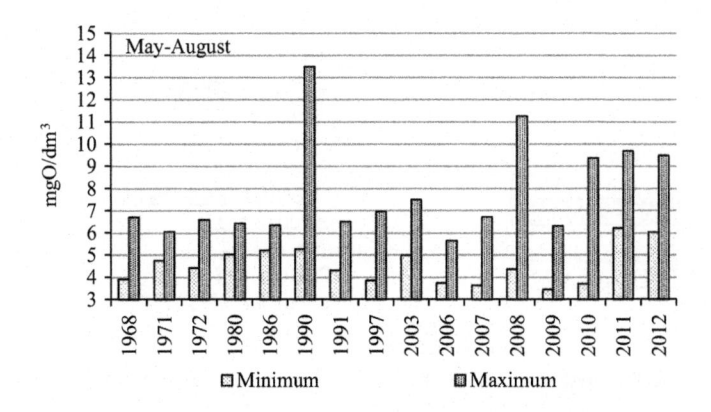

Fig. 3 Generalized dynamics of extreme values of permanganate oxidation in Lake Vistytis

the coast, and for the south, covering the coastal shallow waters, including estuarine areas of flowing rivers and superficial layers of the adjacent deep part of the lake, a higher trophic level called mesotrophic is distinguished. Duck Bay has clearly defined traits of eutrophic reservoir.

4 Zooplankton and Zoobenthos

Available data on zooplankton and zoobenthos of Lake Vistytis includes the data from 1967, 1976–1979, 1990, 1997, 2000, 2003, and 2006 [10–17]. Over the years, studies were conducted with different durations and frequencies, with various numbers of stations. Therefore, in order to compare year-to-year variations of zooplankton and zoobenthos development, we used the data collected in July and August. This period demonstrates the most numerous data, comparable in dates.

Today more than 100 kinds and forms are registered as a part of zooplankton and more than 300 are registered in zoobenthos. Zooplankton includes rotifers, cladocerans, and copepods (Fig. 4). Each of these groups consists of about 30–40 species. Zoobenthos includes nine groups: chironomid larvae, caddis flies, mayflies, dragonflies, crustaceans, leeches, oligochaetes, mollusks, and a group of "others." "Others" include a few water beetles, bugs, mites, and larvae of two wings. The most diverse are chironomid larvae and caddis flies, as well as shellfish.

Most species of zooplankton and zoobenthos are categorized as "rare." The most frequent types of zooplankton are rotifers (*Asplanchna herrickii* (De Guerne), *Asplanchna priodonta* (Gosse), *Kellicottia longispina* (Kellicott), and *Conochilus hippocrepis* (Schrank)), copepods (*Eudiaptomus graciloides* (Lilljeborg), *Mesocyclops leuckarti* (Claus), and *Thermocyclops oithonoides* (G. O. Sars)), and cladocerans (*Bosmina longirostris* (O. F. Müller), *Eubosmina coregoni* (Baird), and *Daphnia cucullata* (G. O. Sars)).

In zoobenthos, the most frequent are crustaceans (*Asellus aquaticus* (L.) and *Gammarus*), mollusks (*Dreissena polymorpha* (Pall.), *Bithynia tentaculata* (L.), and *Pisidium amnicum* (Mull.)), chironomid larvae (*Sergentia longiventris* (Kief.),

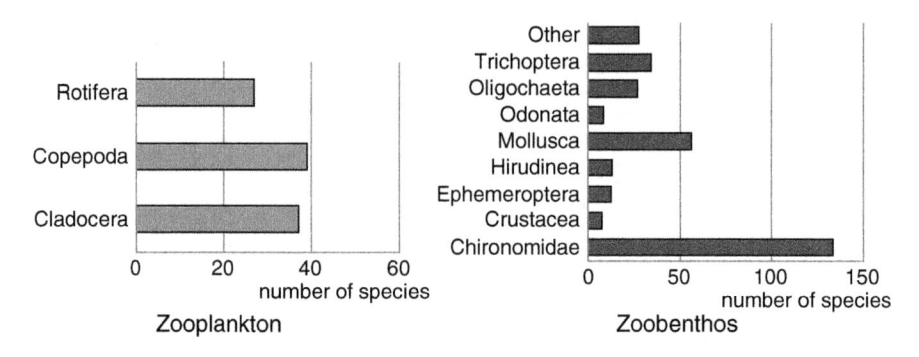

Fig. 4 Species diversity of zooplankton and zoobenthos of Lake Vistytis

Tanytarsus gregarius (Kief.), *Cladotanytarsus mancus* (Walk.), *Microtendipes pedellus* (De Geer), and *Procladius* (Skuse)), mayfly larvae kind (*Caenis* (Stephens)), and oligochaetes (*Potamothrix hammoniensis* (Mich.)).

The lake is home for unique species that are typical of oligotrophic lakes. These species are copepods (*Heterocope appendiculata*) and benthic amphipods (*Pallasiola quadrispinosa* (Sars)). *Heterocope appendiculata* is a pelagic species and is below the thermocline layer. *Pallasiola quadrispinosa* lives in aphytal zone on piled depths and is an important food for the whitefish in the lake.

Spatial distribution of zooplankton is inhomogeneous (Fig. 5). The average number of zooplankton rarely reaches 30,000 species per cubic meter and biomass up to 0.5 g/m^3. The greatest development of zooplankton is characteristic of the Western part of the lake, near Tikhaya Bay. The number of zooplankton in the area is more than 50,000 specimen per cubic meter, and biomass is 2 g/m^3.

The greatest diversity of benthic invertebrates can be observed in the littoral zone – about 200 species (Fig. 6). Chironomids, clams, and caddis flies are widely

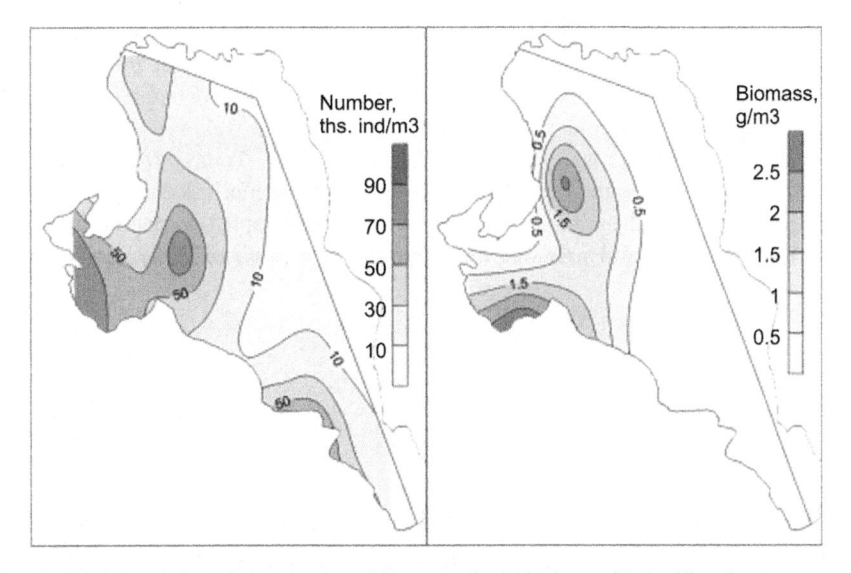

Fig. 5 Spatial variation of abundance and biomass of zooplankton of Lake Vistytis

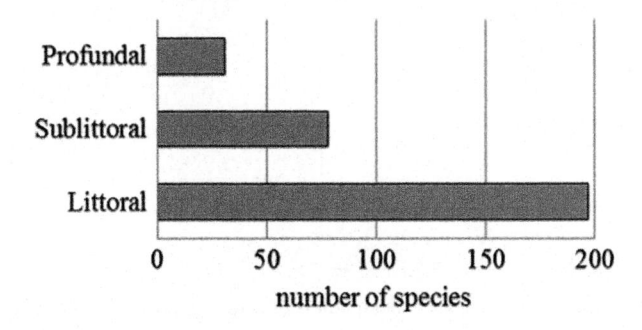

Fig. 6 Species diversity of zoobenthos in different areas of Lake Vistytis

represented. A variety of sublittoral shelf zoobenthos is significantly lower than in littoral, although zoobenthos is represented by all the nine groups. Maximum development of benthic fauna happens in the middle of summer.

Very poor fauna is found in aphytal zone. Permanent inhabitants of this part of the lake are a few Euglesidae shellfish, chironomids, and oligochaetes.

Long-term annual average abundance and biomass of zoobenthos in Lake Vistytis is fairly constant and equal to about 1,400 specimen/m^2 and 70 g/m^2.

There are areas in the littoral zone where the number of zoobenthos often exceeds 3,000 specimen/m^2 (Fig. 7). This is primarily the northwestern part of the lake, at the source of Pissa River, as well as the southwestern coast near the mouth of Chernitsa River. In the central part of the deepwater, the number of zoobenthos rarely exceeds 500 specimen/m^2.

Zoobenthos of the lake consists of four groups: chironomids, mollusks, crustaceans, and oligochaetes. About 40% of them accounts for chironomid larvae. Massive development is seen in the following types: *Paratendipes albimanus*, *Cladotanytarsus mancus*, *Microtendipes pedellus*, and *Stictochironomus crassiforceps*. Permanent dominant of the aphytal zone of all time is *Sergentia longiventris*.

Shellfish account for approximately 20% of the total number of zoobenthos. In some places, up to a depth of 10 m, the bottom of the lake is completely covered with zebra mussel. The role of these species in the lake, as well as its heterogeneity of spatial distribution, was previously described [17]. Today, *Dreissena polymorpha* also continues to play a leading role among the mollusks.

Oligochaetes reach large numbers predominantly in aphytal zone after a mass departure of chironomids. Among them, *Potamothrix hammoniensis* stands out.

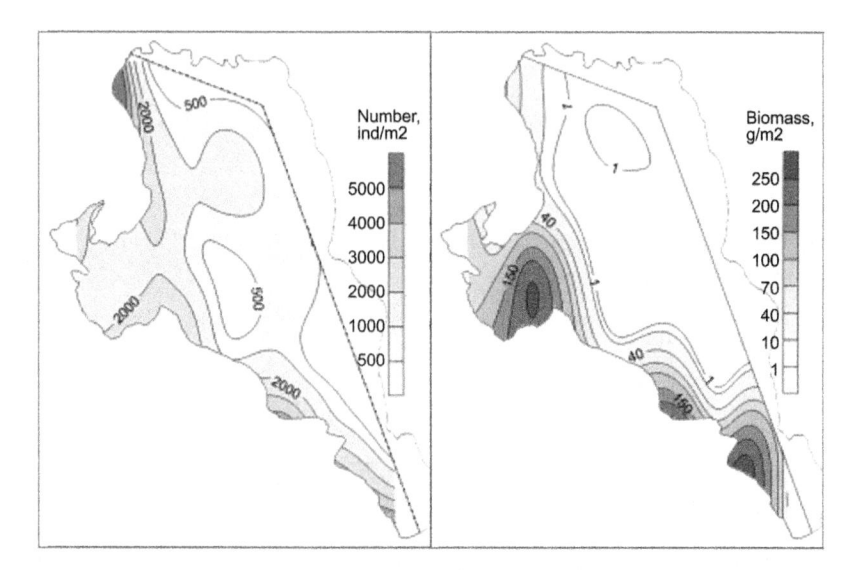

Fig. 7 Spatial variation of abundance and biomass of zoobenthos of Lake Vistytis

Crustaceans account for about 15% of the total number of zoobenthos. They predominate in the intertidal and sublittoral shelf of the north and west parts of the lake. Among them, *Asellus aquaticus* and *Gammarus* dominate.

In some parts of littoral and sublittoral shelf, zoobenthos biomass exceeds 150 g/m^2 (Fig. 7). First of all, these are areas where *Dreissena polymorpha* shellfish are widely spread: Tikhaya Bay area, northwest and southwest coast. These kinds of places can account for up to 95% weight of the benthos. Zoobenthos biomass in the littoral zone except for clams is 10–15 g/m^2.

In aphytal zone, zoobenthos biomass is typically less than 1 g/m^2. *Sergentia longiventris* chironomids mostly dominate here, but after a mass departure of chironomids, oligochaetes and Euglesidae mollusks started to dominate.

The trophic status of the lake according to Kitaev scale [18] is defined as oligotrophic, changing into mesotrophic near Tikhaya Bay and Duck Bay.

An analysis of taxonomic composition of zooplankton and zoobenthos showed that the list of indicator species is broad and can serve as a basis for saprobiological analysis. Saprobic index value zooplankton varies from 1.2 to 1.8 (Fig. 8). Saprobic indicator for Tikhaya Bay, Duck Bay, the west coast right next to the recreation center, at the source of the Pissa near Vištytis, in the eastern part of the lake near the recreation center from the Lithuanian side, is more than 1.5. The status of these areas can be assessed as slightly contaminated (β-mesosaprobic). Saprobic index for most parts of the lake (especially central) is less than 1.5, which corresponds to oligosaprobic water quality.

Saprobic index of zoobenthos is slightly higher compared to zooplankton (1.4–2.8). Saprobic values less than 2 are observed in the coastal area of the lake (Fig. 8). These areas are characterized as slightly polluted (β-mesosaprobic). In the

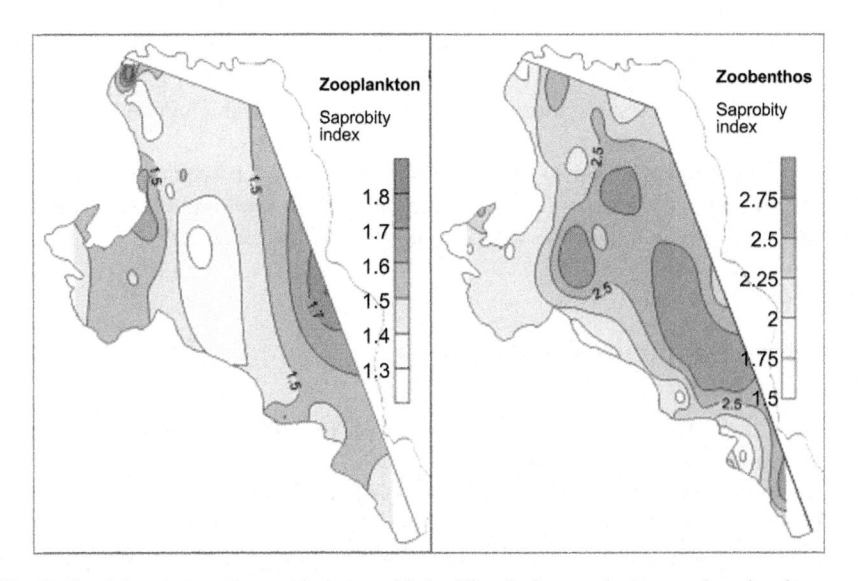

Fig. 8 Spatial variation of saprobic index of Lake Vistytis for zooplankton and zoobenthos

deep zone, saprobic indicator exceeds 2.5, implying a higher content of organic substances. Water quality in bottom layers is contaminated (α-mesosaprobic). It should be borne in mind that in the deep lakes, zoobenthos characterizes only bottom layers of water.

In general, the ecological state of Lake Vistytis can be described as transient from oligosaprobic in the central part to β-mesosaprobic in the coastal part.

In general in terms of hydrobiological indicators, ecosystem of the lake is fairly stable, although compared to the 1970s, some mesotrophic features in the littoral and sublittoral shelf are more evident now. Organic matter is accumulated in the deepwater where the number of Oligochaeta increases from time to time. The least resistant to eutrophication are areas such as Duck Bay and Tikhaya Bay, the coastal areas near recreational center and Vištytis.

5 Ichthyofauna

As part of the fish fauna of Lake Vistytis, 23 species of fish have been noted. Fish that live in the lake are from nine families and a variety of faunal complexes: arctic freshwater (burbot, vendace, whitefish), Ponto-Caspian freshwater (redeye, bream, bleak, white bream), late tertiary freshwater (loach, Western brook lamprey), boreal plain (pike, crucian carp, gudgeon, roach, spiny loach, ruff, perch, tench), boreal foothill (bullhead, loach minnow), and boreal Atlantic (eel) [7, 19–22] (Table 1).

Of particular note is the presence of Coregonidae – vendace and whitefish that are absent in other freshwater bodies of the region. It is known that these species prefer oligotrophic lakes with clean and clear water. At the same time, Lake Vistytis is common with the same types as most other freshwater bodies in the region, mainly represented by carp and perch fish. Moreover, the most diverse among fish fauna are carps – 11 species. Also of interest is the presence in the lake of bull, minnow, and bearded stone loach that prefer lakes with clear water and clean sand and stones in littoral shelf.

The most diverse group in the fish fauna is carps – ten species (about 45%). Some fish and cyclostomes live not in the lake but in the Pissa flowing out of it moving for reproduction to the source of the river from the lake.

6 Fish Community Structure

The structure of the fish communities in Lake Vistytis is characteristic of both general state of the ecosystem and particular fish species. Methods of assessment of fish communities structure in 2011–2013 involved control catches using complex of different gears (bottom gill nets mesh of 14–65 mm in length, beach seine of 18 and 120 m, with 8 mm mesh in purse) with different selective characteristics. Control catches were carried out in the zone of typical habitats so that to obtain reliable

Table 1 Fish species of Lake Vistytis

No	Latin name	Russian name	English name
Whitefish family (Семейство Сиговые) – Coregonidae			
1	*Coregonus albula* (L.)	Ряпушка	*Vendace*
2	*Coregonus lavaretus* (L.)	Сиг обыкновенный	*Whitefish*
Pike family (Семейство Щуковые) – Esocidae			
3	*Esox lucius* L.	Щука обыкновенная	*Pike*
River eel family (Семейство Речные угри) – Anguillidae			
4	*Anguilla anguilla* (L.)	Речной угорь	*European eel*
Carp family (Семейство Карповые) – Cyprinidae			
5	*Abramis brama* (L.)	Лещ	*Bream*
6	*Alburnus alburnus* (L.)	Уклея	*Bleak*
7	*Blicca bjoerkna* (L.)	Густера	*White bream*
8	*Carassius carassius* (L.)	Золотой карась	*Crucian carp*
9	*Gobio gobio* (L.)	Пескарь	*Gudgeon*
10	*Leuciscus leuciscus* (L.)	Елец	*Dace*
11	*Rutilus rutilus* (L.)	Плотва	*Roach*
12	*Scardinius erythrophthalmus* (L.)	Красноперка	*Redeye*
13	*Tinca tinca* (L.)	Линь	*Tench*
Loach family (Семейство Вьюновые) – Cobitidae			
14	*Cobitis taenia* (L.)	Щиповка	*Spiny loach*
15	*Misgurnus fossilis* (L.)	Вьюн	*Loach*
Cod family (Семейство Тресковые) – Lotidae			
16	*Lota lota*	Налим	*Burbot*
Perch family (Семейство Окуневые) – Percidae			
17	*Gymnocephalus cernua* (L.)	Ерш обыкновенный	*Ruff*
18	*Perca fluviatilis* L.	Окунь речной	*River perch*
Cottids family (Семейство Рогатковые) – Cottidae			
19	*Cottus gobio* L.	Бычок подкаменщик	*Bull*
Catfish family (Семейство Сомовые) – Siluridae			
20	*Silurus glanis*	Сом обыкновенный	*Catfish*
Fish in the rivers flowing in the lake and flowing out of it			
Lamprey family (Семейство Миноговые) – Petromyzontidae			
21	*Lampetra planeri* (Bloch)	Минога ручьевая	*Western brook lamprey*
Carp family (Семейство Карповые) – Cyprinidae			
22	*Phoxinus phoxinus* (L.)	Гольян обыкновенный	*Minnow*
23	*Barbatula barbatula* (L.)	Усатый голец	*Bearded stone loach*

characteristics of distribution of species and size composition of fish in the whole water body. All materials on the control catch are standardized and reduced to general indicators: the area of a net setting, 0.5 ha; a value of net fishing effort, a day fishing at 25 m net. As a result, we obtained the value of occurrence (fq), CPUE in numbers (Yn/F), and biomass (Yw/F). Assuming that all things being equal the catch value per unit of fishing effort is determined by the density of the fish population, the figures obtained can be interpreted as indices of abundance and biomass of a particular species, taking into account the fishing net selectivity with a particular mesh size [23, 24].

Considering that there is a clear differentiation of species according to habitats, characteristic of the structure was carried out separately for pelagic and demersal horizons (Table 2).

Perch is currently dominating in the bottom fish community, the incidence reaches 50%, and CPUE fluctuates within 0.6–1.2 kg/day. On the second place in number and biomass is roach and then whitefish. The structure of the bottom ichthyocenosis has been quite stable for 3 years of research, which indicates the absence of a negative impact of fishing or other human factors (Fig. 9). The average CPUE reaches 3.1 kg/day, which is quite high for this type of reservoirs. With a relatively stable structure of ichthyocenosis, the observed fluctuations in number index in some years can be explained by natural interannual changes in the distribution and activity of the behavioral responses of fish, but not by fluctuations in the overall populations. So, in 2010, very high catches of ruff in the standard stations were registered, which led to its majority in the species composition.

In pelagic fish communities, the dominant element is vendace, which is accompanied by ruff in the case when the foot rope is at a distance of less than 1.5 m from the bottom. The incidence of other types – perch, burbot, and whitefish – is significantly less, and they do not play a big role in catches (Table 3 and Fig. 10).

Distribution analysis of the main types allows us to give quantitative characteristics of the confinedness of fish to different habitats. Roach lives mainly in the coastal zone of the lake with depths up to 5 m. Perch occupies a vast water area, starting with the coastal waters up to the depth of 20 m, with few large concentrations of perch noted in the area of the depths of 10–20 m. Coregonidae live mainly in the central deep part of the pond. Whitefish prefer depth of 10–30 m. Vendace live mainly in the pelagic zone of the lake at depths more than 20 m (Fig. 11).

In general, significant changes in the structure of fish population were not observed in recent years, which prove its relative stability.

7 Evaluation of Fish Concentration Density Using Hydroacoustics

Hydroacoustic survey on the quantitative assessment of fish stocks in Lake Vistytis has been carried out using dual frequency hydroacoustic software complex "AsCor" [25] in summer. Network of tacks was used allowing to cover the whole Russian part of the water body at a depth of over 4 m.

Hydroacoustic survey was conducted in May and July in daytime and at night (Figs. 12 and 13). In view of daily migrations of fish in the upper water layers and its distribution and night survey gives bigger density of fish population (five to ten times bigger). In the long term, accumulation of data on the daily distribution of fish will help to find correlation and to unify evaluations.

Table 2 Parameters of demersal community in Lake Vistytis (bottom gill nets)

Species	2012 fq	2012 Yn/F	2012 Yw/F	2013 fq	2013 Yn/F	2013 Yw/F	2014 fq	2014 Yn/F	2014 Yw/F	Average fq	Average Yn/F	Average Yw/F
Perch	47.0	7.10	0.66	37.7	6.70	0.65	45.1	8.22	0.92	43.3	7.34	0.75
Roach	29.8	5.70	0.56	20.8	8.37	0.66	22.0	4.87	0.48	24.2	6.31	0.57
Vendace	19.5	20.12	0.45	20.9	17.12	0.43	28.1	34.14	0.89	22.9	23.79	0.59
Ruff	28.4	8.86	0.14	31.7	13.01	0.21	32.0	14.21	0.22	30.7	12.03	0.19
Pike	17.7	0.30	0.17	5.5	0.13	0.12	7.3	0.15	0.08	10.2	0.19	0.12
White-fish	14.8	0.27	0.11	6.7	0.12	0.06	13.1	0.24	0.12	11.5	0.21	0.10
Tench	1.4	0.08	0.06	4.2	0.13	0.14	1.5	0.02	0.02	2.4	0.07	0.07
Burbot	4.7	0.06	0.03	10.7	0.25	0.11	11.0	0.26	0.08	8.8	0.19	0.07
Bream	2.3	0.24	0.09	3.6	0.14	0.04	2.9	0.03	0.01	2.9	0.14	0.05
Bleak	1.0	0.13	0.00	0.8	0.38	0.01	0.8	0.25	0.01	0.9	0.25	0.01
Eel	0.3	0.01	0.00	0.0	0.00	0.00	0.0	0.00	0.00	0.1	0.00	0.00
Carassius	0.1	0.00	0.00	0.3	0.01	0.00	0.0	0.00	0.00	0.1	0.00	0.00
Total		42.85	2.27		46.36	2.44		62.38	2.83		50.53	2.51

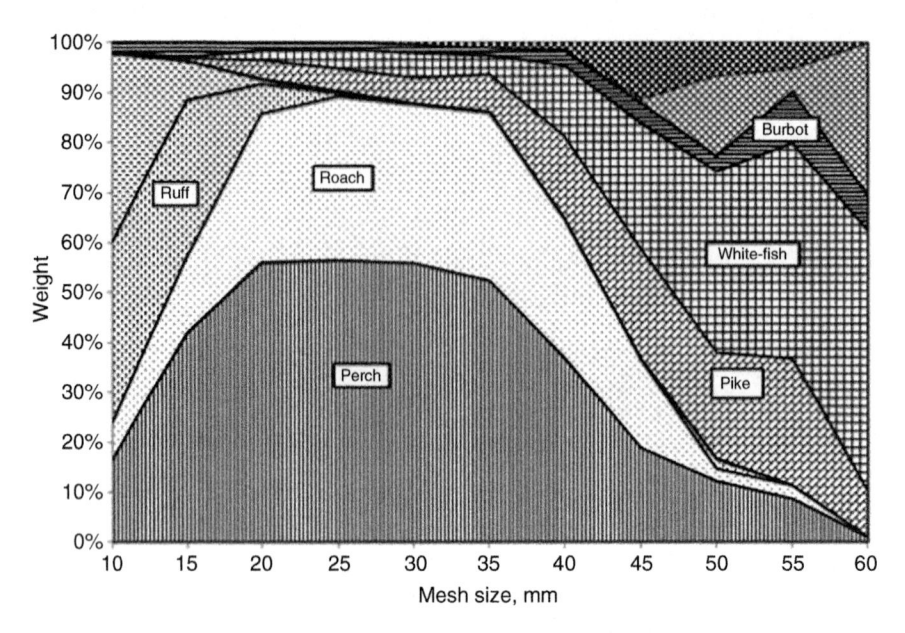

Fig. 9 Structure of demersal fish community of Lake Vistytis (bottom gill nets)

Materials were processed at the operating frequency of 200 kHz; averaging interval was chosen to be 300 sendings at the operating frequency of 200 kHz, which corresponds to approximately 100 s of real time, or 250–300 m of the covered distance.

Total number of fish based on the results of numerical integration of the data obtained turned out to be at night survey 12.7 million species at a density of 1.2 thousand species per hectare. In general, during the years of studies, total number of fish varies within a small range, bearing evidence for steady state of the ichthyofauna of the lake (Fig. 14).

8 Fisheries

After the 1990s, there was a sharp decline in commercial catches, mainly due to socioeconomic reasons. Over the past 15 years, an average catch in the lake was 6.2 tons with fluctuations from 1.1 to 12.4 tons (Fig. 15).

Main fish species in Lake Vistytis are vendace, roach, and perch. Vendace and roach dominate in catches. At a target fishing, whitefish can be caught, and despite the fact that the absolute whitefish catches have decreased in the last decade, its relative value has increased. Over a period, fairly large proportion of commercial catches was taken by eel, by maintaining its population through stocking the pond with baby fish – glass eel. In amateur fishing, pike, perch, and roach are mainly caught.

Table 3 Parameters of pelagic community of Lake Vistytis (pelagic gill nets)

Species	2011 fq	2011 Yn/F	2011 Yw/F	2012 fq	2012 Yn/F	2012 Yw/F	2013 fq	2013 Yn/F	2013 Yw/F	Average fq	Average Yn/F	Average Yw/F
Vendace	62.9	101.44	2.10	81.0	230.95	5.21	90.6	104.13	2.56	78.2	145.51	3.29
Ruff	7.6	0.13	0.00	24.1	0.86	0.01	25.0	2.97	0.05	18.9	1.32	0.02
Burbot	0.0	0.00	0.00	1.7	0.02	0.01	6.3	0.09	0.02	2.7	0.04	0.01
Perch	3.8	0.04	0.00	17.2	0.43	0.01	3.1	0.18	0.03	8.1	0.21	0.01
Roach	1.0	0.00	0.00	0.0	0.00	0.00	0.0	0.00	0.00	0.3	0.00	0.00
Pike	1.0	0.01	0.00	0.0	0.00	0.00	0.0	0.00	0.00	0.3	0.00	0.00
Bleak	1.9	0.15	0.00	0.0	0.00	0.00	0.0	0.00	0.00	0.6	0.05	0.00
Total	105	101.77	2.11	58	232.26	5.24	32	107.37	2.66	195	147.13	3.34

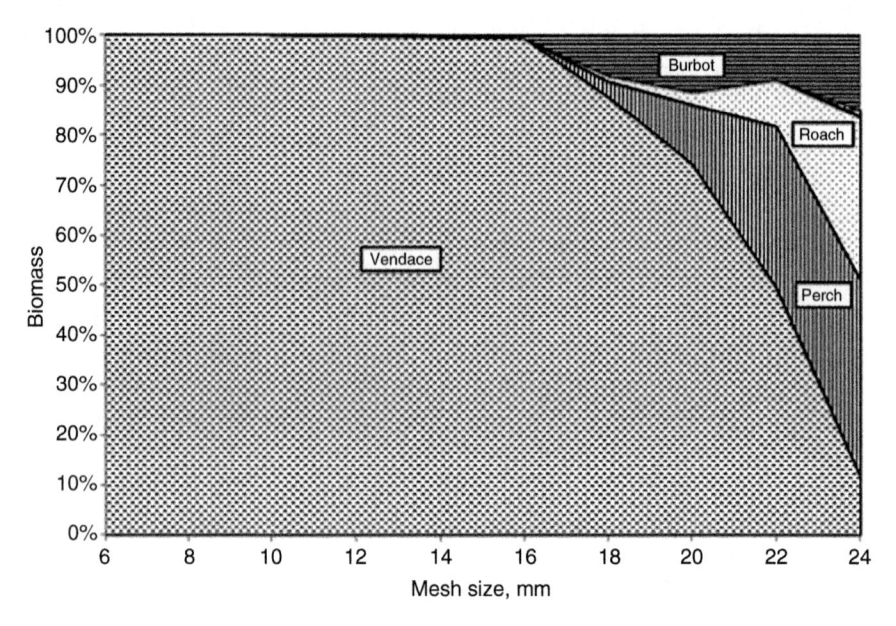

Fig. 10 Structure of pelagic fish community of Lake Vistytis (pelagic gill nets)

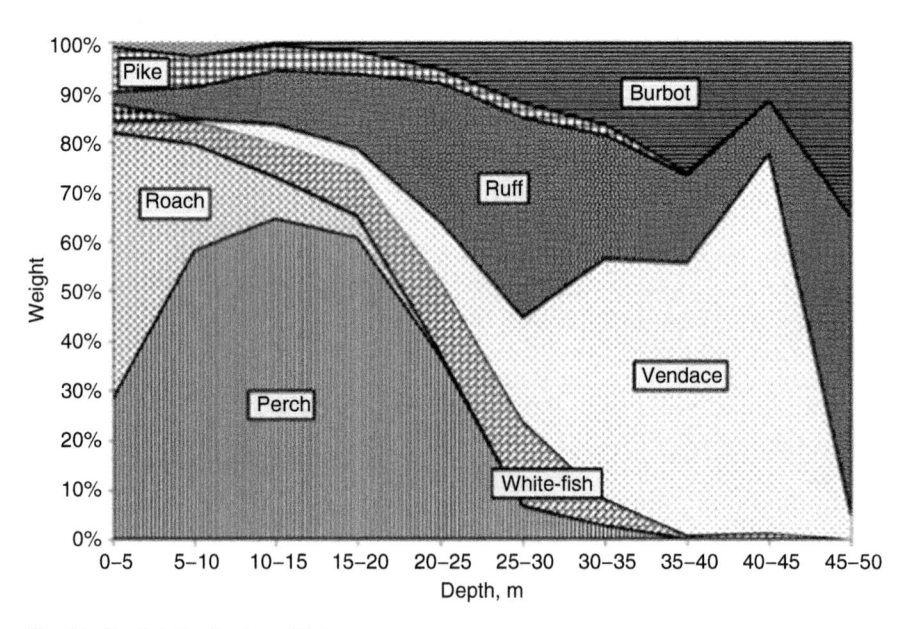

Fig. 11 Spatial distribution of fish

Despite its relatively low productivity, fishing in Lake Vištytis is of great social significance because it provides local employment. In addition, rational management of stocks of valuable fish species may significantly raise their catch and, along

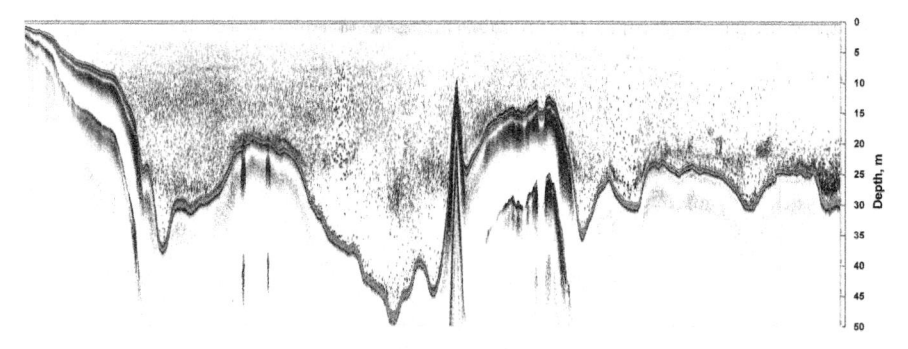

Fig. 12 An example of longitudinal hydroacoustic section of Lake Vistytis (distance, 10 km)

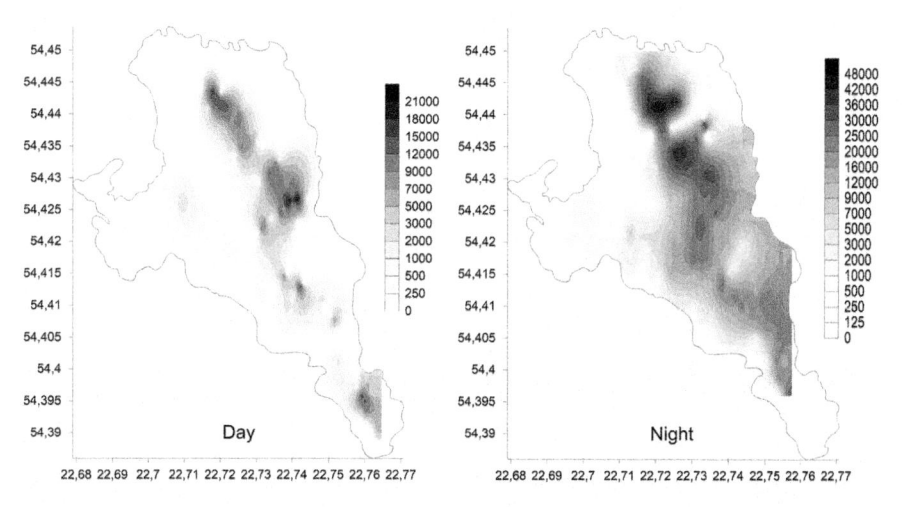

Fig. 13 Map of density distribution of fish abundances on Lake Vistytis in accordance with the data of hydroacoustic survey

with the development of recreational fishing, make a significant contribution to socioeconomic development of the region.

Low results of fishing in the past two decades are the consequence of several factors:

1. Poor development of the fishing base, which is about ten times weaker than during the active use of the reservoir in the 1970s.
2. Insufficiently full record keeping of catch volumes in fishing statistics.
3. As will be shown below, the use of traditional vendace nets with 18 mm mesh bar in changing structure of the population turned out to be ineffective due to low catch per unit effort.

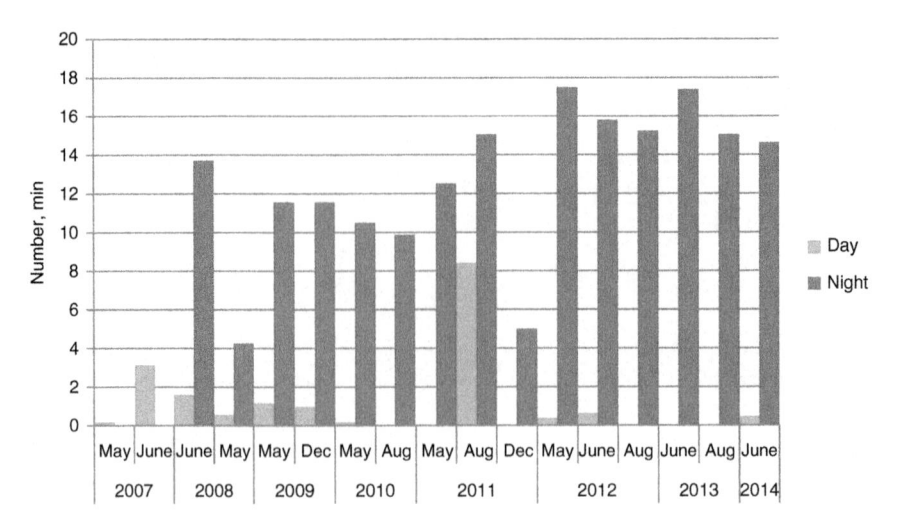

Fig. 14 Summarized data of the number of fish (species) in Lake Vistytis according to the data of hydroacoustic surveys

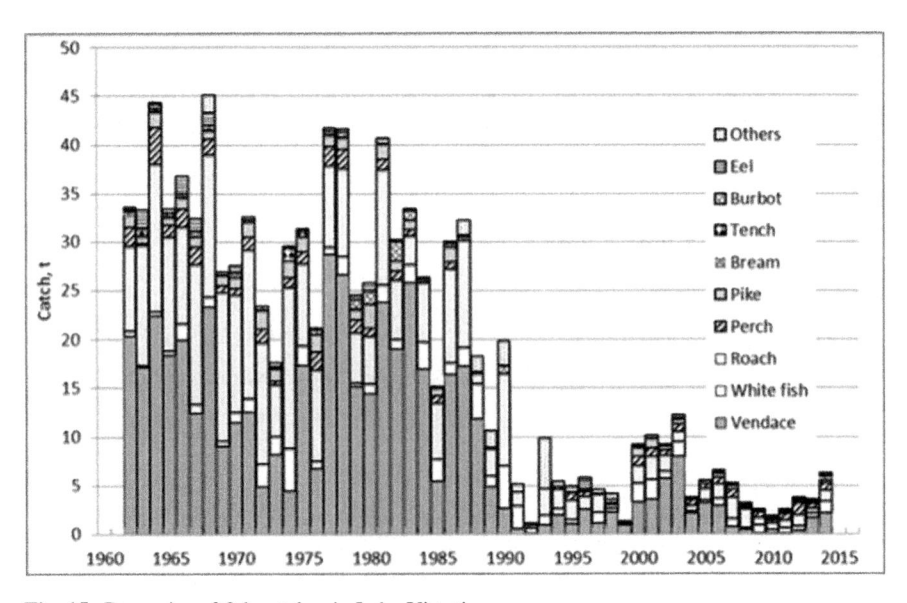

Fig. 15 Dynamics of fish catches in Lake Vistytis

Ongoing development of fisheries in conditions of new market should lead to a gradual increase in catches within the optimal acceptable level.

9 Fishery Characteristics of the Main Fish Species of Lake Vistytis Vendace

The only population of vendace in the Kaliningrad Region is in Lake Vistytis as there are no necessary living conditions in the other water bodies in the region [26].

Hydroacoustic survey showed that the spatial distribution of vendace is directly dependent on the depth of the area (Fig. 16). The highest densities are confined to depths of over 20 m and reach a maximum in the middle of the lake with a depth of 40–50 m. On average, according to the materials of underwater acoustics, concentration of vendace in different years ranged from 9.5 to 47.0 thousand species per hectare. Such fluctuations of the fish population density are connected not only with interannual fluctuations in abundance but with the fact that the surveys were conducted at different dayparts.

The vendace in the lake is characterized by daily vertical migrations: in the daytime, the fish is concentrated in the bottom layers. At the same time, it obviously falls into the "dead zone" and is not registered by echo-sounder. At night, after gathering up in the upper layers of water, vendace appears to be more accessible for survey with underwater acoustics (Fig. 17). In general, it has been noted that night surveys give higher concentrations of vendace than in the daytime. Thus, on average over 3 years of research, the total estimation of number in daily shootings gives a value of 0.8–2 million species and at night 9–14 million species.

Another factor determining spatial distribution of vendace is thermal conditions. Vendace prefers water layers with the temperature no higher than 10°C. In this regard, there is the following seasonal dynamics. During spring homeothermy, main concentrations of vendace are confined to the waters of the depth of more than 8 m, which form dense clusters in the layer of 8–12 m. In summer, in the

Depth map   Biomass distribution

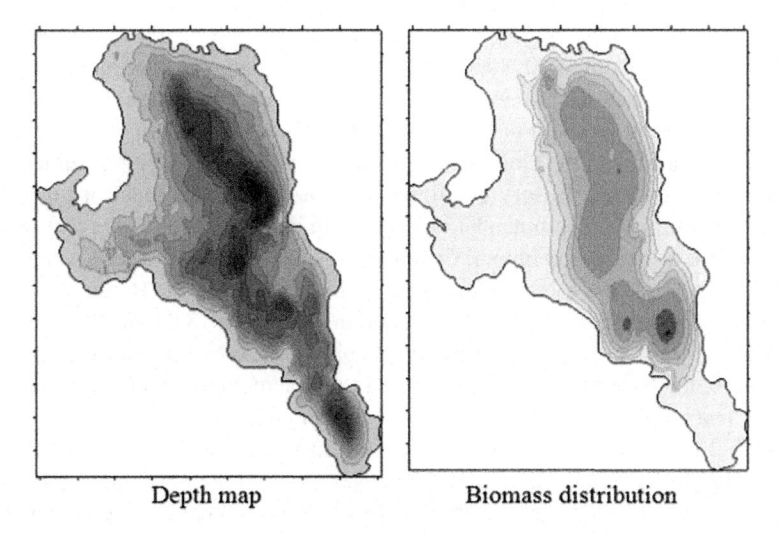

Fig. 16 Horizontal distribution of vendace in Lake Vistytis in correlation with the depth

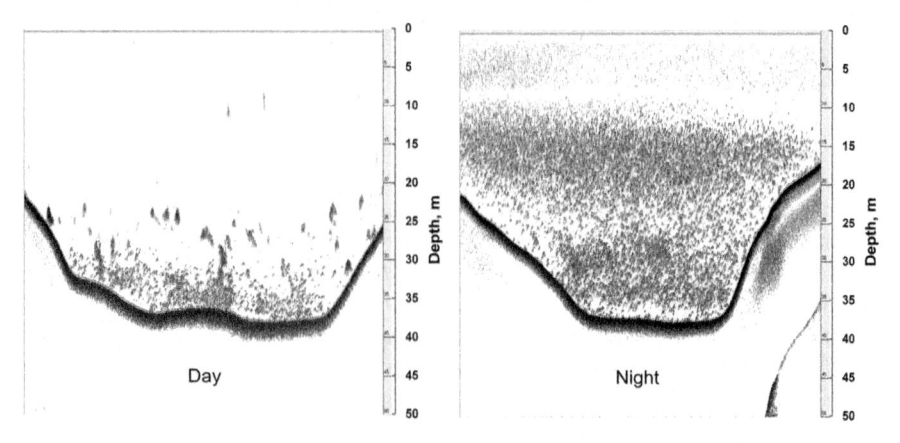

Fig. 17 Vertical distribution of vendace in Lake Vistytis in accordance with the data of hydroacoustic survey (July, 2008)

pelagic zone of the lake, there is a thermocline layer at a depth of about 15 m, and since vendace prefer cooler temperatures, it does not go above this horizon. As a result, at night, concentration of fish is formed in the horizon lower than 15 m.

Size of the vendace in Lake Vistytis in commercial catches range from 12 to 22 cm with an average length of 15 cm, its weight ranges from 18 to 64 g, an average of 39.6. Fluctuations in replenishment number cause changes in size structure of commercial catches, but in recent years, modal group remains the same – 12–13 cm. Its ratio usually ranges between 50 and 60% (Fig. 18).

Four age groups are exploited by fisheries, from 2 to 5 years. The main part of the catch is 2- or 3-year-old fish. Vendace older than 5 years is very rare.

Fishing for vendace is carried out in summer and autumn using nets with mesh size from 16 to 18 mm. Nets are set in the central eastern part of the reservoir in the bottom layer at a depth of 20–30 m. Over the past 40 years, the maximum catches of vendace were registered in 1977 and reached 28.6 tons. Average annual catch is 11 tons, taking into account the period of the 1990s, when fishing was hardly kept.

Relative stability of the biological indicators of vendace allows to assume that its population is in a satisfactory condition. Obtained materials make it possible to provide preliminary recommendations for optimizing vendace fishery in relation to the nature of its spatiotemporal dynamics in Lake Vistytis. It seems appropriate to put into practice the use of pelagic nets with mesh size of 14–16 mm that should be set in spring in the horizon of 8–12 m and in summer in 15–20 m. The use of this mode allows to boost the catches of vendace in the lake by several times.

In 2013, the fishermen at the lake started to implement the scientific recommendations and tried using the nets with 14–16 mm mesh size. It was found that, although 14 mm mesh is the most efficient, disentangling of fish is rather labor consuming. In this connection, preference was given to the nets with mesh size of 16 mm. This allowed to restore the level of vendace catches, which occurred in previous years.

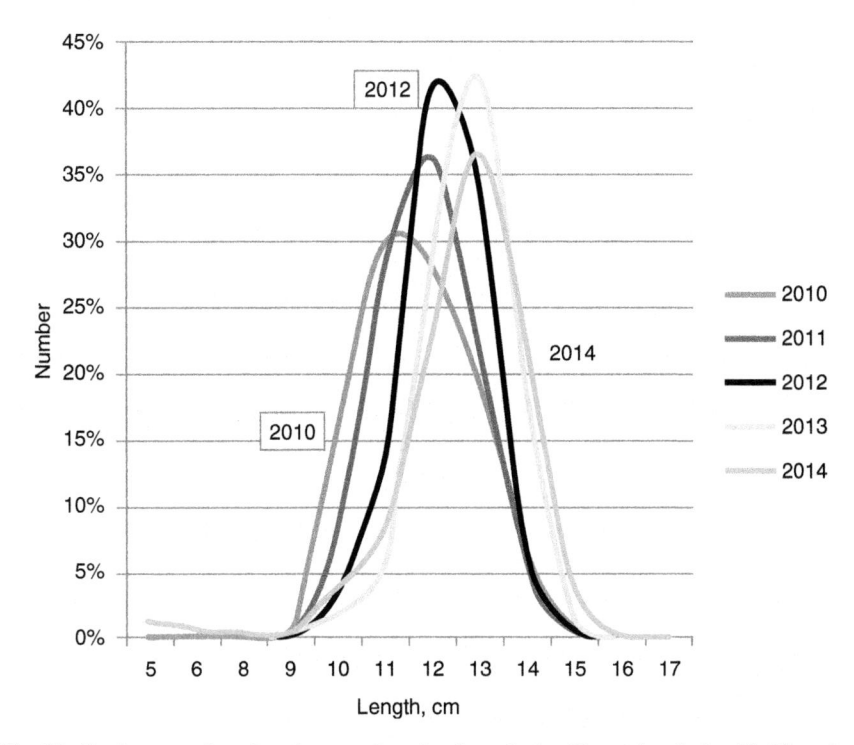

Fig. 18 Catch curve of vendace in control catches by pelagic gill nets (mesh size 10–22 mm)

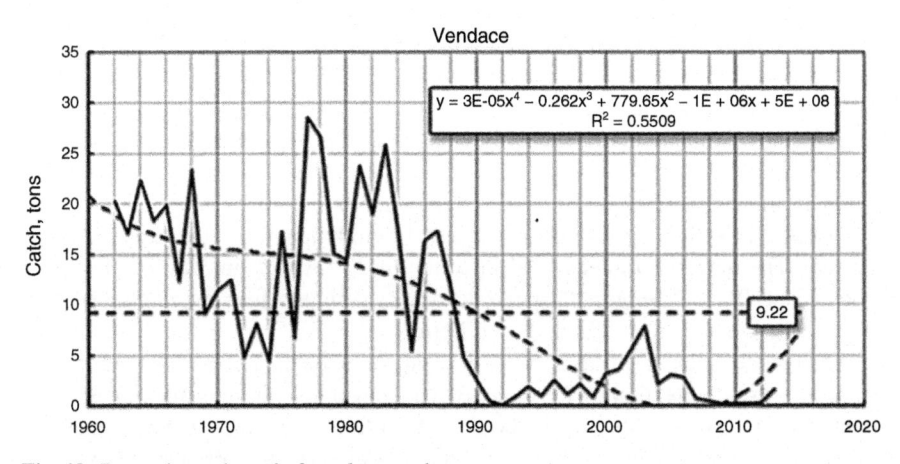

Fig. 19 Dynamics and trend of vendace catches

Taking into account the concept of precautionary approach, the amount of possible catch can be calculated on the basis of the catch trend (Fig. 19). Dynamics of vendace catches is well described by a polynomial of the fourth degree if the correlation coefficient $R = 0.58$, which is acceptable for biological research. Taking into account the observed trends, the TAC for vendace can now be taken at a rate of 12.0 m, which is a little less than the average annual catch level for the period of effective fishing.

10 Parasite Fauna of European Vendace

For several decades starting in 1962 (the year of the start of official fishing), vendace is the main object fishery in Lake Vistytis, providing more than 60% of the catch at the maximum catch of up to 27 tons. The launch of fisheries in the 1990s led to a change in the structure of the population, including a significant increase in the proportion of younger size groups of vendace, which in its turn may be due to an increase in natural mortality of vendace of older age groups. One reason for this may be parasites.

The first studies of parasites fauna of vendace in Lake Vistytis were conducted in the early 1980s and resumed in 2009–2010 [27].

In total, eight kinds of parasites were found (Table 4). Infection by most types was low, except for two. So in 2009 and 2010, of frequent occurrence were high incidence and intensity of invasion by *Proteocephalus exiguus* cestode and *Ergasilus briani* parasitic crustaceans.

Pr. exiguus affects intestines of vendace. It is detected in summer. There are both grown parasites with a well-formed strobila and young ones in the intestine. This is an evidence of constant invasion of the vendace in the lake. Infection occurs by eating first intermediate hosts of the parasite – *Cyclops*, *Eucyclops*, *Mesocyclops* plankton crustacea, and others. Vendace infection in the lake is very extensive. If in the 1980s of the last century it accounts for 64.2%, then in 2010, it reached 100%. Intensity of infection has increased as well with 4–33 species per host in the early years of the study, up to almost 200 species in 2010. High intensity and extent of vendace infestation by a tapeworm indicates the predominance of the copepodid group in the zooplankton of the lake, which is the basis of its food.

Vendace infection begins very early; fish 10 cm in length is 100% infected by a parasite. The maximum intensity of infection by *Pr. exiguus* in May 2010 was for the vendace 10 cm in length, 85 species of a parasite in the host; for 12 cm vendace, 113; for 13 cm vendace, 179; and for 14 cm, 164.

Abundance index in spring 2010 shows that active feeding of the host promotes a sharp increase in infection by *Pr. exiguus*. The longer a vendace is, the higher is the abundance index. At the end of the summer (August) at high extensiveness of infestation, abundance index reduces but remains directly proportional to the length of the host body (Fig. 20).

The parasite is highly toxic and has a significant mechanical impact on the host intestine causing its obstruction, cachexia, and mucositis. The decline in catches of vendace longer than 15 cm provides evidence of either its death because of a parasite or a sharp slowdown in growth under its influence.

Table 4 Parasite fauna of vendace in Lake Vistytis

Parasite	Prevalence (%)	Intensity (ind.)	Abundance	Prevalence (%)	Intensity (ind.)	Abundance	Prevalence (%)	Intensity (ind.)	Abundance
Year	1982			2009			2010		
Myxobolus evdokimovae	21.5			–			–		
Henneguya zschokkei	–			20.0			10.9		
Trichodina sp.	4.3			–			–		
Proteocephalus exiguous	64.2	4–33	13.70	73.0	2–31	9.78	99.5	1–>200	30.71
Piscicola geometra	12.8	1–4	0.52	–	–	–	–	–	–
Glochidia gen. sp.	4.3	2	0.08	–	–	–	–	–	–
Ergasilus briani	43.0	3–14	4.80	100.0	1–76	23.35	97.3	1–173	25.08
Argulus foliaceus	12.9	2–5	0.50	–	–	–	–	–	–

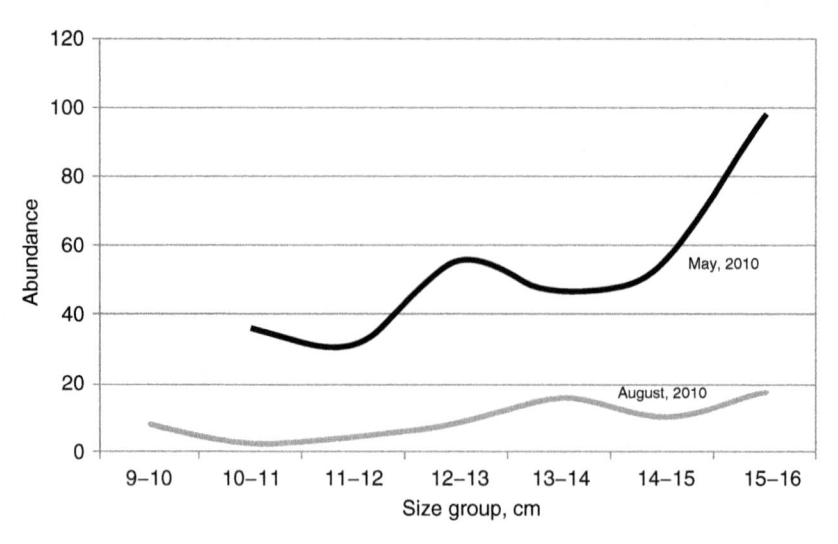

Fig. 20 Abundance dynamics of *Proteocephalus exiguus* with size of vendace

E. briani – a parasite copepod, is parasitizing on several species of fish in Lake Vistytis (perch, ruff, pike, and others) but particularly infectible in vendace. It is located on gills and causes necrosis and destruction of gill tissue. During the 1980s in the last century, infection of vendace with *Ergasilus* in the lake was 43.0% at the intensity of 3–14 species of the parasite in the host. In 2009, both values rose to 100% contamination and 76 species in the host, and in 2010, it reached 90 species of the parasite in the host. Infection of vendace with *Ergasilus* occurs in the exposed parts of the lake, where it was mainly found and where nonsymbiotic parasites live. *E. briani* are present on 10 cm fish, and then infection intensity increases together with its growing. The maximum intensities of infestation by this parasite in May 2010 were as follows: for 10 cm vendace, 24 species of the parasite in the host; for 11 cm, 41; for 12 cm, 51; for 13 cm, 91; and for 14 cm, 60 (Fig. 21).

Ergasilus can cause fish death by eating blood and gill tissue of the host and making toxic effect on it. But more often, they affect the growth and accumulation of weight of the host, reducing it by half. Thus, the analysis of parasite fauna with vendace in Lake Vistytis has showed that two types of parasites can seriously affect its population. These are *Pr. exiguus* tapeworm and *E. briani* parasitic copepod.

11 Whitefish

Whitefish is an important target species. It belongs to fish with an average life cycle. It reaches 60 cm in length, 2.3–2.5 kg in weight, and the age of 15–20 years. Whitefish in Lake Vistytis grow at an average growth rate. By the age of 5, it reaches an average of 32–34 cm and a weight of about 500 (Fig. 22).

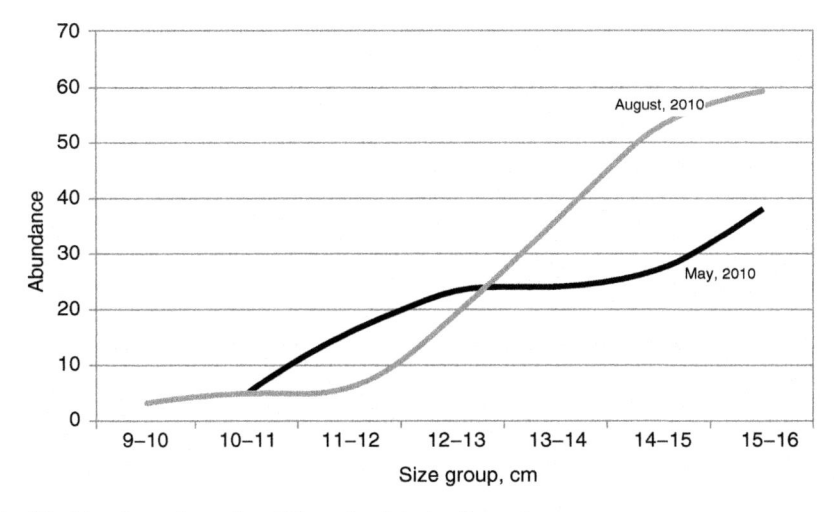

Fig. 21 Abundance dynamics of *Ergasilus briani* with vendace

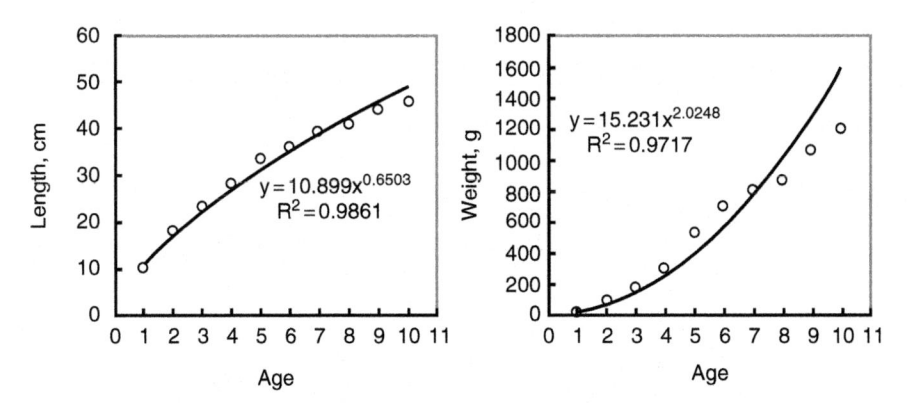

Fig. 22 Curves of linear and weight growth of whitefish in Lake Vistytis

It becomes adult at the age of 4 or 5. Spawning takes place in late autumn in the areas with a sandy bottom. The main spawning areas are located near the Lithuanian coast and in the northern part of the lake at a depth of 3–10 m. In 2008, for the first time ever, whitefish spawning areas were found in the Russian part of the lake at a depth of 10 m. It was previously thought that whitefish reproduce only in the Lithuanian part of the lake that gave Lithuania certain advantages.

The whitefish in commercial catches of the 1960s were from 26 to 50 cm in length. The age composition of the catch was represented by the species from 3 to 11 years inclusive. In the control and commercial catches of the late 1970s, age structure of whitefish varied from 1 to 6 years old, from 16.4 to 50.7 cm in length. In catches of the 1980s, whitefish were from 28 to 54 cm in length, with variations in average length from 38.8 to 43.3 cm, and seven age groups from 3 to 10 years. The basis of the catch was 4–6-year-old fish. In the 1990s, the main part of catches was species 30–40 cm in length. In recent years, the dynamics of the age structure of whitefish population is caused by the emergence of a more or less abundant generation and structure of control catches is largely dependent on the nature of spatial and temporal distribution of whitefish during specific years of the study (Fig. 23). So the dynamics of the size structure of whitefish population relative to aging of strong year class is well marked. As can be seen, in 2013, this generation will be out of fishing, which would entail a reduction of catches. At the same time, it is pointed out that a better year class will replace it that was born in 2010 and have 14–17 cm in length.

In general, biological productivity of whitefish population in Lake Vistytis is relatively low. The maximum catches was recorded in 1974 and reached 4.5 tons. During the observation period, large fluctuations in catches were registered, predominantly related to the instability of the commercial environment and activity rates of the fishery (Fig. 24). Given the biological condition of the whitefish population and the lack of high generations, now it seems rational to maintain the value of potential catch of 1.5 tons. That would be close to the annual average catch given the low intensity of the fishery in recent years.

According to the Lithuanian Centre for Fisheries, in autumn, the Lithuanian part performs annual catches of whitefish producers for harvesting and incubation. The larvae are then used for the stocking of the lakes of Lithuania. Catches in recent years are up to 400 species.

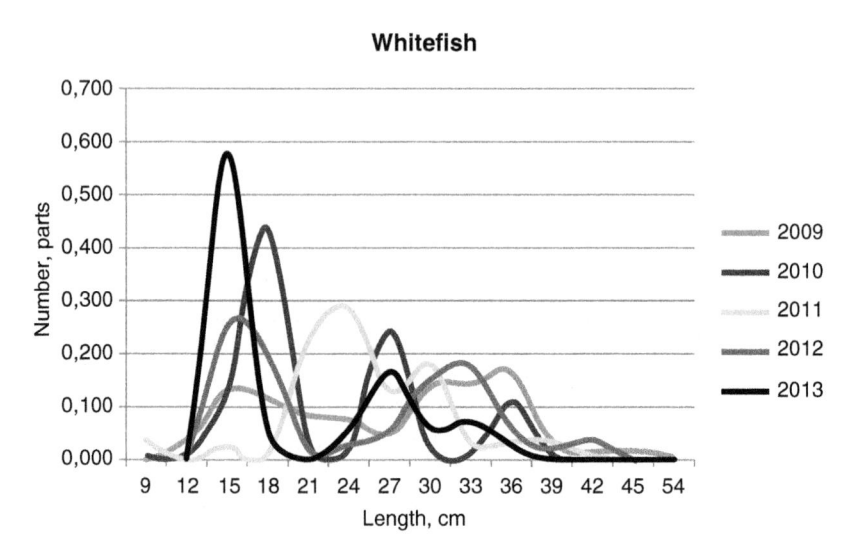

Fig. 23 Whitefish catches by bottom gill nets (mesh size 14–50 mm)

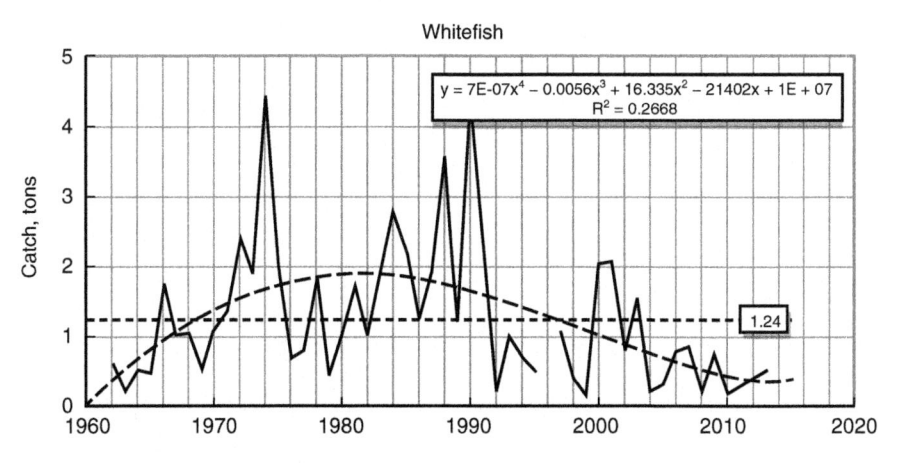

Current Status of the Lake Vistytis in Kaliningrad Region 467

Fig. 24 Dynamics and trend of whitefish catches

12 Roach

Roach is dominant in Lake Vistytis, and in the past, it formed the basis of commercial catches.

In control catches of 2011–2014, roach was registered of up to 32 cm in length, up to 700 g in weight and up to 14 years. Roach become mature at the age of 3–4, with a body length of 12–14 cm. The growth of roach corresponds to the average for the region.

The length of roach in the commercial catches of the 1960s ranged from 14 to 32 cm and from 110 to 450 g in weight. The basis of the catch was species 18–26 cm in length at the age of 6–8 years. In the 1970s, roach caught by fixed nets (40 mm mesh size) was longer, from 22.3 to 30.5 cm, and more weighty – from 245 to 633 g. Species were from 8 to 15 years old rather fat and well fed. The size composition of roach in commercial catches in the 1980s was represented by fish ranging in length from 12 to 30 cm, 70–400 g in weight, aged from 3 to 11 years. The average length was 18–20 cm. In the 1990s, under relative stability of the average weight and length of roach significant fluctuations in the size structure of the catches were observed (Table 5). The latter was due to the variability of methods for observing population, given that in some years, the material was taken either from filtering or enmeshing fishing gear.

In recent years, resulting from the use of a standardized approach to evaluation of biological parameters of roach population in Lake Vistytis, we can talk about its present state. The findings suggest the relative stability of the roach population in growth and also about certain dynamics of the size structure of the population. The last describes the population as highly variable in terms of spatial and temporal distribution of size group population. This is related both to the advent of strong year classes and to migration of certain size groups to a fishing area which causes a change in their relative abundance (Fig. 25).

Table 5 Length structure of roach in commercial catches, % (bottom gill nets, mesh size 24–40 mm)

Length (cm)	Years 1999	2000	2002	2003	2005	2006	2007	Average
<12		0.43				8.80	12.62	3.1
12–13		0.85				4.51	0.96	0.9
13–14		0.57				5.15	0.92	0.9
14–15		0.28				4.72	2.58	1.1
15–16	0.13	1.71				4.94	3.73	1.5
16–17	0.38	1.42				2.79	6.38	1.6
17–18	0.89	0.57				4.51	6.38	2.5
18–19	0.89	0.43				9.87	8.78	3.3
19–20	2.28	1.57			6.67	7.94	12.91	6.0
20–21	4.43	2.42			6.67	12.88	14.68	6.2
21–22	6.33	4.13			6.67	8.80	11.03	7.6
22–23	12.28	6.27			26.67	10.30	9.89	10.0
23–24	15.82	15.53			6.67	6.22	4.43	13.4
24–25	12.15	19.66			6.67	4.51	2.43	7.2
25–26	14.43	16.67				2.15	1.51	7.1
26–27	11.27	12.39			13.33	1.07	0.30	10.8
27–28	9.11	6.41			20.00	0.64	0.26	9.4
28–29	3.92	3.99				0.21	0.15	4.8
29–30	2.91	1.71			6.67		0.04	1.8
30–31	0.76	1.42					0.04	0.3
31–32	1.14	0.14						0.2
32–33	0.76	1.42						0.3
33–34	0.13							0.0
Total	100.0	100.0			100.0	100.0	100.0	100.0
Average (cm)	24.6	24.4			24.3	18.9	18.9	23.0

The maximum catch of roach during observation period reached 16.4 m in 1968 and was stipulated by more or less well-organized common freshwater fishing with beach seine. Since then, the harvest level of roach has decreased monotonically without any connection with the biological condition of its population (Fig. 26). In our times, the value of a possible catch of roach can be at the level of annual average values for the period of 1960–1980s – 5.8 tons.

13 Perch

Perch is a common fish in the coastal zone of the lake and the main subject of amateur fishing. It lives in the coastal zone of Lake Vistytis with depths up to 20 m, but the main concentration forms at the depths of 10–20 m. Species up to 36 cm in length, 1.3 kg in weight, and at the age of 13 years have been registered in catches.

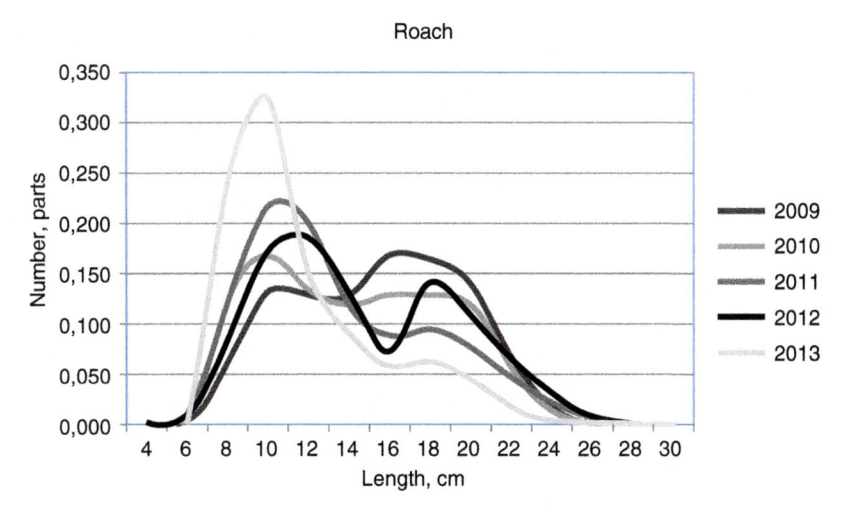

Fig. 25 Catch curves of roach in control catches by bottom gill nets (mesh size 14–50 mm)

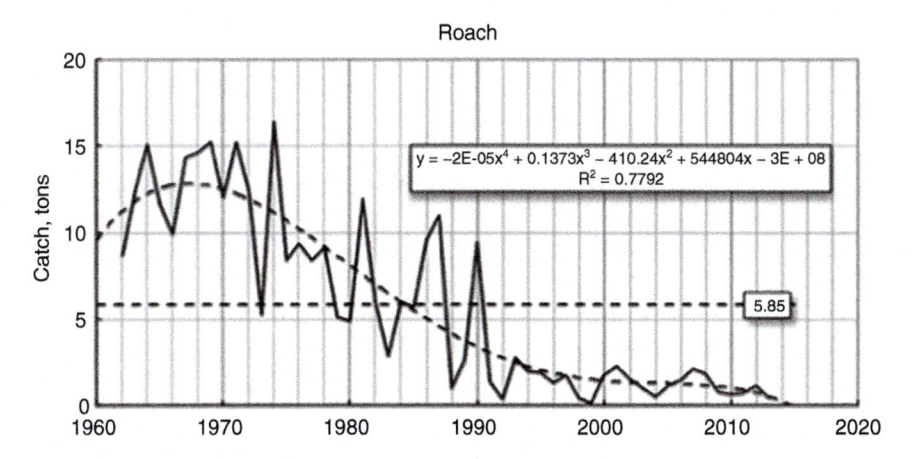

Fig. 26 Dynamics and trend of roach catches

Females become mature at the age of 3 or 4 years with a body length of about 15–17 cm, while males sometimes become mature at 2 years when the body is 10–12 cm long. The growth rate is not high and has a distinctive leap around the age of 4–5 years. Apparently, this is due to the transition to a carnivorous diet.

In commercial catches of the 1960s, perch was represented with a body length of 16–36 cm, with a predominance of size groups 19–23 and 23–27 cm, at the age from 3 to 12 years. In the catches of the 1970s, perch ranged from 12.5 to 30.5 cm with a weight of 29–435 g and aged from 3 to 11 years. Catches of 1987–1988 was characterized by 5–6-year-old perch, with body length ranging from 20.3 to 38 cm and age from 3 to 14 years. Latest figures on perch are from 1999 to 2002. In this period, the length ranged from 10 to 35 cm; most species were from 20 to 26 cm (Table 6), although in 2002, due to the nature of raw material collecting, the structure of the catch was somewhat different.

In the previous 2 years, the size structure of perch population was relatively stable (Figs. 27 and 28).

Due to high nutritional value of perch, especially in Western European countries, it is reasonable to organize a specialized commercial fishing using gill nets and seine. However, the perch may have a value as an object of amateur fishing. Analysis of density of fish population and catch per unit of fishing effort shows the possibility of a significant increase in catches of perch at a level above the average annual value in the period of active fishing – within 6 tons (Fig. 29).

Table 6 Length structure of perch in commercial catches, % (bottom gill nets, mesh size 24–40 mm)

Length (cm)	Years								Average
	1998	1999	2000	2002	2003	2005	2006	2007	
<12			3.81				24.34	28.81	8.1
12–13			6.98				2.92	4.00	2.0
13–14			5.40				3.40	5.36	2.0
14–15			2.54				3.87	5.36	1.7
15–16			3.49				3.68	5.14	1.8
16–17			0.95				5.28	5.31	1.6
17–18		3.24	0.95			6.25	4.72	5.04	2.9
18–19		4.86	0.95			12.50	3.77	5.07	3.9
19–20		6.49	0.63			3.13	5.75	6.58	3.2
20–21		9.73	1.27			4.69	8.21	6.93	4.4
21–22		15.68	3.17			15.63	7.08	6.97	6.9
22–23		15.14	7.30			6.25	8.02	5.59	6.0
23–24		12.97	20.63			18.75	8.49	3.25	10.0
24–25		14.59	21.90			6.25	6.13	2.63	10.0
25–26		7.03	7.94			9.38	2.17	2.18	9.5
26–27		3.24	7.30			4.69	1.04	0.94	4.2
27–28	38.46	1.08	1.90				0.47	0.67	7.0
28–29	23.08	2.16	2.22			4.69	0.19	0.07	5.5
29–30	30.77	2.70	0.63			4.69	0.19	0.02	5.6
30–31	7.69	0.54						0.05	1.2
31–32							0.19	0.03	1.8
32–33		0.54				3.13			0.5
33–34							0.09		0.0
Total	100.0	100.0				100.0	100.0	100.0	100.0
Average (cm)	28.6	22.8	21.7			23.0	18.1	16.9	22.5

Perch

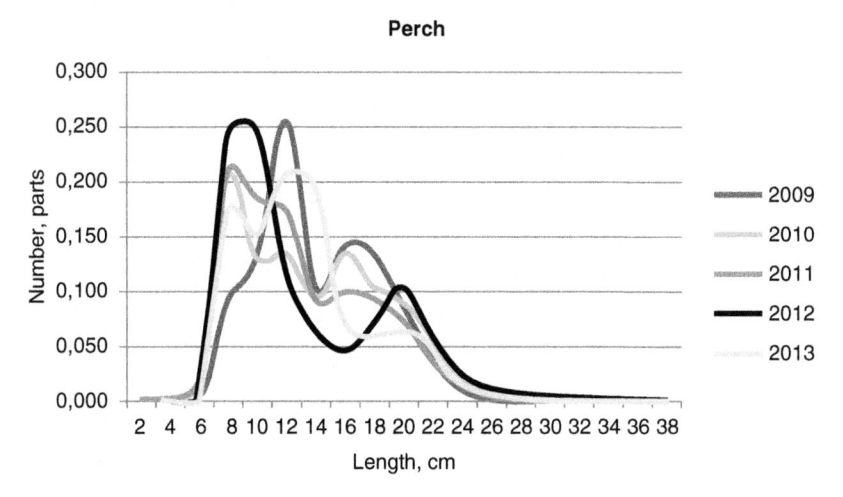

Fig. 27 Catch curve of perch in control catches by bottom gill nets (mesh size 14–50 mm)

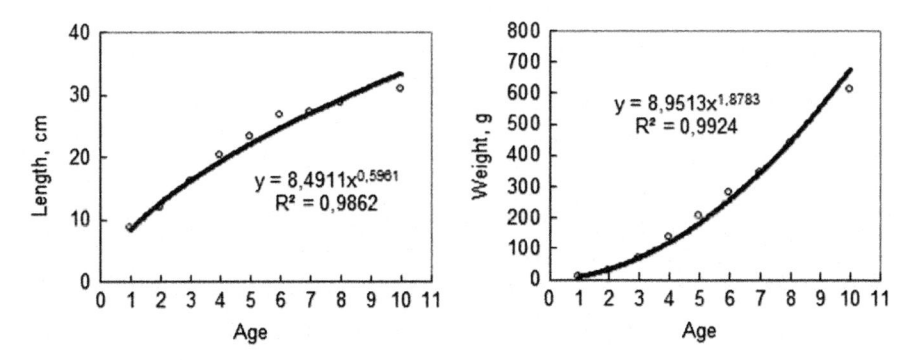

Fig. 28 Growth curves of perch in Lake Vistytis

14 Pike

Target fishing is not carried out as its population is rather small.

In order to save its population target, fishing for pike has been prohibited since 1984. At the same time, pike is regularly found as accidental fish in fishing gears. Feeding areas in spring and summer are coastal zones up to a depth of 15 m.

Pike becomes mature when it is 35–40 cm long and at the age of 3–5 years. Pike live till 25 years or longer. The maximum length is up to 1.5 m and weigh more than 20 kg.

The best living environment for pike is in Tikhaya Bay with vast reed, bulrush, and cattail bed and well-developed submersed aquatic.

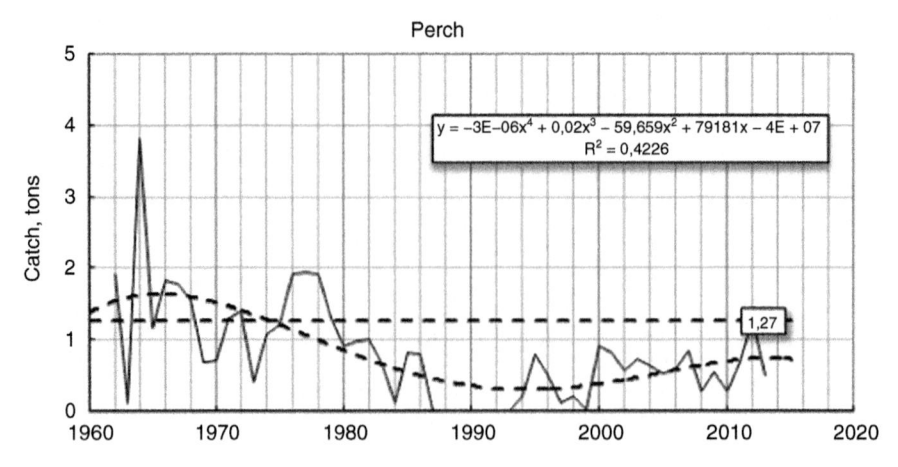

Fig. 29 Dynamics and trend of perch catches

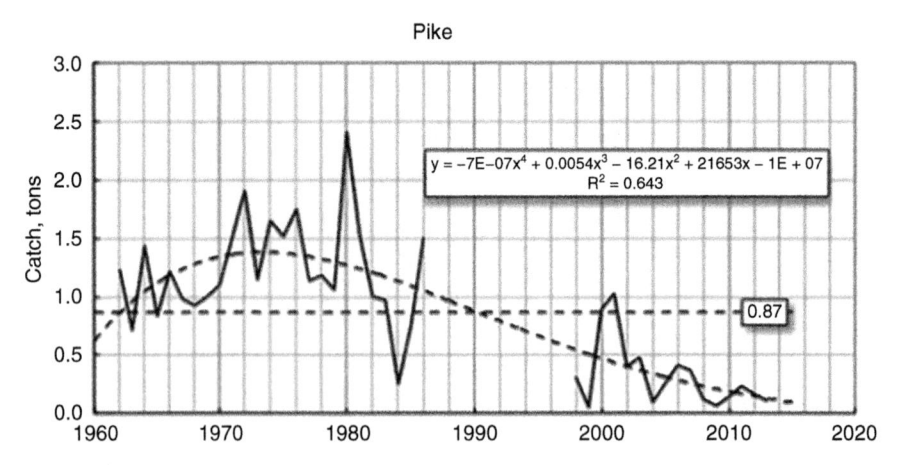

Fig. 30 Dynamics and trend of pike catches

The maximum pike catch was registered in 1980 and amounted to 2.4 tons. Pike is marked by the catch dynamics similar to the other fish: high catches in the 1970s, decline in the 1990s, and a new wave in 2000 (Fig. 30). This tendency is stipulated by better organization of fishing. Considering the fact that being a biological meliorator, pike is an indispensable part of ichthyofauna as well as taking into account its accessibility to fisheries, at present, it is reasonable to set lower catch value of 0.5 tons for this species.

15 Burbot

Burbot is a common predator in the aphytal zone of Lake Vistytis.

It is a bottom-dwelling fish and is the most active in breeding season in winter, late autumn, and early spring.

Burbot becomes mature at the age of 3–4 years. In a water body, burbot grows at a low rate (Fig. 31). By the age of 6, it does not weigh 1 kg, although there have been registered species 112 cm in length and weighing up to 12 kg; however, the most frequent are small-scale burbot no longer than 20 cm.

In control catches, length of burbot varies from 17 to 57 cm, whereas the modal group is 25–43 cm.

In catches by the complex of bottom gill nets with different mesh size, length composition of catches of burbot is characterized by random fluctuations in the older age groups enmeshing at random and high stability of the size group of 17–30 cm, that is caught by the classical scheme (Fig. 32). Attention should be drawn to the fact that catch curves are practically identical for all years, indicating a stable population of burbot in Lake Vistytis.

Target fishing for burbot in Lake Vistytis has not been carried out yet. It is mainly caught as accidental fish in fishing for whitefish and roach.

Catches of burbot are of high variability. The maximum catch was recorded in 1983 and amounted to 2 tons. After "collapse" of fishing in the 1990s, catches reduced. The average annual catch is 0.31 tons (Fig. 33), and it can be the size of regulatory management of the stocks.

16 Tench

Tench is thermophile, for this reason, it avoids open areas of the lake and lives in bays. Tench has been quite rare recently because of its capturing by underwater hunters. Main habitat is Tikhaya Bay. Tench becomes mature in the fourth year of

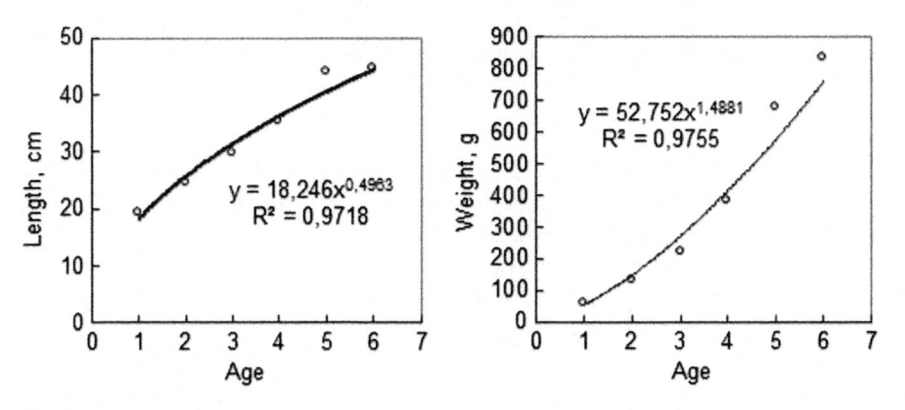

Fig. 31 Curves of linear and weight growth of burbot in Lake Vistytis

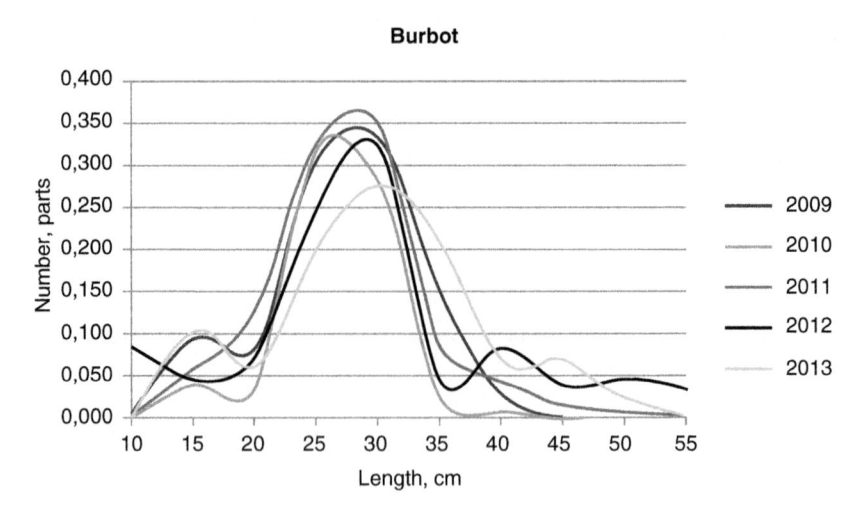

Fig. 32 Catch curves of burbot in control catches (mesh size 12–60 mm)

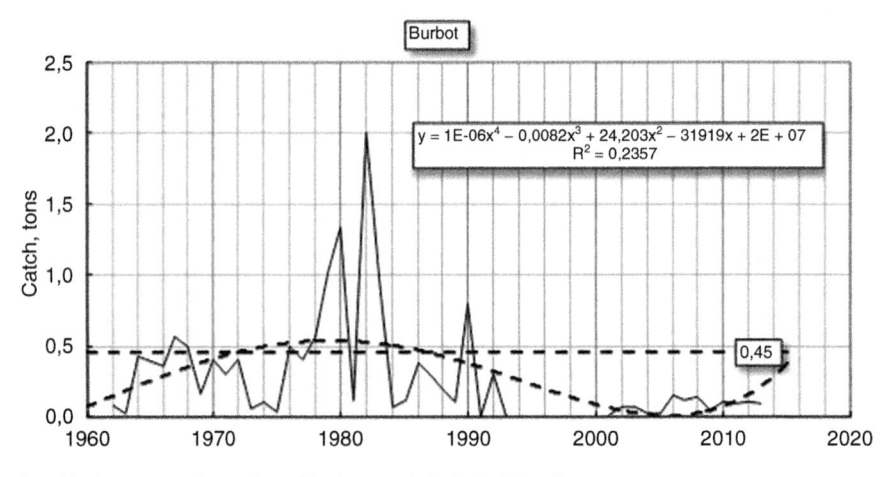

Fig. 33 Long-term dynamics of burbot catch in Lake Vistytis

life with a body length of about 17 cm. The catch is usually represented with the tench 13–45 cm in length, weighing from 44 to 1,750 g. There is tench 56 cm long and weighing up to 3 kg in experimental catches.

The maximum catch of tench was 0.78 tons in 1964 but was generally much lower (Fig. 34). Taking into account the long-term average values, nowadays fishing for tench is hardly reasonable in values exceeding 0.1 tons.

Tench

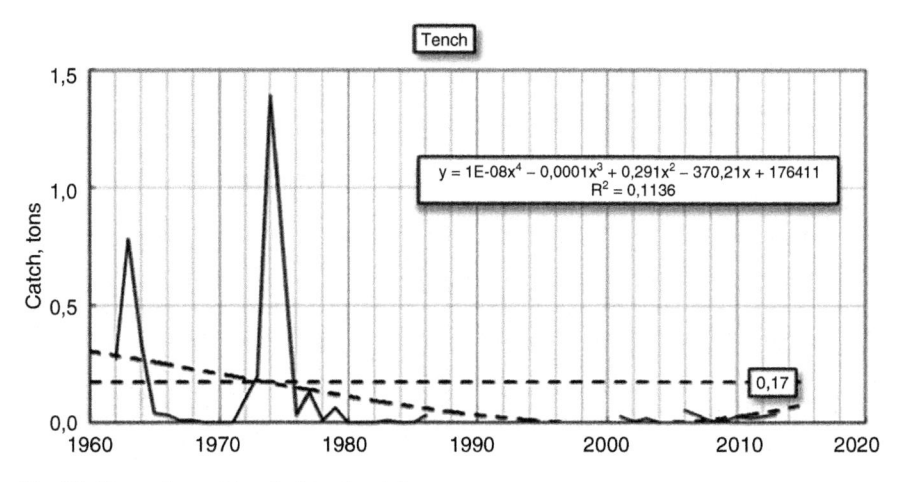

$y = 1E\text{-}08x^4 - 0{,}0001x^3 + 0{,}291x^2 - 370{,}21x + 176411$
$R^2 = 0{,}1136$

0,17

Fig. 34 Dynamics and trend of tench catches

17 Bream

In the period before the 1980s, bream was a target fish and was caught as accidental fish in seine and net fishing. Only in 1958, when a special seining was organized, its catch was 1.75 tons. In the following years, solely because of the low fishing effort, catches of bream fell and did not reflect potential productivity of the species (Fig. 35). In addition to this, in 1990, bream was almost completely out of the catch statistics because of the lack of quotas. In occasional years, e.g., in 2005, bream played a significant role in shaping species composition of commercial catches. Thus, in nets with 45 mm mesh size, it ensured up to 90% of catches. Marked signs of eutrophication indicate that environmental conditions of water may change over a period of time in the way that will be more favorable for bream. This necessitates bream stocks management based on the assessment of possible catch.

At present as the size of possible catch of bream, it is proposed to establish the value slightly higher than long-term average – 0.5 tons.

18 Eel

Eel is one of the valuable species found in Lake Vistytis.

It is a migratory fish performing catadromous migrations. Eel can live in the lake for a long time from 9 to 15 years. Its feeding grounds are almost the entire water area of the lake, but it prefers places with muddy bottom and aquatic vegetation, such as Tikhaya Bay and specific coastal areas.

Bream

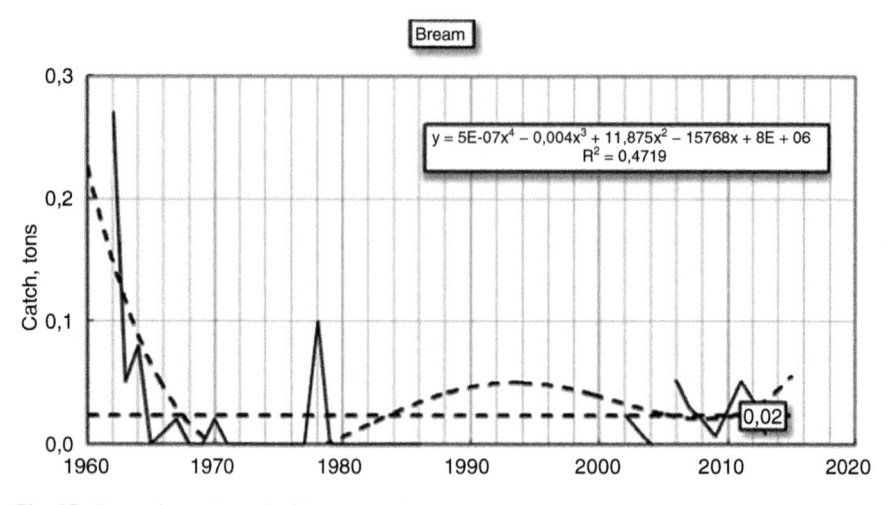

$y = 5\text{E-}07x^4 - 0{,}004x^3 + 11{,}875x^2 - 15768x + 8\text{E} + 06$
$R^2 = 0{,}4719$

Fig. 35 Dynamics and trend of bream catches

In commercial catches of the 1960s, eel was represented with eight age groups from 6 to 13 years, with the size of 86–105 cm and weighing from 1.3 to 1.85 kg. Its growth rate in freshwater is relatively low. The annual growth is about 4 cm. In catches of 1986, eel was from 2 to 10 years, with a length of 51–97 cm and a weight of 273–1,700 g, and in 1987, eel was represented with age groups only from 3 to 7 years, 47–68 cm in length, and weighing 110–480 g. In 2000, eel was from 40 to 80 cm in length, mostly from 46 to 66 cm.

In the 1970s and 1980s, Lake Vistytis was stocked with glass eel, which made catches much higher. Lake Vistytis is a favorable habitat for eel; however, regulation of Masurian Lakes in Poland and the Pregolya and the Pissa in the Kaliningrad Region has led to decrease in the level of replenishment of stocks and a drop in catches. Artificial stocking can effectively solve the problem of maintaining stable exploited population of eel. In addition, by providing high-quality food products, eel is a valuable biological meliorator. High fish-holding density of eel could make ruff population significantly thinner, although its population in the lake is big.

The maximum catch of eel amounted to 2 tons in 1963. In the following years, catches were decreasing to the end of the 1970s, when a new artificially "stocked" generation has appeared in fisheries. In recent years, there has been some increase in eel catch, which is probably due to more efficient fishing. Given the trends observed (Fig. 36), possible eel catch is now recommended to be set at 0.5 tons.

The Department of Ichthyology and Ecology began to attract international funds for development and implementation of the program of stocking Lake Vistytis with glass eel larvae. Together with AtlantNIRO and Zapbaltrybvod plant, a Russian program is designed to restore eel stocks in transboundary water bodies, which is presented to the Second Session of the Joint Commission on the Baltic Sea of Russia and the EU. The program is currently under consideration in DG MARE

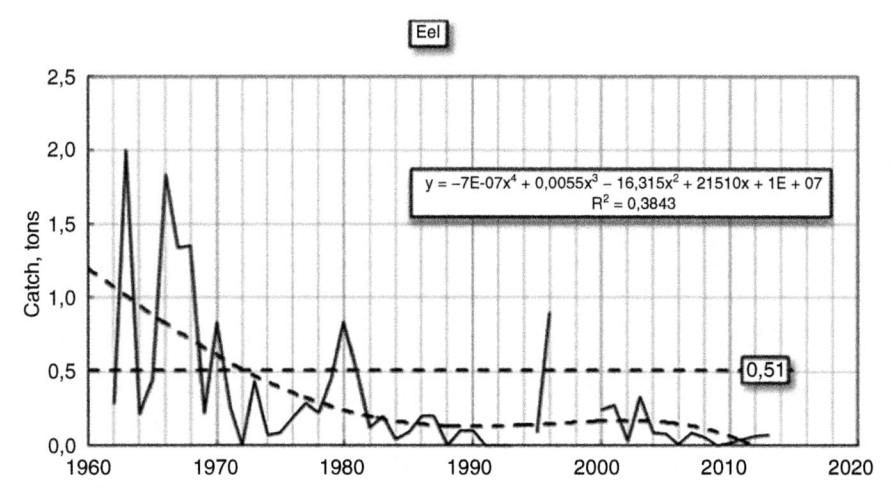

Fig. 36 Dynamics and trend of eel catches

(Fisheries Directorate of the EU), and it is supposed to issue a decision on Russian access to the procurement of vitreous eel for the purpose of stocking.

A special map is designed to show possible migration paths of eel in the Baltic Sea in the event of artificial stocking of inner waters of Lithuania, Poland, and the Kaliningrad Region with glass eel. It has been found out that Lake Vistytis can be the most advantageous for stocking due to the following reasons:

1. There is a unique habitat for eel in the lake.
2. The only migration path of eel to the Baltic Sea is the Pissa, where in Soviet times, an eel trap was set allowing to register all the eel migrating down. Using it today enables solving two problems – strict regulations of commercial catch and evaluation of commercial return of eel which is impossible in other waters.
3. Eel can easily migrate from Lake Vistytis to the Vistula Lagoon through the Pissa and the Pregolya, gain weight there for some time, or continue migrating to the Baltic Sea.

In 2013, Poland initiated the development of cross-border (Russia-Poland) plan for replenishment of eel stocks in the Prokhladnaya River. The first version of the plan was considered and approved at the 13th session of the Russian-Polish Joint Commission for Fisheries (Kaliningrad, December 2013). Now the plan is being refined, and afterward, it will be submitted for approval to the Ministries of Agriculture of Russia and Poland. If it is approved, the Polish side is going to submit it to the European Commission for funding. In this case, Russia can receive access to the purchase of glass eel larvae, which export is currently prohibited by CITES and start implementing its program.

19 Conclusions

Lake Vistytis is a unique cross-border water body in terms of origin, hydrological features, and productivity of all trophic levels and fish fauna composition.

It is a deepwater body of oligotrophic type with clean and clear water.

In terms of both hydrochemical and hydrobiological characteristics, the trophic status of Lake Vistytis is stable.

However, in some areas of the lake, mainly belonging to Vistytis settlement and recreation centers on both sides of the border, signs of mesotrophy and in some cases eutrophy are present. Nowadays, they are still not distinct, but monitoring of the observed trend is required.

Twenty-three fish species are registered within the ichthyofauna of the lake including Coregonidae (whitefish and vendace) absent in most of other freshwater reservoirs of the region.

When analyzing distribution of the main fish species according to the depths, their belongingness to different habitats was clearly defined. Roach lives mainly in the coastal zone of the lake with depths up to 5 m. Perch occupies a vast water area, from the coastal waters up to a depth of 20 m, with a few large concentrations of perch registered at the depths of 10–20 m. Whitefish live mainly in the deep central part of the lake. Whitefish prefer depths of 10–30 m. Vendace inhabits pelagic zone of the lake at depths greater than 20 m.

Commercial fishery importance is determined by both the existing traditional fishing and well-developed amateur fishing. The lake is also directly used for recreational purposes.

In Soviet times, catches in Lake Vistytis were about 30 tons over a long period. The basis of the catch was vendace and roach. After the 1990s, fishing intensity decreased significantly, and during the past 15 years, catches are at the level of 6 tons. The fishery is currently conducted mainly in Russian territorial waters. The basis of the species composition of catches is as before vendace and roach.

Ichthyological activities of recent years do not allow drawing the conclusion that commercial fishery importance of Lake Vistytis has decreased. Therefore, reduction in catches after the 1980s is not subject to ecological and biological factors but rather to the general deterioration of socioeconomic situation in the country and is generally typical for the Russian fishing industry, as well as the acquisition of the status of transboundary lake. However, as far as certain fish species (pike, tench) are concerned, one should speak of quite strained condition of their stocks.

Due to a unique character of Lake Vistytis, activities undertaken on it must be precautionary. In view of this, conservation actions should have priority over recreational trend. It appears that the current level of recreational loading should be fixed and does not increase thereafter.

Sustainable use of fish stocks in Lake Vistytis should be based on the following principles:

- Application of precautionary approach to management of living aquatic resources in the unique lake and eliminating any fishery changes (e.g., artificial reproduction) without critical justification for it
- Preservation of traditional character of the existing commercial fishery with a primary focus on the fishing of Coregonidae
- Development of managed recreational and sport fishery, based on conservation of the coastal fish (roach, perch, pike, and others) without expanding the existing infrastructure (recreation centers and camping) to prevent water pollution
- Coordination with Lithuania of mutually acceptable measures for management and rational use of water bioresources of the lake, excluding unilateral actions aimed at fishery transformation of the lake and the impact on aquatic resources, including stocking, reproduction, acclimatization, and removal of aquatic biological resources for these purposes

References

1. Bernikova TA (1999) Lakes. The Kaliningrad Region: essays about nature. Yantarnyi skaz, Kaliningrad, pp 84–91
2. Bernikova TA (2013) Evaluation of the trophic status of a water body according to permanganate value (as exemplified by Lake Vistytis of the Kaliningrad Region). In: Bernikova TA, Nagornova NN, Tsupikova NA (eds) Vestnik of PFUR, vol 3. PFUR, Moscow, pp 12–23
3. Orlenok VV, Barinova GM, Kucheryavyi PP, Ulyashev GL (eds) (2001) Lake Vistytis: nature, history, ecology. KGU, Kaliningrad, p 212
4. Geographical atlas of the Kaliningrad Region (2002) Kaliningrad
5. Bernikova TA (2013) In: Bernikova TA, Nagornova NN, Tsupikova NA (eds) Long-term changes of organic content of water as an indicator of the trophic status of Lake Vistytis, vol 28. Izvestiya KGTU, Kaliningrad, pp 70–78
6. Alekseev NK, Demidova AG, Bernikova TA et al (1976) Lake Vistytis. Kaliningrad Publishing House, Kaliningrad, p 47
7. Tylik KV, Shibaev SV (2008) Lake Vistytis. IE Mishutkina, Kaliningrad, p 144
8. Baranov IV (1962) Limnological types of the lakes of USSR. Gidrometeoizdat, Leningrad, p 276
9. Алекин О. А. Основы гидрохимии. Л.: Гидрометеоиздат, 1970. 443 с (Alekin OA (1970) Basics of hydrochemistry. Gidrometeoizdat, Leningrad, p 443)
10. Mordukhai-Boltovskaya ED (1975) Some features of Vistytis lakes of the Kaliningrad Region. In: Mordukhai-Boltovskaya ED, Bernikova TA, Demidova AG (eds) Basics of biological productivity of inland water of the Baltic states: proceedings of XVII scientific conference on inland water of the Baltic states, Vilnius, pp 155–158
11. Mordukhai-Boltovskaya ED, Ivanov PI, Mashinets IP (1971) Zooplankton and zoobenthos of Lake Vistytis. In: Biology of fish and aquatic invertebrates of seas and inland water bodies: proceedings of Kaliningrad Technical Institute for Fisheries, vol 26. Kaliningrad, pp 48–56
12. Masyutkina EA (2014) Evaluation of ecological condition of Lake Vistytis with application of different hydrobiological indexes, vol 7. Vestnik of Immanuel Kant Baltic Federal University, Kaliningrad, pp 66–76

13. Shibaeva MN (2011) In: Shibaeva MN, Masyutkina EA, Matveeva EP (eds) Zoobenthos diversity, bioindication and ecological condition of the lakes of the Kaliningrad Region, vol 7. Vestnik of I. Kant BFU, Kaliningrad, pp 91–96
14. Shibaeva MN (2008) Zoobenthos. In: Tylik KV, Shibaev SV (eds) Lake Vistytis. IE Mishutkina, Kaliningrad, pp 50–55
15. Shibaeva MN (2008) Zoological plankton. In: Tylik KV, Shibaev SV (eds) Lake Vistytis. IE Mishutkina, Kaliningrad, pp 44–50
16. Shcherbina GK, Shilova AI (1983) On the fauna of chironomids in some lake of the Kaliningrad Region (Diptera, Chironomidae). In: Biology of inland water: newsletter, vol 59. Leningrad, pp 26–31
17. Shcherbina GK (1993) Role of Dreissena Polymorpha in bottom communities of Lake Vistytis. Zoocenosis of the upper Volga under human intervention. Gidrometeoizdat, Saint Petersburg, pp 145–159
18. Kitaev SP (2007) Basics of limnology for aquatic biologists and ichthyologists. Karelian Research Center of Russian Academy of Science, Petrozavodsk, p 395
19. Alekseev NK (1972) Biogeographical profile of Lake Vistytis. In: Biology of the commercial fish in inland water of the Kaliningrad Region, vol 40. Kaliningrad, pp 54–56
20. Alekseev NK, Probatov AN (1969) Zoogeographical profile of the freshwater ichthyofauna of the Kaliningrad Region. Technical regulations, vol XXIY. KTI, Kaliningrad, pp 7–17
21. Shibaev SV, Khlopnikov MM, Sokolov AV et al (2008) Fishery cadaster of cross-border water bodies in Russia (the Kaliningrad Region) and Lithuania. IE Mishutkina, Kaliningrad, p 200
22. Probatov AN (1969) Fish of the Kaliningrad Region. In: Essays about nature. Kaliningrad, pp 166–170
23. Shibaev SV (2002) Theoretical basics of systematic approach in fisheries research and information support of water bioresources management of inland water bodies: extended abstract of Dr. Sci. (Biology). Dissertation, Kaliningrad, p 42
24. Shibaev SV, Sokolov AV (2013) Analysis method of ichthyocenosis of the small lakes in the Kaliningrad Region based on control catches with nets. In: Proceedings of VNIRO, vol 151
25. Degtev AI, Ivanter DE (2002) Computer-assisted system for quantitative assessment of fish stocks using hydroacoustic method ASKOR-2. Fishery 4
26. Tylik KV (1988) Biological features of vendace in Lake Vistytis. In: All-Union meeting on salmonoids, Tolyatti, pp 15–16
27. Zaostrovtseva SK (2009) Parasite fauna of vendace (Coregonus albula (L.)) in Lake Vistytis. In: Zaostrovtseva SK, Evdokimova EB (eds) Proceedings of 7th anniversary scientific conference "innovations in science and education", part 1, KSTU, Kaliningrad, pp 57–59

Specially Protected Natural Areas of the Kaliningrad Region

I.I. Volkova, T.V. Shaplygina, and E.S. Bubnova

Abstract The main categories of specially protected natural areas introduced by Russian environmental legislation are shown in this chapter. The historical prerequisites, the current state, and the prospects for the development of the Kaliningrad Region conservation areas are considered.

Keywords Natural monument, Natural park, Partial nature reserve, Protected area

Contents

1 Introduction 482
2 The National Park "Curonian Spit" 486
3 The Natural Park "Vishtynetsky" 486
4 State Partial Nature Complex (Landscape) Reserves "Dyunny" and "Gromovsky" 490
5 State Partial Nature Geological Reserves 490
6 Natural Monuments 491
7 Conclusions 491
References 494

The original version of this chapter was revised. The erratum to this chapter is available at DOI 10.1007/698_2017_198.

I.I. Volkova (✉) and T.V. Shaplygina
Department of Geography, Nature Management and Spatial Development, Immanuel Kant Baltic Federal University, Kaliningrad, Russia
e-mail: volkova.bfu@yandex.ru; tshaplygina@gmail.com

E.S. Bubnova
Immanuel Kant Baltic Federal University, Kaliningrad, Russia

P.P. Shirshov Institute of Oceanology, Russian Academy of Sciences, 36, Nakhimovsky Pr., Moscow 117997, Russia
e-mail: bubnova.kat@gmail.com

1 Introduction

The main approach to conserve a biological and landscape diversity is to create and develop specially protected natural areas (SPNA). For such territories, a special regime of protection and natural resource use is established, including full or partial, permanent or temporary restriction of use of natural resources, and in some cases – the practice of special methods of nature management or carrying out restoration activities.

The demand to preserve the biodiversity and landscapes in the Kaliningrad Region is based on a number of features: a small area of preserved natural habitats, rapid rates of their transformation and degradation, intensively growing recreational loads, a high concentration of sites of increased environmental risk, and insufficient quantity of conservation areas with their small law protected status.

In the Russian environmental legislation, the following categories are included in specially protected natural areas: state nature reserves, national parks, nature parks, state partial nature reserves, natural monuments, dendrological parks, and botanical gardens [1].

In the hierarchy of protected areas, state nature reserves have the highest status and, accordingly, the strictest legal regime for protection. They belong to specially protected natural areas of federal importance. Within the boundaries of state nature reserves, the environment is preserved in its natural conditions and all economic or other activities are completely prohibited, except for cases stipulated by the federal law "On Specially Protected Natural Territories" [1].

National parks (NP) belong to the protected areas of federal importance too. Within their boundaries two types of zones are distinguished. Zones in which the natural environment is preserved in natural conditions and any activities, which are not allowed by federal law, are prohibited, and zones in which economic and other activities are restricted in order to preserve natural and cultural heritage objects and their use for recreational purposes. Functional zoning of national parks provides the following zones: reserved zones, specially protected zones, recreational zones, zones with protection of cultural heritage sites (historical and cultural monuments) of the nations of the Russian Federation, zones with economic purposes, zones with traditional extensive nature management [1].

Natural parks are specially protected natural areas of regional importance, there are zones of ecological, cultural, or recreational purposes within parks' boundaries. Such zones have prohibitions and restrictions on economic and other activities. Natural parks use various modes of special protection and land management depending on the ecological and recreational value of natural sites. In this regard, natural parks are divided into environmental, recreational, agro-economic, and other functional zones, including zones for the protection of historical and cultural complexes and objects [1].

State partial nature reserves include territories (water areas), holding particular importance for the conservation or remediation of natural regions or their components and maintaining the ecological balance. They can be federally or regionally important. State partial natural reserves can be of a various type, including: complex (landscape), biological (botanical and zoological), paleontological, hydrological (marsh, lake, river, marine), geological [1].

Natural monuments are unique, irretrievable, valuable in ecological, scientific, cultural, and aesthetic means natural sites, as well as objects of natural and artificial origin.

Dendrological parks and botanical gardens are specially protected natural areas created to form special collections of plants in order to preserve the plant diversity. The territory of dendrological parks and botanical gardens can be divided into different functional zones, including: exposition, scientific and experimental, administrative [1].

Natural monuments, dendrological parks, and botanical gardens can be of either federal or regional importance.

The opportunity to develop protected areas in the Kaliningrad Region is quite high and it is possible to guarantee the preservation of a part of landscapes in a natural homeostasis. Through analysis of the region's spatial organization, it is possible to identify key areas, such as the Vishtynets Hills, the Seshupy-Neman interfluve, the Curonian and Vistula Spits [2], the marshy natural site of the Baltic glaciolacustrine plane. The Vishtynets Hills (frontal moraine ridges and water-glacial lakes, the largest of which is Vishtynetskoe) and the Curonian and Vistula Spits (high coastal dunes) are unique natural sites, that represent natural diversity and ecological stabilization in the Southeastern Baltic.

At the present time five out of six categories of SPNA, defined by the federal law "On Specially Protected Natural Territories" are represented in the Kaliningrad Region. They are: a national park, a partial natural reserve, a natural park, a natural monument, and a botanical garden. There are no state nature reserves and dendrological parks on the territory of the region.

It is also possible for the subjects of the Russian Federation to establish other categories of specially protected natural areas of regional and local importance. So, for example, in the structure of SPNA in the Kaliningrad Region a category of wetlands (peatlands) has been introduced.

Historically, the first "reserved" areas appeared in the region in the tenth century AD. Prussians had to protect sacred groves (with rivers, stones, animals) in which all economic activities were prohibited, in particular, timber harvesting. At the beginning of the eighteenth century hunting land needed a protection from poaching. The "Royal Forest" nature reserve at the Curonian Spit was established due to the fact that this place was a site for royal hunting [3].

In 1890, Kaiser Wilhelm declared the Romincka Forest as the hunting area, famous for the spread of red deer.

The beginning of twentieth century was also the beginning of protected areas expansion:

- In 1937 the nature reserve "German Elk Forest" was founded, which included land to the east and southeast from the Curonian Lagoon with a total area of 46,550 ha. The forest districts of the Curonian Spit – Rositen and Schwarzort were added to this territory in 1939;
- In 1941 the district "Elk Forest" directly subordinated to the Imperial Service of Forestry and Hunting in Germany;
- On the peat bog Tselaubrukh a state nature reserve was established [3].

The modern structure of the SPNA in the region began to form in the 1960s. Between 1963 and 1976 the network of state zoological partial reserves is being created in the region: The Vistula Spit (1963), Kamensky (1963), The Mayko-Krasnopolyansky (1963), The Curonian Spit (1967), Novoselovsky (1976). Some objects received the status of a natural monument, including Lake Vishtynetskoe and the Krasnaya River.

The state natural national park "Curonian Spit" was established in 1987 – the first SPNA in the region of the federal level. In order to preserve valuable natural complexes in their natural state, protect and enrich the wildlife, integrated nature partial reserves of regional importance were established: "Vishtynetsky," "Gromovsky," "Dyunny" in 1994.

The idea of international, across boundary specially protected natural areas occurred in the late 1990s of the twentieth century. In this regard, the SPNA have been defined as a single natural system under different national jurisdictions in order to maintain the ecological equilibrium with the achievement of the greatest possible ecological and socio-economic effect, in the context of coordinated management of the administrations of neighboring countries. The Vishtynetsky, Vistula Spit, Dyunny and Novoselovsky partial reserves as well as the national park "Curonian Spit" were considered to be promising as across boundary SPNA.

Natural partial reserves "Vishtynetsky," "Gromovsky," "Dyunny" became haunting reserves in 1998, while the partial reserve "Vistula Spit" has lost the status of protected area. In order to preserve the moose population, the Zapovedny zoological partial reserve was established in 1999. Thirty-two wetlands (peat deposits) with a total area of 264.9 km^2 have also been granted the status of specially protected areas.

Thus, the system of specially protected natural areas of the Kaliningrad Region was actively formed from 1963 to 1999. These SPNA represented a big deal of the valuable and unique ecosystems of the region. In the following years, significant changes occurred in the system of protected areas, which occurred owing to the fact that documents, designated natural reserves functioning expired in 2004. The work of restoring the network of protected areas of regional importance was resumed in 2012 – the natural park "Vishtynetsky" was established as well as the state partial nature complex (landscape) reserves "Gromovsky" and "Dyunny." Ten state partial nature geological reserves were created in 2013–2014.

Resorts "Svetlogorsk-Otradnoye" and "Zelenogradsk," formerly part of the specially protected natural areas of federal importance, were excluded from the specially protected natural areas in 2013 by federal law.

Currently, 67 specially protected natural areas are located on the territory of the Kaliningrad Region, occupying about 5% of the region's total area. The "Curonian Spit" national park is one of federal importance and 66 SPNA are regionally important, including the natural park Vishtynetsky, the state partial nature complex (landscape) reserves "Gromovsky" and "Dyunny," the state partial nature geological reserves "Dunaevskoe," "Romanovskoe," "Pionerskoe," "Filino," "Shatrovskoe," "Tihorechenskoe," "Mayskoe," "Mogaykino," "Nadezhdinskoe 2," "Pokrovskoe," and 53 natural monuments [4] (Fig. 1).

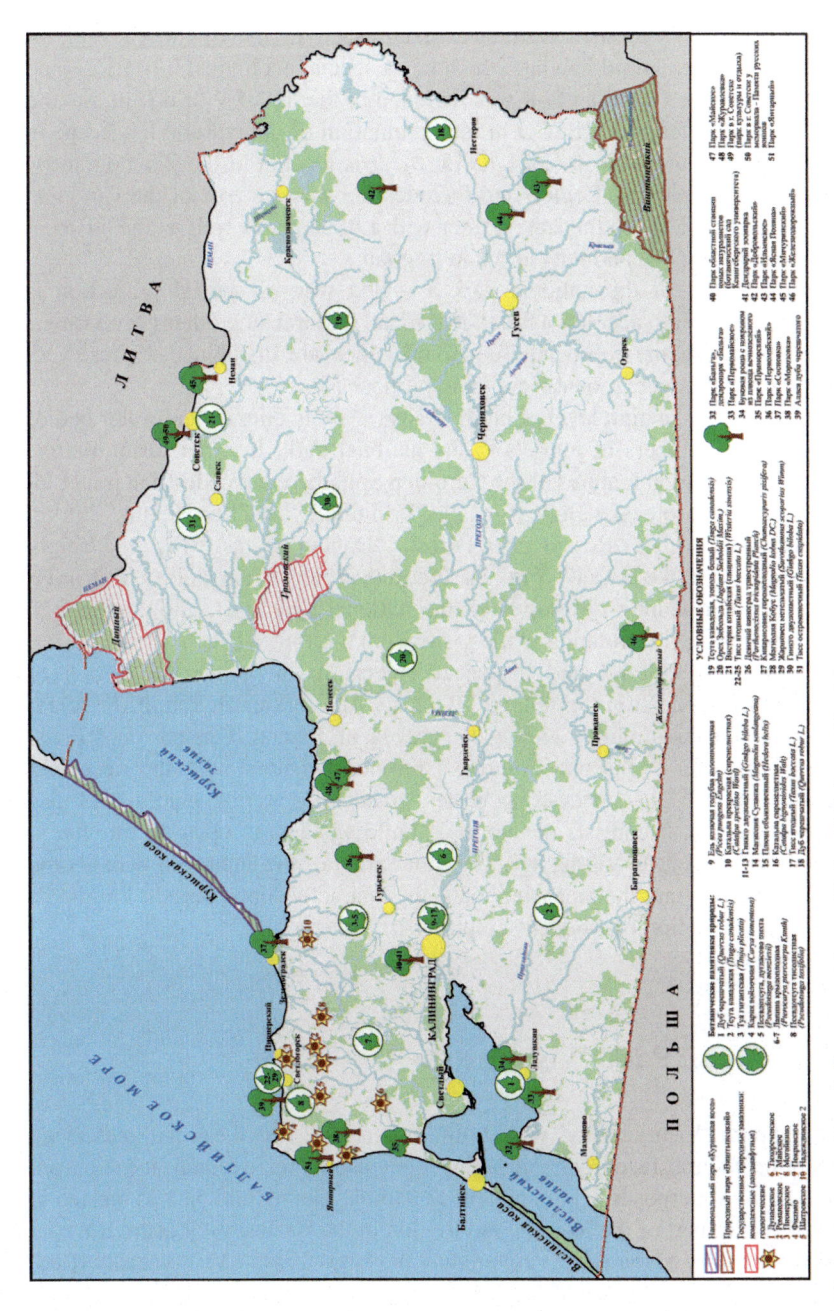

Fig. 1 Current system of specially protected areas in the Kaliningrad Region

2 The National Park "Curonian Spit"

The national park "Curonian Spit" was established in 1987 with the aim of preserving a unique natural habitat, which has a special ecological, historical, and aesthetic value. The Curonian Spit was included in the list of World Cultural and Natural Heritage sites of UNESCO in the nomination of the cultural landscape in 2000. The area of the park is 6,621 ha [5, 6]. It is located in the Zelenogradsky district of the Kaliningrad Region and includes the Russian part of the Curonian Spit from Zelenogradsk to the state border with Lithuania, as well as the adjacent waters of the Baltic Sea and the Curonian Lagoon.

On the grounds of the national park a scaled way of special protection is established, taking into account natural, historical, cultural, and other peculiarities, according to which the following functional zones are distinguished: reserved, specially protected, recreational, economic purpose [6].

Dune landscapes, a high level of biodiversity, friendly climate and water bodies annually attract numbers of tourists to the national park. In order to familiarize tourists with the natural features of the Curonian Spit, there are pedestrian routes all over the park: "Korolevsky Bor" (Fig. 2), "Fringilla," "Mueller Height," "From Rossittena to Rybachiy," "Rossittensky Forest," "Dancing Forest" (Fig. 3), "Height of Ef" (Fig. 4), "Lake Swan" (Fig. 5). Free electronic audio guides with GPS function for smartphones, developed by the national park staff, allow tourists to get all the necessary information while visiting hiking routes [6].

There are also cultural and historical objects in the park: the church of St. Sergius of Radonezh in the village. Rybachiy (the former church of the Evangelical Lutheran community, 1872), the graves of Johann Tineman and Franz Ef. The visit-center "Museum Complex" provides visitors with the necessary information about sights, pedestrian routes, transport, catering places, and also offers guided tours. It consists of two main buildings and an open-air museum – there is the village of the Viking age "Ancient Sambia," the enclosure with hoofed animals, the arboretum, the pier with boats, the interactive complex, the children's town and the collection of wooden sculptures in the area.

3 The Natural Park "Vishtynetsky"

The natural park "Vishtynetsky" was established in 2012 with the aim of preserving and restoring natural resources, managing their use in recreational and environmental education purposes [7]. Its area is 22,935 ha. The natural park is located in the south-eastern part of the Kaliningrad Region in the Nesterovsky district. About 75% of the territory of the park is covered by the large forest – Vishtynetsky (Red) forest.

The natural system includes the Vishtynetskoe lake (Fig. 6), several small lakes (Marinovo, Goldap, Rybnoye, Protochnoe, Kamyshevoye, Chistoe, Ostrovnoe,

Fig. 2 Hiking trail "Korolevsky Bor"

Dorozhnoe, Utinoe), valleys of the Krasnaya River (Fig. 7) and Pissa River in their upper parts, numerous swamps of the transitional type and also meadow and leafy biotopes adjacent to the woodland [4].

Bicycle ("To the ruins of the Castle of Wilhelm," "To the Mountain Dozor") and hiking trails ("Raiffe's Forest Trail," "The Trail of Three Reindeers," "The Path of Emperor Wilhelm II") are situated in the park along with the museum [8].

Fig. 3 Hiking trail "Dancing Forest"

Fig. 4 Hiking trail "Height of Ef"

Fig. 5 Hiking trail "The Swan Lake"

Fig. 6 Lake Vishtynetskoe

Fig. 7 Krasnaya River

4 State Partial Nature Complex (Landscape) Reserves "Dyunny" and "Gromovsky"

State partial nature complex (landscape) reserves "Dyunny" and "Gromovsky" were created in 2012 with the aim of preserving and restoring ecosystems (natural landscapes) and providing biological diversity of flora and fauna [7]. Both partial reserves are situated in the Slavsky district. The area of the partial nature reserve "Dyunny" is 18,600 ha [4]. It is located in the delta of the Neman River. Its territory includes wetlands: Kozie, Dyunnoe, Zelentsovka. The area of the partial nature reserve "Gromovsky" is 9,900 ha [4]. It is located in the interfluve of the Rzhevka River, Lugovaya River, and Golovkinsky Canal in the low lying area of ancient flat bogs, in the southern part of the Neman Lowland. It includes the largest swamp in the region – the Great Moss with an area of about 5,000 ha and Gromovsky forest [9].

5 State Partial Nature Geological Reserves

The main reason for creating such partial reserves is presence of amber deposits at the Sambia Peninsula. The majority of the partial geological reserves was founded in 2013 – "Dunaevskoe," "Romanovskoe," "Pionerskoe," "Shatrovskoe,"

"Tihorechenskoe," "Mayskoe," "Mogaykino," "Nadezhdinskoe 2" (Zelenogradsky district); Filino (Svetlogorsky district); the Pokrovskoe (Yantarny district) was opened in 2014 [4]. The main goal of their creation is the preservation of valuable objects and complexes of inorganic nature (deposits, amber yields, and related landscape elements) [7].

Amber yields are linked with the "blue earth" layer of the Prussian formation and are covered by Neogene and Quaternary sediments. The thickness of amber deposits ranges from 0.5 m in the Tikhorechenskoe partial reserve to 8.2 m in the Pionerskoe partial reserve; while amber wealth varies from 1,415 tons in the "Tikhorechenskoye" to 52,110.4 tons in the "Maiskoe" [4].

6 Natural Monuments

On the territory of the Kaliningrad Region 53 natural monuments of regional importance are represented. The list of natural monuments includes two hydrological objects – Lake Vishtynetskoe and the section of the River Krasnaya is 18 km long (Figs. 6 and 7) and 51 botanical objects – 16 parks, the oak alley, the beech grove and 33 single plants of various types (Figs. 8 and 9). The total area of botanical monuments of nature is 180 ha [4]. The largest number of natural monuments is located in Kaliningrad itself (11), including the park of young naturalists (the University of Konigsberg botanical garden) and the zoo. In the city of Svetlogorsk there are seven natural monuments, but all of them are single plants. Five natural monuments were listed as in Bagrationovsky District (including the Balga park, the Balga arboretum, the Pervomaiskoye park, the beech grove with evergreen ivy), in Guryevsky District (including the Pervomaisky park) and also in Zelenogradsk District (including the "Sosnovka" park and "Morozovka" park).

7 Conclusions

Specially protected natural areas play an important role in biodiversity and landscape preserving in the Kaliningrad Region. However, further development of the SPNA in the region has to fight several problems. In particular, Kaliningrad should improve the regional environmental management and control; increase the level of supply and technical support for protected areas, deepen the system of international and, above all, cross-border cooperation in the environmental sphere.

International collaboration in the environmental sphere should be focused on sustainable development of the entire Baltic region and the Kaliningrad Region in particular.

Remarkable landscapes, cultural, historical, and recreational advantages of the Kaliningrad Region make it valuable even for the whole Europe. Several specially protected natural areas in the Kaliningrad Region are located along the borders with

Fig. 8 Natural monument "European yew," Svetlogorsk

Lithuania and Poland (the Curonian Spit National Park, the Vishtynetsky Nature Park, the Dyunny Partial Reserve), they play an important role in the system of long-term cross-border cooperation, especially the national park "Curonian Spit," which was included in the list of World Natural Heritage by UNESCO along with the Lithuanian national park "Kurshu Neria."

The cross-border SPNA compose a system of specially protected natural areas under different national jurisdictions that share the same ecosystem and is capable to ensure the preservation of ecological balance at a level that gives the maximum ecological and socio-economic effect.

Fig. 9 Natural monument "Boston ivy," Svetlogorsk

The cross-boundary specially protected natural areas are an integral part of the European strategy for the conservation of biological and landscape diversity.

The basic principles of the formation of cross-border SPNA:

- the principle of unity and mutual complementation, meaning all SPNA located on the territories of different countries should be linked into a single multifunctional system.
- creation of legal, management and socio-economic conditions for cross-border SPNA.

- the principle of step-by-step foundation, carried out taking into account the possibilities of financing, legal support, the availability of project documentation.
- the principle of creative initiatives, in which the initiatives of the population and various organizations of the adjacent territory become important.
- the principle of efforts coordination to preserve natural resources, natural, cultural, and historical heritage.
- the principle of ecological consciousness training.
- the principle of continuous improvement [10].

The tasks of cross-border SPNA with top priority are:

- studying of unique landscapes and their components;
- conservation, remediation, and maintenance of biological diversity;
- international cooperation in science and environmental management [10].

Consequently, the cross-border specially protected natural areas are an effective form of international environmental cooperation for solving different problems of improving the pan-European environment, creating a framework for the Eastern Europe long-term sustainable development.

The leading direction for the regional environmental compliance at the current time is improving SPNA system not only by new territories involving, but also by improving the law protecting status of existing areas. For example, it makes sense to give the status of specially protected nature areas to the Vistula (Baltic) Spit, the area around the Balga Castle with the Vistula Lagoon, to the unique intact bog in the Pravdinsky District, to the areas along the Curonian Lagoon in the Slavsky and Polessky Districts, and so on. At the same time, even more important is to improve the system of environment protection activities and to make the control over economic entities within existing protected areas stronger in order to encourage initiatives with positive ecological effect.

References

1. Federalniy zakon ot 14.03.1995 No 33-FZ (red. 31.12.2014) "On specially protected natural territories" (Federal law No 33-FL 14.03.1995 (ed. 28.12.2016) "On specially protected natural areas"). http://base.consultant.ru
2. Volkova II (2002) Vislinskaya (Baltiyskaya) kosa. Potencial vozmozhnostey (Vistula (Baltic) spit. Future opportunities) (Volkova II, Korneevets VS, Fedorov GM, eds). KGU, Kaliningrad, p 72
3. Novova EE (2007) Prigranichnye osobo okhranyaemye prirodnye territorii Kaliningradskoy oblasti: ecologo-geograficheskiy analiz (Near-border specially protected natural areas in the Kaliningrad region: ecological and geographical analysis). Diss. kand nauk: 25.00.36, Kaliningrad, p 198
4. Gosudarstvenny doklad "O ecologicheskoy obstanovke v Kaliningradskoy oblasti v 2015 godu" (The Kaliningrad region national environmental report 2015) (2015) Kaliningrad, p 200. http://minprirody.gov39.ru/
5. Zotov SI, Volkova II, Zhindarev LA, Shaplygina TV, Sokolov AA, et al. (2008) Osobo okhranyaemae prirodnye territorii (Specially protected natural areas). In: Kadjoyan YS,

Moscow NS (eds) Neft' i okruzhayuschaya sreda Kaliningradskoy oblasti (Susha). Tom 1, vol 2. Yantany skaz, Kaliningrad, pp 106–131. Chap. 2
6. Federal State Institution "National park Curonian Spit". http://www.park-kosa.ru/
7. Medvedev VA, Alexeev FE (eds) (2013) Priroda Kaliningradskoy oblasti/Landshafty. Osobo okhranyaemay prirodnye territorii (spravochnoe posobie) (Kaliningrad region nature. Landscapes. Specially protected natural areas). Istok, Kaliningrad, p 192
8. Kaliningrad Region State Public Institution "National park Vishtynetskiy". http://vishtynec.gov39.ru
9. Kropinova EG (2015) Prirodnye, istoricheskie I kulturnye perekrestki Baltiki (uchebnoe posobie) (Natural, historical and cultural crossroads in Baltic region). Kaliningrad, p 144
10. Shaplygina TV (2010) Geoecologicheskaya ocenka sostoyaniya prirodnyh kompleksov Kurshscoi I Vislinskoy kos (Geoecological assessment of ecosystems condition at Corinian and Vistula spits). Diss. kand. geogr. nauk: 25.00.36, Kaliningrad, p 198

The Reduction in the Beach Area as the Main Limiting Factor for Sustainable Tourism Development (Case for the Kaliningrad Oblast)

Elena G. Kropinova

Abstract *The aim of this chapter is* to study the trend of alteration of the beach area of the marine coast of the Kaliningrad Oblast (region) in 1995–2015 as a limiting factor for sustainable development of tourism in the region. Both Russian and foreign approaches to defining the notion of "area carrying capacity" were studied. The historical perspective of shore protection attitude was investigated. The paper focuses on natural factors of development of the Kaliningrad Oblast. The recommendations for federal and regional authorities for sustainable tourism are developed. The Kaliningrad Oblast possesses lengthy sand beaches with different capacities. However, if the current trend of destruction of the waterfront continues, the most popular resort areas will be limited in their development by the natural carrying capacity of the territory. The development of the Kaliningrad Oblast as a tourism centre should include a programme for conducting sand-entrapping and shore protection works.

Keywords Baltic region, Carrying capacity, Coastal zone, Ecology, Kaliningrad Oblast, Sustainable tourism

Contents

1	Introduction	498
2	The Approaches to the Definition of the Notion "Area Carrying Capacity"	499
3	Sustainable Development of Coastal Territories	500
4	Natural Factors Influencing the Development of Tourism and Recreation in the Region	502
5	The Recreation Capacity of Beaches and Coastal Habitats	502
6	Conclusions	508
	References	510

E.G. Kropinova (✉)
Immanuel Kant Baltic Federal University, Kaliningrad 236000, Russia

Museum of the World Ocean, Kaliningrad 236000, Russia
e-mail: kropinova@mail.ru

1 Introduction

The coastal zone is a complex mix consisting of geographical, ecological, economic and social systems. It is the coastal zone which is linked to potentially dangerous trends of global climate change. There is a lot of research done on its complex management.

At the same time, it is well known that tourism and recreation are smokeless industries. Moreover, the utilization of oceanic and coastal beaches for recreation and tourism is technologically simple, is profitable and causes the least damage in ecological terms when compared to other forms of use. That is why they are more and more involved in the economic development.

The attractiveness of coastal territories connected with an enhancement of development generates a need for a comprehensive study of potential opportunities of the natural complexes of coastal areas in order to secure their sustainable development. According to the estimation by Armand [1], littoral zone (horizontally up to 200 m, 18% of the total area of the earth surface) was inhabited by approximately 60% of the Earth's population; two-thirds of the cities with the population of over 1.6 mln are located in the zone [1]. As UNESCO projects, by the year 2025, the figure can increase to 75%. 16 out of 23 metropolitan cities on the Earth are situated in the littoral zone.

The decisions of the UN international conference on environmental conservation and sustainable development (it was conducted in Rio de Janeiro in 1992) determine the necessity to establish in the countries and regions, possessing a seacoast, an efficient system of integrated management of the processes of the coastal territories [2]. A number of documents, serving as guidelines and aimed at the recognition by the administrating authorities and general public of the main tasks and approaches to the development of coastal areas, such as "Model Law on Sustainable Management of Coastal Zones" and "European Code of Conduct for Coastal Zones", were elaborated on the initiative of the European Commission [3]. An integrated management of coastal areas is at present a focus of studies of the entire research institutes, and the issues connected with the problems of development of coastal areas are among priorities of numerous EU-funded programmes. For example, the research conducted by the Ministry of Economic Development and Trade of Russia in 2011–2012 within the measures of the public contracting authority of the federal targeted programme "World Ocean" resulted in the elaboration of proposals on the structure and content of the said coastal-maritime component of model draft strategies and programmes of the socio-economic development of maritime subjects of the Russian Federation and development programmes for coastal municipal entities, on an indicative set of target indicators for monitoring the course of implementation of this component, and also draft guidelines for the elaboration of this coastal-maritime component [4].

At the same time, the economic activity and – to a considerable degree – inactivity lead to a degradation of seashores and a loss of such an important part as beaches. The beach area is the most significant factor of attractiveness of any beach resort. Its presence and qualitative composition (sandy or pebbly-sandy) determine the ranking of tourist destinations. That is why the monitoring of alteration in the width of beaches (which means their area) can be considered as one of the indicators of sustainability for coastal territories. And in determining the prospective number of tourists, the monitoring can be viewed as a limiter of permissible carrying capacity of the area. Works on studying carrying capacity of coastal territories have been carried out for more than 40 years. At the same time, each section of the coastal zone is unique due to its geomorphological, hydrological, geological and other physical and geographic conditions and also owing to a different functional purpose and the history of development. This virtually excludes the possibility of applying a unified approach. For example, researchers E. Jurado, A. Dantas and C. Pereira da Silva, who elaborated tools for establishing a set of indicators to assess the beach carrying capacity (Costa del Sol, Spain), devoted considerable attention, besides the ecological component of the carrying capacity, to the social aspect, such as how many of those staying on the beach make other holidaymakers feel comfortable [5].

The said issues are especially urgent for the area of the Kaliningrad Oblast of the RF which, on the one hand, possesses an extended line of the seacoast (147 km), and where, on the other hand, tourism is considered as one of the priorities of the regional development, which means that the further development of the coastal territory is expected for the purposes of tourism and recreation. That is why the issues of coast protection and widening of the beach line in the region should be given more and more attention.

2 The Approaches to the Definition of the Notion "Area Carrying Capacity"

Every territory has its limits on permissible loads. In this regard, the tolerance range (perseverance) of ecological systems (which any biogeocenosis, including anthropogenous, is) to loads varies depending on the kind of the territory and the type of loads. In order to determine the limits of the load carrying capacity (i.e. the impact when irreversible processes have not come yet), the notion of "carrying capacity" is introduced. This notion is analogous to the notion "the limits to growth" within the concept of sustainable development and to a considerable degree – to the notion of "maximum allowable concentration" in natural resource management – in the Russian practice. According to the definition proposed by L. Ortolano, carrying capacity is defined as "the growth limits an area can accommodate without violating environmental capacity goals" [6]. It should be noted that the notion of "limits", unlike the ecology where it was derived from initially (it was used in the estimation of the growth of population productivity), can be applied in the social environment regarding both physical-geographic and socio-economic indicators of human activity depending on the research purposes.

For example, in relation to the population, the term "area carrying capacity" was introduced by P. P. Semenov-Tyan-Shansky in 1871 as a "maximum number of inhabitants capable of obtaining means of subsistence residing in the area and making use of its resources" [7]. The area carrying capacity is determined by the greatest possible population density and depends on the level of development of production forces, the type of economy (including the degree of involvement in the territorial division of labour) and natural conditions.

When studying economic activities of people, not only physical-geographic and demographic indicators are considered but also infrastructural and financial. Then the definition acquires an economic character: *"the area carrying capacity (economic) –* a possibility of enlargement of economic activity on the area both without significant ancillary costs for its improvement, largely by means of intensification, integrated use of reclaimed resources and with additional capital costs on space planning and involvement of new resources in the economic use (potential area carrying capacity)" [8].

Besides static measures, characterizing the "capacity" at a time, for further area development, it is important to understand to what extent this "capacity" enables the system to develop (natural or sociocultural, economic, etc.) in the future. Thus, a conclusion can be drawn that the measure "area carrying capacity" can be considered as one of the indicators of the possibility of its sustainable development [9].

3 Sustainable Development of Coastal Territories

In spite of national, subregional, regional and global efforts, current approaches to harmonious exploitation of marine and coastal resources have not always led to sustainable development, and many areas in the world see a rapid process of deterioration of coastal resources, degradation and deterioration in the conditions of the environment of coastal areas [2]. In the past century, the significance of beaches has changed completely: from being ineffective inhospitable ground areas, they have turned into a driving force of economic well-being. A textbook example that one can quite often find in research papers is Cannes that used to be an unknown and a poor fishermen's village and that within a short period of time turned into a worldwide famous seaside resort.

At the same time, the population pressure and overuse of both territories located depthward of the coastal margins (dikes on the rivers, farming and tourism) and the coasts as such (sewage dump, solid waste disposal) resulted in a general worsening of sedimentation in the coastal zone. The issues of studying the carrying capacity of the maritime territory within the context of its sustainable development are considered in the works for various subregions of the world. Given that, a special attention was devoted to studying the carrying capacity of beaches. Thus, studies in South Africa are considered in the work by DeRuyck (1997), in Australia by

Williams and Lemckert (2007), in Spain by Roca and Villares (2008), in Maldives by Brown (1997), in Great Britain by Nelson (2000) and also in the Mediterranean by Valdemoro and Jeimenez (2006) [10]. The research gives an opportunity not only to carry out an analysis of the current situation but also to make long-term forecasts necessary for the area planning and designing.

In the meantime, an unsustainable utilization of resources of the coastal territory and the lack of reasonable, scientifically well-founded management approaches frequently lead to such negative effects as follows:

- The loss of economically valuable land resources
- The disappearance of property (of residential and utility buildings, infrastructure facilities)
- The loss of land resources having a high natural and aesthetic value
- The disappearance of sea and terrestrial species of animals and plants
- The loss of sites of historical, cultural and archaeological value
- The limitation of public access to coastal resources
- The noise pollution and increased human pressure on the environment
- The air contamination [11]

The problems that exist in the coastal areas of Russia are similar to those of foreign states and are connected with the collision of interests of natural resource users. In Russia, the problems have reached a critical level and can be resolved through an integrated coastal area management. This means finding reasonable ways for overcoming the following difficulties: jurisdictional conflicts of central and regional government authorities, competition between resource users, inefficiency of dispute-settlement mechanism and the lack of relevant process of decision-making at the federal and local levels regarding coastal area management [9].

One of the most important documents for studying and development of coastal areas is the "European Code of Conduct for Coastal Zones" [12]. The conceptual basis of the code is made up by the following legally relevant provisions: the coastal area is a valuable natural resource; this resource is nonrenewable; the coastal zone is of great economic, cultural, aesthetic and historical significance; the exploitation of the coastal zone resources, also by means of enlargement of urban development and areas of country cottages along the coast, leads to the loss of natural coastal landscape and natural habitats. One of the factors that is supposed to ensure the sustainable development of the coastal zone is the harmonization of users' interests by means of establishing its legal bases, including the limitations on the use of coastal areas.

The Russian Federation Water Code is in force in Russia which determines the parameters of water conservation zones as well as coastal protective belts and specifies the rules of exploitation of water conservation zones [13].

Article 50 of the code should be especially noted which immediately regulates the design, construction and use of buildings, facilities, and water bodies for recreational purposes – leisure, tourism and sport – and also the development of beaches.

4 Natural Factors Influencing the Development of Tourism and Recreation in the Region

The coastline of Kaliningrad fringes the southeastern part of the Baltic Sea and extends over a distance of 147 km (in some sources – 157 km). This coastline encompasses the northern part of the Vistula Spit, the western and northern sections of the Sambia Peninsula and the southern section of the Curonian Spit. Cliffs dominate the coast of the Sambia Peninsula, and sand beaches of various widths occur between the numerous cliffed headlands (Table 1). These contrast sharply with the dune-backed beaches of the Kaliningrad coast spit formations [14].

In comparison to other Russian and Baltic coast resorts, the climate of the region is favourable. The warmest month is July with a mean temperature of 17°C for Svetlogorsk, and winter temperatures are somewhat 4–8°C higher than on the coast of the Gulf of Finland. The average period of sunshine is 1,830 h/year, although in 1968 a peak of 2,200 h was recorded [15].

These warm and sunny conditions favour sunbathing and also favour bathing in the sea from mid-June until mid-September. This length of season is similar to that observed in Poland and Germany but is much longer than the seabathing period for Estonia and St. Petersburg (Russia). In spite of this, heated saltwater swimming pools are still provided.

The presence of sandy beaches, warm sea water and a mild and sunny climate are all factors which have favoured the development of the area as a centre for tourism, recreation and health care (convalescence). This is reflected by the presence of numerous well-established resorts along the northern coast of the Sambia Peninsula, including Zelenogradsk, Pionerskii, Svetlogorsk, Otradnoye and Primorie (Fig. 1). In contrast, the beaches of the western coast of the peninsula are mainly used by local inhabitants, while little development has occurred along the Curonian Spit. The beaches of the Vistula Spit in particular are very infrequently used.

5 The Recreation Capacity of Beaches and Coastal Habitats

One of the most important natural factors which can limit the number of visitors to the coast is the capacity of beaches. The average width of beaches on the coast of the Sambia Peninsula is 20 m (as of the year 2014) (Table 1).

The minimum width (5–10 m) is at the capes, and the maximum (up to 100 m) is in the area of Yantarny (Fig. 1) – these are primarily artificially washed beaches. In order to ensure complete extinction of waves in case of severe storms, the beach width should be not less than 70 m. The beach width is a nonconstant value. Thus, since 1995 the average beach width on the Vistula Spit has reduced from 30 to 25 m, that on the Curonian Spit from 40 to 30 m, and the maximum beach width on the Spits has decreased to 35 and 55 m, respectively. The most unfavourable issue is

Table 1 Trend of key indicators of the beaches of the Kaliningrad Oblast (1995, 2015) (compiled by the author)

Number	Beach name	Length (km)	Average width (m) 1995	Average width (m) 2015	Width range (m) 1995	Width range (m) 2015	Area (ha) 1995	Area (ha) 2015	Theoretical visitor capacity (th. people)	Substrate
1	Vistula	25	30	25	15–45	15–35	75	62.5	62.5	Sand
2	Baltyisk	9	30	30	25–50	20–70	27	27	30	Sand
3	Khmelevka-Obzorny	10	40	50	25–80	40–100	40	50	66.7	Sand
4	Yantarny	10	23	60	20–50	40–100	23	60	83.3	Sand
5	Bakalny	0.5	23	30	20–50	20–70	1.15	1.5	1.7	Boulder/pebble
5	Bakalny	0.5	23	30	20–50	20–70	1.15	1.5	1.7	Sand/pebble
6	Donskoy	3.6	23	20	20–50	15–30	8.28	7.2	6	Sand
7	Cape Taran	0.5	6	6	5.0–7.0	5.0–7.0	0.3	0.3	0	Sand/pebble
7	Cape Taran	1	6	6	5.0–7.0	5.0–7.0	0.6	0.6	0	Boulder/pebble
7	Cape Taran	0.5	6	6	5.0–7.0	5.0–7.0	0.3	0.3	0	Sand/pebble
8	Filino	1.5	8	8	5.0–15.0	5.0–15.0	1.2	1.2	0	Boulder/pebble
9	Primorie	3.3	70	50	60–80	40–60	23.1	16.5	22	Sand
10	Otradnoye	1.5	11	10	8.0–20.0	5.0–20.0	1.65	1.5	0	Boulder/pebble
10	Otradnoye	0.5	11	10	8.0–20.0	5.0–20.0	0.55	0.5	0	Sand/pebble
11	Svetlogorsk	4.5	33	10	30–45	5.0–25.0	14.85	4.5	0	Sand
12	Pionerskii	6.4	23	10	20–45	2.0–20.0	14.72	6.4	0	Boulder/pebble
13	Cape Gvardeysky	0.5	11	10	8.0–15.0	7.0–15.0	0.55	0.5	0	Sand/pebble
13	Cape Gvardeysky	0.5	11	10	8.0–15.0	7.0–15.0	0.55	0.5	0	Boulder/pebble
14	Cape Gvardeysky – Kulikovo	2.2	11	15	8.0–15.0	10.0–25.0	2.42	3.3	1.8	Sand/pebble
15	Kulikovo-Zelonogradsk	11	30	20	25–45	10.0–40.0	33	22	18.3	Sand
16	Edge of the Curonian Spit	3.8	33	15	30–50	10.0–30.0	12.54	5.7	3.2	Sand

(continued)

Table 1 (continued)

Number	Beach name	Length (km)	Average width (m) 1995	Average width (m) 2015	Width range (m) 1995	Width range (m) 2015	Area (ha) 1995	Area (ha) 2015	Theoretical visitor capacity (th. people)	Substrate
17	Lesnoye	3.8	33	25	30–50	20.0–30.0	12.54	9.5	9.5	Sand
18	Curonian	34.4	40	30	35–60	15–55	137.6	103.2	114.7	Sand
	Total	134.5					432.05	386.2	421.4	
	West coast	60.6					176.78	210.9	251.9	
	North coast	73.9					255.27	175.3	169.5	

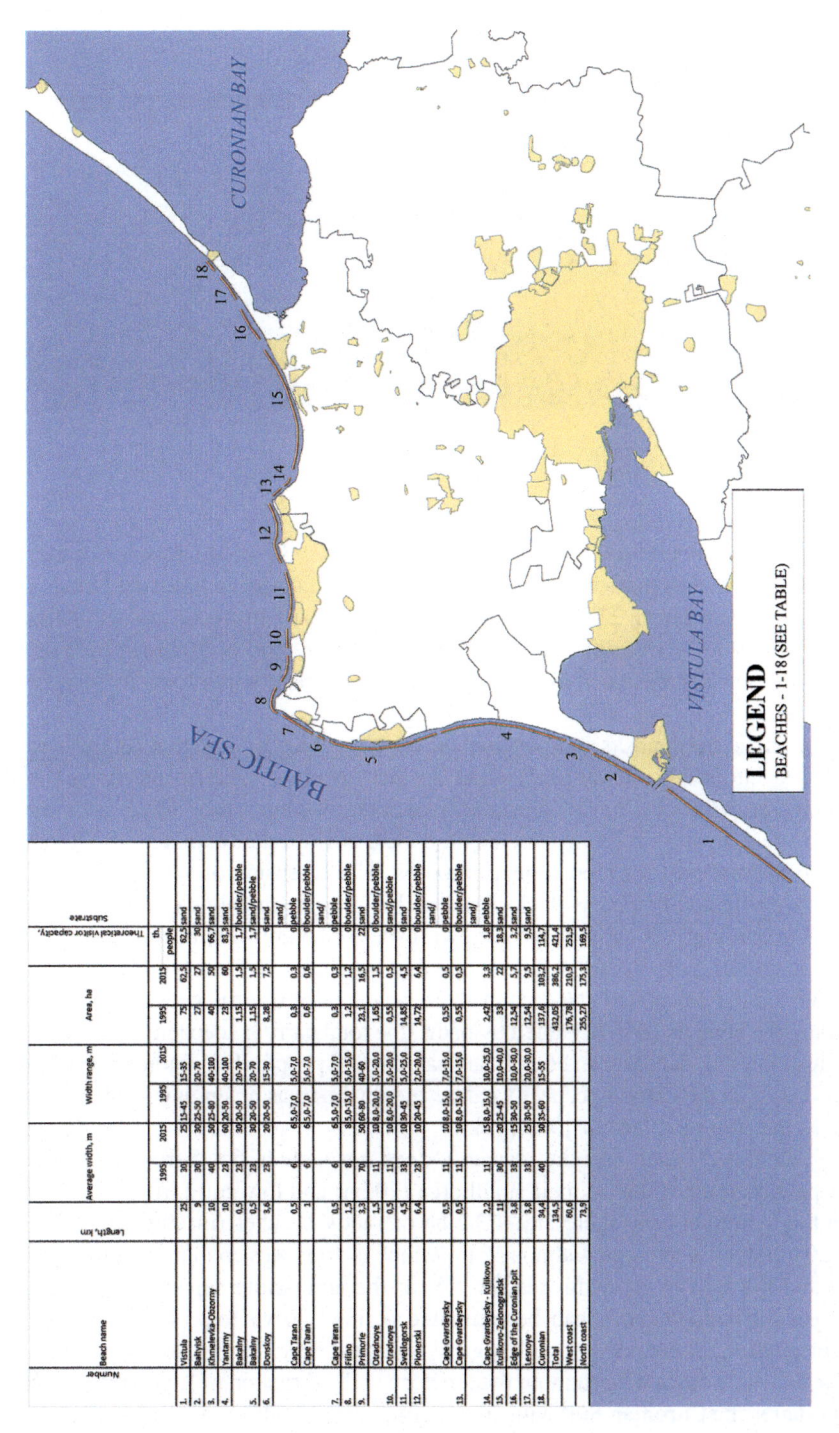

Fig. 1 Schematic map of beaches of the sea coast of the Kaliningrad Oblast

Fig. 2 The promenade in Svetlogorsk, 2016

that the beach in Svetlogorsk (the resort of the federal meaning and the most popular among inbound tourists) yet in 1995 was about 30 m and now it is partly less than 10 m or even there is no beach at all (Fig. 2). At the same time, the Bakalny beach has increased from 23 m of its average width to 30 m; its maximum width has grown from 50 to 70 m (with the exception of a narrow strip of 200 m, where the beach width has decreased from 20 to 25 m of the average width to 5–7 m on the cape proper).

According to the deposits composition, the beaches of the Kaliningrad coastal area are divided into sandy, sandy-pebbly and bouldery-pebbly. Most of the shores are fringed by sandy beaches made up primarily of average-grained sand (0.25–0.5 mm) with the average width of 20 m, the length of 136 km and the total area of 386 ha. They are mainly used for recreational purposes. The extension of scarcely utilized bouldery-pebbly and bouldery-blocky beaches building up mainly capes and shoulders and, accordingly, occupying the narrowest sites is 3.5 km with the area of 9.4 ha, and the length of partially utilized sandy-pebbly beaches is 4.7 km with the area of 11 ha. Given the calculated standard of the beach area of 6 m^2 per visitor [16] and excluding the zone of 10 m occupied by paths, the total carrying capacity of beaches of the Kaliningrad seacoast can be estimated at 410,000 people. This figure is 100,000 visitors less than the carrying capacity calculated on the basis of similar standards for the areas that were in place in 1995 [17].

The calculated beach carrying capacity does not include bouldery-pebbly and sandy-pebbly beaches due to their small area (20 ha) and low comfort level.

On the Curonian Spit, there are 103.2 ha of sandy beaches (in 1995 the figure was 150 ha), whose average width is 30 m (which is 10 m less compared to the year 2015) and the maximum width reaches 55 m; the admissible capacity according to the established standards can be estimated at 114,700 visitors (Table 1). That notwithstanding, in connection with a good transport accessibility, a motorway of federal significance runs throughout the territory in the direction of Lithuania which Russia shares the Curonian Spit with, and high-quality (sandy) beaches are actively used throughout their extension. This, on the one hand, furthers spreading of beachgoers thus reducing the loads. But, at the same time, this creates an effect of an unorganized, spontaneous use of the area that is associated with an additional negative impact on vulnerable natural complexes of the national park (fires, trips to the beach, etc.). At the same time, accommodating more than 10,000 visitors is unreasonable for environmental reasons, since the area is unique, sensitive and has a status of a national nature park.

The total area of beaches of the Vistula Spit is 75 ha with the rated capacity of more than 80,000 people, but here for the same reasons, the admissible number of visitors should be limited to 10,000 people. Within the Spit's shoreline, there are only three recreational outlets of 500 m: in the area of Shchukinsky lighthouse, the 23rd km and the 15th–16th km. Notably, this is virtually the only territory where the current tourist flow is much lower than the potential carrying capacity of the area, which is connected with its isolated character (due to the presence of a strait between the continental part of the Sambia Peninsula and the Spit).

The highest regular in the summer season actual load is carried by the beaches in Zelenogradsk, Svetlogorsk, Pionerskii, Sokolniky and Kulikovo. This is determined by their rapid accessibility (it takes 30–40 min to get there by car/bus) for the residents of the regional centre along the Primorskoe ring. The load on the beaches in Baltiisk and Yantarny is lower. However, in high season the loading level of these beaches also exceeds their optimal carrying capacity. Thus, in the summer season, the 4 km of Zelenogradsk beach with the average width of 20 m receives over 40,000–50,000 visitors with the standard for the area of 22 ha of only 18,000. The beach of Svetlogorsk with the length of 4.5 km with the average width of 10 m hosts 10,000–15,000 beachgoers having the rated standard of only 45,000. It should be noted that in 1995, the total beach area in Svetlogorsk was threefold higher than the present day. This is to a considerable degree connected with the then-existed practice of artificial beach alluvion in the area of Yantarny (during pulp transfer in the course of exploitation of sandy amber-bearing strata in quarries) and transport of part of those sandy deposits by currents to the area of Primorie – Svetlogorsk. In addition to that, a positive role was played by the well-preserved until the 1990s German coast-protecting structures (breakwarters logs and groins).

Modern coast-protecting structures of the Soviet period, made according to the model popular for the Black Sea resorts (reinforced concrete structures), fell short of expectations. The comparative table of beach behaviour pattern that we elaborated in the course of annual monitoring in 1995–2015 clearly demonstrates this (Table 1).

Prospective are the beaches featuring all the necessary elements of comfort (sand deposits, a considerable width, flat slopes) on the sites from Khmelevka as far as Pokrovskoe, in the area of Yantarny, Primorie. At the same time, they are the most distant from the regional centre, and the tourism infrastructure is the least fitted for the 24-h stay of beachgoers. In case of their development, these beaches can to a considerable extent lessen the load of the currently used beaches. Altogether, the sandy beaches of the Kaliningrad Oblast sea coast can receive without damaging the natural environment (in condition of uniform distribution of holidaymakers along the entire extension of the coast) up to 410,000 people simultaneously. However, in order to make this happen, it is necessary to develop the tourism infrastructure at distant and undeveloped at present areas.

6 Conclusions

The processes of beach disruption are connected with both natural phenomena (currents, high water level, intensive wave action) and human activities. Recreation is considered 1 among 40 factors having an impact on the processes of shore working [18].

It should be noted that the following kinds of influence can be identified on the part of tourism and recreation, placed in the order of decreasing of the degree of their impact on beach degradation and coast disruption:

1. The construction of temporary objects (cafes, camp sites, etc.) immediately on the beach or within the dunes-palve
2. Driving jeeps on the beach or within the dunes-palve, unauthorized parking on the beach
3. Laying new paths through dunes
4. Beach trampling

All this has an impact on the dislocation of sand, which in its turn leads to the erosion of shores, dunes sloughing and the sea advance. This requires additional shore protection. Neglecting abrasion processes can also cause shore sloughing which jeopardizes the buildings (also those of recreational purpose).

Consequently, the following factors should be considered when planning and constructing new tourism and recreation complexes:

1. The natural carrying capacity of the available beaches must be taken into account, with new recreation and tourism facilities being located in the areas having sufficient capacity (on the west coast of the Kaliningrad Oblast).
2. The shore protection should be carried out according to a plan taking into consideration traditional modes of coast protection that have already proved efficient in the Baltic Sea region ("willow cells" for dune fixing, groins, etc.).
3. In case of the lack of financial possibilities for the shore protection activities, new facilities of tourism infrastructure should be located taking into account the proposed in the paper carrying capacity of beaches and at a sufficient distance from coastal slopes (since it is obvious that the potential visitors during the season will increase the standards of beach occupancy that have already been exceeded).

It should be noted that with a view to retaining tourism attractiveness of the town of Svetlogorsk as a seaside resort, the Government of the Russian Federation within the Federal Target Programme "The development of domestic and inbound tourism in the Russian Federation (2011–2018)" allocated 1.66 billion rubles to the

Fig. 3 Construction of the promenade and conducting shore protection works in Svetlogorsk, 2017

Kaliningrad Oblast for the implementation of the project of establishment of a tourism cluster "Rauschen" (the name of Svetlogorsk resort before 1946). 1.2 billion rubles will be spent on the construction of a promenade and conducting shore protection works in the area of the resort (Fig. 1, number 11). "The facilities that are supposed to be constructed in the coastal area include a promenade involving a possibility for vehicular traffic, a pedestrian bridge, a sea beach with the length of 4.5 km, a quay pier with the extension of approximately 320 m and also beach-consolidating facilities (groins) with the length of 150 m" [19]. The construction has started in 2016 (Fig. 3). However, this will only cover 4.5 km out of the total 140 km of beaches and is well known for the Black Sea not for the Baltic, where it is still to be tested.

At the beginning of 2016, the other regional programme on beach protection started in Zelenogradsk (named Kranz before 1946) (Fig. 1, number 15). There another technology is used: breakwarters from larch logs at every 50 m. It is more common for the Baltic region technology, historically used in the former East Prussia. But it will help to protect just another 5 km. For over 40 km of sand beaches, of 140 km are most in demand in the summer season in the Kaliningrad Oblast, so the issue is still urgent and requires a comprehensive solution and joining efforts of federal and regional authorities. To develop an ICZM program (Integrated Coastal Zone Management) by analogy with ICZM, proposed by N. Plink for St. Petersburg [20], could be the best solution.

Acknowledgements I am indebted to the support and wisdom, both professional and personal, of Prof. Eugeniy Krasnov and Prof. Lars Ryden, people who inspired me with a love of environmental science and armed with knowledge on sustainable development.

References

1. Armand AD (1988) Self-organization and self-regulation of geographical systems. Moscow University Publishing House, Moscow
2. Agenda for the 21st century (1992) Section II. Conservation and harmonious exploitation of resources for development purposes. Chapter 17 protection of oceans and all kinds of seas including enclosed seas and semi-enclosed seas, and coastal areas and conservation, harmonious exploitation and exploration of their living resources. Programme areas. A. Integrated harmonious exploitation and sustainable development of coastal and marine areas, including exclusive economic zones
3. Sychev SL (2006) The integrated development of the black sea coastal zone – a most important factor of its sustainable development. Extended abstract of PhD dissertation, Krasnodar. http://www.kazedu.kz/referat/88724. Accessed 15 Oct 2013
4. Guidelines for the elaboration of the coastal-marine entry of the strategy of the socio-economic development of the coastal entity of the Russian Federation, sent to the heads of executive authorities of coastal entities of the Russian Federation. Ministry of Economy of RF. http://economy.gov.ru/minec/activity/sections/fcp/doc20131025_6. Accessed 15 Oct 2013
5. Jurado E, Dantas A, Pereira da Silva C (2009) Coastal zone management: tools for establishing a set of indicators to assess beach carrying capacity (Costa del Sol-Spain). Proceedings for the 10th international coastal symposium, Lisbon, Portugal. J Coast Res SI 56:1125–1129
6. Ortolano L (1984) Environmental planning and decision making. Wiley, New York. http://www.coastalwiki.org/wiki/Carrying_capacity_analysis#Tourism_carrying_capacity. Accessed 10 Mar 2015
7. Lopatina EB, Pokshishevsky VV (1961) On the history of the notion 'area carrying capacity' in Russian literature on human geography. In: Voprosy geografii nasleniya SSSR, Moscow. http://demography.academic.ru/1666/Area_carrying_capacity. Accessed 17 Feb 2015
8. Human Ecology (2005) In: Prokhorov BB (ed) Dictionary of terms and notions, Rostov-on-the Don. http://human_ecology.academic.ru/1464/. Accessed 21 Sept 2013
9. Kropinova EG, Afanasieva EP (2014) The sustainable development of coastal areas as a basis for an integrated management of coastal areas. Vestnik BFU im. I. Kanta, Estestvennye nauki, vol 1. IKBFU, Kaliningrad, pp 152–156
10. Pessoa RMC, Pereira LCC, Sousa RC, Magalhães A, da Costa RM (2013) Recreational carrying capacity of an Amazon macrotidal beach during vacation periods. Proceedings 12th international coastal symposium (Plymouth, England). J Coast Res Special Issue 65:1027–1032. http://ics2013.org/papers/Paper4471_rev.pdf. Accessed 10 Mar 2015
11. Programme of remote learning in the sphere of integrated coastal area management "CoastLearn". http://www.biodiversity.ru/coastlearn/planning-rus/sustainable_development.html. Accessed 6 Oct 2015
12. European Code of Conduct for Coastal Zones. Strasbourg, 16 Dec 1998. http://campusdomar.es/observatorio/_documentos/ordenacion_del_litoral/documentacion/europea/cje/de08_cje.pdf. Accessed 1 Jul 2017
13. The Water Code of the Russian Federation, No. 74-FZ of 03/06/2006
14. Boldyrev VL, Ryabkova OI (1989) The establishment of artificial sandy beaches on the northern coast of the Sambian Peninsula. Problemy geomorfologii i chetvertichnoy geologii shelfovyh morey, Kaliningrad, pp 5–10

15. Fedorov GM (ed) (2016) Kaliningrad Region. In: Natural conditions and resources: rational use and protection. Monograph. I. Kant Baltic Federal University, p 224
16. Guidance on the Development of Resort and Recreation Systems (1984) Stroiizdat Press, Moscow, p 180
17. Kropinova H, Litvin V (1996) Factors influencing the development of the recreation and tourism industry on the coastline of Kaliningrad Region. Studies in european coastal management. Samara Publishing, Cardigan, pp 69–75
18. Fokina NA (2008) Recreational activities and abrasion processes. In: Stroitelstvo i tehnogennaya bezopasnost, vol 23, pp 88–92. http://tourlib.net/statti_tourism/fokina.htm. Accessed 21 Sept 2013
19. Federal Targeted Programme "The development of domestic and inbound tourism in the Russian Federation (2011–2018)"
20. Plink NL (2000) The concept of integrated management of St. Petersburg coastal zone. Collected volume "Issledovanie i podgotovka kadrov v oblasti morskih nauk". RGGMU Press, St. Petersburg, pp 37–57

Rare and Protected Macrophytes and Semiaquatic Plants of Flora of the Kaliningrad Region

M.A. Gerb and A.A. Volodina

Abstract The list of rare and endangered aquatic and semiaquatic plants basing on surveys of inland waters of the Kaliningrad Region in 2000–2015 is presented. Some of the 28 species of rare plants revealed are recorded for the first time in the Kaliningrad Region since 1945, three species in the region (*Aegagropila linnaei, Audouinella chalybea, Hildenbrandia rivularis*) and one in European Russia (*Chara inconnexa*). Nine species of vascular plants are supposed to be category of rarity 3 ("rare").

Keywords Kaliningrad Region, Rare macrophytes, Semiaquatic plants

Contents

1 Introduction 514
2 Materials and Methods 514
3 Results and Discussion 514
4 Conclusion 525
References 525

M.A. Gerb (✉)
P.P. Shirshov Institute of Oceanology, Russian Academy of Sciences, 36, Nakhimovski Prospect, Moscow 117997, Russia
e-mail: marikegerb@gmail.com

A.A. Volodina
P.P. Shirshov Institute of Oceanology, Russian Academy of Sciences, 36, Nakhimovski Prospect, Moscow 117997, Russia

Immanuel Kant Baltic Federal University, 14, A. Nevskogo Str., Kaliningrad 236041, Russia
e-mail: volodina.alexandra@gmail.com

1 Introduction

The geographical position of the Kaliningrad Region on the coast of the Baltic Sea, a dense hydrographic network with abundant diverse water courses and specific hydrological regime of rivers in areas that are below sea level, contributes to the greater presence of hygrophilous flora component in the vegetation of the region. The set of semiaquatic and aquatic plants in various water bodies and water courses or in flooded riparian areas may be similar. A number of rare species protected in the Kaliningrad Region are adapted to more specific environmental conditions defined by water quality. They are preserved only in certain habitats that serve as refugiums for them.

Identifying recent floristic composition of water bodies at the Kaliningrad Region and documenting the distribution of various aquatic and wetland species are relevant as they are understudied. Nowadays climatic fluctuations influence the distribution of species, so does the constantly increasing anthropogenic as well as recreational load on water bodies. These factors put forward additional tasks: documenting the changeability in composition of floras and certain vegetation limits and raising the question of monitoring rare and endangered species, in particular aquatic species. This work is aimed at recording the distribution of rare and wetland plants and macrophyte in inland waters of the Kaliningrad Region for the period 2000–2015.

2 Materials and Methods

The materials used for this work are the results of field research done by the authors in 2000–2015; literary sources on the flora of the region [1–3]; herbarium specimens from the collection of the Marine Ecology Laboratory of the Atlantic Branch of P.P. Shirshov Institute of Oceanology, Russian Academy of Sciences (AB IORAS); and herbarium stock of the Immanuel Kant Baltic Federal University (BFU).

The studies were conducted in compliance with the standard methods of studying vegetation of water bodies adopted in the Russian hydrobotany [4–7]. The collected material is kept in the collection of the Marine Ecology Laboratory of the AB IORAS.

3 Results and Discussion

Available information about rare and protected aquatic and helophytic plants in the region remains insufficient. The data is very scattered and confined to several publications. These are (1) the summary of the rare vascular plants [8]; (2) the

Red Data Book [9], where the descriptions are provided only for the species of category of rarity 1 ("endangered"); and (3) the publication of the findings of rare macrophytes [10–12].

Sokolov [13] assigns 59 species (15.7%) of the total amount of rare vascular plants of the region (375 species) to the aquatic eco-phytocenotic type, 30 of which come from the aquatic eco-phytocenotic group and 21 belong to the helophytic group and 8 to the littoral.

Information about macroalgae were not included in the Red Data Book [9] due to their insufficient study. The review of the recent composition of macroalgae in inland water bodies is absent in the scientific literature at all.

There are 243 species in total [14] on the list of hydrophilic flora, which is about 17% of the composition of the flora of higher vascular plants [3]. According to the Red Data Book [9], 28 species from this list are given the status of historical species (not found in the postwar period), and 6 species (*Acorus calamus* L., *Elodea canadensis* Michx., *Bidens connata* Muehl. ex Willd., *B. frondosa* L., *Nasturtium officinale* R. Br., *Suaeda maritima* (L.) Dumort are considered as element of adventive flora.

Ten species of aquatic and wetland plants of the category of rarity 1 (*Alisma gramineum* Lej., *Juncus gerardii* Loisel., *Carex buxbaumii* Wahl., *Potamogeton obtusifolius* Mert. et Koch, *P. praelongus* Wulf., *P. rutilus* Wolfg., *Hippuris vulgaris* L., *Nymphoides peltata* (S. G. Gmel) O. Küntze., *Utricularia minor* L., *Ranunculus reptans* L.) are included in the Red Data Book [9]. The book did not include the information about the plants of categories 2–5, but the list of species requiring special attention and control is provided, of which 22 species are helophytic and aquatic plants. Among the latter we have found new habitats for the three species (*Equisetum telmateia*, *Batrachium eradicatum* (Laest.) Fries, *Batrachium trichophyllum* (Chaix) Bosch.), and the growth of five vegetation species has been confirmed: *Zannichellia major* Boenn, *Z. palustris* L., *Batrachium fluitans* (Lam.) Wimm. (Vistula Lagoon), *Potamogeton friesii* Rupr. (Pregolya River), and *Alisma gramineum* Lej. (Vishtytis Lake).

The habitats of rare sensitive species of algae *Hildenbrandia rivularis* (Liebmann) J. Agardh (Fig. 1) and *Audouinella halybaea* (Roth) Bory de Saint-Vincent (Fig. 2) have been discovered for the first time ever.

Besides, *Aegagropila linnaei* Kütz., *Chaetophora elegans* (Roth) C. Agardh. (Fig. 3), *C. lobata* Schrank, *Draparnaldia glomerata* (Vaucher) C. Agardh (Fig. 4), *Vaucheria frigida* (Roth.) C. Agardh., *V. sessilis* (Vaucher) De Candolle, *V. canaulicularis* (L.) T. A. Christensen, *Chroodactylon ornatum* (C. Agardh) Basson are revealed in the region for the first time in the postwar period. The status of the recent findings is unclear due to poor knowledge of algal flora of the region available.

During the observation period, 28 species of rare plants are found in water bodies of the region. Some of them are recorded in the Kaliningrad Region for the first time in the postwar period. In the below synopsis of rare macrophyte findings of the Kaliningrad Region, the following information is given in brackets: the coordinates of the habitat, harvest date, collector, and herbarium specimen storage (the

Fig. 1 *Hildenbrandia rivularis* (Liebmann) J. Agardh; Photo: A. Volodina

Fig. 2 *Audouinella halybaea* (Roth) Bory de Saint-Vincent; Photo: A. Volodina

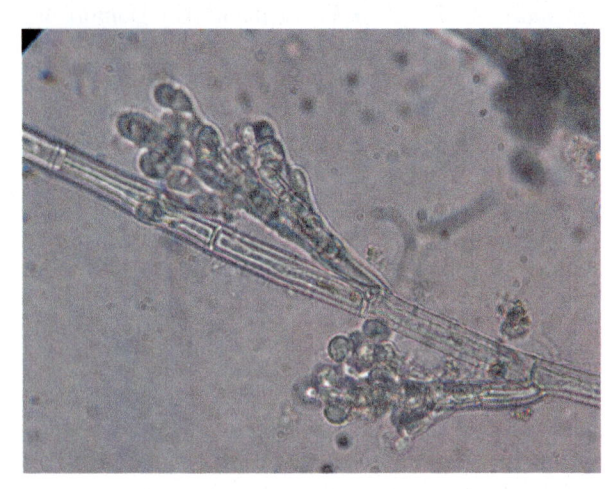

collection of hydrobiological samples, Marine Ecology Laboratory (MEL) of AB IORAS; Herbarium of the Immanuel Kant Baltic Federal University (BFU); photograph collection of the authors).

Hildenbrandia rivularis (Liebmann) J. Agardh (Fig. 1) – the first findings of the species in the region. In the shaded areas on the rocks and boulders in the shallow water of Pissa River (54°27.006″N, 22°40.069″E; 09.07.2014, Volodina A., MEL), Angrapa River (54° 22.523″N, 21°59.850″E; 07.07.2014, Volodina A., MEL), and Krasnaya River (54°26.922″N, 22°19.693″E; 07.07.2014, Volodina A., MEL).

Audouinella chalybea (Roth) Bory de Saint-Vincent (Fig. 2) – the first findings of the species in the region. Shallow water with a rocky bottom and flow rate of 1–1.4 m/c in Angrapa River (54°36.265″N, 21°57.947″E, 14.05.2014, Volodina A., MEL).

Fig. 3 *Chaetophora elegans* (Roth) C. Agardh.; Photo: A. Volodina

Fig. 4 *Draparnaldia glomerata* (Vaucher) C. Agardh.; Photo: A. Volodina

Chaetophora elegans (Roth) C. Agardh. (Fig. 3) – the first findings of the species in the region. Instrutch River, on the older leaves of aquatic plants (54°39.266″N, 21°47.851″E, 14.05.2014, A. Volodina); in the reed, on the old leaves of aquatic plants at Utinaya Bay, Vishtynetskoe Lake 54°25′49.9″N, 22°41′18.1″E, 09.07.2014, A. Volodina, MEL; Reed community in Curonian Spit (55°09′09.4″N, 20°51′39.4″E. A. Volodina 21.05.2015, MEL).

Draparnaldia glomerata (Vaucher) C. Agardh (Fig. 4) – on the rocks at the source of the Pissa River (54°41.728″N, 20°36.806″E, 12.05.2015, A. Volodina, MEL); on the rocks at Uzkaya River, Slavsk District (55°10′58.0″N, 21°14′29.7″E, 19.05.2011, A. Volodina, MEL). On the rocks at Sinya River near Krasnolesye village (54°23′26.8″N, 22°21′34.5″E, 25.05.2015, A. Volodina, MEL).

Aegagropila linnaei Kütz. – the first findings of the species in the region. At the source of the Pissa River (54°41.728″N 20°36.806″E, 15.05.2014; A. Volodina, M. Rogotnev, MEL).

Batrachospermum turfosum Bory (Fig. 5) – polder drainage canal in the vicinity of the village of Khrustalnoe (no coordinates, Feschenko J., 24.06.2013, KLGU), lake complex of Zehlau peat bog (54°31′58.7″N, 20°55′33.5″E; 09.1995; Napreenko M., Volodina A., KLGU).

Batrachospermum gelatinosum (L.) DC – in the shallow water of Angrapa River with a rocky bottom and fast current (54°36.265″N, 21 57.947″E, 14.05.2014, Volodina A., MEL) (Fig. 6).

Chroodactylon ornatum (C. Agardh) Basson – epiphyte on macroalgae, Curonian Lagoon, Lesnoje village (55°00′47.5″N, 20°37′11.7″E, 03.07.2015,

Fig. 5 *Batrachospermum turfosum* Bory; Photo: A. Volodina

Fig. 6 *Batrachospermum gelatinosum* (L.) DC; Photo: A. Volodina

A. Volodina, MEL); Vishtynetskoe Lake at the source of the Pissa River (54°41.728″N, 20°36.806″E, 15.05.2014; 15.05.2014, A. Volodina).

Vaucheria frigida (Roth.) C. Agardh. – on wet soil of Pissa River shore (54°39.298″N, 21°58.153″E, 06.05.2015, A. Volodina, MEL).

V. sessilis (Vaucher) De Candolle – on wet soil at the source of the Pissa River (54°41.728″N 20°36.806″E, 15.05.2014; A. Volodina, MEL). Reed community in Curonian Spit (55°09′09.4″N, 20°51′39.4″E., A. Volodina 21.05.2015, MEL). Golubaya River (54°36′43.71″N, 21°31′09.91″E, 28.11.2014, A. Volodina, MEL).

V. canaulicularis (L.) T. A. Christensen. – on wet soil of Pissa River shore (54°27′04.8″N, 22°40′03.3″E, 06.05.2015, A. Volodina, MEL).

Six eurybiontic species of the genus *Chara* are identified; only three species are in common with the findings of the nineteenth century [15]: *Chara vulgaris* L., *Chara globularis* Thuill., *Chara inconnexa* Allen, *Chara contraria* A. Br. ex Kütz z. s. str., and *Chara virgata* Kütz. The last two for the first time are identified in Kaliningrad Region, and *Ch. inconnexa* is indicated for the first time in European Russia. *Ch. inconnexa* is likely to be the form of *Chara contraria* [16]. The list of identified Charophyta is given below.

Chara vulgaris L. – 4 km south-westward of Sosnovka village (54°37′34.4″N, 20°39′55.8″N) overgrowing pit. 07.07.1987, Kamjeber, Davydova, KLGU; Gromovskoye forestry, 68 quarter, sandpit 54°58′44.9″N, 21°24′23.81″N, 29.09.1995, A. Volodina, KLGU; lake in Oktyabrsky village in Kaliningrad City (54°42′10.8″N, 20°35′25.2″E, 21.07.2003, J. Parfenova, BFU).

Chara globularis Thuill. – sandpits at Ozerki village, 1 km ha northwest of Ozerki village (54°38′23.0″N, 20°53′51.6″E, 20.06.1989, I. Gubareva, KLGU); Curonian Lagoon, 55°00′47.5″N, 20°37′11.7″E, Lesnoje village, 29.05.2014, M. Gerb, MEL; Curonian Lagoon, 55°01′32.9″N, 20°38′08.0″E, 27.05, 2014, M. Gerb, MEL.

Chara contraria A. Braun ex Kütz. s. str. – western part of Vishtynetskoe Lake, no coordinates, in water. 03.07.1990, A. Sokolov, BFU.

Chara cf. *contraria* A. Braun ex Kütz. s. str. – juvenile plant, drainage ditch PK 71 + 89 (54°35′45.9″N, 21°53′59.1″E, 18.11.2014, A. Volodina; MEL).

Chara contraria vel *inconnexa* – at the source of the Pissa River, plants seem to be whipstick and sterile (54°41.728″N, 20°36.806″E, 15.05.2014; A. Volodina, M. Rogotnev, MEL).

Chara inconnexa Allen. – Vishtynetskoe Lake at the source of the Pissa River (54°41.728″N, 20°36.806″E, 15.05.2014; 10.07.2014, A. Volodina, M. Rogotnev, MEL) (Fig. 7); Curonian Lagoon, Lesnoje village (55°00′47.5″N, 20°37′11.7″E, 23.07.2014, M. Gerb, MEL); Curonian Lagoon (55°01′32.9″N, 20°38′08.0″E, 27.05.2014, M. Gerb, MEL).

Chara cf. *inconnexa* Allen – Lake Sinyavinskoye, clear water, black mud, sand (54°53′42.8″N, 19°57′00.8″E, 05.07.2014, E. Ezova, A. Volodina, MEL (Fig. 8); Neman River (55°03′312N, 22°12′048″E, 08.07.2014, A. Volodina, MEL).

Chara virgata Kütz – at the source of the Pissa River. Among the thallome there are fragments of filaments *Aegagropila linnaei* Kütz (54°41.728″N, 20°36.806″E, 15.05.2014; A. Volodina, M. Rogotnev, MEL).

Fig. 7 *Chara* cf. *inconnexa* Allen, Sinyavinskoye Lake; Photo: A. Volodina

Fig. 8 The community of *Chara inconnexa* and *Potamogeton pectinatus*, Vishtynetskoe Lake; Photo: A. Volodina

Chaetophora lobata Schrank – the first findings of the species in the region. In the reed, on the old leaves of aquatic plants at Utinaya Bay, Vishtynetskoe Lake (54°25′49.9″N, 22°41′18.1″E, 16.05.2014, A. Volodina, MEL).

Hippuris vulgaris L. (Fig. 9) – category of rarity 1. The confirmation of the location of the species habitat in the northwestern part of the Vishtynetskoe Lake known since 1870 [1], but not indicated for the region in the recent period [17]. Fruiting specimens in coastal washouts in the northwestern part of this lake (54°27′10.9″N, 22°42′11.6″E, 08.07.2014, Gerb M., Volodina a., MEL; 05.10.2014 Sokolov A.; 15.10.2014, Volodina A., MEL). The previous findings on the coast of the Curonian Lagoon [17] have not been confirmed in recent years due to the bank protection works by beach aggradation being carried out.

Alisma gramineum Lej. (Fig. 10) – category of rarity 1. The confirmation of the location of species habitat in the northwestern part of the Vishtytis Lake, known

Fig. 9 *Hippuris vulgaris* L.; Photo: A. Volodina

Fig. 10 *Alisma gramineum* Lej.; Photo: A. Volodina

from the prewar sources [1], not indicated in the recent period in this lake [18] (54°27.095″N, 22°42.070″E, 15.10.2014, A. Volodina, MEL). Six specimens in the status of fruiting are recorded at a depth of 0.5–0.8 m, biotope with *Dreissena polymorpha* and *Elodea canadensis*. The previous finding within the city of Kaliningrad [18] in the overgrown Melnichny pond in 2013 was not confirmed.

Nymphoides peltata (S. G. Gmel.) O. Küntze (Fig. 11) – category of rarity 1. Confirmation of findings on the eastern coast of Curonian Lagoon in 1953 [19] near Mysowka village (55°12.115″N, 21°16.374″E, 12.05.2012, Volodina A., Gerb M., MEL), not indicated in the Red Data Book of the Kaliningrad Region [20].

Batrachium eradicatum (Laest.) Fries (Fig. 12) – species requiring attention. New habitat in the region. Pregolya River (54°38.636″N 20°58.767″E, 20.05.2014, Gerb M., MEL).

Batrachium trichophyllum (Chaix) Bosch (Fig. 13) – species requiring attention. New habitat in the region. Krasnaya River (54°24′59.0″N, 22°23′49.9″E, 24.10.2014, Volodina A., Gerb M., MEL).

Fig. 11 *Nymphoides peltata* (S. G. Gmel.) O. Küntze; Photo: A. Volodina

Fig. 12 *Batrachium eradicatum* (Laest.) Fries; Photo: A. Volodina

Fig. 13 *Batrachium trichophyllum* (Chaix) Bosch; Photo: A. Volodina

Fig. 14 *Potamogeton praelongus* Wulf.; Photo: A. Volodina

Potamogeton praelongus Wulf. (Fig. 14) – category of rarity 1. New habitat in the region. The Sheshupe River at the confluence with the Neman River (55°03.286″N 22°12.069″E; 15.05.2014, Volodina A., MEL). Previous findings revealed for the City of Kaliningrad [21].

Wolffia arrhiza (L.) Horkel ex Wimm – species of category of rarity 0, previously considered extinct [22]. The only recent finding – in a roadside drainage ditch at the base part of Curonian Spit [23] (54°57′57.1″N, 20°29′51.2″E, 2004, Gubareva I., BFU).

Equisetum telmateia L. (Fig. 15) – category of rarity 1 [16]. New habitat in the region. Angrapa River (54°22′30.8″N, 21°59′51.2″E, 07.07.2014, Kuzmin A., Volodina A., MEL).

Tripolium pannonicum (Jacq.) Dobrocz (Fig. 16) – species considered extinct [3]; however, it was recorded to be found on the coast of Vistula Lagoon in the Vistula Spit since 2005 (54°36′56.02″N, 19°53′01.25″E, A. Volodina, 08.2005; A. Volodina, M, Gerb 2010–2016). At present, there is an active distribution of this species along the lagoon coast.

Fig. 15 *Equisetum telmateia* L.; Photo: A. Volodina

Fig. 16 *Tripolium pannonicum* (Jacq.) Dobrocz. Photo: A. Volodina

Zannichellia major Boenn. – species requiring attention [22]. Western shore of Curonian Lagoon (55°01′34.0″N, 20°38′10.0″E, 30.09.2016, M. Gerb, MEL), Vistula Lagoon (54°37′31.8″N, 19°53′14.0″E; 54°37′28.7″N, 19°52′59.5″E, 16.07.2014, M. Gerb, MEL).

Z. palustris L. – species requiring attention [22]. Vistula Lagoon, western shore of Curonian Lagoon (55°01′34.0″N, 20°38′10.0″E, 23.08.2014; 55°02′01.9″N, 20°39′11.0″E, M. Gerb, MEL).

4 Conclusion

Nine species of macrophytes need to be assigned to the category of rarity 3 ("rare"), in particular *Batrachium eradicatum* (Laest.) Fries, *B. fluitans* (Lam.) Wimm., *B. trichophyllum* (Chaix) Bosch, *Callitriche hermaphroditica* L., *Ceratophyllum submersum* L. *Zannichellia palustris* L., *Z. major* Boenn, *Potamogeton acutifolius* Linc, and *P. friesii* Rupr [12].

Acknowledgments The authors express their gratitude to Elena Ezova, Irina Gubareva, and Alexey Sokolov for valuable comments during the preparation of the article and to Roman Romanov and Irina Gubareva for the identification of some macrophytes.

References

1. Abromeit J, Neuhoff W, Steffen H, Jentzsch A, Vogel G (1898–1940) Flora von Ost- und WestpreuBen. Berlin, H.l-2, p 1248 (in German)
2. Pobedimova EG (1955) Composition, distribution by area and economic significance of the Kaliningrad Region flora. In: Proceedings of the Botanical Institute of V.L. Komarov, Academy of Sciences of the USSR, Series III (Geobotany), vol 10. Publishing House of the USSR Academy of Sciences, Moscow, Leningrad, pp 225–329 (in Russian)
3. Gubareva IY, Dedkov VP, Napreenko MG et al (1999) In: Dedkov VP (ed) Checklist of vascular plants of the Kaliningrad Region. Kaliningrad University Press, Kaliningrad, p 107 (in Russian)
4. Katanskaya VM (1981) Higher aquatic vegetation inland waters of the USSR. Methods of study. Nauka, Leningrad, p 278 (in Russian)
5. Papchenkov VG (2001) The vegetation cover of reservoirs and waterways of the middle Volga. CMP MUBiNT, Yaroslavl, p 214 (in Russian)
6. Papchenkov VG (2003) Dominant and determinant classification of aquatic vegetation. In: Hydrobotany: methodology, methods. Materials of School on Hydrobotany (n. Borok, 2003). OAO Rybinsk Printing House, Rybinsk, pp 126–131 (in Russian)
7. Papchenkov VG (2006) Different approaches to the classification of plants ponds and streams. In: Proceedings of the 6th all Russian conference school on aquatic weeds (Hydrobotany 2005), Borok, 11–16 Oct 2005. OAO Rybinsk Printing House, Rybinsk, pp 16–23 (in Russian)
8. Sokolov AA (2003) Summary of rare vascular plants of the Kaliningrad Region. In: Scientific notes of the Russian Geographical Society (Kaliningrad branch), text electronic edition. KSU, RGO, Kaliningrad, 5C-1-5C-116 (in Russian)
9. Dedkov VP, Grishanov GV (eds) (2010) The Red Data Book of the Kaliningrad Region. Immanuel Kant University, Kaliningrad, p 334 (in Russian)
10. Parfenowa YV (2008) Rare species of hydrophilic plants Kaliningrad ponds. In: Some problems ichthyopathology and hydrobiology: first steps in science. Kaliningrad, pp 105–120 (in Russian)
11. Gubareva IY (2000) Rare and protected plants in the Kaliningrad Region of the Baltic spit. In: Problems of geographical, biological and chemical sciences. Materials permanent scientific seminars. Kaliningrad University Press, Kaliningrad, pp 73–75 (in Russian)
12. Volodina AA, Gerb MA, Gubarewa IY, Sokolov AA (2015) About records of rare species of macrophytes in water bodies and rivers of the Kaliningrad Region. In: Hydrobotany 2015 materialy 8th all Russian conference with international participation on the aquatic weeds, Borok, Filigran, Yaroslavl, 16–20 Oct 2015, pp 88–90 (in Russian)

13. Sokolov AA (2000) Rare vascular plants of the Kaliningrad Region and their protection. Diss ... cand. biol. sciences, Kaliningrad State University Press, Kaliningrad, p 238 (in Russian)
14. Gubareva IY (2015) List of hydrophilic flora of reservoirs of the Kaliningrad Region. In: Alekseev FA et al (eds) Nature of the Kaliningrad Region. Water features. Istok, Kaliningrad, pp 94–98 (in Russian)
15. Romanov RE, Volodina AA (2015) Study of Chara algae (Streptophyta: Charophyceae, Charales) in Kaliningrad Region. In: Problems of taxonomy and geography aquatic plants: proceedings of the international conference, Borok, Russia, 21–24 Oct 2015. Filigree, Yaroslavl, pp 35–37 (in Russian)
16. Gubareva IY (2010) Equisetum telmateia. In: Dedkov VP, Grishanov GV (eds) The red book of the Kaliningrad Region. Immanuel Kant University, Kaliningrad, p 203 (in Russian)
17. Sokolov AA (2010) Hippuris vulgaris. In: Dedkov VP, Grishanov GV (eds) The red book of the Kaliningrad Region. Immanuel Kant University, Kaliningrad, p 159 (in Russian)
18. Sokolov AA (2010) Alisma gramineum. In: Dedkov VP, Grishanov GV (eds) The red book of the Kaliningrad Region. Immanuel Kant University, Kaliningrad, p 126 (in Russian)
19. Minkyavichus A, Pippinis J (1959) Overview of flora and vegetation of the Curonian Lagoon. Kursiu mares. The results of a complex study. Lithuanian SSR Academy of Sciences, Institute of Biology, Vilnius, pp 109–116 (in Russian)
20. Gubareva IY (2010) Nymphoides peltata. In: Dedkov V, Grishanov G (eds) The red book of the Kaliningrad Region. Immanuel Kant University, Kaliningrad, p 173 (in Russian)
21. Sokolov AA (2010) Potamogeton praelongus. In: Dedkov VP, Grishanov GV (eds) The red book of the Kaliningrad Region. Immanuel Kant University, Kaliningrad, p 188 (in Russian)
22. Gubareva IY, Sokolov AA (2010) List of taxa not included in the red book of the Kaliningrad Region, need special attention and control. List of animal and plant species have disappeared from the territory of the Kaliningrad Region for the last 50 years. In: Dedkov VP, Grishanov GV (eds) The red book of the Kaliningrad Region. Immanuel Kant University, Kaliningrad, pp 307–311, 318–324 (in Russian)
23. Gubareva IY (2009) Botanical objects. In: Ryl'kov O, Zhukovskaya I (eds) Especially valuable natural and cultural sites (Curonian Spit) National park: sat. scientific article. Immanuel Kant University, Kaliningrad, p 19 (in Russian)

Conclusions

Vladimir A. Gritsenko, Vadim V. Sivkov, Artem V. Yurov, and Andrey G. Kostianoy

Abstract This concluding chapter briefly describes the results of the research presented in the book, which has 22 chapters including the present Conclusions and Introduction written by volume editors of the book. This book entitled "Terrestrial and Inland Water Environment of the Kaliningrad Region" is the first one in the series of four volumes which will be published in the coming years under the general title "Environmental Studies in the Kaliningrad Region." This first volume deals with physicogeographical and bio-geo-ecological conditions and environmental problems of the Kaliningrad Region focusing on terrestrial and inland water environment. This book is addressed to the specialists working in various fields of environmental problems and ecology, water resources and management, land reclamation and agriculture, and international cooperation in the Baltic Sea Region.

Keywords Bioecology, Geoecology, Inland water, Kaliningrad Region, Terrestrial environment, Southeastern Baltic Sea Region

---

V.A. Gritsenko (✉)
Shirshov Institute of Oceanology Russian Academy of Sciences, Immanuel Kant Baltic Federal University, Kaliningrad, Russia
e-mail: gritsenko_vl@mail.ru

V.V. Sivkov
Shirshov Institute of Oceanology Russian Academy of Sciences, Immanuel Kant Baltic Federal University, Moscow, Russia
e-mail: sivkov@kaliningrad.ru

A.V. Yurov
Immanuel Kant Baltic Federal University, Kaliningrad, Russia
e-mail: AIUrov@kantiana.ru

A.G. Kostianoy
Shirshov Institute of Oceanology Russian Academy of Sciences, Moscow, Russia
e-mail: kostianoy@gmail.com

The chapters presented in this book are devoted to various aspects of the Kaliningrad Region environment, which allows readers to make their generalized opinion on environmental features and problems. Some chapters are compilations and generalizations of the works published mainly in Russian editions which are hardly accessible to western scientists, while others contain unpublished data and custom approaches, sometimes debatable.

The main results and aspects of the entire book are:

1. Two structural stages are distinguished in the geological section of the Kaliningrad Region: the lower one is composed of gneisses, crystalline schists, and amphibolites of the *Archaean-Proterozoic age* (platform basement), and the upper one is the *Phanerozoic platform* cover, represented by poorly dislocated and slightly metamorphosed sediments. The Kaliningrad Region is rich in diverse mineral resources. Actually all economic minerals discovered within Kaliningrad Region are spatially and genetically associated with the deposits of the platform cover. They are represented by oil, drinking and mineral underground water, amber, building materials, which are actively exploited [1]. Potassium-magnesium salts, numerous occurrences of which are known here, are of the greatest economic interest. The main treasure of the region is the Baltic amber, which has been the research object for almost 150 years. One might assume that after so many years of research, there are only a few unresolved issues of fundamental importance. Publications in recent years still prove the opposite. Fundamental issues, such as the resin source, the formation time of the Baltic amber, and the position and period of the legendary "amber forest," are still a matter of debate. The updated taxonomic composition of amber inclusions was published for the first time in English [2]. The data was obtained as a result of studying the primary material from the quarries of the Kaliningrad Amber Combine. Unfortunately, comprehensive studies of amber in Russia have not been carried out since the 1970s. The existence of the world's largest amber deposit still allows us to hope for such research [2].

2. The lithogenic basis of the Kaliningrad Region landscapes is the Quaternary deposits. The surface of pre-Quaternary deposits in the region is characterized by complex relief, including denudation uplands ("remnants"). Erosion-denudation pre-Quaternary relief was transformed by glacial and fluvioglacial processes. The incision system exists in the northern and western parts of the Sambia Peninsula, continuing on the bottom of the Baltic Sea. Along with the uplifts and depressions that have been inherited from the pre-Quaternary surface, there are inversion forms of the modern relief [3]. At the same time, there are some geological facts, proving the fact that the main strata of the Pleistocene sediments are not continental (glacial, water-glacial, and separating alluvial facies), but were accumulated in the sea with the participation of icebergs and fast ice. According to the author [4], the Pleistocene glacier was so thin and slow-moving that it produced virtually no damage, so the tectonic factor had much more significant effect on the relief than the glacier. Even

some doubts about the existence of a blanket Pleistocene glaciation within the Kaliningrad Region occurred.
3. The modern appearance of the Kaliningrad Region landscapes shares similarities with the flat European landscapes, which lies within the accumulation zone of the last Quaternary glaciation. Based on the features of Quaternary sediments and relief, various landscapes were distinguished: landscapes of glacial origin, landscapes of fluvial origin, and landscapes of marine and lagoon origin. They relatively are divided into plains of the main moraine, finely moraine elevations, lacustrine-glacial plains, seaside landscapes, ancient delta lowlands, valley landscapes, and ancient alluvial plains undergoing aeolian processing [5].

The soil and vegetation cover, evolved in the postglacial period from tundra to forest, underwent significant changes in the last millennium due to active human activity. On the most part of the territory, the lithogenic base was formed in the Pleistocene period. Some of the younger landscape complexes (marsh landscapes, river terraces) were formed in the Holocene. The youngest dynamic landscapes – modern floodplains of rivers and seashores – are still in genesis; channel and marine accumulation and abrasion processes are in action. Soils and vegetation cover have passed several stages of development along with climatic trends [5].

Conducted recently in the region, targeted palynological studies for the first time made it possible to clarify the main pattern of forest and peatland formation in the Kaliningrad Region during Holocene [6]. According to studied pollen spectra, the territory of the Kaliningrad Province doesn't form an entire palaeoenvironmental district and is to be divided, in this respect, onto two different parts, each of those could be united with the neighboring regions of Poland and Lithuania. A boundary between these two identified palynological districts is stretched along the rivers Deima, Pregolya, and Pissa. They are palynologically similar in the Early and the Middle Holocene. The differences between them appeared in the Late Holocene when conifers became dominating in the northeastern part of the region having gradually replaced communities of the "nemoral complex." The latter remained common in the south and especially in the southwestern parts of the Kaliningrad Region where they were key components of forest vegetation in the Late Holocene together with *Carpinus* and *Fagus*. Having been affected by human activity, vegetation structure suffered an essential change on the territory of Pregolya glaciolacustrine plain since the beginning of the seventeenth century (400 BP), while in Lower Neman Lowland, such alterations are recorded only since the middle of the eighteenth century (250 years BP), apparently due to its hard approachability that restricted human impact onto pristine forests in this area. Intensive land use and clear cutting resulted into a large reduction of broad-leaved and spruce-broad-leaved forests in the area and, simultaneously, caused an increase of agricultural areas and synanthropic habitats as well as secondary pine and birch stands in both study landscapes [6].

The analysis of the modern natural complexes stability on the Curonian and Vistula Spits has shown some similarities. In particular, the distribution of potential sustainability stages – the "weakly stable" landscapes – is about 1/3 in area expression. The remaining categories show a certain variability, which is primarily due to the structural features of natural complexes of spits. There are 19% more natural complexes classified as "stable" on the Curonian Spit than on the Vistula Spit, which is owing to the structure of the deflationary-accumulative plain. Unstable natural complexes are 18% more on the Vistula Spit, which can be mainly associated with a large number of complexly oriented dune ridges; highly unstable are about 11% more on the Curonian Spit, which is due, first of all, to the presence of not fixed or weakly fixed dune arrays. The main trend of the natural complexes digression of spits is the leading role of the first stage (55% and 52% of the Curonian and Vistula Spit areas, respectively). The highest level of digression is more common for coastal zone of the spits (beach, seaside dune ridge complex) [7].

4. The main factors of landscape genesis at the Kaliningrad Region are land use and settlement systems. The modern system of settlement is to a certain extent formed by landscape conditions, but also, in many cases, settlement system is associated with other factors – primarily political, socioeconomic, and demographic. The greatest contrast in the settlement structure in comparison with prewar time is possessed by three natural regions: the ancient Neman Delta, an array of ancient alluvial sands of the Neman and Sheshupe interfluvial areas, and end-moraine hills [5].

At current time, the landscapes of the Kaliningrad Region represent a system of territorial complexes that vary in transformation degree and are at different stages of their development. The modern settlement system of the Kaliningrad Region is determined by their system before 1945, which was strongly connected with the hydrographic network and landscape structure of the territory. After World War II, the new administrative division occurred, and the transport network of the region has changed, resulting in railway network reduction. Now the region is less populated than before the war, and the vast majority of the present population is concentrated in the regional center and around it. After World War II and the restructuring in the 1990s, the modern land use of the Kaliningrad Region is characterized by agriculture development, which reflects use of previously abandoned lands for agricultural needs. At the same time, the restorative succession also takes place in the region. Many of such territories have become sort of "natural reserves," which affects the biodiversity of the territory. The main feature of the region is the landscape mosaic pattern. The state borders of the Kaliningrad Region are particularly important for modern landscape genesis [8].

Cities of the Kaliningrad Region include two or more types of natural landscapes. Most of the cities in the region are located on the riverbanks: the Pregolya River and its tributaries. Features of inner landscapes define not only the planning features of cities but also the development of some negative processes. Thus, the coastal position of cities determines the restriction of their development by the seashore, as well as the risk of coast destruction. Cities in the lower course of the Pregolya River experience floods due to wind surges. The cities' transport system is the main factor determining the functionality of their modern spatial structure [9].

5. The river network is well developed and very dense [10]. The watercourses are classified as plain streams and belong to the Baltic Sea catchment area. Over 95% of watercourses are "small," characterized by a large short-term variability of hydrological and hydrochemical parameters comparable to the seasonal changes. Much of the Kaliningrad Region's territory is occupied by the Pregolya River basin, as well as by the basins of its several major tributaries.

As a consequence of small slopes, the estuarine and adjacent parts of the rivers are in the backwater of the receiving water bodies and very dependent on the wind-driven surges. Most small rivers are channelized, and the water pumped out from the polder lands is dumped therein. Almost all the rivers in the Neman Delta and at the Curonian Lagoon coast are connected with channels and form the united drainage system. Hydrometric parameters fluctuate significantly and are closely associated with both the weather conditions and the aggregate natural peculiarities of the drained area. Ice conditions are very unstable, and the formation of two-layer ice is possible.

According to the chemical composition of water, the studied streams belong to bicarbonate class, calcium group, mainly of the first water type $[HCO_3^- < (Ca^{2+} + Mg^{2+})]$. Pursuant to the total hardness value, the water is "moderately hard." Oxygen conditions, the content of organic matter, and nutrients in most watercourses do not comply with maximum concentration limits for fishery water bodies, especially during low-water periods.

The Neman River enters the Kaliningrad Region already quite polluted, and the major watercourses in the delta plain (the Matrosovka Canal, the Nemonin River) determine the load on the Curonian Lagoon. The main source of nutrient load on the Vistula Lagoon is the Pregolya River. The Pregolya River is exposed to intensive pollution. The condition of some small rivers is indicative of their catastrophic contamination. Small rivers emptying directly into the Baltic Sea build up a load on the sea mostly only during wet periods.

Transboundary rivers arrive in the Kaliningrad Region already quite polluted. The proportion of nutrient runoff received from the territory of the neighboring states ranges from 10 to 80% or more of the total nutrient runoff of the transboundary watercourses discharged from the Kaliningrad Region. The amount of nitrite nitrogen, mineral phosphorus, and total iron discharged into the Curonian Lagoon by the Neman River differs little from what is incoming from the adjacent territories [10].

6. The Pregolya River runoff from the area of its catchment is directed to two receiving reservoirs – the Vistula Lagoon and the Curonian Lagoon. The watershed of the Pregolya River was divided into 42 interconnected subbasins in the model installation, the allocation of which was carried out taking into account major tributaries, hydrological stations, as well as the existing state border between the segments of catchment areas in Poland and Russia. The

calculation of the volume of water flow rate from the Pregolya River catchment was made by tools of numerical simulation [11]. The flows to the Vistula and Curonian Lagoons constitute 1.96 and 1.2 km^3/year, respectively, which in total give 3.16 km^3/year from the Pregolya River catchment toward the Baltic Sea through both lagoons. In general, 1.46 km^3 of water per year comes to the Kaliningrad Region from the territory of watershed of the Pregolya River in the neighboring countries (Poland and Lithuania). In addition, next 1.26 km^3 more of water is formed in the catchment before the division of the Pregolya River into arms. Each individual subbasin of the Pregolya River major tributaries provides through outlet section at the confluence: the Lyna-Lava River, 1.37 km^3/year; the Wengorapa-Angrapa River, 0.69 km^3/year; the Instruch River, 0.3 km^3/year; and the Golubaya River, 0.14 km^3/year. Thus, out of the annual average of 760 mm of precipitation, 530 mm comes back to the atmosphere by means of evaporation and 230 mm flows with surface runoff [11].

7. More than 100 species of zooplankton are found only in the largest water bodies of region such as the Neman River and Lake Vishtynetskoe (Vištytis) [12]. In other water bodies, diversity of zooplankton is lower (60–80 species). Abundance and biomass of zooplankton in lakes are significantly higher than in rivers. Small lakes had the highest number of indicators and weights. In the Lake Vishtynetskoe, these indices are 2.5–3 times lower. With the increase in river flow, the quantity of zooplankton tends to be zero. The number and biomass in almost all the groups of water bodies are dominated by copepods. The exception is the Neman River, where this group is inferior in numbers to rotifers and in weight to Cladocera.

   In the zoobenthos there were identified 450 species [12]. The most diverse zoobenthos is of the Neman River and fleeting rivers. A smaller variety of zoobenthos is characteristic for small lakes. A large number of zoobenthos was observed at the Neman River, but 80% of it was shellfish. The high number of organism is also noted in slow rivers shorter at length. The highest biomass of zoobenthos is registered in the Neman River. This rate is significantly lower than in small rivers and Lake Vishtynetskoe. In general, with increasing flow rate of the river the biomass of benthos reduce on 15%. Zoobenthos biomass is also low in small lakes. The biomass basis in all groups of water bodies is consisted of shellfish.

8. Comparing the data for 1982–1998 shows a gradual depletion of the river plankton and benthos communities in the Pregolya River by 1995, which is explained by chemical pollution and anthropogenic eutrophication [13]. At the same time, since 1997, there has been a sharp increase in the quantitative characteristics and qualitative diversity of communities. Analysis of interannual changes in biodiversity, abundance, and biomass of components of the river biota in 2000–2011 shows improvement of environmental situation in comparison with the 1990s. The boundaries of biotic communities with the dominance of filter-feeding mollusk and rooted aquatic vegetation are gradually moved downstream, with the appearance of benthic species which did not occur in the lower reaches in the 1980s and 1990s. Currently, despite the fact that

contamination level of the Pregolya River is still high, the biota of the river in the lower reaches is in the best condition ever recorded in the scientific press [13].

9. During the vegetation period, the structure of phytoplankton of the small rivers of the Pregolya River system is mainly determined by diatoms—benthic forms and fouling species [14]. The phytoplankton structure in the Pissa and Angrapa rivers is largely determined by the fact that they flow from the Vyshtynetskoye and Mamry lakes, respectively. Phytocenosis of these rivers is characterized by higher taxonomic diversity, domination of cyanobacteria in terms of numbers and biomass, in some cases, and the highest phytoplankton productivity. The Pregolya River is more susceptible to the influence of the Angrapa River, whose water flow rate is noticeably higher than the second tributary—the Instruch River. In some watercourses of the second order, the mass development of *Euglenids* was noted, which indicated the contamination of waters with organic substances [14].

In the river coasts, the dominant species of vegetation are *P. australis*, *S. fragilis*, *S. alba*, *A. glutinosa*, *S. sagittifolia*, *S. sylvaticus*, and *P. arundinacea*, and frequently occurring are *C. acuta*, *R. amphibia*, *S. erectum*, and *B. umbellatus*. Aquatic vegetation is represented predominantly by communities with *P. pectinatus*, *P. nodosus*, *S. emersus*, *N. lutea*, and *M. spicatum*. Data on the growth of rare sensitive algal species in the region (*H. rivularis*, *A. chalybaea*) and findings of *C. elegans*, *D. glomerata*, *V. frigida*, *V. bursata*, and *V. canalicularis* are published for the very first time [15]. In the Krasnaya River, *B. trichophyllum* was found, recorded in the Red Book of the Kaliningrad Region. A new habitat for *E. telmateia*, a vanishing species, also listed in the Red Book of the Kaliningrad Region, has been identified [15].

10. From the point of view of nature management and protection, underground waters, which are the main source of household and drinking water supply in the Kaliningrad Region, are essential [16]. Fresh underground waters of the Kaliningrad Region are the main source of the regional water supply: 53% of people in the cities and 99% in rural are supplied with water from underground sources. Water is provided from the Upper Cretaceous, Paleogene, and Quaternary aquifers: the Moscow-Valdai (formerly known as mid-Russian-Valdai) and Oka-Dnieper (previously Lithuanian-Central Russian) intermoraine aquifers. The quaternary aquifers with 63% of water consumption are of the aquifers with the highest operational importance. The highest anthropogenic burden falls on the upper intermoraine (the Moscow-Valdai, mid-Pleistocene) aquifer, which is widespread in the southern part of the region. The aquifer distribution area is more than 9,300 km^2 (70% of the regional territory).

The study of the features of the geological structure, relief, and hydrogeological conditions allowed us to typify the conditions for the protection of groundwaters of the upper intermoraine horizon. Since the protection of groundwater is understood as the degree of their isolation from pollution sources, for the first classification level of protection conditions, the nature of the overlapping and underlying aquifer of sediments was adopted. The most dangerous is a combination of security conditions when the upper water-resistant horizon is absent and a hydraulically uniform aquifer of groundwater and pressure water is formed [16].

According to the data on the distribution of the upper aquiclude reduced capacity values, a map of isochrones of pollutant penetration from the surface to the Moscow-Valdai aquifer was designed [16]. It should be noted that this map does not display the exact time intervals, but their approximate values allow us to estimate the rate of pollutant infiltration. The disintegration time of pollutants varies widely. Such pollutants as solutions of some mineral salts (chlorides, sulfates, nitrates, etc.) or long-lived radioactive isotopes are very persistent and very slowly disintegrate. Other pollutants are also quite resistant but with a limited lifetime. A large group of pesticides is characterized with a large interval of disintegration time – from several months to 5–10 years.

The joint analysis of various environmental protection factors in relation to the Moscow-Valdai aquifer in the Kaliningrad Region identified three categories of protection of groundwater. The groundwater of an area of 1,700 km^2 (18% of the area of the aquifer) is considered unprotected. The territories which are characterized as partly protected occupy 3,500 km^2 (38% of the area of the aquifer). The 44% of the territory (4,100 km^2) of the Moscow-Valdai aquifer distribution is characterized as protected from contamination [16].

Potential hazards are located in the vicinity or directly on unprotected areas, on mineral deposit developments, municipal solid waste landfills, and stocks of mineral and organic fertilizers. The hydraulic connection between the rivers and the waters of the Moscow-Valdai aquifer is manifested particularly actively in the areas with weak protection, and it threatens with the penetration of pollutants from the rivers Pregolya, Lava, Instruch, Sheshupe, Neman, and Deima. Particular attention should be focused on the intakes which receive the water of the Moscow-Valdai aquifer and located in the areas characterized as unprotected or partly protected. The greatest potential hazard to the natural systems and groundwater in particular is created by the oil fields. One-third of the region's oil fields are located in the areas where the groundwater of the studied aquifer is not protected, and it increases the risk of contamination. Developments of construction material deposits are less dangerous compared to oil extraction, which is due to the relative geochemical inertness of these minerals. However, removing the overburden rock and subsequently operating the useful clay formation reduce the capacity of impermeable clays rocks overlying the aquifer [16].

11. The Lake Vishtynetskoe (Vištytis) is a unique transboundary reservoir in terms of origin, hydrological features, productivity of all trophic levels, and composition of the ichthyofaunal [17]. It is a deep pond of oligotrophic type with clear, transparent water. According to hydrochemical and hydrobiological indicators, the trophic status of the Lake Vishtynetskoe remains at a stable level. At the same time, for individual zones of the lake on both sides of the border, there are signs of mesotrophy, in some cases eutrophic. The fishery importance of the lake is determined by the existing traditional fish fishery and developed amateur fishing. The lake is also directly used for touristic purposes.

The main reasons for the catch decline after the 1980s are non-ecological and biological, and the general deterioration of socioeconomic conditions in the country and, in general, is typical for the fishing industry in Russia, as well as the acquisition of the status of a transboundary basin by Lake Vishtynetskoe. The conclusion is that the achieved level of recreational load should be fixed and not be increased in the future, and rational use of fish stocks in the lake should be coordinated with Lithuania [17].

The processes of beach disruption are connected with both natural phenomena (currents, high water level, intensive wave action) and human activities. The following kinds of influence can be identified on the part of tourism and recreation, placed in the order of decreasing degree of their impact on beach degradation and coast disruption: the construction of temporary objects (cafes, camp sites, etc.) immediately on the beach or within the dunes-palve; driving jeeps on the beach or within the dunes-palve, laying new paths through dunes; and also beach trampling. This requires additional shore protection. Neglecting abrasion processes can also cause shore sloughing which jeopardizes the buildings. Consequently, the natural carrying capacity of the available beaches should be considered when planning and constructing new tourism and recreation complexes. The good news are that the Government of the Russian Federation within the Federal Target Programme "The Development of Domestic and Inbound Tourism in the Russian Federation (2011–2018)" allocated money for shore protection works in Svetlogorsk. However, this will only cover 4.5 km out of the total 140 km of beaches. For over 40 km of sand beaches, which are very popular in the Kaliningrad Region especially during the summer season, the issue is still urgent and requires a comprehensive solution and joining efforts of federal and regional authorities [18].

Specially protected natural areas (SPNA) play an important role in biodiversity and landscape preserving in the Kaliningrad Region. However, further development of the SPNA in the region has to fight several problems. In particular, Kaliningrad should improve the regional environmental management and control, increase the level of supply and technical support for protected areas, and deepen the system of international and, above all, cross-border cooperation in the environmental sphere. International collaboration in the environmental sphere should be focused on sustainable development of the entire Baltic Region and the Kaliningrad Region in particular.

The Kaliningrad region is valuable for the whole Europe in many regards owing to its remarkable landscapes, cultural, historical, and recreational places of interest. Several specially protected natural areas in the Kaliningrad Region are located along the borders of Lithuania and Poland (the Curonian Spit National Park, the Vishtynetsky Nature Park, the Dyunny Partial Reserve); they play an important role in the system of long-term cross-border cooperation, especially the national park "Curonian Spit," which was included in the list of World Natural Heritage by UNESCO along with the Lithuanian National Park "Kurshu Neria." The cross-border SPNA compose a system of specially protected natural areas under different national jurisdictions that share the same ecosystem and are capable to ensure the preservation of ecological balance at a level that gives the maximum ecological and socio-economic effect. The cross-boundary specially protected natural areas are an integral part of the European strategy for the conservation of biological and landscape diversity.

The leading direction for the regional environmental compliance at the current time is improving SPNA system not only by involving new territories but also by improving the law protecting the status of existing areas. For example, it makes sense to give the status of specially protected nature areas to the Vistula (Baltic) Spit, the area around the Balga Castle with the Vistula Lagoon, to the unique intact bog in the Pravdinsky District, to the areas along the Curonian Lagoon in the Slavsky and Polessky Districts, and so on. At the same time, even more important is to improve the system of environmental protection activities and to make the control over economic entities within existing protected areas stronger in order to encourage initiatives with positive ecological effect [19].

And, finally, another proposal in the context of environmental activities in the region is based on the results of macrophyte study in the reservoirs of the region; nine species were proposed to be granted the status of "rare species," in particular *Batrachium eradicatum* (Laest.) Fries, *B. fluitans* (Lam.) Wimm., *B. trichophyllum* (Chaix) Bosch, *Callitriche hermaphroditica* L., *Ceratophyllum submersum* L., *Zannichellia palustris* L., *Z. major* Boenn. ex Reichenb., *Potamogeton acutifolius* Link, and *P. friesii* Rupr [20].

This book presents a brief systematization and description of the knowledge on the terrestrial environment and inland water resources in the Kaliningrad Region. The publication is based on observational data, scientific literature mainly published in Russian editions, and long-standing experience of authors of the chapters in the scientific research on the Kaliningrad Region environment. This book is addressed to the specialists working in various fields of environmental problems and ecology, water resources and management, land reclamation and agriculture, and international cooperation in the Baltic Sea Region. This is the first book in the series of four volumes, which will be published in the coming years under the general title "Environmental Studies in the Kaliningrad Region." The other three volumes will be devoted to physical oceanography, geoecology, and bioecology of the Southeastern Baltic Sea.

Acknowledgments V.A. Gritsenko was partially supported by Immanuel Kant Baltic Federal University Program "Integrated Geographical Surveys of the Kaliningrad Region and the Baltic Region." A.G. Kostianoy was partially supported by the Russian Science Foundation Grant N 14-50-00095. As volume editors of this book, we would like to thank Immanuel Kant Baltic Federal University for the investment in our book project by funding translation of most of the chapters. Also, we are very grateful to Springer-Verlag and *The Handbook of Environmental Chemistry* book series in particular, for support of our four volume book project "Environmental Studies in the Kaliningrad Region."

References

1. Zhamoida VA, Sivkov VV, Nesterova EN (2017) Mineral resources of the Kaliningrad Region. In: Gritsenko VA, Sivkov VV, Yurov AV, Kostianoy AG (eds) Terrestrial and Inland water environment of the Kaliningrad Region. Environmental studies in the Kaliningrad Region. The handbook of environmental chemistry, Springer, Heidelberg
2. Sivkov VV, Zhamoida VA (2017) Amber deposits in the Kaliningrad Region. In: Gritsenko VA, Sivkov VV, Yurov AV, Kostianoy AG (eds) Terrestrial and Inland water environment of the Kaliningrad Region. Environmental studies in the Kaliningrad Region. The handbook of environmental chemistry. Springer, Heidelberg
3. Mikhnevich GS (2017) Composition of pre-quaternary surface and quaternary sediments allocation on the territory of the Kaliningrad Region. In: Gritsenko VA, Sivkov VV, Yurov AV, Kostianoy AG (eds) Terrestrial and Inland water environment of the Kaliningrad Region. Environmental studies in the Kaliningrad Region. The handbook of environmental chemistry, Springer, Heidelberg
4. Kolesnik TB (2017) Pleistocene deposits in the Kaliningrad Region. In: Gritsenko VA, Sivkov VV, Yurov AV, Kostianoy AG (eds) Terrestrial and Inland water environment of the Kaliningrad Region. Environmental studies in the Kaliningrad Region. The handbook of environmental chemistry. Springer, Heidelberg
5. Romanova EA, Vinogradova OL, Frizina IV (2017) Modern landscapes in the Kaliningrad Region. In: Gritsenko VA, Sivkov VV, Yurov AV, Kostianoy AG (eds) Terrestrial and Inland water environment of the Kaliningrad Region. Environmental studies in the Kaliningrad Region. The handbook of environmental chemistry. Springer, Heidelberg
6. Napreenko-Dorokhova TV, Napreenko MG (2017) The history of forest and peatland formation in the Kaliningrad Region during the Holocene. In: Gritsenko VA, Sivkov VV, Yurov AV, Kostianoy AG (eds) Terrestrial and Inland water environment of the Kaliningrad Region. Environmental studies in the Kaliningrad Region. The handbook of environmental chemistry. Springer, Heidelberg
7. Volkova II, Shaplygina TV, Belov NS, Danchenkov AR (2017) Eolian coastal-marine natural systems in the Kaliningrad Region. In: Gritsenko VA, Sivkov VV, Yurov AV, Kostianoy AG (eds) Terrestrial and Inland water environment of the Kaliningrad Region. Environmental studies in the Kaliningrad Region. The handbook of environmental chemistry. Springer, Heidelberg
8. Romanova EA, Vinogradova OV, Sergeeva DV (2017) Factors and patterns of current development of territorial units in the Kaliningrad Region. In: Gritsenko VA, Sivkov VV, Yurov AV, Kostianoy AG (eds) Terrestrial and Inland water environment of the Kaliningrad Region. Environmental studies in the Kaliningrad Region. The handbook of environmental chemistry. Springer, Heidelberg
9. Romanova EA, Vinogradova OV, Danishevskyi VV, Frizina IV (2017) Specific features of urban geosystems in the Kaliningrad Region. In: Gritsenko VA, Sivkov VV, Yurov AV, Kostianoy AG (eds) Terrestrial and Inland water environment of the Kaliningrad Region. Environmental studies in the Kaliningrad Region. The handbook of environmental chemistry. Springer, Heidelberg
10. Bernikova TA, Nagornova NN, Tsoupikova NA, Shibaev SV (2017) Environmental features of watercourses in the Kaliningrad Region. In: Gritsenko VA, Sivkov VV, Yurov AV, Kostianoy AG (eds) Terrestrial and Inland water environment of the Kaliningrad Region. Environmental studies in the Kaliningrad Region. The handbook of environmental chemistry. Springer, Heidelberg
11. Domnin DA, Chubarenko BV, Capell R (2017) Formation and re-distribution of the river runoff in the catchment of Pregolya River. In: Gritsenko VA, Sivkov VV, Yurov AV, Kostianoy AG (eds) Terrestrial and Inland water environment of the Kaliningrad Region. Environmental studies in the Kaliningrad Region. The handbook of environmental chemistry. Springer, Heidelberg

12. Shibaeva MN, Masyutkina EA, Shibaev SV (2017) Hydrobiological characteristics of water bodies of Kaliningrad Region. In: Gritsenko VA, Sivkov VV, Yurov AV, Kostianoy AG (eds) Terrestrial and Inland water environment of the Kaliningrad Region. Environmental studies in the Kaliningrad Region. The handbook of environmental chemistry. Springer, Heidelberg
13. Ezhova EE, Gerb MA, Kocheshkova OV, Lange EK, Polunina JJ, Molchanova NS (2017) The structure and composition of biological communities in the Pregolya River (Vistula Lagoon, the Baltic Sea). In: Gritsenko VA, Sivkov VV, Yurov AV, Kostianoy AG (eds) Terrestrial and Inland water environment of the Kaliningrad Region. Environmental studies in the Kaliningrad Region. The handbook of environmental chemistry. Springer, Heidelberg
14. Lange EK (2017) Phytoplankton community of small rivers of the Pregolya River basin. In: Gritsenko VA, Sivkov VV, Yurov AV, Kostianoy AG (eds) Terrestrial and Inland water environment of the Kaliningrad Region. Environmental studies in the Kaliningrad Region. The handbook of environmental chemistry. Springer, Heidelberg
15. Volodina AA, Gerb MA (2017) Flora and vegetation of the small rivers of the Pregolya River system in the Kaliningrad Region. In: Gritsenko VA, Sivkov VV, Yurov AV, Kostianoy AG (eds) Terrestrial and Inland water environment of the Kaliningrad Region. Environmental studies in the Kaliningrad Region. The Handbook of environmental chemistry. Springer, Heidelberg
16. Mikhnevich GS (2017) The protection conditions of the groundwater against pollution in the Kaliningrad Region. In: Gritsenko VA, Sivkov VV, Yurov AV, Kostianoy AG (eds) Terrestrial and Inland water environment of the Kaliningrad Region. Environmental studies in the Kaliningrad Region. The handbook of environmental chemistry. Springer, Heidelberg
17. Shibaev SV, Sokolov AV, Tylik KV, Bernikova TA, Shibaeva MN, Nagornova NN, Aldushin AV (2017) Current status of the Lake Vistytis in Kaliningrad Region. In: Gritsenko VA, Sivkov VV, Yurov AV, Kostianoy AG (eds) Terrestrial and Inland water environment of the Kaliningrad Region. Environmental studies in the Kaliningrad Region. The handbook of environmental chemistry. Springer, Heidelberg
18. Kropinova EG (2017) The reduction in the beach area as the main limiting factor for sustainable tourism development (case for the Kaliningrad Oblast). In: Gritsenko VA, Sivkov VV, Yurov AV, Kostianoy AG (eds) Terrestrial and Inland water environment of the Kaliningrad Region. Environmental studies in the Kaliningrad Region. The handbook of environmental chemistry. Springer, Heidelberg
19. Volkova II, Shaplygina TV, Bubnova ES (2017) Specially protected natural areas of the Kaliningrad Region. In: Gritsenko VA, Sivkov VV, Yurov AV, Kostianoy AG (eds) Terrestrial and Inland water environment of the Kaliningrad Region. Environmental studies in the Kaliningrad Region. The handbook of environmental chemistry. Springer, Heidelberg
20. Gerb MA, Volodina AA (2017) Rare and protected macrophytes and semi-aquatic plants of flora of the Kaliningrad Region. In: Gritsenko VA, Sivkov VV, Yurov AV, Kostianoy AG (eds) Terrestrial and Inland water environment of the Kaliningrad Region. Environmental studies in the Kaliningrad Region. The handbook of environmental chemistry. Springer, Heidelberg

Erratum to: Mineral Resources of the Kaliningrad Region

Vladimir Zhamoida, Vadim Sivkov, and Elena Nesterova

V.A. Gritsenko et al. (eds.), *Terrestrial and Inland Water Environment of the Kaliningrad Region - Environmental Studies in the Kaliningrad Region*, Hdb Env Chem, DOI 10.1007/698_2017_115,
© Springer International Publishing AG 2017

Erratum to: Hdb Env Chem
10.1007/698_2017_115

The names of the authors were mentioned incorrectly as Zhamoida Vladimir, Sivkov Vadim, and Nesterova Elena, instead of rightly mentioning them as

---

The updated online version for this chapter can be found under
DOI 10.1007/698_2017_115

V. Zhamoida (✉)
A.P. Karpinsky Russian Geological Research Institute (VSEGEI), 74, Sredny Prospect, St. Petersburg, Russian Federation

St. Petersburg State University, Universitetskaya Embankment, 13B, St. Petersburg 199034, Russian Federation
e-mail: vladimir_zhamoida@vsegei.ru

V. Sivkov
P.P. Shirshov Institute of Oceanology RAS, Nakhimovsky Pr., 36, Moscow 117997, Russia

Immanuel Kant Baltic Federal University, 14, A. Nevskogo Street, Kaliningrad, Russia
e-mail: sivkov@kaliningrad.ru

E. Nesterova
A.P. Karpinsky Russian Geological Research Institute (VSEGEI), 74, Sredny Prospect, St. Petersburg, Russian Federation

Vladimir Zhamoida, Vadim Sivkov, and Elena Nesterova, as there was an inadvertent interchange between the family name and given name of all the authors of this chapter. This has now been updated.

Erratum to: Specially Protected Natural Areas of the Kaliningrad Region

I.I. Volkova, T.V. Shaplygina, and E.S. Bubnova

V.A. Gritsenko et al. (eds.), *Terrestrial and Inland Water Environment of the Kaliningrad Region - Environmental Studies in the Kaliningrad Region*, Hdb Env Chem, DOI 10.1007/698_2017_98,
© Springer International Publishing AG 2017

Erratum to: Hdb Env Chem
10.1007/698_2017_98

Inadvertently, the second affiliation of the co-author E.S. Bubnova was mentioned incorrectly in this chapter. This has now been updated.

The updated online version for this chapter can be found under
DOI 10.1007/698_2017_98

I.I. Volkova (✉) and T.V. Shaplygina
Department of Geography, Nature Management and Spatial Development, Immanuel Kant Baltic Federal University, Kaliningrad, Russia
e-mail: volkova.bfu@yandex.ru; tshaplygina@gmail.com

E.S. Bubnova
Immanuel Kant Baltic Federal University, Kaliningrad, Russia

P.P. Shirshov Institute of Oceanology, Russian Academy of Sciences, 36, Nakhimovsky Pr., Moscow 117997, Russia
e-mail: bubnova.kat@gmail.com

Index

A
Aegagropila linnaei, 513, 515, 518
Agathis australis, 36
Agriculture, 100, 109, 114, 186, 206, 323, 527, 536
Airfields, 171, 194
Air masses, 210
Air pollution, 211–213
Air temperatures, 7, 35, 104, 233, 272, 276
Alisma gramineum, 334, 515, 521
Alnus
    A. glutinosa, 137, 142, 150, 406
    A. incala, 150
Amber, 3, 6, 13, 25, 33
Ammonia, 241, 244, 253, 256, 259, 433
Angrapa River, 208, 270, 373
Apatite, 25
Aquatic plants, 320, 332, 513
Aquifers, depth, 422
    nitrite, 242, 432
    pollutant penetration, 419
Archaean-Proterozoic, 6, 16
Areal-networking components, 189
Arsenic, 323
Assessment, 147
Asteraceae, 388
Audouinella chalybaea, 513, 516
Autumn floods, 224

B
Baltic Naval Fleet, 194
Baltic Sea, southeastern, 1, 527
Baltiysk, 191, 207

Batrachium
    B. eradicatum, 334, 515, 522, 525, 536
    B. trichophyllum, 389, 408, 523
Batrachospermum
    B. gelatinosum, 390, 402, 518
    B. turfosum, 518
Beaches, 10, 497, 502, 509, 535
Beef, 189
Belligerent landscapes, 196
Benzopyrene, 211
Betula
    B. pendula, 150
    B. pubescens, 150
Bioecology, 1, 527
Blue earth, 15, 25, 33, 34, 39, 41, 491
Bol'shoye Mokhovoye raised bog, 125
Bombing, 181, 209, 213
Borders/borderlines, 196
    state/administrative, 179
Botanical garden, 482, 483, 491
Bream, 475
Brick clay, 26
Bridges, 115, 190, 219, 249, 270
Brines, industrial, 27
Brown coal, 15, 20, 22, 41, 47
Building materials, 25, 28, 52, 70, 435, 538
Building percentage, 201
Burbot, 473

C
Cadmium, 23
Cambrian system, 17
Carbon-14, 126

Carnallite, 24
Carpinus betulus, 133, 137, 142, 406
Carrying capacity, 497, 500, 506–508, 535
Catchment, transboundary, 269
Cattle breeding, 188
Cedrus atlantica, 36
Celestine, 24
Cellulose factories, 323
Cereal crops, 188
Chaetophora elegans, 517
Chalk marl, 26
Chara inconnexa, 513, 520
Chernozem, 209
Chernyakhovsk, 191, 208, 270
Chironomids, 296, 309–312, 321, 344, 446
Chlorococcales, 375
Chlorophyll a, 223, 254, 443
Cladocera, 288, 294, 313, 321, 336, 363, 446, 532
Clay, 8, 15–28, 47, 57, 81, 108, 285, 415, 534
 red glacio-lacustrine, 108, 195
Climate, 7, 18, 36, 59, 103, 122, 203, 211, 486, 502
 change, 73, 272, 498
Clinoptilolite, 25
Coal, brown, 15, 20, 22, 41, 47
Coastal defensive-industrial subzone, 185
Coastal zones, 15, 47, 112, 175, 497, 530
Cobalt, 23
Construction, 190
 debris, 208, 534
 materials, 9, 83, 195, 206, 411, 435
Copepods, 288, 294–296, 313, 321, 336, 362, 446, 464, 532
Copper, 23
Cretaceous system, 19
Crops, 187–189
Crustaceans, 289, 296, 312, 336, 446, 462
Cryptophytes, 375
Curonian Lagoon, 2, 7, 35, 48, 121, 270, 373, 483, 531
Curonian Spit, 3, 10, 147, 149, 318, 373
 natural park, 486

D
Dachas, 181, 192
Dams, 182, 194, 210, 251
Deima River, 57, 73, 139, 194, 373, 435, 529, 534
Demographic processes, 185
Devonian system, 17
Diatoms, 326–332, 375–382, 533

Digression, 147, 153, 174, 530
Dikes, 16, 187, 209, 210, 249, 500
Dinoflagellates, 375, 380
Drainage, 26, 69, 105, 109, 187, 225, 262, 531
 canals, 229, 244, 255–262
Draparnaldia glomerata, 517
Dreissena polymorpha (zebra mussel), 309, 345, 346, 348, 446, 448
Drinking water, 6, 69, 109, 426, 533
Dyunny, 490

E
Eels, 475
Endangered plants, 334, 385, 390, 402, 515
Eolian coastal-marine natural complexes (NC), 147
Equisetum telmateia, 524
Euglenophytes, 375
Eutrophication, 323, 351, 445, 450, 475, 532
Exclave, 2, 201
Explosive materials, 187

F
Fagus sylvatica, 133, 142
Ferrous iron, 243
Fisheries, 454
Fishing, 442, 457–479, 535
Floods, 224, 233, 236–257, 273–283, 287, 427, 531
Fontinalis antipyretica, 333
Forage production, 188
Forests, 36, 98, 104–115, 121–144, 150, 187, 387, 483, 529
 amber, 36, 38, 41, 528
 coniferous, 104, 141, 390, 529
 history, 133
Formaldehyde, 211

G
Gardens, public, 211
Gas, 20
 fields, 22
Geoecology, 1, 527
Geology, 5, 13
Glaciation, 5, 25, 49, 61, 67, 72, 81, 98, 250, 529
Glaciers, 7, 33, 82, 106, 427, 528
Glaciodislocations, 94
Glaciolacustrine plains, 7, 97, 107, 115, 141, 208, 428, 483, 529

Glauconite, 24, 25
Grain crops, 189
Gromovsky, 484, 490
Groundwater, 27, 411
  level (GWL), 151
  pollution, 9, 412
Guryevka River, 208
Guryevsk, 208, 432
Gusev, 192
Gvardeisk/Gvardeysk, 208, 270

H
Hayfields, 189
Heavy metals, 210
Hildenbrandia rivularis, 513, 516
Hippuris vulgaris, 521
Holocene, 121
Household waste, 208
Hydrobiology, 8
Hydrochemistry, 223, 237
Hydrogen sulfide, 38, 244, 322, 324, 346, 351
Hydrology, 223, 441
Hydrometry, 223
Hydrophytes, 399
Hydropower plants, 195
Hypnum moss communities, 135

I
Ice, marine sediments, 81
  two-layer, 234
Ichthyofauna, 441, 450
Industrial brines, 27
Industrial effluents, 325
Industrial pollution, 348
Industrial production, 204, 323
Industrial waste, 154, 210
Industrial wastewater, 238, 242, 323
Industrial zones, 195, 215, 324, 334
Inland water, 1, 527
Instruch River, 7, 72, 102, 191, 208, 228, 270, 378–382, 387–392, 398, 435, 532–534
Invasive species, 334, 389
Iron, in water, 243
  ore, 23

J
Juglans sp., 133, 142
Jurassic system/deposits, 18, 55, 57

K
Kaliningrad Oblast, 497
Keramzit clays, 26
Kozye raised bog, 123–125, 133, 136
Krasnaya River, 193, 245, 260, 374, 379, 400, 484, 490, 533
Krasnoznamensk, 72, 192, 208, 211, 415, 433

L
Ladushkin, 208, 242, 428
Lagoons, 224
Lake Vistytis, 9, 441
Landgraben (Pityevoy), 210
Landscape-dependent settlement, 110
Landscapes, 7, 97, 179
  current, 179
  genesis, 97
Land use, 97, 179
Laser scanning, 156
Lava River, 208, 270
Lead, 23
Leeches, 296–298, 308, 314, 344, 446
Legumes, 189
Lignin, 325
Lignite, 6, 14, 20, 84, 88–93
Limestones, 13
Littoral zone, 498
Livestock production, 188
Lugovaya River, 193, 194, 490

M
Macroalgae, 332, 388, 401, 515
Macrophytes, 10, 317, 320, 385, 388
  rare, 513
Macrozoobenthos, 321, 350
Magnesium, 6, 24, 325
Maloye Olenye Lake, 125
Mamonovka River, 208, 228, 237, 242, 258, 427
Mamonovo, 15, 73, 208
Marls, 13, 18, 26, 57, 76, 84, 426
Masurian Lakes, 373, 377, 381, 442, 476
Matrosovka River, 111, 194, 248
Meltwater, 41, 51, 74, 106, 239
Mesozoic, 13, 18, 61
Metals, 6, 187, 203
  ferrous, 6
  heavy, 210
  nonferrous, 6, 13, 20
  rare, 23

Milk, 188
Minerals, metallic, 23
　nonmetallic, 24
　resources, 5, 13, 20, 539
Mineral water, 22, 27, 69
Mire peat deposits, 121
Molybdenum, 23
Moraine plains, 97, 102, 106, 243, 260, 287
Morphometry, 223
Moscow-Valdai aquifer, 9, 411, 421
Motorways, 109, 193, 506

N
Natural complex digression, 154
Natural geological reserves, 490
Natural monuments, 481, 483, 491
Natural parks, 481, 482
Nature complex (landscape) reserves, 490
Nature management/protection, 9
Nature reserves, 189
　partial, 481
Neman River, 7, 9, 25, 55, 101, 122, 223, 247, 434, 490, 529–534
Nemonin, 194
Neogene, 19, 33
Nesterovsky district, 23
Nitrate, 242–244, 253, 256, 259, 325, 376, 428, 433, 534
Nitrite, 242, 259, 432, 531
Nitrogen, 324, 377, 432, 531
　compounds, water, 241, 253, 435
Nitrogen dioxide, 211
Numerical simulation, 269
Nutrients, in water, 243
Nymphoides peltata, 522

O
Oil, 6, 13, 20, 195, 411, 433, 528, 538
　deposits, 15, 195
　extraction, 9, 411, 534
　fields, 15, 20, 435, 534
Oil-recovery facilities, 195
Oka-Dnieper aquifer, 421
Oligochaetes, 296, 304, 308, 364, 446–450
Oligosaprobes, 296, 312
Oolites, 23
Ordovician system, 17
Organic substances, 223, 239, 253, 256, 325, 445, 450, 533
Ozerkovskyi canal, 194

P
Palaeogeography, 121
Paleogene, 19, 33
Paleoincisions, 45, 61
Paleozoic, 17
Palynology, 121. *See also* Pollen analysis
Paper mill, 321
Pasture grounds, 8, 110, 186–189, 259, 323
Peat, 6, 13, 20, 23, 26, 85, 103, 121–144, 249, 425, 483
　bog, 108, 483, 518
　composition, 121
Peatland, 7, 23, 121–144, 529
Perch, 468
Permanganate index, 239, 253–258, 262, 445
Permeability, 9, 411
Permian system, 17
Permutite, 25
Petrophytum sp., 186
Phanerozoic, 6
Phosphates, 19, 39, 84, 243, 253
Phosphorites, 13, 19, 24, 25, 28, 39
Phosphorus, 243, 257, 260, 323, 377, 531
Phosphorus pentoxide, 25
Phragmites australis, 334
Phytoplankton, 317, 319, 326, 373
　seasonal dynamics, 373
Picea abies, 137, 150
Pike, fishing, 471
Pinus
　P. succinifera, 36
　P. sylvestris, 129, 150
Pionersk, 239, 242, 507
Pionersky, 47, 194, 196, 207, 502
Pipelines, 195
Pissa River, 60, 75, 139, 208, 270, 442, 476, 477, 487, 516–519, 529, 533
Planktothrix agardhii, 373
Pleistocene, 6, 20, 35, 41, 47, 76, 81, 115, 422, 528
Poaceae, 388
Podlipovsky peatland, 126
Polders, 101, 108, 111, 187, 210, 228, 249, 262, 531
Polessk, 57, 186, 208, 432
Polesskaya valley, 74, 262, 421
Polessky canal, 194, 227, 248
Pollen analysis, 7, 121, 126–133, 137, 529
Pollution, 9, 209–262, 411, 445, 531
　air, 212, 213
　groundwater, 9, 411, 417, 428, 435
　industrial, 348

transboundary, 258
Polygalite, 24
Population density, 10, 109–113, 181, 182, 186, 213, 258, 459, 500
　rural, 111, 184, 207
Populus tremula, 150, 409
Potamogeton
　P. acutifolius, 358, 525, 536
　P. lucens, 334, 358, 401
　P. nodosus, 389, 391, 398, 404
　P. pectinatus, 358, 389, 396, 520
　P. praelongus, 523
Potamothrix hammoniensis, 297, 305, 308, 310, 365, 447
Potassium, 6, 18, 20, 24, 28, 36, 39, 238, 528
Power plants, electric, 195
Pravdinsk, 208
Pregolya River, 8, 244, 269, 317, 373, 385
　phytoplankton, 373, 375
Pre-Quaternary surfaces, 6, 18, 40, 45, 528
Primorsk, 192, 239, 429
Primorskaya River, 228, 234, 237, 239, 247, 258, 427
Primorsky amber deposits, 6, 15, 25, 35, 67
Primorsky canal, 194, 248
Prokhladnaya River, 234
Protected areas, 215, 481, 493, 533
Protection, 9, 411, 513
　forests, 174
　shore/beaches, 165, 208, 497, 509
　vs. vulnerability, 412
Proterozoic, 6, 16
Pumping stations, 187

Q

Quarries, 109, 195, 287, 412, 528
　amber, 19, 35, 41, 47, 507
　dumps, 195
Quartz sands, 15, 24, 25, 84, 88
Quaternary aquifers, 421, 533
Quaternary deposits/sediments, 6, 14, 45, 148, 181, 195, 491
Quaternary glaciation, 81, 529
Quaternary system, 15, 20

R

Radiocarbon dating, 126
Railways, 109, 193, 201, 530
Ranunculaceae, 388
Rapeseed, 189

Reclamation, 52, 110, 180, 197, 235, 527, 536
Recreation, 8, 114, 150, 168, 203, 441, 498, 502, 535
Red Data Book, 515
Relief, 45, 208
Reservoirs, 204, 210, 269, 416, 442, 452, 478, 531, 534, 536
　transboundary, 534
Residential neighborhoods, 185
Riparian vegetation, 389
Rivers, basin, 223
　small, 373
　transboundary, 237, 241, 259–262, 531
　vegetation, 389
Roach, 467
Rock salt, 6, 24
Rotifers, 288–295, 313, 336–340, 360, 446, 532
Runoff, 211, 269
　coefficients, 229
　groundwater, 204
　meltwater, 51
　river, 250, 269, 276–283, 531
Rural settlements, 184, 187, 192

S

Salinity, 237, 252, 336, 418
Salts, 6, 13, 28, 237, 428, 528, 534
Sambia Peninsula, 6, 34, 81, 210, 228, 240, 258, 429, 490, 502, 528
Sandstones, 17–23, 26, 57, 63, 70, 76
Sandy gravel, 25, 28
Sapropel, 26–28, 85, 425
Seaside resort subzone, 185
Secondary succession, 97
Sediments, 45, 249, 323, 341, 346, 351, 413, 416, 425, 434, 491, 528, 533
　alluvial, 6, 20, 55, 60, 425
　bog, 425
　ice-marine, 81
　marine, 18, 81, 425
　Pleistocene, 528
　Quaternary, 46, 491, 529
Semiaquatic plants, 513
Settlement systems, 97, 179
Severnaya River, 248
Sewage, 244, 323, 335, 500
　discharge, 323, 325
　irrigation, 323
　municipal, 211, 352
　systems, 203, 211, 244, 325

Sewage (cont.)
　treatment, 215
Shale oil, 23
Sheshupe River, 108, 208, 255
Soils, 7, 101, 105–115, 148, 153, 175, 208, 529
　acidity, 210
　artificial, 204
　podzolic, 105, 150–153, 210
　urban, 209
Sovetsk, 191
Specially protected natural areas (SPNA), 482
Species, endangered/rare, 334, 385, 390, 402, 514, 515
　invasive, 334, 389
Sphagnum, moss/peat/bog, 121, 128, 133, 135, 137, 143
Spits, 3
　Curonian, 3, 10, 147, 149, 318, 373, 486
　forests, 153
　transformation/stability, 151
　Vistula, 3, 8, 102, 147, 149, 175, 422, 426, 483, 484, 502, 507, 530
Stratigraphy, 16, 121
Strontium ores, 24
Succession, secondary, 97, 100, 109, 187, 198
Succinite, 36
Sulfates, 24, 325, 428, 534
Surface laser scanning (SLS), 147, 156
Surface runoff, 211, 279
Sustainable development, 189, 491, 498, 535
Svetlogorsk, 27, 62–72, 85, 88, 101, 196, 207, 421, 429, 434, 491, 506, 535
Svetly, 72, 192, 194, 208, 228, 429, 435
Sylvite, 24

T
Technozem, 209
Tench, 473
Terrestrial environment, 1, 527
Therapeutic muds, 27
Thermal waters, 27
Tilia platyphyllos, 133, 142
Total hardness, 239, 257, 262, 444, 531
Tourism, 5, 10, 502, 507, 535
　sustainable, 497
Town status–town size, 207
Traffic, 219, 509
　network, 190
Transboundary catchment, 269
Transboundary pollution, 258
Transmission lines, 196

Transportation system/network, 185, 216
Transport connectivity, 201
Triassic system/deposits, 18, 23
Tripolium pannonicum, 523, 524
Two-layer ice, 234
Typha
　T. angustifolia, 334, 359, 404
　T. latifolia, 334, 359, 390, 393, 404

U
Urban environment, 201, 214
Urban frames, 205
Urban geosystem, 202
Urban green spaces, 212
Urban settlement, 201
　evolution, 206
Urban vegetation, 211
Urban zoning concept, 215
Urbozem, 209

V
Valdai Glacier, 7, 72, 81, 88, 98, 422
Vanadium, 23
Vegetation, 104, 385
　cover, 104, 150, 168, 171, 181, 183, 211, 529
　period, 104, 320, 327, 350, 375, 382, 533
　riparian, 389
Vendace, 454
Vishtynetskoe Lake, 286–298, 307–314, 392, 489, 517, 535
Vishtynetsky Natural Park, 486, 535
Vistula Lagoon, 2, 8, 19, 84, 102, 113, 150, 229, 236, 270, 318, 375, 442, 531, 536
Vistula Spit, 3, 8, 102, 147, 149, 175, 422, 426, 483, 484, 502, 507, 530
Vistytis Lake, 9, 10, 441–479
VNIGRI, 16
Vulnerability, 10, 212, 411, 417

W
Wastewater, 224, 324, 412, 416
　discharge, 247, 418
　disposal, 174
　household/domestic, 239, 241
　industrial, 238, 242, 323
　municipal, 240, 242
Water, balance, 269, 283
　drinking, 6, 69, 109, 426, 533

groundwater, 27, 411
hardness, 239, 254–257, 262, 444, 531
meltwater, 41, 51, 74, 106, 239
mineral, 22, 27, 69
pollution, 211, 324, 382, 428, 445
quality, 224, 241, 254, 286, 296, 322, 335, 449, 514
routes, 194
temperatures, 225, 233, 241, 252, 338, 443
thermal, 27
use conflicts, 434
wastewater, 224, 324, 412, 416
Watercourses, 5, 59, 223, 374, 422, 531
transboundary, 224, 258–262, 270
Watersheds, 59, 89, 143, 204, 228–281, 324, 431, 531
transboundary, 245
Wetlands, 123, 142, 225, 239, 243, 483, 490
formation, history, 133
Whitefish, 464

Wind power, 195
Winds, 7, 103, 160, 236, 245, 270
surges, 228, 318, 329, 337, 375, 531

Y
Yuzhnaya River, 248

Z
Zebra mussels (*Dreissena polymorpha*), 309, 345, 346, 348, 446, 448
Zehlau raised bog, 124
Zelenogradsk, 72, 101, 160, 184, 207, 287, 434, 484, 502, 509
Zeolites, 6, 20–25
Zinc, 23
Zoobenthos, 8, 285, 296, 317, 341, 441
Zooplankton, 8, 285, 317, 320, 336, 441